Other Titles in This Series

Volume

3 **Albrecht Böttcher, Aad Dijksma, Heinz Langer, Michael A. Dritschel, James Rovnyak, M. A. Kaashoek**
Peter Lancaster, Editor
Lectures on operator theory and its applications
1996

2 **Victor P. Snaith**
Galois module structure
1994

1 **Stephen Wiggins**
Global dynamics, phase space transport, orbits homoclinic to resonances, and applications
1993

Lectures on Operator Theory and Its Applications

FIELDS INSTITUTE MONOGRAPHS

THE FIELDS INSTITUTE FOR RESEARCH IN MATHEMATICAL SCIENCES

Lectures on Operator Theory and Its Applications

Albrecht Böttcher
Aad Dijksma and Heinz Langer
Michael A. Dritschel and James Rovnyak
M. A. Kaashoek

Peter Lancaster, Editor

American Mathematical Society
Providence, Rhode Island

The Fields Institute for Research in Mathematical Sciences

The Fields Institute is named in honour of the Canadian mathematician John Charles Fields (1863–1932). Fields was a remarkable man who received many honours for his scientific work, including election to the Royal Society of Canada in 1909 and to the Royal Society of London in 1913. Among other accomplishments in the service of the international mathematics community, Fields was responsible for establishing the world's most prestigious prize for mathematics research—the Fields Medal.

The Fields Institute for Research in Mathematical Sciences is supported by grants from the Ontario Ministry of Education and Training and the Natural Sciences and Engineering Research Council of Canada. The Institute is sponsored by McMaster University, the University of Toronto, and the University of Waterloo and has affiliated universities in Ontario and across Canada.

The author received support for this volume from the Ontario Ministry of Education and Training and the Natural Sciences and Engineering Research Council of Canada.

1991 *Mathematics Subject Classification.* Primary 47-02;
Secondary 15-A18, 34G10, 47A58, 47B50, 47E05, 47N70, 65F15, 93C35.

Library of Congress Cataloging-in-Publication Data

Lectures on operator theory and its applications/Böttcher, Albrecht... [et al.]; edited by Peter Lancaster.
p. cm. —(Fields Institute monographs, ISSN 1069-5273; 3)
At head of title: The Fields Institute for Research in Mathematical Sciences.
Includes bibliographical references and index.
ISBN 0-8218-0457-X
1. Operator theory. I. Böttcher, Albrecht. II. Lancaster, Peter, 1929– . III. Series.
QA329.L43 1995
515′.724–dc20 95-43465
CIP

This publication was prepared by the Fields Institute.

10 9 8 7 6 5 4 3 2 1 01 00 99 98 97 96

Contents

Preface xi

Lecture Series 1. Infinite Matrices and Projection Methods
Albrecht Böttcher

1. Matrix representations of operators 3
2. Three problems for infinite matrices 7
3. The finite section method 14
4. Selfadjoint and compact operators 16
5. Toeplitz matrices with continuous symbols 20
6. Toeplitz operators: algebraization of stability 25
7. Toeplitz operators: localization 30
8. Block case and higher dimensions 35
9. Banach space phenomena 43
10. Norms of inverses and pseudospectra 48
11. Toeplitz determinants 59
12. More general projection methods 64

Bibliography 69

Lecture Series 2. Operator Theory and Ordinary Differential Operators
Aad Dijksma and Heinz Langer

Introduction 75

1. Definitizable operators in Kreĭn spaces 77
   1. Basic concepts 77
   2. Definitizable operators 80
   3. The spectral function of a definitizable operator 82
   4. Critical points 84
   5. The critical point ∞ 85

2. Boundary eigenvalue problems for Sturm-Liouville operators and related holomorphic functions 87
1. The Sturm–Liouville differential operator 87
2. The Titchmarsh–Weyl coefficient 90
3. The boundary eigenvalue problem 93
4. The linearization of the boundary eigenvalue problem 94
5. Special cases 96

3. Operator representations of holomorphic functions 99
1. Colligations 99
2. Linearizations and Štraus extensions 101
3. Models for the Titchmarsh–Weyl coefficients 106

4. Sturm–Liouville operators with indefinite weight 109
1. The problem 109
2. Full range and half range completeness 113
3. The Titchmarsh–Weyl coefficient 115

5. Interface conditions and singular potentials 117
1. Interface conditions 117
2. A singular potential continuous in the graph norm of $\hat{A}$ 120
3. A singular potential discontinuous in the graph norm of $\hat{A}$ 122
4. Strings with point masses and dipoles 125

6. Operator pencils 128
1. Basic notations 128
2. Operator polynomials 128
3. A λ-rational pencil 130
4. The Klein–Gordon equation 131

Bibliography 133

Symbols used in Lecture Series 2 139

Lecture Series 3. Operators on Indefinite Inner Product Spaces
Michael A. Dritschel and James Rovnyak

Preface 143

Introduction: Preliminaries and notation 144
A. Prerequisites 144
B. Inner product spaces 144
C. Kreĭn spaces and fundamental decompositions 145
D. Example 147
E. Books and expositions 148

Lecture 1. Kreĭn spaces and operators 149

Lecture 2. Julia operators and contractions 158
Addendum: Adjoints of contractions 168

Lecture 3. Extension and completion problems 171

Lecture 4. The Schur algorithm 181

Lecture 5. Reproducing kernel Pontryagin spaces and colligations 192

Lecture 6. Invariant subspaces 207
A. The commutant of a selfadjoint operator 207
B. Dual pairs of invariant subspaces for definitizable operators 213
C. Compact perturbations of definitizable operators 221
Addendum: Nondefinitizable operators 224

Bibliography 227

Lectures Series 4. State Space Theory of Rational Matrix Functions and Applications
M.A. Kaashoek

Introduction 235

Chapter 1. Canonical factorization and the state space method 237
1.1. Elements of mathematical system theory 237
1.2. Inverting systems of Wiener–Hopf integral equations 242
1.3. Canonical factorization of rational matrix functions 244
1.4. Wiener–Hopf integral equations revisited 248
1.5. Notes 250

Chapter 2. J-unitary rational matrix functions 252
2.1. J-unitary rational matrix functions 252
2.2. J-inner rational matrix functions 253
2.3. J-inner-outer factorization in $\mathcal{R}at(H_\infty)$ 258
2.4. Notes 262

Chapter 3. Analysis of zeros 263
3.1. Eigenvectors and canonical systems of null functions 263
3.2. Canonical right and left null pairs at a point 268
3.3. Right and left null pairs on a set 271
3.4. The analytic case 274
3.5. Notes 276

Chapter 4. Inverse problems involving null pairs 277
4.1. An inverse problem in $\mathcal{R}at(H_\infty)$ 277
4.2. J-inner matrix functions in $\mathcal{R}at(H_\infty)$ with prescribed null pairs 279
4.3. Left tangential Nevanlinna–Pick interpolation 281
4.4. Notes 287

Chapter 5. Analysis of zeros and poles 288
5.1. Pole pairs and null-pole subspace 288
5.2. Coupling transformation and null-pole triples 290
5.3. More about the coupling transformation 294
5.4. Null-pole triples from minimal realizations 297
5.5. Another construction of a null-pole triple 299
5.6. Null-pole triples and the Smith–McMillan form 304
5.7. Notes 308

Chapter 6. Inverse problems involving null-pole triples 309
6.1. Rational matrix functions with prescribed null-pole triples 309
6.2. J-inner matrix functions with prescribed null-pole triples 316
6.3. Bitangential Nudelman interpolation 320
6.4. Notes 324

Bibliography 327

Index 335

Preface

Much of the importance of mathematics lies in its ability to provide theories which are useful in widely different fields of endeavour. The large and amorphous body of knowledge known as "the theory of linear operators" or "operator theory" is a good example of this and came to life about a century ago as a theory to encompass properties common to matrix, differential, and integral operators, in particular. Thus, it is a primary purpose of operator theory to provide a coherent body of knowledge which can explain phenomena common to the enormous variety of problems in which such linear operators play a part. The theory is a vital part of "functional analysis" whose methods and techniques are one of the major advances of twentieth century mathematics, and now play a pervasive role in the modelling of phenomena in probability, imaging, signal processing, systems theory, etc., as well as the more traditional areas of theoretical physics and mechanics.

In the latter half of the twentieth century advances have been made in the theory initiated earlier (as in Lecture Series 1, for example) and, in addition, new areas of application have stimulated the growth of new branches of the subject (see Lecture Series 4, for example). In the fall of 1994 two linked meetings on "Operator Theory and its Applications" were organized in Canada. The first was held under the auspices of the Fields Institute and the purpose was to provide instructional lectures giving introductions to some of the methods being used, and problems being tackled, in current research involving the use of operator theory. The lectures were aimed at an audience of graduate students and more senior researchers in universities and industry, and the present volume has been developed from four of the five lecture series presented at the Fields Institute in the last week of September, 1994. The organizers were fortunate to enlist authors (and lecturers) who are leaders in their field and, as the reader will discover, excellent expositors. The second meeting referred to above was an international research conference on "Applications of Operator Theory" organized by the Institute of Industrial Mathematical Sciences of the University of Manitoba, and held in the following week. The proceedings of that conference are to be published by Birkhäuser Verlag in their series on "Operator Theory: Advances and Applications".

Each of the four lecture series presented here is essentially a self-contained creation of its author, or authors, and in preparing this volume it soon became apparent that the imposition of uniform format and stylistic conventions would be counter-productive. Consequently, the reader will find different equation numbering and sectioning conventions used in each series. They are all easily understood and widely used strategies and should create no problems for the attentive reader. Each series has its own bibliography, but a unified list of contents and a unified index have been prepared for the reader's convenience. Concerning subject matter, each series has its own preface or introduction and it would be unnecessary and presumptuous to outline their contents here. The reader is invited to consult pages 7, 45, 143 and 235 for this information.

This volume has been prepared with the conviction that the subject matter will be useful and interesting to a wide audience in mathematics and its applications and owes its existence to several groups of people. First and foremost the authors who, as well as delivering excellent and stimulating lectures, have devoted much time and energy to the production of the TeX masters of their lectures, and have been very patient with editorial requests concerning the merging of their files. The programs of the two meetings, and hence the contents of this volume, were the responsibility of the Program Committee. The committee members were:

M.A Dahleh (M.I.T.),
P.A. Fillmore (Dalhousie),
B.A. Francis (Toronto),
F. Ghahramani (Manitoba),
K. Glover (Cambridge),
I. Gohberg (Tel Aviv),
T. Kailath (Stanford),
P. Lancaster (Calgary), Chair,
H. Langer (Vienna),
P. N.Shivakumar (Manitoba),
A.A. Shkalikov (Moscow),
B. Simon (Cal.Tech.),
H. Widom (Santa Cruz).

The staff of the Fields Institute have been helpful and supportive throughout and, in particular, we are grateful to Sue Embro and Sandra Valeriote for assistance in the preparation of this volume.

Finally, my personal thanks to Israel Gohberg for assistance and wise counsel from the inception of the project, and to Joanne Mellard for dedication and enthusiasm in the preparation of the final LaTeX master copies of this work.

Calgary,
June, 1995.

Peter Lancaster
Editor.

LECTURE SERIES 1

INFINITE MATRICES AND PROJECTION METHODS

Albrecht Böttcher
Technische Universität Chemnitz-Zwickau
Fakultät für Mathematik
09107 Chemnitz
Germany

1 Matrix representations of operators

Many problems of mathematics, physics, or engineering lead to the consideration of equations of the form

$$Ax = y \quad \text{or} \quad Ax - \lambda x = y,$$

where A is a linear and bounded operator on some infinite-dimensional separable Hilbert space H. Operator theory provides strategies and general recipes for tackling such equations, but without additional information about the structure of the operator A concrete results are not available. We here study the above equations under the assumption that we are able to represent A by an infinite matrix on the Hilbert space l^2 of all square-summable complex-valued sequences.

A matrix representation for A may be obtained in the usual way. Choose two orthonormal bases $\{e_k\}_{k=1}^\infty$ and $\{f_k\}_{k=1}^\infty$ of H. Then

$$x = \sum_{k=1}^{\infty} x_k e_k, \quad y = \sum_{k=1}^{\infty} y_k f_k$$

with

$$\sum_{k=1}^{\infty} |x_k|^2 = \|x\|^2 < \infty, \quad \sum_{k=1}^{\infty} |y_k|^2 = \|y\|^2 < \infty,$$

and the equation $Ax = y$ may be written as

$$\begin{pmatrix} a_{11} & a_{12} & a_{13} & \dots \\ a_{21} & a_{22} & a_{23} & \dots \\ a_{31} & a_{32} & a_{33} & \dots \\ \dots & \dots & \dots & \dots \end{pmatrix} \begin{pmatrix} x_1 \\ x_2 \\ x_3 \\ \vdots \end{pmatrix} = \begin{pmatrix} y_1 \\ y_2 \\ y_3 \\ \vdots \end{pmatrix} \tag{1}$$

with $a_{jk} := (Ae_k, f_j)$. We so arrive at an infinite linear algebraic system for $x = \{x_k\}_{k=1}^\infty \in l^2$. The structure of the infinite matrix in (1) depends on A itself and on the choice of the bases $\{e_k\}$ and $\{f_k\}$. It should not be hidden that in practice the evaluation of the scalar products (Ae_k, f_j) may cause serious problems.

Example 1.1 Let $H = L^2(0, \infty)$ and let A be the integral operator given by

$$(Ax)(t) = \int_0^\infty \alpha(t, s) x(s)\, ds \quad (t > 0).$$

1991 *Mathematics Subject Classification.* Primary 15–01, 47–01, 65–01; Secondary 15A06, 15A15, 15A18, 15A60, 45E05, 45E10, 45L05, 46N05, 46N55, 47A10, 47A30, 47A58, 47B35, 47C15, 47N40, 65F15, 65F30, 65F40.

Research supported by the Alfried Krupp Förderpreis für junge Hochschullehrer of the Krupp Foundation.

If

$$\int_0^\infty \int_0^\infty |\alpha(t,s)|^2 \, dt \, ds < \infty, \tag{2}$$

then A is bounded (even Hilbert–Schmidt and thus compact). An orthonormal basis in $L^2(0,\infty)$ is constituted by the functions

$$\lambda_n(t) = \frac{1}{n!} L_n(t) e^{-t/2} \quad (t > 0),$$

where

$$\begin{aligned} L_0(t) &= 1, \quad L_1(t) = 1 - t, \quad L_2(t) = 2 - 4t + t^2, \quad \dots \\ L_n(t) &= e^t \frac{d^n}{dt^n} (t^n e^{-t}) \end{aligned}$$

are the Laguerre polynomials. Hence, on choosing $e_k = f_k = \lambda_{k-1}$ for $k = 1, 2, \dots$ we get

$$a_{jk} = (Ae_k, f_j) = \int_0^\infty \int_0^\infty \alpha(t,s) \lambda_{k-1}(t) \lambda_{j-1}(s) \, dt \, ds.$$

Without additional information about the kernel $\alpha(t,s)$ all we can say is that $\sum_{j,k=1}^\infty |a_{jk}|^2 < \infty$.

Suppose now that the kernel depends only on the difference of its arguments:

$$(Ax)(t) = \int_0^\infty \alpha(t-s) x(s) \, ds \quad (t > 0). \tag{3}$$

In this case A is referred to as a *Wiener–Hopf integral operator*. One easily checks that (2) is never satisfied unless α is identically zero. However, if

$$\int_{-\infty}^\infty |\alpha(t)| \, dt < \infty, \tag{4}$$

then A is bounded on $L^2(0,\infty)$ and one can show (see e.g. Gohberg and Feldman [1971, Chap. III., Sec. 3]) that with respect to the bases $\{e_k\} = \{f_k\} = \{\lambda_{k-1}\}$ the matrix of A is given by

$$a_{jk} = (Ae_k, f_j) = \frac{1}{2\pi} \int_{-\infty}^\infty \hat{\alpha}(\xi) \left(\frac{\xi + i/2}{\xi - i/2} \right)^{j-k} \frac{d\xi}{\xi^2 + 1/4}, \tag{5}$$

where

$$\hat{\alpha}(\xi) := \int_{-\infty}^\infty \alpha(t) e^{i\xi t} \, dt \quad (\xi \in \mathbf{R})$$

stands for the Fourier transform of the kernel α. Thus, the matrix $(a_{jk})_{j,k=1}^\infty$ now has a special structure: it is constant along the parallels to its main diagonal. Such matrices are called *Toeplitz matrices* . ■

Example 1.2 Let $\mathbf{T} = \{z \in \mathbf{C} : |z| = 1\}$ denote the complex unit circle and put $H = L^2(\mathbf{T})$. The operator $A = M(a)$ of multiplication by the function a is bounded on $L^2(\mathbf{T})$ if (and only if) $a \in L^\infty(\mathbf{T})$. Define

$$\chi_n(e^{i\theta}) = \frac{1}{\sqrt{2\pi}} e^{in\theta} \quad (n \in \mathbf{Z}).$$

Then $\{\chi_n\}_{n=-\infty}^{\infty}$ is an orthonormal basis of $L^2(\mathbf{T})$. A matrix representation of $M(a)$ that is of the form $(a_{jk})_{j,k=1}^{\infty}$ may be obtained by choosing

$$e_1 = f_1 = \chi_0,\ e_2 = f_2 = \chi_1,\ e_3 = f_3 = \chi_{-1},\ e_4 = f_4 = \chi_2, \ldots,$$

but such a representation is inconvenient for many purposes. We rather put up with working with a "doubly infinite" matrix $(a_{jk})_{j,k=-\infty}^{\infty}$, i.e. we let $e_k = f_k = \chi_k$ for all $k \in \mathbf{Z}$ so that $M(a)$ is represented by the matrix

$$\left(\begin{array}{ccc|cccc} \ldots & \ldots & \ldots & \ldots & \ldots & \ldots & \ldots \\ \ldots & a_0 & a_{-1} & a_{-2} & a_{-3} & a_{-4} & \ldots \\ \ldots & a_1 & a_0 & a_{-1} & a_{-2} & a_{-3} & \ldots \\ \hline \ldots & a_2 & a_1 & a_0 & a_{-1} & a_{-2} & \ldots \\ \ldots & a_3 & a_2 & a_1 & a_0 & a_{-1} & \ldots \\ \ldots & a_4 & a_3 & a_2 & a_1 & a_0 & \ldots \\ \ldots & \ldots & \ldots & \ldots & \ldots & \ldots & \ldots \end{array}\right) \tag{6}$$

with

$$a_n = \frac{1}{2\pi}\int_0^{2\pi} a(e^{i\theta})e^{-in\theta}\,d\theta \quad (n \in \mathbf{Z}).$$

Matrices of the form (6) are usually referred to as *Laurent matrices.*

The *Cauchy singular integral operator* S is defined by

$$(Sx)(t) = \frac{1}{\pi i}\int_{\mathbf{T}} \frac{x(s)}{s-t}\,ds \quad (t \in \mathbf{T}),$$

the integral understood in the principal value sense. It is easily seen that

$$S\chi_n = \begin{cases} \chi_n & \text{for} \quad n \geq 0 \\ -\chi_n & \text{for} \quad n < 0 \end{cases},$$

which implies that S is bounded on $L^2(\mathbf{T})$ and that its matrix representation with respect to the bases $\{e_k\} = \{f_k\} = \{\chi_k\}$ is

$$\left(\begin{array}{ccc|cccc} \ldots & \ldots & \ldots & \ldots & \ldots & \ldots & \ldots \\ \ldots & -1 & 0 & 0 & 0 & 0 & \ldots \\ \ldots & 0 & -1 & 0 & 0 & 0 & \ldots \\ \hline \ldots & 0 & 0 & 1 & 0 & 0 & \ldots \\ \ldots & 0 & 0 & 0 & 1 & 0 & \ldots \\ \ldots & 0 & 0 & 0 & 0 & 1 & \ldots \\ \ldots & \ldots & \ldots & \ldots & \ldots & \ldots & \ldots \end{array}\right).$$

Thus, the *singular integral equation*

$$a(t)x(t) + b(t)\frac{1}{\pi i}\int_{\mathbf{T}} \frac{x(s)}{t-s}\,ds = y(t) \quad (t \in \mathbf{T}) \tag{7}$$

may be converted into the infinite linear system

$$\left(\begin{array}{cccc|cccc} \cdots & \cdots & \cdots & \cdots & \cdots & \cdots \\ \cdots & a_0 - b_0 & a_{-1} - b_{-1} & a_{-2} + b_{-2} & a_{-3} + b_{-3} & \cdots \\ \cdots & a_1 - b_1 & a_0 - b_0 & a_{-1} + b_{-1} & a_{-2} + b_{-2} & \cdots \\ \hline \cdots & a_2 - b_2 & a_1 - b_1 & a_0 + b_0 & a_{-1} + b_{-1} & \cdots \\ \cdots & a_3 - b_3 & a_2 - b_2 & a_1 + b_1 & a_0 + b_0 & \cdots \\ \cdots & a_4 - b_4 & a_3 - b_3 & a_2 + b_2 & a_1 + b_1 & \cdots \\ \cdots & \cdots & \cdots & \cdots & \cdots & \cdots \end{array}\right) \left(\begin{array}{c} \vdots \\ x_{-2} \\ x_{-1} \\ \hline x_0 \\ x_1 \\ x_2 \\ \vdots \end{array}\right) = \left(\begin{array}{c} \vdots \\ y_{-2} \\ y_{-1} \\ \hline y_0 \\ y_1 \\ y_2 \\ \vdots \end{array}\right).$$

Since this matrix is composed by two halves of Laurent matrices, it is called a *paired matrix.* ■

Example 1.3 Let $L^2(\mathbf{D})$ denote the Hilbert space of all functions x on the complex unit disk $\mathbf{D} = \{z \in \mathbf{C} : |z| < 1\}$ that are square-integrable with respect to (normalized) area measure:

$$\|x\|^2 = \int_{\mathbf{D}} |x(z)|^2 \, dA(z) = \frac{1}{\pi} \int_0^1 \int_0^{2\pi} |x(re^{i\theta})|^2 r \, dr \, d\theta < \infty.$$

The *Bergman space* $A^2(\mathbf{D})$ is the set of all functions in $L^2(\mathbf{D})$ that are holomorphic on $\mathbf{D}$. It is well known (see e.g. Gaier [1987] or Zhu [1990]) that $A^2(\mathbf{D})$ is a closed subspace of $L^2(\mathbf{D})$ and that the orthogonal projection P of $L^2(\mathbf{D})$ onto $A^2(\mathbf{D})$ acts by the rule

$$(Px)(z) = \int_{\mathbf{D}} \frac{x(\omega)}{(1 - z\overline{\omega})^2} \, dA(\omega) \quad (z \in \mathbf{D}).$$

The *Bergman space Bergman space Toeplitz operator* $T^B(a)$ induced by a function $a \in L^\infty(\mathbf{D})$ is the operator that sends $x \in A^2(\mathbf{D})$ to $P(ax) \in A^2(\mathbf{D})$. An orthonormal basis in $A^2(\mathbf{D})$ is built up by the functions $\{e_n\}_{n=1}^\infty$ defined by

$$e_n(z) = \sqrt{n} z^{n-1}.$$

The matrix of $T^B(a)$ with respect to the bases $\{e_k\}_{k=1}^\infty = \{f_k\}_{k=1}^\infty$ is $(a_{jk})_{j,k=1}^\infty$ with

$$a_{jk} = (T^B(a)e_k, e_j) = \sqrt{kj} \int_{\mathbf{D}} a(z) z^{k-1} \overline{z}^{j-1} \, dA(z). \tag{8}$$

In case $a \in L^\infty(\mathbf{D})$ is the harmonic extension of some function $b \in L^\infty(\mathbf{T})$,

$$a(re^{i\theta}) = \sum_{n \in \mathbf{Z}} b_n r^{|n|} e^{in\theta}, \quad b_n = \frac{1}{2\pi} \int_0^{2\pi} b(e^{i\theta}) e^{-in\theta} \, d\theta,$$

the integrals in (8) can be evaluated, and what results is

$$a_{jk} = \frac{2\sqrt{kj}}{|j-k| + j + k} b_{j-k}. \tag{9}$$

The matrix with the entries (9) is not a Toeplitz matrix unless b (and thus a) is constant. However, if b is continuous on $\mathbf{T}$, we have the decomposition

$$\begin{pmatrix} a_{11} & a_{12} & a_{13} & \dots \\ a_{21} & a_{22} & a_{23} & \dots \\ a_{31} & a_{32} & a_{33} & \dots \\ \dots & \dots & \dots & \dots \end{pmatrix} = \begin{pmatrix} b_0 & b_{-1} & b_{-2} & \dots \\ b_1 & b_0 & b_{-1} & \dots \\ b_2 & b_1 & b_0 & \dots \\ \dots & \dots & \dots & \dots \end{pmatrix} + \begin{pmatrix} c_{11} & c_{12} & c_{13} & \dots \\ c_{21} & c_{22} & c_{23} & \dots \\ c_{31} & c_{32} & c_{33} & \dots \\ \dots & \dots & \dots & \dots \end{pmatrix}$$

and it can be shown (see e.g. Böttcher [1990]) that $(c_{jk})_{j,k=1}^{\infty}$ is the matrix of some compact operator. ■

2 Three problems for infinite matrices

In what follows we will not distinguish between an operator and its matrix representation with respect to a pair of orthonormal bases. Thus, we think of A as given on l^2 by the infinite matrix

$$A = \begin{pmatrix} a_{11} & a_{12} & a_{13} & \dots \\ a_{21} & a_{22} & a_{23} & \dots \\ a_{31} & a_{32} & a_{33} & \dots \\ \dots & \dots & \dots & \dots \end{pmatrix}. \tag{10}$$

What have we won by the transition from operators to infinite matrices? The point is that we may replace the matrix (10) by its $n \times n$ principal section

$$A_n = \begin{pmatrix} a_{11} & a_{12} & \dots & a_{1n} \\ a_{21} & a_{22} & \dots & a_{2n} \\ \vdots & \vdots & & \vdots \\ a_{n1} & a_{n2} & \dots & a_{nn} \end{pmatrix}, \tag{11}$$

having hopes that for large n the properties of the finite matrices (11) mimic those of the infinite matrix (10). We will exemplify what we have in mind by three concrete questions.

Problem I Before solving the equations $Ax = y$ or $Ax - \lambda x = y$ we should find out whether A or $A - \lambda I$ is invertible. Hence, it is desirable to know the *spectrum* of A,

$$\operatorname{sp} A := \{\lambda \in \mathbf{C} : A - \lambda I \text{ is not invertible}\}.$$

Assuming that we are able to determine the spectrum (eigenvalues) of the matrices A_n, we naturally arrive at the following problem: If n is large, is $\operatorname{sp} A_n$ close (in some sense) to $\operatorname{sp} A$?

We remark that this problem is of great interest also from the "reverse" point of view: frequently the spectrum of A is known and what one is interested in is the spectrum of A_n for large n (where n might be 100 in numerical analysis and 10^{23} in statistical physics).

Example 2.1 Let

$$A = \begin{pmatrix} 0 & 1 & 0 & 0 & \dots \\ 1 & 0 & 1 & 0 & \dots \\ 0 & 1 & 0 & 1 & \dots \\ 0 & 0 & 1 & 0 & \dots \\ \dots & \dots & \dots & \dots & \dots \end{pmatrix}.$$

Then $\operatorname{sp} A = [-2, 2]$, and it is a nice exercise to show that

$$\operatorname{sp} A_n = \left\{ 2\cos\frac{\pi}{n+1},\ 2\cos\frac{2\pi}{n+1},\ \dots,\ 2\cos\frac{n\pi}{n+1} \right\}.$$

Consequently, $\operatorname{sp} A_n$ mimics $\operatorname{sp} A$ as n approaches infinity. ■

Example 2.2 Consider now the Toeplitz matrix

$$A = \begin{pmatrix} 0 & 0 & 0 & 0 & \dots \\ 1 & 0 & 0 & 0 & \dots \\ 0 & 1 & 0 & 0 & \dots \\ 0 & 0 & 1 & 0 & \dots \\ \dots & \dots & \dots & \dots & \dots \end{pmatrix}.$$

We have $\operatorname{sp} A = \overline{\mathbf{D}} = \{\lambda \in \mathbf{C} : |\lambda| \le 1\}$, while $\operatorname{sp} A_n = \{0\}$ for all n. Thus, we cannot say that $\operatorname{sp} A_n$ is close to $\operatorname{sp} A$ if n is large. ■

In the two previous examples, we encountered Toeplitz matrices (operators) of very simple structure. The *Toeplitz operator* $T(a)$ onl^2 generated by a function $a \in L^\infty(\mathbf{T})$ is defined via the matrix

$$T(a) = \begin{pmatrix} a_0 & a_{-1} & a_{-2} & \dots \\ a_1 & a_0 & a_{-1} & \dots \\ a_2 & a_1 & a_0 & \dots \\ \dots & \dots & \dots & \dots \end{pmatrix} \tag{12}$$

where

$$a_n = \frac{1}{2\pi}\int_0^{2\pi} a(e^{i\theta})e^{-in\theta}\,d\theta \quad (n \in \mathbf{Z}). \tag{13}$$

Note that we were led to such operators also in the examples of Section 1. The function a is called the *symbol* of the operator (12). Since $T(a)$ may be identified with the south-east block of the matrix (6), we have

$$\|T(a)\| \le \|M(a)\| = \|a\|_\infty$$

for every $a \in L^\infty(\mathbf{T})$. The symbols of the operators A in Examples 2.1 and 2.2 are $a(t) = t^{-1} + t$ and $a(t) = t \quad (t = e^{i\theta} \in \mathbf{T})$, respectively. The following two (by now classical) theorems describe $\operatorname{sp} T(a)$ in two important cases.

Theorem 2.3 *The operator $T(a)$ is normal, i.e. it commutes with its adjoint, if and only if the essential range $\mathcal{R}(a)$ of the symbol a lies on some line segment. In that case* $\operatorname{sp} T(a)$ *equals* $\operatorname{conv} \mathcal{R}(a)$, *the (closed) convex hull of $\mathcal{R}(a)$.*

The adjoint of $T(a)$ is $T^*(a) = T(\overline{a})$, where $\overline{a}(t) = \overline{a(t)}$ $(t \in \mathbf{T})$. In particular, $T(a)$ is self-adjoint if and only if its symbol is real-valued. In that case Theorem 2.3 tells us that

$$\operatorname{sp} T(a) = [\operatorname*{ess\,inf}_{t\in\mathbf{T}} a(t),\ \operatorname*{ess\,sup}_{t\in\mathbf{T}} a(t)]. \tag{14}$$

Since a function $a \in L^\infty(\mathbf{T})$ is defined only almost everywhere, some care is in order when speaking about the range of a. The *essential range* $\mathcal{R}(a)$ is the set of all $\lambda \in \mathbf{C}$ such that $\{t \in \mathbf{T} : |a(t) - \lambda| < \varepsilon\}$ has positive (Lebesgue) measure for every $\varepsilon > 0$.

Theorem 2.4 *If a is continuous on $\mathbf{T}$, $a \in C(\mathbf{T})$, then*

$$\operatorname{sp} T(a) = a(\mathbf{T}) \cup \{\lambda \in \mathbf{C} \setminus a(\mathbf{T}) : \ \operatorname{wind}(a - \lambda) \neq 0\},$$

where wind $(a - \lambda)$ *is the winding number (index) of the naturally oriented curve $a(\mathbf{T})$ about the point λ.*

A few curves and their winding numbers are plotted in Figure 1. If, in particular, $a(\mathbf{T})$ is the solid curve in the lower part of Figure 1, then $\operatorname{sp} T(a)$ is the union of this curve and the dotted regions.

Proofs of the preceding two theorems can be found in the books Halmos [1967], Gohberg and Feldman [1971], or Böttcher and Silbermann [1990], for example. ■

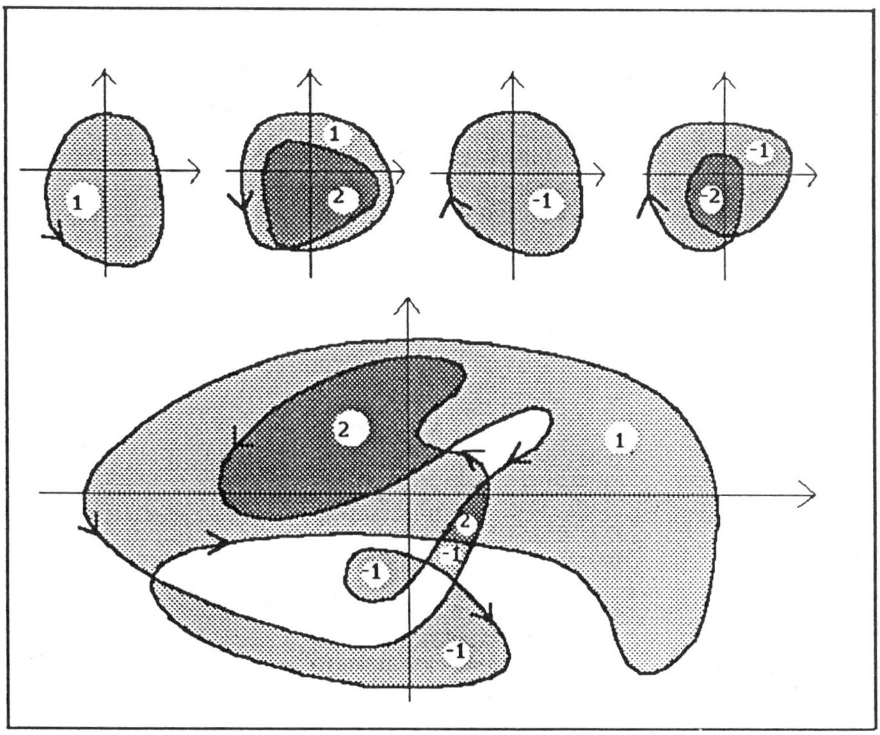

Figure 1

Example 2.5 The Figures 2 to 7 show the ranges $a(\mathbf{T})$ for six (randomly chosen) trigonometric polynomials a and the spectrum (the eigenvalues) of $T_{60}(a)$, the 60×60 principal section of $T(a)$. The situation is the same for $T_n(a)$ with $n > 60$: it follows from Schmidt and Spitzer [1960] that if $T(a)$ is a band matrix, then the eigenvalues of $T_n(a)$ cluster along a certain "thin" set as n goes to infinity. This "thin" set can be completely described and it is in no obvious way related to $\operatorname{sp} T(a)$ in case this spectrum has a nonempty interior. ■

Problem II We know from numerical analysis that the question "Is A invertible?" is not the right question. A better question reads: What is $\|A^{-1}\|$? In this context one puts $\|A^{-1}\| = \infty$ if A is not invertible.

Given A and A_n by (10) and (11), we now ask: Is $\|A_n^{-1}\|$ close to $\|A^{-1}\|$ if n is large? This is a very difficult question in general, and even relaxed versions of it are of interest. Such a baby version is: If $\|A^{-1}\| < \infty$, does it follow that $\|A_n^{-1}\| < \infty$ for all sufficiently large n? Equivalently: If A is invertible, are the operators A_n invertible for all n that are large enough? A more intrinsic baby version is: If $\|A^{-1}\| < \infty$, is there an M such that $\|A_n^{-1}\| \le M$ for all sufficiently large n? We will have more to say about this problem in the forthcoming sections. At the present moment, let us leave it with two simple examples.

Example 2.6 Let A be an infinite matrix that contains exactly one unit in every row and every column. Then A is an unitary operator on l^2 and thus $\|A\| = \|A^{-1}\| = 1$. If A is given by

$$A = \begin{pmatrix} 0 & 1 & 0 & 0 & 0 & 0 & \dots \\ 1 & 0 & 0 & 0 & 0 & 0 & \dots \\ 0 & 0 & 0 & 1 & 0 & 0 & \dots \\ 0 & 0 & 1 & 0 & 0 & 0 & \dots \\ 0 & 0 & 0 & 0 & 0 & 1 & \dots \\ 0 & 0 & 0 & 0 & 1 & 0 & \dots \\ \dots & \dots & \dots & \dots & \dots & \dots & \dots \end{pmatrix},$$

then A is such matrix, but the last row of A_{2k+1} consists only of zeros. Hence, although A is invertible, the matrices A_n are noninvertible whenever n is odd. ■

Example 2.7 We now slightly modify the matrix in the previous example: let $\{\varepsilon_n\}_{n=1}^{\infty}$ be a sequence of positive numbers converging to zero and put

$$A = \begin{pmatrix} \varepsilon_1 & 1 & 0 & 0 & 0 & 0 & \dots \\ 1 & \varepsilon_1 & 0 & 0 & 0 & 0 & \dots \\ 0 & 0 & \varepsilon_2 & 1 & 0 & 0 & \dots \\ 0 & 0 & 1 & \varepsilon_2 & 0 & 0 & \dots \\ 0 & 0 & 0 & 0 & \varepsilon_3 & 1 & \dots \\ 0 & 0 & 0 & 0 & 1 & \varepsilon_3 & \dots \\ \dots & \dots & \dots & \dots & \dots & \dots & \dots \end{pmatrix},$$

Assume also that $\varepsilon_n \le 1/2$ for all n. Since

$$\left\| \begin{pmatrix} \varepsilon & 1 \\ 1 & \varepsilon \end{pmatrix}^{-1} \right\| = \frac{1}{1-\varepsilon^2} \left\| \begin{pmatrix} \varepsilon & -1 \\ -1 & \varepsilon \end{pmatrix} \right\| \le \frac{(2+2\varepsilon^2)^{1/2}}{1-\varepsilon^2} < \frac{4^{1/2}}{1-1/4} = \frac{8}{3},$$

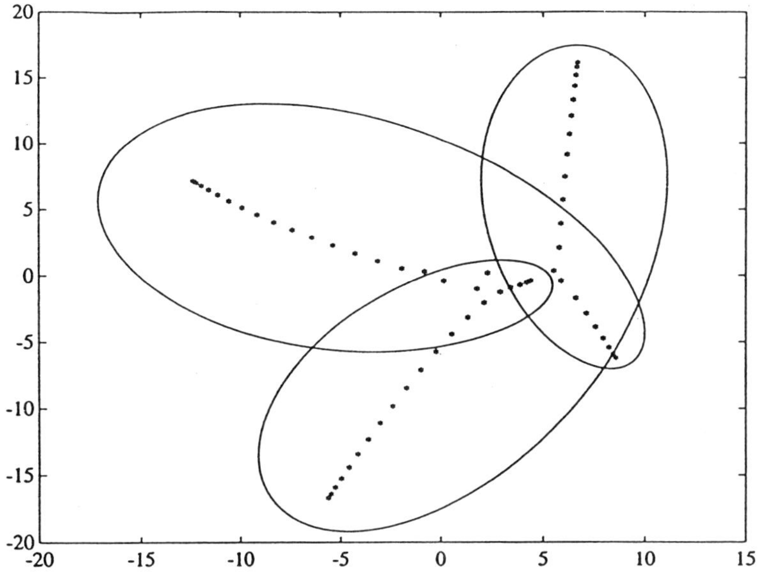

Figure 2

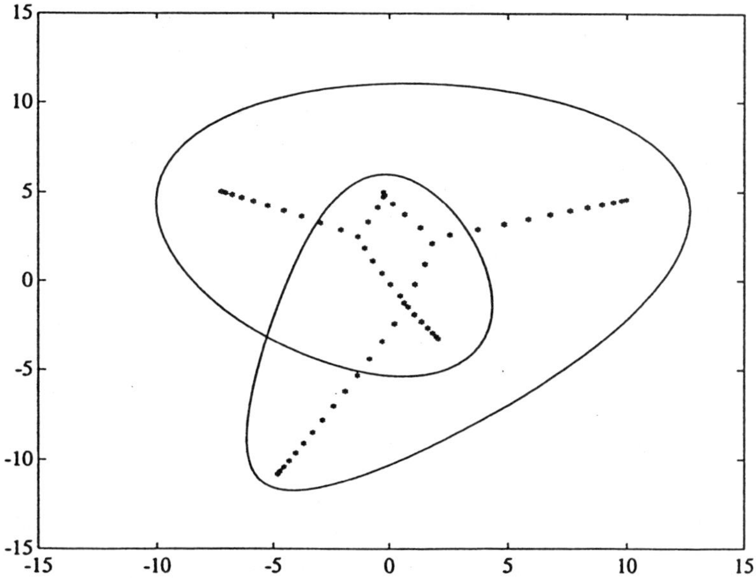

Figure 3

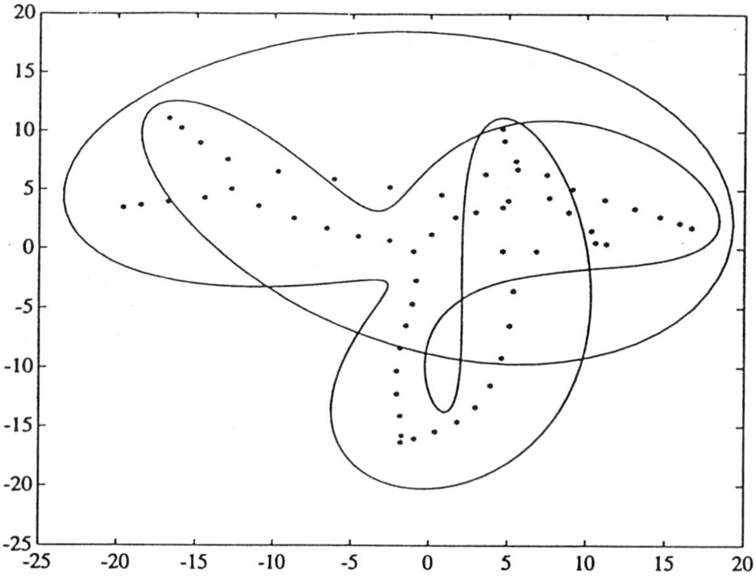

Figure 4

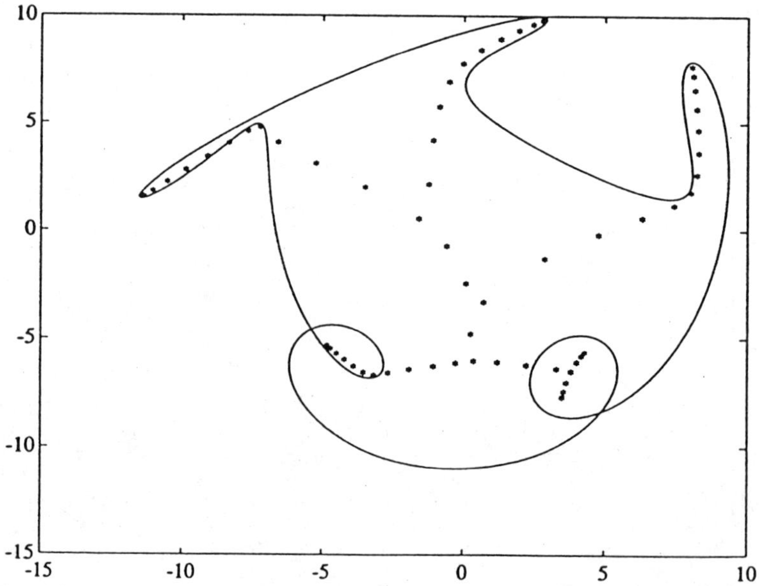

Figure 5

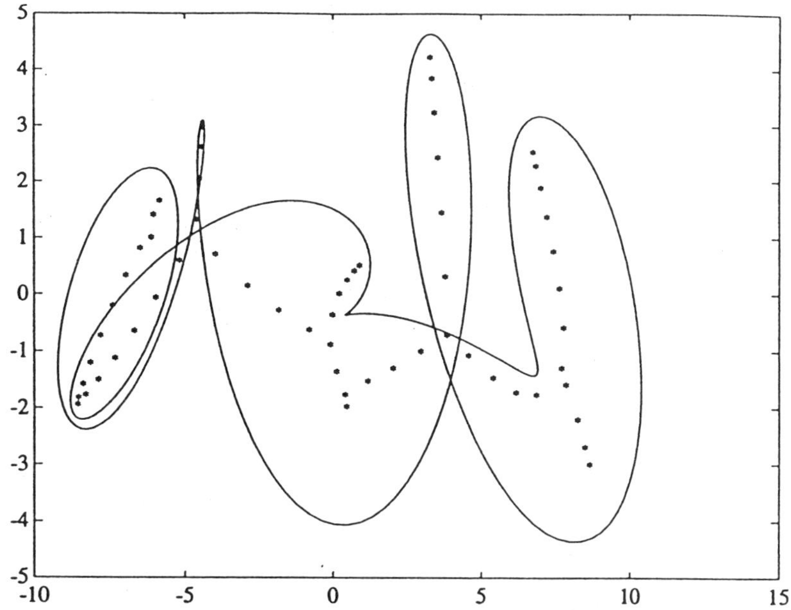

Figure 6

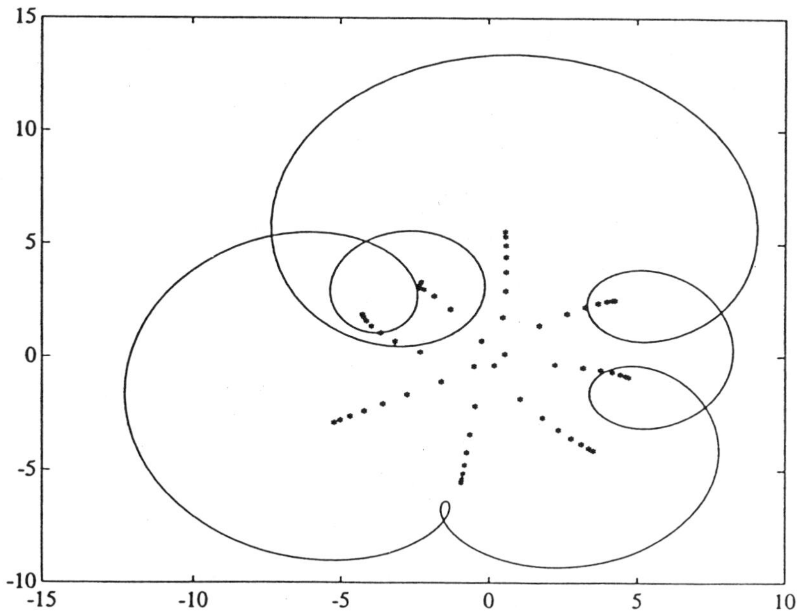

Figure 7

the operator A is invertible and $\|A^{-1}\| \leq 8/3$. Clearly, A_n is invertible for every $n \geq 1$, and if n is even then $\|A_n^{-1}\| \leq 8/3$, while if $n = 2k-1$ is odd then $\|A_{2k-1}^{-1}\| \geq 1/\varepsilon_k$. Thus, although all A_n are invertible, the norms $\|A_n^{-1}\|$ are not uniformly bounded. ■

Problem III Suppose we know that A is invertible and moreover, assume we can estimate $\|A^{-1}\|$ and have thus an idea of the sensitivity of the dependence of x on $y = Ax$. In order to solve the equation $Ax = y$, i.e. the infinite system

$$\begin{pmatrix} a_{11} & a_{12} & a_{13} & \cdots \\ a_{21} & a_{22} & a_{23} & \cdots \\ a_{31} & a_{32} & a_{33} & \cdots \\ \cdots & \cdots & \cdots & \cdots \end{pmatrix} \begin{pmatrix} x_1 \\ x_2 \\ x_3 \\ \vdots \end{pmatrix} = \begin{pmatrix} y_1 \\ y_2 \\ y_3 \\ \vdots \end{pmatrix}, \tag{15}$$

we consider the "truncated" systems

$$\begin{pmatrix} a_{11} & \cdots & a_{1n} \\ \vdots & & \vdots \\ a_{n1} & \cdots & a_{nn} \end{pmatrix} \begin{pmatrix} x_1^{(n)} \\ \vdots \\ x_n^{(n)} \end{pmatrix} = \begin{pmatrix} y_1 \\ \vdots \\ y_n \end{pmatrix}. \tag{16}$$

To abbreviate notation, we introduce the projections P_n that act on l^2 by the rule

$$P_n : \{x_1, x_2, x_3, \ldots\} \mapsto \{x_1, x_2, \ldots, x_n, 0, 0, \ldots\}. \tag{17}$$

The system (16) may then be written as the equation

$$P_n A x^{(n)} = P_n y \quad (x^{(n)} \in \operatorname{Im} P_n), \tag{18}$$

where, of course, $\operatorname{Im} P_n$ denotes the image (range) of P_n. Since $P_n x^{(n)} = x^{(n)}$ for $x^{(n)} \in \operatorname{Im} P_n$, we might also write

$$P_n A P_n x^{(n)} = P_n y \quad (x^{(n)} \in \operatorname{Im} P_n)$$

for (18). The matrix A_n given by (11) may therefore be identified with the restriction of $P_n A P_n$ to $\operatorname{Im} P_n$:

$$A_n = P_n A P_n | \operatorname{Im} P_n.$$

The basic question is now as follows: Are the operators A_n invertible for all sufficiently large n and do the solutions $x^{(n)}$ of (18) converge in l^2 to the solution $x \in l^2$ of the equation $Ax = y$?

Example 2.6 shows that the answer to this question may be negative. Nevertheless, things are not as bad as it might seem in view of this example. In the following sections, we will establish a number of "positive" results for large classes of operators A.

3 The finite section method

Let A be given on l^2 by the infinite matrix (10), define A_n by (11), and let P_n stand for the projections given by (17). We say that the *finite section method converges* for A and write $A \in \Pi\{P_n, P_n\}$ if

(i) the matrices $A_n = P_n A P_n|\text{Im}\, P_n$ are invertible for all sufficiently large n, for $n \geq n_0$ say;

(ii) for each $y \in l^2$ the (unique) solutions $x^{(n)} \in \text{Im}\, P_n$ of the equations $P_n A P_n x^{(n)} = P_n y$ $(n \geq n_0)$ converge in l^2 to a solution of the equation $Ax = y$.

A sequence of operators B_n on l^2 is said to *converge strongly* to an operator B on l^2 if $B_n x \to Bx$ in l^2 for every $x \in l^2$. The following simple proposition does two things: it first shows that A must be necessarily invertible if $A \in \Pi\{P_n, P_n\}$ and it secondly relates the finite section method to strong convergence.

Proposition 3.1 *We have $A \in \Pi\{P_n, P_n\}$ if and only if A is invertible, the matrices A_n are invertible for all sufficiently large n, and $A_n^{-1} P_n$ converges strongly to A^{-1}.*

Proof We only show that A is invertible whenever $A \in \Pi\{P_n, P_n\}$. So suppose $A \in \Pi\{P_n, P_n\}$. Then for every $y = l^2$ the sequence $\{A_n^{-1} P_n y\}_{n \geq n_0}$ is convergent and hence, by the Banach–Steinhaus theorem (also known as the "uniform boundedness principle"),

$$M := \sup_{n \geq n_0} \|A_n^{-1} P_n\| < \infty.$$

Given any $x \in l^2$, we therefore have for $n \geq n_0$

$$\begin{aligned} \|P_n x\| &= \|A_n^{-1} A_n P_n x\| = \|A_n^{-1} P_n A P_n x\| \\ &\leq \|A_n^{-1} P_n\| \, \|A P_n x\| \leq M \|A P_n x\|, \end{aligned}$$

and since P_n converges strongly to I, it follows that $\|x\| \leq M\|Ax\|$. Thus, A is injective. Because A is surjective due to condition (ii), we arrive at the conclusion that A is invertible. ∎

From the basic course in numerical analysis, we know the formula

$$\text{convergence} = \text{approximation} + \text{stability}. \tag{19}$$

In the situation at hand we need not be worried about approximation: $A_n P_n = P_n A P_n$ converges strongly to A and thus A_n "approximates" A. What about stability? We say that the sequence $\{A_n\}_{n=1}^{\infty}$ is *stable* if A_n is invertible for all n large enough, for $n \geq n_0$ say, and

$$\sup_{n \geq n_0} \|A_n^{-1} P_n\| < \infty.$$

Here now is what formula (19) states for the finite section method.

Proposition 3.2 *The finite section method converges for A, that is $A \in \Pi\{P_n, P_n\}$, if and only if A is invertible and $\{A_n\}$ is a stable sequence.*

Proof Let $A \in \Pi\{P_n, P_n\}$. Then A is invertible and $A_n^{-1} P_n \to A^{-1}$ due to Proposition 3.1, whence $\sup \|A_n^{-1} P_n\| < \infty$ by virtue of the Banach–Steinhaus theorem.

Conversely, suppose A is invertible and $\{A_n\}$ is stable. Then for each $y \in l^2$,

$$\|A_n^{-1}P_n y - A^{-1}y\| \le \|A_n^{-1}P_n y - P_n A^{-1}y\| + \|P_n A^{-1}y - A^{-1}y\|.$$

The second term on the right goes to zero since $P_n \to I$ strongly, and for the first term on the right we have

$$\begin{aligned}&\|A_n^{-1}P_n y - P_n A^{-1}y\| = \|A_n^{-1}(P_n y - P_n A P_n A^{-1}y)\|\\ &\le M\|P_n y - P_n A P_n A^{-1}y\| = o(1),\end{aligned}$$

again because $P_n \to I$ strongly. ■

Propositions 3.1 and 3.2 do not tell us anything that might solve the question whether the finite section method converges for a given concrete operator. However, what we learn from Proposition 3.2 is that Problem III of Section 2 is equivalent to the baby version of Problem II that asks whether $\|A_n^{-1}\| \le M$ for all sufficiently large n if $\|A^{-1}\| < \infty$.

4 Selfadjoint and compact operators

To study the finite section method for concrete operators, let us begin with two classes of operators that enjoy great popularity in almost every text on functional analysis: the selfadjoint and compact operators.

An operator A on l^2 is *selfadjoint* if it coincides with its adjoint operator, that is, if $A = A^*$. In case A is given by the matrix $(a_{jk})_{j,k=1}^{\infty}$ on l^2, the selfadjointness of A is equivalent to requiring that $a_{jk} = \overline{a_{kj}}$ for all j and k. An operator A on l^2 is said to be *positive definite* if it is selfadjoint and there exists an $\varepsilon > 0$ such that

$$(Ax, x) \ge \varepsilon\|x\|^2 \quad \text{for all} \quad x \in l^2. \tag{20}$$

Notice that (20) implies that A is invertible and $\|A^{-1}\| \le 1/\varepsilon$.

Theorem 4.1 *If A is positive definite then $A \in \Pi\{P_n, P_n\}$.*

Proof Since $P_n^* = P_n = P_n^2$, we have

$$(P_n A P_n x, P_n x) = (A P_n x, P_n x) \ge \varepsilon\|P_n x\|^2,$$

which implies that $\|(P_n A P_n)^{-1}P_n\| \le 1/\varepsilon$ for all $n \ge 1$. The assertion is now immediate from Proposition 3.2. ■

To be positive definite is of course an extremely strong property and so the preceding result (does it deserve the title "theorem"?) should not come as a surprise. But what happens if A is merely supposed to be selfadjoint? Then the finite section method may converge or may not converge. For instance, if

$$A = \begin{pmatrix} 0 & 1 & 0 & 0 & 0 & \dots \\ 1 & 0 & 0 & 0 & 0 & \dots \\ 0 & 0 & 1 & 0 & 0 & \dots \\ 0 & 0 & 0 & 1 & 0 & \dots \\ 0 & 0 & 0 & 0 & 1 & \dots \\ \dots & \dots & \dots & \dots & \dots & \dots \end{pmatrix}$$

then the finite section method converges although A is not positive definite. On the other hand, Examples 2.6 and 2.7 (in conjunction with Proposition 3.2) reveal that the finite section method need not converge if A is selfadjoint but not positive definite. In case $A = T(a)$ is a selfadjoint Toeplitz operator (matrix), we infer from (14) that $T(a)$ is invertible if and only if A or $-A$ is positive definite. Thus, combining Proposition 3.1 and Theorem 4.1 we arrive at the rather puzzling conclusion that

> *the finite section method converges for a selfadjoint Toeplitz operator if and only if this operator is either positive definite or negative definite.*

A useful generalization of Theorem 4.1 states that $A \in \Pi\{P_n, P_n\}$ whenever there is an $\varepsilon > 0$ such that $\operatorname{Re}(Ax, x) \geq \varepsilon \|x\|^2$ for all $x \in l^2$, or equivalently, whenever $A = G + iH$ with a positive definite operator G and a selfadjoint operator H (see e.g. Gohberg and Feldman [1971, Chap. II, Sec. 2]).

Finally, we remark that if A is any invertible operator, then A^*A is positive definite. Thus, in case we are able to effectively compute the entries of the matrix A^*A, we could replace the equation $Ax = y$ by the equivalent equation $A^*Ax = A^*y$ and try solving the latter equation by having recourse to the finite section method.

Let us now consider compact operators . Recall that an operator K on l^2 is said to be *compact* if it maps bounded sets to sets whose closure is compact. Since l^2 is a separable Hilbert space, an operator K on l^2 is compact if and only if there is a sequence $\{K_n\}_{n=1}^{\infty}$ of finite-rank operators K_n such that $\|K - K_n\| \to 0$ as $n \to \infty$.

Examples 4.2 (a) Let K be given by the infinite matrix $(a_{jk})_{j,k=1}^{\infty}$ on l^2. If $\sum\limits_{j,k=1}^{\infty} |a_{jk}|^2 < \infty$, then K is compact. Indeed, the matrix

$$K_n = \begin{pmatrix} a_{11} & a_{12} & a_{13} & \dots \\ \dots & \dots & \dots & \dots \\ a_{n1} & a_{n2} & a_{n3} & \dots \\ 0 & 0 & 0 & \dots \\ \dots & \dots & \dots & \dots \end{pmatrix}$$

induces a finite-rank operator and we have

$$\|K - K_n\|^2 \leq \sum_{j=n+1}^{\infty} \sum_{k=1}^{\infty} |a_{jk}|^2 = o(1) \quad (n \to \infty).$$

(b) An infinite diagonal matrix $\operatorname{diag}(\lambda_1, \lambda_2, \lambda_3, \dots)$ defines a compact operator on l^2 if and only if $\lambda_n \to 0$ as $n \to \infty$.

(c) Every function $a \in L^\infty(\mathbf{T})$ generates two *Hankel operators (matrices)* $H(a)$ and $H(\tilde{a})$ on l^2: define the Fourier coefficients a_n $(n \in \mathbf{Z})$ by (13) and put

$$H(a) = \begin{pmatrix} a_1 & a_2 & a_3 & \dots \\ a_2 & a_3 & \dots & \dots \\ a_3 & \dots & \dots & \dots \\ \dots & \dots & \dots & \dots \end{pmatrix}, \quad H(\tilde{a}) = \begin{pmatrix} a_{-1} & a_{-2} & a_{-3} & \dots \\ a_{-2} & a_{-3} & \dots & \dots \\ a_{-3} & \dots & \dots & \dots \\ \dots & \dots & \dots & \dots \end{pmatrix}. \quad (21)$$

The matrices $H(a)$ and $H(\tilde{a})$ may be identified with the south-west and north-east blocks of the matrix (6), respectively, which implies that

$$\|H(a)\| \le \|M(a)\| = \|a\|_\infty, \ \|H(\tilde{a})\| \le \|M(a)\| = \|a\|_\infty.$$

and thus that $H(a)$ and $H(\tilde{a})$ are bounded for every $a \in L^\infty(\mathbf{T})$. We claim that $H(a)$ and $H(\tilde{a})$ are compact if $a \in C(\mathbf{T})$. To see this, denote by $\sigma_n a$ ($n \ge 0$) the n-th Fejer–Cesaro mean of a,

$$(\sigma_n a)(t) = \sum_{k=-n}^{n} \Big(1 - \frac{|k|}{n+1}\Big) a_k t^k \quad (t \in \mathbf{T}),$$

and notice that $\|a - \sigma_n a\|_\infty \to 0$ if $a \in C(\mathbf{T})$. It follows that

$$\|H(a) - H(\sigma_n a)\| = \|H(a - \sigma_n a)\| \le \|a - \sigma_n a\| = o(1),$$

and since $H(\sigma_n a)$ has finite rank, we deduce that $H(a)$ is compact. The same argument gives the compactness of $H(\tilde{a})$.

(d) Finite-rank operators are compact. If $\varphi_1, \dots, \varphi_n, \psi_1, \dots, \psi_n \in l^2$, then the operator defined by

$$Kx = \sum_{j=1}^{n} (x, \psi_j)\varphi_j$$

has finite rank. Conversely, every finite-rank operator on l^2 is of this form. ■

Compact operators on l^2 are never invertible and hence the equation $Kx = y$ with a compact operator K is not a "nice" equation. Equations involving a compact operator are often of the form $(\lambda I + K)x = y$ or $(A + K)x = y$ with some "nice" operator A. If A is invertible and K is compact, then the problem of whether $A + K$ is invertible may be tackled ("halved") by having recourse to the *Fredholm alternative*:

$$A + K \text{ is invertible} \iff A + K \text{ is surjective} \iff A + K \text{ is injective.}$$

The following lemma is of great importance in the theory of projection methods and therefore we will give two proofs of it.

(Basic) Lemma 4.3 *If K is compact and $B_n \to B$ strongly, then $B_n K \to BK$ uniformly, i.e. $\|B_n K - BK\| \to 0$.*

First proof Let $\varepsilon > 0$. Since K maps the unit ball of l^2 to a set whose closure is compact, there are $x_1, \dots, x_N$ in the unit ball such that for every x on the unit sphere (sic !) we can find an x_j satisfying $\|Kx - Kx_j\| < \varepsilon$. We have

$$\begin{aligned} &\|B_n Kx - BKx\| \\ &\le \|B_n\| \, \|Kx - Kx_j\| + \|B_n Kx_j - BKx_j\| + \|B\| \, \|Kx_j - Kx\| \\ &\le \|B_n\|\varepsilon + \|B_n Kx_j - BKx_j\| + \|B\|\varepsilon, \end{aligned}$$

the Banach–Steinhaus theorem implies that $\|B_n\| \le M$ for all n, and because B_n converges strongly to B, the norm $\|B_n Kx_j - BKx_j\|$ is less than ε for all sufficiently large n. Thus, if n is large enough, then

$$\|B_nKx - BKx\| \le (M+1+\|B\|)\varepsilon = (M+1+\|B\|)\varepsilon\|x\|$$

whenever $\|x\| = 1$. ∎

Second proof Given $\varepsilon > 0$, there is a finite rank operator

$$K_0 : l^2 \to l^2,\ x \mapsto \sum_{j=1}^{N} (x, \psi_j)\varphi_j$$

such that $\|K - K_0\| < \varepsilon$. Since $\|B_n\| \le M$ by the Banach–Steinhaus theorem, we obtain

$$\begin{aligned}&\|B_nK - BK\| \\ &\le \|B_n\|\,\|K - K_0\| + \|B_nK_0 - BK_0\| + \|B\|\,\|K - K_0\| \\ &\le M\varepsilon + \|B_nK_0 - BK_0\| + \|B\|\varepsilon.\end{aligned}$$

Further,

$$\begin{aligned}\|B_nK_0x - BK_0x\| &= \left\| \sum_{j=1}^{N} (x, \psi_j)(B_n\varphi_j - B\varphi_j) \right\| \\ &\le \left(\sum_{j=1}^{N} \|\psi_j\|\,\|B_n\varphi_j - B\varphi_j\| \right) \|x\| < \varepsilon\|x\|\end{aligned}$$

if only n is large enough. ∎

The next theorem reveals that convergence of the finite section method is stable under compact perturbations.

Theorem 4.4 *Let $A \in \Pi\{P_n, P_n\}$ and suppose K is compact. Then $A + K \in \Pi\{P_n, P_n\}$ if and only if $A + K$ is invertible.*

Proof Proposition 3.1 shows that $A + K$ is invertible if the finite section method converges for it.

To prove the reverse implication, write $A_n = P_nAP_n|\mathrm{Im}\, P_n$ and note first that

$$P_n(A+K)P_n = P_nAP_n(I + A_n^{-1}P_nK)P_n.$$

Since $A_n^{-1}P_n \to A^{-1}$ strongly due to Proposition 3.1, we infer from Lemma 4.3 that $A_n^{-1}P_nK$ converges uniformly to $A^{-1}K$. The operator $I + A^{-1}K = A^{-1}(A+K)$ is invertible; put $\|(I + A^{-1}K)^{-1}\| = 1/\varepsilon$. We then have for every $x \in l^2$ the estimate

$$\begin{aligned}&\varepsilon\|P_nx\| \le \|(I + A^{-1}K)P_nx\| \\ &\le \|(I + A_n^{-1}P_nK)P_nx\| + \|A_n^{-1}P_nK - A^{-1}K\|\,\|P_nx\|\end{aligned}$$

and the second term on the right is less than $(\varepsilon/2)\|P_nx\|$ if n is large enough. For these n,

$$\begin{aligned}(\varepsilon/2)\|P_nx\| &\le \|(I + A_n^{-1}P_nK)P_nx\| \\ &\le \|A_n^{-1}\|\,\|(A_n + P_nKP_n)P_nx\| = \|A_n^{-1}\|\,\|P_n(A+K)P_nx\|,\end{aligned}$$

and since $\|A_n^{-1}\| \le M$ with some M (Proposition 3.2), we arrive at the inequality

$$(\varepsilon/2M)\|P_n x\| \le \|P_n(A+K)P_n x\|. \tag{22}$$

It follows that $P_n(A+K)P_n|\mathrm{Im}\, P_n$ is injective and thus invertible (the Fredholm alternative on finite-dimensional spaces). Moreover, (22) also shows that

$$\left\|(P_n(A+K)P_n)^{-1}P_n\right\| \le 2M/\varepsilon,$$

implying that $\{P_n(A+K)P_n\}$ is a stable sequence. Proposition 3.2 now completes the proof. ■

The previous theorem was established by Gohberg and Feldman [1971], who attribute it to S. G. Mikhlin in the case where A is required to be positive definite.

Here are two simple consequences of this theorem. If $A = I + K$ where K is as in Example 4.2(a) (notice that such matrices arise naturally from the integral operators considered in Example 1.1), then $A \in \Pi\{P_n, P_n\}$ if and only if A is invertible. Or if $T^B(a)$ is the Bergman space Toeplitz operator generated by the harmonic extension of a positive function $b \in C(\mathbf{T})$ (recall Example 1.3), then $T^B(a) = T(b) + K$ with some compact operator K and hence, since $T(b)$ is positive definite, Theorems 4.1 and 4.4 imply that $T^B(a) \in \Pi\{P_n, P_n\}$.

It turns out that the range of application of Theorem 4.4 is much larger than it might appear at the first glance. In order to take advantage of this theorem, we have to learn the art of decomposing a given operator into a compact operator and an operator for which the finite section method can be shown to converge. More about this will be said in the forthcoming sections.

5 Toeplitz matrices with continuous symbols

Toeplitz matrices are, in a sense, the best understood non-selfadjoint and non-compact operators. We already know that the Toeplitz matrix $T(a)$ induces a selfadjoint operator on l^2 if and only if a is a real-valued function. Clearly, this is in general not the case. What about compactness? Here is what Gohberg [1952] says about this question.

Proposition 5.1 *The operator $T(a)$ is not compact unless $a = 0$ identically.*

Proof In addition to the projections P_n, we consider the complementary projections Q_n,

$$Q_n : l^2 \to l^2,\ \{x_1, x_2, x_3, \ldots\} \mapsto \{0, \ldots, 0, x_{n+1}, x_{n+2}, \ldots\}. \tag{23}$$

Obviously, $Q_n \to 0$ strongly. Assume now that $T(a)$ is compact. Then, by Lemma 4.3, $\|Q_n T(a) Q_n\| \to 0$. But $Q_n T(a) Q_n|\mathrm{Im}\, Q_n$ results from $T(a)$ by erasing the first n rows and columns, and since $T(a)$ is a Toeplitz matrix, it follows that $Q_n T(a) Q_n|\mathrm{Im}\, Q_n$ has the same matrix as $T(a)$. Consequently,

$$\|T(a)\| = \|Q_n T(a) Q_n|\mathrm{Im}\, Q_n\| = \|Q_n T(a) Q_n\| = o(1),$$

implying that $T(a)$ is the zero operator. ■

We remark that $T(a) - I = T(a-1)$, so that $T(a)$ is of the form identity plus compact operator only in the trivial case where $a = 1$ identically. Thus, the results of the previous section are not of great use for Toeplitz matrices. To tackle the finite section method for Toeplitz matrices we will (of course) take heavy advantage of the special structure of these matrices.

This is perhaps the right place to say a few words about the history. About 30 years ago, almost nothing was known about the finite section method for non-selfadjoint and non-compact operators. The first deep result on the finite section method for Toeplitz matrices was established in the early sixties and is in Baxter [1963]. In that pioneering time, Gohberg and Feldman also started working on projection methods. After an astonishingly short period, they presented final results for block Toeplitz matrices with continuous symbols (see Gohberg and Feldman [1965]), and a little bit later they accomplished an impressive theory of projection methods for convolution operators, including block Toeplitz operators and their continuous analogues with continuous symbols on l^p and L^p, paired operators and singular integral operators with continuous matrix-valued coefficients on l^p and L^p, as well as all these classes of operators with piecewise continuous symbols/coefficients on l^2 and L^2: see the book Gohberg and Feldman [1971]. They so built up a theory that attracted other people, and many mathematicians have been working on the finite section method since that time (see the books Böttcher and Silbermann [1983b] and [1990] for recent accounts of the topic). A few of the results that emerged from this fascinating development will be discussed in what follows. But let us start at the beginning ...

The following simple result is something like a magic wand. It has been well known for a long time and has been employed by several authors in various contexts (see e.g. Devinatz and Shinbrot [1969], Böttcher and Silbermann [1983a], Bart et al. [1984], Speck [1985]).

Proposition 5.2 *Let X be a linear space, let P and Q be complementary projections on X (i.e. $P^2 = P$, $Q^2 = Q$, $P + Q = I$) and let A be an invertible linear operator on X. Then the compression $PAP|\mathrm{Im}\, P$ is invertible on* $\mathrm{Im}\, P$ *if and only if the compression $QA^{-1}Q|\mathrm{Im}\, Q$ is invertible on* $\mathrm{Im}\, Q$. *In that case*

$$(PAP)^{-1}P = PA^{-1}P - PA^{-1}Q(QA^{-1}Q)^{-1}QA^{-1}P. \tag{24}$$

Proof We have

$$\begin{aligned}
&PAP(PA^{-1}P - PA^{-1}Q(QA^{-1}Q)^{-1}A^{-1}P)\\
&= PA(I-Q)A^{-1}P - PA(I-Q)A^{-1}Q(QA^{-1}Q)^{-1}A^{-1}P\\
&= P - PAQA^{-1}P - 0 + PAQA^{-1}P = P,
\end{aligned}$$

and similarly we get $(PA^{-1}P - PA^{-1}Q(QA^{-1}Q)^{-1}A^{-1}P)PAP = P$. ■

Here is a trick with our magic wand: we show that the finite section method always converges for matrices that are the inverses of Toeplitz matrices ! Notice in this connection that the inverse of a Toeplitz matrix is in general not a Toeplitz matrix; I do not even know a nice (and useful) description of those matrices that are the inverse of some Toeplitz matrix.

Theorem 5.3 *If $a \in L^\infty(\mathbf{T})$ and $T(a)$ is invertible, then $T^{-1}(a)$ belongs to $\Pi\{P_n, P_n\}$.*

Proof By Proposition 3.2, we are left with showing that $\{P_n T^{-1}(a) P_n\}$ is a stable sequence. With Q_n defined by (23), we deduce from Proposition 5.2 that $P_n T^{-1}(a) P_n | \mathrm{Im}\, P_n$ is invertible if and only if so is $Q_n T(a) Q_n | \mathrm{Im}\, Q_n$. But the latter operator has the same matrix as $T(a)$, hence it is invertible for all $n \geq 1$, and (24) implies that

$$\begin{aligned}&\|(P_n T^{-1}(a) P_n)^{-1} P_n\| \\ &\leq \|P_n T(a) P_n\| + \|P_n T(a) Q_n (Q_n T(a) Q_n)^{-1} Q_n T(a) P_n\| \\ &\leq \|T(a)\| + \|T(a)\|\, \|T^{-1}(a)\|\, \|T(a)\|\end{aligned}$$

for all $n \geq 1$. ■

The next proposition is a key result for understanding many properties of Toeplitz matrices.

Proposition 5.4 *If $a, b \in L^\infty(\mathbf{T})$, then*

$$T(ab) = T(a)T(b) + H(a)H(\tilde{b}). \tag{25}$$

Proof Computing the jk entries of both sides of (25) one sees that (25) is equivalent to the obvious identity

$$\sum_{n=-\infty}^{\infty} a_{j-1-n} b_{-k+1+n} = \sum_{n=0}^{\infty} a_{j-1-n} b_{-k+1+n} + \sum_{n=-\infty}^{-1} a_{j-1-n} b_{-k+1+n}. \blacksquare$$

Thus, while $M(a)M(b) = M(ab)$ and therefore the product of two Laurent matrices is again a Laurent matrix, it turns out that the product of two Toeplitz matrices is a Toeplitz matrix minus the product of two Hankel matrices. We know that Hankel matrices generated by continuous functions are compact, and this is the observation that disposes of the finite section method for Toeplitz operators with continuous symbols.

Theorem 5.5 *Let $a \in C(\mathbf{T})$. Then $T(a) \in \Pi\{P_n, P_n\}$ if and only if $T(a)$ is invertible.*

Proof If $T(a)$ is invertible, then $a^{-1} \in C(\mathbf{T})$ due to Theorem 2.4, and formula (25) with $b = a^{-1}$ gives

$$I = T(a)T(a^{-1}) + H(a)H((a^{-1})^{\sim}). \tag{26}$$

Again from Theorem 2.4 we infer that $T(a^{-1})$ is also invertible. Hence,

$$T(a) = T^{-1}(a^{-1}) - H(a)H((a^{-1})^{\sim})T^{-1}(a^{-1}),$$

and since $H(a)$ is compact and the product of a compact operator and a bounded operator is anew a compact operator, we have

$$T(a) = T^{-1}(a^{-1}) + K$$

with some compact operator. As $T^{-1}(a^{-1}) \in \Pi\{P_n, P_n\}$ by Theorem 5.3, we deduce from Theorem 4.4 that $T(a) \in \Pi\{P_n, P_n\}$. ■

We remark that a result like Theorem 5.5 was already established in Baxter [1963]. Theorem 5.5 as it is stated here appeared in Gohberg and Felman [1965] for the first time. It was independently obtained also by Devinatz and Shinbrot [1969]. The proof given above is in spirit of the book Böttcher and Silbermann [1983b].

By combining Theorem 5.5 with Theorem 4.4 one can prove the convergence of the finite section method for a lot of operators.

For example, let A be the singular integral operator defined by the left-hand side of equation (7). The matrix of A is then a paired matrix and if the coefficients a and b are continuous, then this matrix is of the form

$$\begin{pmatrix} C & 0 \\ 0 & D \end{pmatrix} + \begin{pmatrix} 0 & K \\ L & 0 \end{pmatrix}$$

where C and D are Toeplitz matrices with continuous symbols and K and L are compact Hankel matrices. Suppose A as well as the two Toeplitz operators C and D are invertible. Define $\tilde{P}_n$ on $l^2(\mathbf{Z})$ by

$$\tilde{P}_n : \{x_k\}_{k=-\infty}^{\infty} \mapsto \{\dots, 0, x_{-n}, \dots, x_0, \dots, x_{n-1}, 0, \dots\}.$$

Then

$$\tilde{P}_n \begin{pmatrix} C & 0 \\ 0 & D \end{pmatrix} \tilde{P}_n$$

is the direct sum of $T_n(c)$ and $T_n(d)$ with $c, d \in C(\mathbf{T})$, and since the corresponding infinite matrices C and D are invertible Toeplitz operators, we deduce from Theorems 5.5 and 4.4 that

$$A = \begin{pmatrix} C & 0 \\ 0 & D \end{pmatrix} + \begin{pmatrix} 0 & K \\ L & 0 \end{pmatrix} \in \Pi\{\tilde{P}_n, \tilde{P}_n\}.$$

One can show that in fact the invertibility of the three operators A, C, D is not only sufficient but also necessary for the applicability of the finite section method to the operator A (see Gohberg and Feldman [1971, Chap. VI, Theorem 2.1]).

Or let b be a continuous function on $\mathbf{T}$ and let a denote the harmonic extension of b to $\mathbf{D}$. Then the matrix of the Bergman space Toeplitz operator $T^B(a)$ is of the form $T(b) + K$ with some compact operator K. It is well known (see e.g. Böttcher [1990] or Zhu [1990]) that $T(b)$ must be invertible if $T^B(a)$ is an invertible operator, and consequently, Theorems 5.5 and 4.4 imply that $T^B(a) \in \Pi\{P_n, P_n\}$ if and only if $T^B(a)$ is invertible.

On the other hand, Theorem 5.5 produces a new class of invertible operators for which the finite section method need not converge. We have already made the acquaintance of such matrices in Examples 2.6 and 2.7. Let now $A = M(a)$ be a Laurent matrix of the form (6), say

$$A = \begin{pmatrix} \dots & \dots & \dots & \dots & \dots & \dots & \dots & \dots & \dots & \dots \\ \dots & 0 & 1 & 4 & 2 & 0 & 0 & 0 & 0 & \dots \\ \dots & 0 & 0 & 1 & 4 & 2 & 0 & 0 & 0 & \dots \\ \dots & 0 & 0 & 0 & 1 & 4 & 2 & 0 & 0 & \dots \\ \dots & 0 & 0 & 0 & 0 & 1 & 4 & 2 & 0 & \dots \\ \dots & \dots & \dots & \dots & \dots & \dots & \dots & \dots & \dots & \dots \end{pmatrix}.$$

Which diagonal of this matrix is the main diagonal? According to I. Gohberg, "we have democracy, and thus every diagonal has the right to claim that it be the main diagonal." Hence, to solve the equation $Ax = y$ by the finite section method, we may consider the finite truncations of this equation with the $n \times n$ matrices

$$A_n = \begin{pmatrix} 0 & 1 & 4 & 2 & 0 & \dots & 0 \\ 0 & 0 & 1 & 4 & 2 & \dots & 0 \\ \dots & \dots & \dots & \dots & \dots & \dots & \dots \\ 0 & 0 & 0 & 0 & 0 & \dots & 0 \end{pmatrix} \tag{27}$$

or

$$A_n = \begin{pmatrix} 1 & 4 & 2 & 0 & \dots & 0 \\ 0 & 1 & 4 & 2 & \dots & 0 \\ \dots & \dots & \dots & \dots & \dots & \dots \\ 0 & 0 & 0 & 0 & \dots & 1 \end{pmatrix} \tag{28}$$

or

$$A_n = \begin{pmatrix} 4 & 2 & 0 & \dots & 0 \\ 1 & 4 & 2 & \dots & 0 \\ \dots & \dots & \dots & \dots & \dots \\ 0 & 0 & 0 & \dots & 4 \end{pmatrix} \tag{29}$$

or

$$A_n = \begin{pmatrix} 2 & 0 & 0 & \dots & 0 \\ 4 & 2 & 0 & \dots & 0 \\ \dots & \dots & \dots & \dots & \dots \\ 0 & 0 & 0 & \dots & 2 \end{pmatrix} \tag{30}$$

etc. The operator A is invertible, since we may write $A = M(a)$ with $a(t) = 4 + t + 2t^{-1}$ ($t \in \mathbf{T}$). However, the finite section method for A with the matrices (27) to (30) is equivalent to the finite section method for the Toeplitz matrix $T(a)$ with

$$\begin{aligned} a(t) &= t^{-2}(t + 4 + 2t^{-1}), \\ a(t) &= t^{-1}(t + 4 + 2t^{-1}), \\ a(t) &= t + 4 + 2t^{-1}, \\ a(t) &= t(t + 4 + 2t^{-1}), \end{aligned}$$

respectively, and from Theorem 2.4 we know that $T(a)$ is invertible exactly if $a(t) = t+4+2t^{-1}$. Thus, the finite section method converges in the case (29) but diverges in all other cases.

6 Toeplitz operators: algebraization of stability

Having settled the theory of the finite section method for Toeplitz operators with continuous symbols, we now turn to Toeplitz operators with discontinuous symbols. Consideration of discontinuous symbols may be philosophically motivated by saying that every beautiful theory is interesting and of relevance for practice and that Toeplitz operators with discontinuous symbols (and singular integral operators with discontinuous coefficients) form one of the most beautiful pieces of present-day concrete operator theory. Here are two examples of concrete operators that are Toeplitz operators with a discontinuous symbol.

Example 6.1 Let A be the Wiener–Hopf integral operator given by (3) (Example 1.1) on $L^2(0,\infty)$. We know that the matrix of A with respect to the Laguerre functions $\{\lambda_n\}_{n=0}^{\infty}$ is a Toeplitz matrix $T(a)$. A bijective and orientation preserving map of $\mathbf{R}$ onto $\mathbf{T}$ is given by $\xi \mapsto (\xi - i/2)/(\xi + i/2)$, and hence (5) implies that the nth Fourier coefficient of a is equal to

$$a_n = \frac{1}{2\pi}\int_0^{2\pi} \hat{\alpha}\left(\frac{i}{2}\frac{1+e^{i\theta}}{1-e^{i\theta}}\right)e^{-in\theta}\,d\theta.$$

If the kernel α of the Wiener–Hopf integral operator (3) is subject to the condition (4), then $\hat{\alpha}$ and thus a is continuous. However, if $\alpha(t) = -1/(\pi i t)$, in which case A is the *Cauchy singular integral on the positive half-line*,

$$(Ax)(t) = \frac{1}{\pi i}\int_0^{\infty} \frac{x(s)}{s-t}\,ds \quad (t > 0),$$

then $\hat{\alpha}(\xi) = -\operatorname{sign}\xi$ is a piecewise continuous function with jumps at the origin and at infinity, implying that the function

$$a(e^{i\theta}) = \hat{\alpha}\left(\frac{i}{2}\frac{1+e^{i\theta}}{1-e^{i\theta}}\right)$$

has jumps at $e^{i\theta} = \pm 1$. Moral: such a fundamental operator as the Cauchy singular integral on the half-line is (unitarily equivalent to) a Toeplitz operator with a piecewise continuous symbol. ■

Example 6.2 The so-called *Hilbert Toeplitz matrix*

$$\big(1/(j-k-1/2)\big)_{j,k=1}^{\infty}$$

generates a bounded operator on l^2 whose norm is π. Why? Because it is the Toeplitz operator $T(a)$ with

$$a(e^{i\theta}) = \pi i e^{i\theta/2} \quad (0 \le \theta < 2\pi).$$

Notice that a has a jump at $e^{i\theta} = 1$. ■

The reasoning of the proof of Theorem 5.5 breaks down hopelessly in case a is a piecewise continuous function: one can show that $H(a)$ is definitely not compact if a has a jump (see e.g. Power [1982]). The purpose of this section is to illustrate how

Banach algebra techniques may be employed in order to study Toeplitz operators with piecewise continuous symbols.

Let $\mathcal{L}(l^2)$ denote the Banach algebra of all bounded linear operators on l^2 and let $\mathcal{K}(l^2)$ stand for the collection of all compact operators on l^2. It is well known that $\mathcal{K}(l^2)$ is a closed two-sided ideal of $\mathcal{L}(l^2)$. An operator $A \in \mathcal{L}(l^2)$ is said to be *Fredholm* if its image (= range) $\mathrm{Im}\, A$ is closed and the two spaces

$$\mathrm{Ker}\, A := \{x \in l^2 : Ax = 0\}, \ \mathrm{Coker}\, A := l^2/\mathrm{Im}\, A$$

have finite dimension; in that case the *index* of A is defined by

$$\mathrm{Ind}\, A = \dim \mathrm{Ker}\, A - \dim \mathrm{Coker}\, A.$$

Fredholm operators are "almost invertible": what is a hundred-dimensional kernel or cokernel in comparision with the infinite dimension of l^2! A deep result of Fredholm theory says that an operator is Fredholm if and only if it is invertible modulo compact operators. This result reduces the study of Fredholmness of an operator $A \in \mathcal{L}(l^2)$ to the invertibility of the coset $A + \mathcal{K}(l^2)$ in the quotient algebra $\mathcal{L}(l^2)/\mathcal{K}(l^2)$. For instance, if $a \in C(\mathbf{T})$ has no zeros on $\mathbf{T}$, then $a^{-1} \in C(\mathbf{T})$ and Proposition 5.1 in conjunction with the compactness of $H(a)$ and $H(\tilde{a})$ implies that both

$$T(a^{-1})T(a) - I = -H(a^{-1})H(\tilde{a}), \quad T(a)T(a^{-1}) - I = -H(a)H((a^{-1})^{\sim})$$

are compact and hence that $T(a)$ is Fredholm. Notice the ease with which we are able to conclude that $\mathrm{Ker}\, T(a)$ and $\mathrm{Coker}\, T(a)$ are of finite dimension in case $a \in C(\mathbf{T})$ has no zeros!

Banach algebra techniques for the investigation of the finite section method are based on the construction of a Banach algebra $\mathcal{B}$ such that

$$A \in \Pi\{P_n, P_n\} \Longleftrightarrow \text{something is invertible in } \mathcal{B}.$$

To begin with, let $\mathcal{F}$ be the set of all sequences $\{A_n\}_{n=1}^{\infty}$ of operators $A_n \in \mathcal{L}(\mathrm{Im}\, P_n)$ for which

$$\|\{A_n\}\| := \sup_{n \geq 1} \|A_n\| < \infty. \tag{31}$$

The set $\mathcal{F}$ with the algebraic operations

$$\{A_n\} + \{B_n\} := \{A_n + B_n\}, \ \lambda\{A_n\} := \{\lambda A_n\}, \ \{A_n\}\{B_n\} := \{A_n B_n\},$$

and the norm (31) is easily seen to be a Banach algebra. An element $\{A_n\} \in \mathcal{F}$ is invertible in $\mathcal{F}$ if and only if A_n is invertible for all $n \geq 1$ and $\sup_{n \geq 1} \|A_n^{-1}\| < \infty$. Clearly, invertibility in $\mathcal{F}$ is not equivalent to the convergence of the finite section method, but it has something to do with it.

Now denote by $\mathcal{N}$ the subset of $\mathcal{F}$ consisting of all sequences $\{C_n\} \in \mathcal{F}$ such that $\|C_n\| \to 0$ as $n \to \infty$. It is not difficult to show that $\mathcal{N}$ is a closed two-sided ideal of $\mathcal{F}$ and hence we may consider the quotient algebra $\mathcal{F}/\mathcal{N}$.

Proposition 6.3 *A sequence $\{A_n\} \in \mathcal{F}$ is stable if and only if $\{A_n\} + \mathcal{N}$ is invertible in $\mathcal{F}/\mathcal{N}$.*

Proof If $\{A_n\}+\mathcal{N}$ is invertible in $\mathcal{F}/\mathcal{N}$, then there exists a sequence $\{B_n\}\in\mathcal{F}$ such that $B_nA_n = P_n + C_n$ with $\|C_n\| \to 0$ as $n \to \infty$. If $\|C_n\| < 1$, then $(P_n + C_n)|\mathrm{Im}\, P_n$ is invertible and $(P_n + C_n)^{-1}B_n$ is the inverse of A_n. Since

$$\|(P_n + C_n)^{-1}B_n\| \leq (1 - \|C_n\|)^{-1}\|\{B_n\}\|,$$

it follows that $\{A_n\}$ is stable.

Conversely, let A_n be stable. Suppose A_n is invertible for $n \geq n_0$ and denote the inverse by B_n. Then the sequence $\{D_n\}$ defined by $D_n = P_n$ for $n < n_0$ and $D_n = B_n$ for $n \geq n_0$ lies in $\mathcal{F}$ and $\{D_n\}\{A_n\} - \{P_n\}$ as well as $\{A_n\}\{D_n\} - \{P_n\}$ belong to $\mathcal{N}$. ■

Thus, we have "algebraized" the concept of stability. This is not yet sufficient for our purposes. We still need know something like a "finite section analogue" of Proposition 5.1. Let W_n $(n \geq 1)$ stand for the operator on l^2 acting by the rule

$$W_n : \{x_k\}_{k=1}^{\infty} \mapsto \{x_n, \ldots, x_1, 0, 0, \ldots\}. \tag{32}$$

Obviously, $W_n^2 = P_n$. The following formula was established by Widom [1976].

Proposition 6.4 *If $a, b \in L^\infty(\mathbf{T})$, then*

$$T_n(ab) = T_n(a)T_n(b) + P_nH(a)H(\tilde{b})P_n + W_nH(\tilde{a})H(b)W_n. \tag{33}$$

Proof Define Q_n by (23). Formula (25) gives

$$P_nT(ab)P_n = P_nT(a)P_nT(b)P_n + P_nT(a)Q_nT(b)P_n + P_nH(a)H(\tilde{b})P_n$$

and we have

$$P_nT(a)Q_nT(b)P_n = W_nW_nT(a)Q_nQ_nT(b)W_nW_n.$$

A simple computation shows that

$$W_nT(a)Q_n = P_nH(\tilde{a})V^{(-n)}, \; Q_nT(b)W_n = V^nH(b)P_n, \tag{34}$$

where the "unilateral shifts" $V^{(\pm n)}$ are defined by

$$\begin{aligned} V^n &: \{x_1, x_2, x_3, \ldots\} \mapsto \{0, 0, \ldots, 0, x_1, x_2, x_3, \ldots\} \quad (n \text{ zeros}) \\ V^{(-n)} &: \{x_1, x_2, x_3, \ldots\} \mapsto \{x_{n+1}, x_{n+2}, \ldots\}. \end{aligned}$$

Because $V^{(-n)}V^n = I$, we obtain

$$P_nT(a)Q_nT(b)P_n = W_nH(\tilde{a})H(b)W_n. \blacksquare$$

Now we are coming to one of the most exciting points of our excursion into the finite section method for Toeplitz operators: to the train of thought B. Silbermann was inspired to in the early eighties by Widom's formula (33). Suppose $a \in C(\mathbf{T})$ has no zeros on $\mathbf{T}$. Then $a^{-1} \in C(\mathbf{T})$, formula (33) gives

$$T_n(a^{-1})T_n(a) = P_n - P_nH(a^{-1})H(\tilde{a})P_n - W_nH((a^{-1})^{\sim})H(a)W_n, \tag{35}$$

and all occuring Hankel operators are compact. It would follow from Proposition 6.3 that $\{T_n(a)\}$ is stable if sequences of the form $\{P_nKP_n+W_nLW_n\}$ with compact operators K and L would belong to $\mathcal{N}$. This is, however, not the case: $P_nKP_n +$

W_nLW_n converges strongly to K (we remark that W_n converges weakly to zero, implying that W_nLW_n converges strongly to zero). So let us replace $\mathcal{N}$ by the set

$$\mathcal{J} = \Big\{\{P_nKP_n + W_nLW_n + C_n\}_{n=1}^{\infty} : K, L \in \mathcal{K}(l^2),\ \{C_n\} \in \mathcal{N}\Big\}.$$

Then $\{T_n(a^{-1})\}$ is an inverse of $\{T_n(a)\}$ modulo $\mathcal{J}$ and it would be nice to construct a Banach algebra $\mathcal{S}$ that has the following three properties:

$\mathcal{J}$ is a closed two-sided ideal of $\mathcal{S}$;

$\{T_n(a)\} \in \mathcal{S}$ for all $a \in L^{\infty}(\mathbf{T})$;

invertibility in the quotient algebra $\mathcal{S}/\mathcal{J}$ has something to do with the invertibility in $\mathcal{F}/\mathcal{N}$, i.e. with stability.

Silbermann succeeded in constructing such a Banach algebra. Here it is. Let $\mathcal{S}$ be the set of all $\{A_n\}_{n=1}^{\infty} \in \mathcal{F}$ for which there are two operators A and $\tilde{A}$ in $\mathcal{L}(l^2)$ such that

$$A_nP_n \to A, \quad A_n^*P_n \to A^*, \quad W_nA_nW_n \to \tilde{A}, \quad W_nA_n^*W_n \to \tilde{A}^*,$$

where the asterisk refers to the adjoint operator and $\to$ denotes strong convergence.

Lemma 6.5 *$\mathcal{J}$ is a closed two-sided ideal of $\mathcal{S}$.*

A **proof** is in Silbermann [1982] or Böttcher and Silbermann [1983b], p. 67. We only remark that if

$$\{A_n\} = \{P_nKP_n + W_nLW_n + C_n\} \in \mathcal{J},$$

then $A_n \to K$ and $W_nA_nW_n \to L$ strongly, which shows that

$$\|K\| \le \liminf_{n\to\infty} \|A_n\|, \quad \|L\| \le \liminf_{n\to\infty} \|A_n\|$$

and thus that $\mathcal{J}$ is closed. Further, if $\{B_n\}$ is any element of $\mathcal{S}$, then

$$\begin{aligned} B_nA_n &= B_nP_nKP_n + B_nW_nLW_n + B_nC_n \\ &= P_nB_nP_nKP_n + W_nW_nB_nW_nLW_n + B_nC_n \end{aligned}$$

and since B_nP_nK and $W_nB_nW_nL$ converge in the norm to BK and $\tilde{B}L$, respectively (Lemma 4.3), it follows that

$$B_nA_n = P_nBKP_n + W_n\tilde{B}LW_n + C_n' \quad \text{with} \quad \{C_n'\} \in \mathcal{N}$$

and hence that $\{B_n\}\{A_n\} \in \mathcal{J}$ whenever $\{B_n\} \in \mathcal{S}$ and $\{A_n\} \in \mathcal{J}$. ■

For a function $a \in L^{\infty}(\mathbf{T})$, denote by $\tilde{a} \in L^{\infty}(\mathbf{T})$ the function given by

$$\tilde{a}(t) = a(1/t) \quad (t \in \mathbf{T}).$$

Then $(\tilde{a})_n = a_{-n}$ for all $n \in \mathbf{Z}$. Notice that with this definition at hand the Hankel operator $H(\tilde{a})$ may also be defined as $H(b)$ with $b = \tilde{a}$. In particular, we have $H((a^{-1})^{\sim}) = H(\tilde{a}^{-1})$ etc.

Lemma 6.6 *If $a \in L^{\infty}(\mathbf{T})$, then $\{T_n(a)\} \in \mathcal{S}$.*

Proof Clearly, $T_n(a) \to T(a)$ and $T_n^*(a) = T_n(\overline{a}) \to T(\overline{a}) = T^*(a)$. It is readily verified that

$$W_n T_n(a) W_n = T_n(\tilde{a}), \tag{36}$$

whence $W_n T_n(a) W_n \to T(\tilde{a})$ and $W_n T_n^*(a) W_n \to T^*(\tilde{a})$. ∎

The next theorem, which was established in Silbermann [1982], relates invertibility in $\mathcal{S}/\mathcal{J}$ to the convergence of the finite section method.

Theorem 6.7 *Let $\{A_n\} \in \mathcal{S}$ and denote the strong limits of A_n and $W_n A_n W_n$ by A and $\tilde{A}$, respectively. Then the following are equivalent:*
(i) A is invertible and $\{A_n\}$ is stable;
(ii) A and $\tilde{A}$ are invertible and $\{A_n\} + \mathcal{J}$ is invertible in $\mathcal{S}/\mathcal{J}$.

A **proof** can be found in Silbermann [1982] or Böttcher and Silbermann [1983b, p. 68], so we here confine ourselves to two points. First, if A_n is invertible, then so also is $W_n A_n W_n$ and in case $\|A_n^{-1}\| \le M$ for $n \ge n_0$, we have

$$\|(W_n A_n W_n)^{-1} P_n\| = \|W_n A_n^{-1} W_n\| \le \|A_n^{-1}\| \le M,$$

which implies that $\|P_n x\| \le M \|W_n A_n W_n x\|$ for $n \ge n_0$ and thus $\|x\| \le M \|\tilde{A} x\|$ for all $x \in l^2$. A similar argument may be applied to the adjoint of $\tilde{A}$ and so gives the invertibility of $\tilde{A}$ if $\{A_n\}$ is stable. Secondly, suppose $A, \tilde{A}$ and $\{A_n\} + \mathcal{J}$ are invertible. Let $\{R_n\} + \mathcal{J}$ denote the inverse of $\{A_n\} + \mathcal{J}$ and let R and $\tilde{R}$ stand for the strong limits of R_n and $W_n R_n W_n$, respectively. One can show that then

$$\{R_n + P_n(A^{-1} - R)P_n + W_n(\tilde{A}^{-1} - \tilde{R})W_n\} + \mathcal{N}$$

is the inverse of $\{A_n\} + \mathcal{N}$ in $\mathcal{F}/\mathcal{N}$. ∎

Here is what the previous theorem does for Toeplitz operators.

Corollary 6.8 *Let $a \in L^\infty(\mathbf{T})$. Then $T(a) \in \Pi\{P_n, P_n\}$ if and only if $T(a)$ is an invertible operator and $\{T_n(a)\} + \mathcal{J}$ is an invertible element of $\mathcal{S}/\mathcal{J}$.*

Proof Theorem 6.7 immediately implies that $T(a) \in \Pi\{P_n, P_n\}$ if and only if $T(a), T(\tilde{a})$ as well as $\{T_n(a)\} + \mathcal{J}$ are invertible (also recall (36)). But $T(\tilde{a})$ is the transposed operator of $T(a)$, i.e. $T(\tilde{a}) = CT(\overline{a})C$ with

$$C : l^2 \to l^2,\ \{x_k\}_{k=1}^\infty \mapsto \{\overline{x}_k\}_{k=1}^\infty,$$

and consequently, $T(\tilde{a})$ is invertible if and only if so is $T(a)$. ∎

To get a first feeling of the usefulness of Corollary 6.8 notice that this corollary in conjunction with formula (35) implies at once Theorem 5.5. But Corollary 6.8 does much more. It reduces the problem of the convergence of the finite section method for Toeplitz operators to invertibility in the algebra $\mathcal{S}/\mathcal{J}$, and this algebra is so nice that it may be easily studied with the help of so-called local principles.

7 Toeplitz operators: localization

When learning basic Banach algebra theory, we make the acquaintance of Gelfand theory: given a *commutative* Banach algebra $\mathcal{A}$ with identity element, one has the maximal ideal space Ω of $\mathcal{A}$ and the Gelfand map $\Gamma : \mathcal{A} \to C(\Omega)$, and an element $a \in \mathcal{A}$ is invertible in $\mathcal{A}$ if and only if $(\Gamma a)(\tau) \neq 0$ for all $\tau \in \Omega$.

Local principles may be viewed as far-reaching extensions of Gelfand theory to *non-commutative* Banach algebras. Their philosophy is to associate with an element a of a Banach algebra $\mathcal{A}$ a family $\{a_\tau\}_{\tau\in\Omega}$ of "simpler objects" (where a_τ may also belong to $\mathcal{A}$ or may belong to some individual algebra $\mathcal{A}_\tau$) such that a is invertible in $\mathcal{A}$ if and only if a_τ is invertible for every $\tau \in \Omega$. For example, if $\mathcal{A}$ is commutative and has the identity element e, we may put $a_\tau = (\Gamma a)(\tau)e \in \mathcal{A}$ or may define $a_\tau = (\Gamma a)(\tau) \in \mathbf{C}$; Gelfand theory then says that a is invertible in $\mathcal{A}$ if and only if a_τ is invertible for all τ in the maximal ideal space Ω.

For lack of space we cannot give a detailed presentation of local principles here. We refer to Gohberg and Krupnik [1973], Douglas [1972], Böttcher and Silbermann [1990] in this connection and restrict ourselves to stating a theorem that is fully sufficient for our purposes.

Theorem 7.1. *Let $\mathcal{A}$ be a Banach algebra with identity element e and let $\gamma : L^\infty(\mathbf{T}) \to \mathcal{A}$ be a linear bounded operator with the following properties:*

(i) $\gamma(1) = e$;

(ii) if $a \in L^\infty(\mathbf{T})$ and $\varphi \in C(\mathbf{T})$, then $\gamma(a\varphi) = \gamma(a)\gamma(\varphi) = \gamma(\varphi)\gamma(a)$.

Suppose $a \in L^\infty(\mathbf{T})$ and for each $\tau \in \mathbf{T}$ we are given a function $a_\tau \in L^\infty(\mathbf{T})$ such that

$$\inf\{\|(a - a_\tau)\varphi\| : \varphi \in C(\mathbf{T}),\ \varphi(\tau) = 1\} = 0. \tag{37}$$

If $\gamma(a_\tau)$ is invertible in $\mathcal{A}$ for all $\tau \in \mathbf{T}$, then $\gamma(a)$ is also invertible.

This theorem can be easily **proved** by using the local principle of Gohberg and Krupnik (see Gohberg and Krupnik [1973] or Böttcher and Silbermann [1990]). ∎

Example 7.2 Let $\mathcal{A} = \mathcal{L}/\mathcal{K} := \mathcal{L}(l^2)/\mathcal{K}(l^2)$ and define

$$\gamma : L^\infty(\mathbf{T}) \to \mathcal{L}/\mathcal{K},\ a \mapsto T(a) + \mathcal{K}.$$

It is clear that γ is a linear and bounded operator ($\|\gamma\| = 1$) and that $\gamma(1)$ is the identity element of $\mathcal{L}/\mathcal{K}$. Proposition 5.4 and the compactness of $H(\varphi)$ and $H(\tilde{\varphi})$ for $\varphi \in C(\mathbf{T})$ imply that (ii) is satisfied. ∎

Example 7.3 Now put $\mathcal{A} = \mathcal{S}/\mathcal{J}$ and let

$$\gamma : L^\infty(\mathbf{T}) \to \mathcal{S}/\mathcal{J},\ a \mapsto \{T_n(a)\} + \mathcal{J}.$$

If K is any compact operator, then $\{P_nKP_n\} \in \mathcal{J}$ and hence

$$\begin{aligned}\|\{T_n(a)\} + \mathcal{J}\| &\le \sup_{n\ge1} \|T_n(a) + P_nKP_n\| \\ &= \sup_{n\ge1} \|P_n(T(a) + K)P_n\| \le \|T(a) + K\|.\end{aligned}$$

It follows that

$$\|\{T_n(a)\} + \mathcal{J}\| \le \inf_{K \in \mathcal{K}} \|T(a) + K\| = \|T(a) + \mathcal{K}\| \le \|T(a)\| \le \|a\|_\infty$$

and consequently, γ is bounded. Condition (i) is obvious, and condition (ii) is a consequence of Proposition 6.4 and the compactness of $H(\varphi)$ and $H(\tilde{\varphi})$ for $\varphi \in C(\mathbf{T})$. ■

Let us now turn to (37). If $a \in C(\mathbf{T})$, then (37) clearly holds for the function a_τ that is identically equal to $a(\tau)$. So Theorem 7.1 applied to the situation of Example 7.2 implies that $T(a)$ is Fredholm whenever $a \in C(\mathbf{T})$ and $a(\tau) \neq 0$ for all $\tau \in \mathbf{T}$, while Theorem 7.1 combined with Example 7.3 and Corollary 6.8 shows that $T(a) \in \Pi\{P_n, P_n\}$ if $a \in C(\mathbf{T})$ has no zeros and $T(a)$ is invertible. From Theorem 2.4 we know that a cannot have zeros if $T(a)$ is invertible.

The conclusions of the preceding paragraph are not new. But let us now consider discontinuous symbols. We denote by $PC(\mathbf{T})$ the closed algebra of all *piecewise continuous functions* on $\mathbf{T}$, i.e. $a \in PC(\mathbf{T})$ if and only if $a \in L^\infty(\mathbf{T})$ and the one-sided limits

$$a(t \pm 0) := \lim_{\varepsilon \to 0} a(e^{i(\theta \pm \varepsilon)}) \quad (t = e^{i\theta})$$

exist for every $t \in \mathbf{T}$. Notice that functions in $PC(\mathbf{T})$ have at most countably many discontinuities (jumps) and that for each $\delta > 0$ the set

$$\{t \in \mathbf{T} : |a(t+0) - a(t-0)| > \delta\}$$

is finite. For $a \in PC(\mathbf{T})$, denote by $a^\#$ the closed, continuous, and naturally oriented curve that results from the essential range of a by joining the endpoints of each jump by a straight line segment.

Theorem 7.4 *Let $a \in PC(\mathbf{T})$. Then $T(a)$ is Fredholm if and only if $0 \notin a^\#$. In that case the index of $T(a)$ equals minus the winding number of $a^\#$ with respect to the origin. The operator $T(a)$ is invertible if and only if it is Fredholm of index zero.*

This theorem is well known and was discovered by several mathematicians, including H. Widom, I.B. Simonenko, A. Devinatz, and I. Gohberg (see Gohberg and Krupnik [1973] or Böttcher and Silbermann [1990]). Here is a **proof outline.** Suppose $0 \notin a^\#$. For $\tau = e^{i\theta_0} \in \mathbf{T}$, define $a_\tau \in PC(\mathbf{T})$ by

$$a_\tau(e^{i\theta}) = \begin{cases} a(\tau - 0) & \text{for} \quad \theta_0 - \pi < \theta < \theta_0, \\ a(\tau + 0) & \text{for} \quad \theta_0 < \theta < \theta_0 + \pi. \end{cases}$$

Then (37) is satisfied if we apply Theorem 7.1 in the setting of Example 7.2. The essential range of a_τ is the point $\{a(\tau)\}$ if a is continuous at τ and the doubleton $\{a(\tau - 0), a(\tau + 0)\}$ in case a has a jump at τ. Since $0 \notin a^\#$, Theorem 2.3 implies that $T(a_\tau)$ is Fredholm (even invertible) for each $\tau \in \mathbf{T}$, and Theorem 7.1 then gives the Fredholmness of $T(a)$. The assertion concerning the index follows from a standard homotopy argument and Theorem 2.4. If $0 \in a^\#$, then $T(a)$ cannot be Fredholm, since in that case we could slightly perturb a to obtain operators with different indices. Finally, the invertibility criterion follows from a more general theorem by L. Coburn, which says that if a is any function in $L^\infty(\mathbf{T})$, then $T(a)$ is invertible if and only if $T(a)$ is Fredholm of index zero. ■

Theorem 7.5 *Let $a \in PC(\mathbf{T})$. Then $T(a) \in \Pi\{P_n, P_n\}$ if and only if $T(a)$ is invertible.*

Proof By virtue of Corollary 6.8 we are left with proving that $\{T_n(a)\} + \mathcal{J}$ is invertible in $\mathcal{S}/\mathcal{J}$ if $T(a)$ is invertible. For $\tau \in \mathbf{T}$, define $a_\tau \in PC(\mathbf{T})$ as in the preceding proof.

Assume for a moment that a_τ is real-valued. If $a(\tau \pm 0)$ have the same sign, then $T(a_\tau)$ is positive or negative definite and hence $T(a_\tau) \in \Pi\{P_n, P_n\}$ (Theorem 4.1), implying that $\{T_n(a_\tau)\} + \mathcal{J}$ is invertible in $\mathcal{S}/\mathcal{J}$ (Corollary 6.8). If $a(\tau \pm 0)$ have different signs, then $T(a_\tau)$ is not Fredholm due to Theorem 7.4. But in that case $\{T_n(a_\tau)\} + \mathcal{J}$ cannot be invertible in $\mathcal{S}/\mathcal{J}$: if $\{R_n\} + \mathcal{J}$ were an inverse, then the strong limit of R_n were an inverse of $T(a_\tau)$ modulo compact operators. Conclusion: if a_τ is real-valued, then the spectrum of $\{T_n(a_\tau)\} + \mathcal{J}$ in $\mathcal{S}/\mathcal{J}$ is the convex hull of the essential range of a_τ, $\operatorname{conv} \mathcal{R}(a_\tau)$, which is the singleton $\{a(\tau)\}$ if a is continuous at τ and the segment $[a(\tau - 0), a(\tau + 0)]$ if a has a jump at τ.

In the general case we can find complex numbers α_τ and $\beta_\tau \neq 0$ such that $\alpha_\tau + \beta_\tau a_\tau$ is real-valued. Since, by the spectral mapping theorem,

$$\alpha_\tau + \beta_\tau \operatorname{sp}(\{T_n(a_\tau)\} + \mathcal{J}) = \operatorname{sp}(\{T_n(\alpha_\tau + \beta_\tau a_\tau)\} + \mathcal{J})$$

and the spectrum on the right was already shown to be equal to

$$\operatorname{conv} \mathcal{R}(\alpha_\tau + \beta_\tau a_\tau) = \alpha_\tau + \beta_\tau \operatorname{conv} \mathcal{R}(a_\tau),$$

it follows that $\operatorname{sp}(\{T_n(a_\tau)\} + \mathcal{J}) = \operatorname{conv} \mathcal{R}(a_\tau)$. Finally, Theorem 7.4 shows that $\operatorname{conv} \mathcal{R}(a_\tau)$ does not contain the origin and thus that $\{T_n(a_\tau)\} + \mathcal{J}$ is invertible for every $\tau \in \mathbf{T}$. Theorem 7.1 in the context of Example 7.3 completes the proof. ■

The previous theorem is due to Gohberg [1967], who proved it by different (and more elementary) methods; also see Gohberg and Feldman [1971, Chap. IV, Theorem 4.1]. The present proof is essentially the one of Silbermann [1982]. Although the proof given here seems to be unnecessarily complicated, the idea underlying this proof (namely, the application of local principles to the algebra $\mathcal{S}/\mathcal{J}$) may be extended to plenty of other situations. We will return to this question in Sections 8 and 9.

It turns out that one may prove the equivalence "$T(a) \in \Pi\{P_n, P_n\} \Longleftrightarrow T(a)$ is invertible" for classes of symbols that are much larger than $PC(\mathbf{T})$. Given $a \in L^\infty(\mathbf{T})$, denote by $\mathcal{R}(a|U)$ the essential range of the restriction of a to the open subset $U \subset \mathbf{T}$. For $\tau \in \mathbf{T}$, the set $\mathcal{R}_\tau(a)$ of the essential limit values of a at τ is defined by

$$\mathcal{R}_\tau(a) = \bigcap_{U \ni \tau} \mathcal{R}(a|U).$$

If a is continuous at τ, then $\mathcal{R}_\tau(a) = \{a(\tau)\}$, while if a has a jump at τ, we have $\mathcal{R}_\tau(a) = \{a(\tau - 0), a(\tau + 0)\}$. A function $a \in L^\infty(\mathbf{T})$ is said to be *locally sectorial* if for each $\tau \in \mathbf{T}$ the set $\mathcal{R}_\tau(a)$ is contained in some open half-plane whose boundary passes through the origin (this half-plane may depend on τ). Taking into account that the finite section method converges not only for positive-definite operators but also for operators A whose real part $\operatorname{Re} A$ is positive-definite, one can employ the reasoning of Theorem 7.5 to show that if a is locally sectorial, then $T(a) \in \Pi\{P_n, P_n\}$ if and only if $T(a)$ is invertible.

What we have discussed so far is what is nowadays called "localization over C". Much recent work is devoted to "localization over QC". Without attempting to go into details, I want at least to introduce a few of the definitions one encounters in this connection.

A crucial role in our investigations was played by compact Hankel operators. We know that the Hankel operator $H(a)$ is compact whenever $a \in C$ (we now suppress the $\mathbf{T}$). Are there discontinuous functions a for which $H(a)$ is compact? This happens of course if $a \in L^\infty$ has no Fourier coefficients with positive indices, in which case $H(a)$ is simply the zero operator. The set of all such functions is denoted by $\overline{H^\infty}$. Does $\overline{H^\infty}$ contain discontinuous functions? Yes, it does: examples are (the complex conjugates of) infinite Blaschke products or such functions as

$$a(t) = \exp\left(\frac{1+t}{1-t}\right) \quad (t \in \mathbf{T}).$$

Notice, however, that the functions in $\overline{H^\infty}$ cannot have jumps. The set of all $a \in L^\infty$ for which $H(a)$ is compact is described by *Hartman's theorem*:

$$H(a) \text{ is compact} \Longleftrightarrow a \in C + \overline{H^\infty}. \tag{38}$$

Here $C + \overline{H^\infty}$ is what it ought to be, namely

$$C + \overline{H^\infty} = \{f + g : f \in C, g \in \overline{H^\infty}\}.$$

It turns out that $C + \overline{H^\infty}$ is a closed subalgebra of L^∞; moreover, it is the minimal closed subalgebra of L^∞ properly containing $\overline{H^\infty}$: if a closed subalgebra of L^∞ contains $\overline{H^\infty}$ then this algebra is either equal to $\overline{H^\infty}$ or it contains $C + \overline{H^\infty}$. These and many more mysteries of $C + \overline{H^\infty}$ are presented in Douglas [1972] and Sarason [1978] in a very readable fashion.

We know that if $a \in C$, then

$$T(a) \text{ is Fredholm} \Longleftrightarrow a \text{ has no zeros on } \mathbf{T}. \tag{39}$$

Equivalently, if $a \in C$, then

$$T(a) \text{ is Fredholm} \Longleftrightarrow a \text{ is invertible in } C. \tag{40}$$

The equivalence (39) has no nice analogue for symbols in $C + \overline{H^\infty}$, because it is a little bit troublesome to define what it means that a function in $C + \overline{H^\infty}$ has a zero on $\mathbf{T}$. However, extending (40) does not cause any problems: Douglas [1972] showed that if $a \in C + \overline{H^\infty}$, then

$$T(a) \text{ is Fredholm} \Longleftrightarrow a \text{ is invertible in } C + \overline{H^\infty}. \tag{41}$$

Once (38) and (41) are available, it is easy to deduce from Corollary 6.8 and formula (35) that for $a \in C + \overline{H^\infty}$ the following equivalence is true:

$$T(a) \in \Pi\{P_n, P_n\} \Longleftrightarrow T(a) \text{ is invertible}.$$

Since functions in $\overline{H^\infty}$ cannot have jumps, one is led to the consideration of the smallest closed subalgebra of L^∞ containing PC and $\overline{H^\infty}$. Unfortunately, this algebra is too intricate to be well understood. However, it contains a sufficiently interesting subalgebra, the algebra PQC of all *piecewise quasicontinuous functions*, which has attracted considerable attention since the paper Sarason [1977].

Let H^∞ stand for the set of all functions in L^∞ whose Fourier coefficients with negative indices vanish. From (38) we easily infer that

$$H(\tilde{a}) \text{ is compact} \Longleftrightarrow a \in C + H^\infty, \tag{42}$$

and combining (38) and (42) we arrive at the equivalence

$$H(a) \text{ and } H(\tilde{a}) \text{ are compact} \Longleftrightarrow a \in (C + \overline{H^\infty}) \cap (C + H^\infty). \tag{43}$$

While $\overline{H^\infty} \cap H^\infty = \mathbf{C}$ (= the set of constant functions), it turns out that $(C + \overline{H^\infty}) \cap (C + H^\infty)$ is larger than C ! This intersection is denoted by QC,

$$QC = (C + \overline{H^\infty}) \cap (C + H^\infty),$$

and is referred to as the algebra of all *quasicontinuous functions.* Besides (43), there is another important characterization of QC: Douglas [1978] proved that if $A \in \mathcal{L}(l^2)$, then

$$AT(a) - T(a)A \in \mathcal{K}\ \forall a \in L^\infty \Longleftrightarrow A = T(\varphi) + K \text{ with } \varphi \in QC \text{ and } K \in \mathcal{K}. \tag{44}$$

The implication "$\Longleftarrow$" in (44) is a straightforward consequence of (43); the hard part of (44) is the implication "$\Longrightarrow$". Using only the implication "$\Longleftarrow$" of (44) one can extend Theorem 7.1 to the following result.

Theorem 7.6 *Let $\mathcal{A}$ be a Banach algebra with identity element e and let $\gamma : L^\infty \to \mathcal{A}$ be a linear bounded operator mapping 1 to e and satisfying*

$$\gamma(a\varphi) = \gamma(a)\gamma(\varphi) = \gamma(\varphi)\gamma(a) \quad \text{for } a \in L^\infty \text{ and } \varphi \in QC.$$

Suppose that $a \in L^\infty$ and that for each ξ in the maximal ideal space $M(QC)$ of QC we are given a function $a_\xi \in L^\infty$ such that

$$\inf\{\|(a - a_\xi)\varphi\| : \varphi \in QC,\ (\Gamma\varphi)(\xi) = 1\} = 0.$$

If $\gamma(a_\xi)$ is invertible in $\mathcal{A}$ for all $\xi \in M(QC)$, then $\gamma(a)$ is also invertible.

This is an example of what is called "localization over QC". For theorems like the previous one we refer to Böttcher and Silbermann [1990]. The next theorem may be derived from Theorem 7.6.

Theorem 7.7 *Let a be in PQC, the smallest closed subalgebra of L^∞ containing PC and QC. Then $T(a) \in \Pi\{P_n, P_n\}$ if and only if $T(a)$ is invertible.*

Thus, we now know a lot of symbols a in L^∞ for which the convergence of the finite section method for the Toeplitz operator $T(a)$ is equivalent to the invertibility of $T(a)$. Of course, all these results would turn out to be superfluous if one day someone would come and would tell us that for **every** $a \in L^\infty$ the invertibility of $T(a)$ is equivalent to the convergence of the finite section method. Fortunately, in 1987 (but only in 1987!) Treil was able to dispose of this unlovely expectation. Here is his theorem (see Treil [1987]).

Theorem 7.8 *There exist $a \in L^\infty$ such that $T(a)$ is invertible but $T(a)$ is not in $\Pi\{P_n, P_n\}$.*

8 Block case and higher dimensions.

We know from the previous section that non-convergence of the finite section method for Toeplitz matrices is only the result of exotic discontinuities of the symbol. Things are more complicated for non-Toeplitz matrices.

A Toeplitz matrix is constant along the parallels to the main diagonal. Matrices whose entries in the parallels to the main diagonal form periodic sequences (with the same period) are referred to as *block Toeplitz matrices.* Equivalently, A is a block Toeplitz matrix if and only if

$$A = \begin{pmatrix} a_0 & a_{-1} & a_{-2} & \dots \\ a_1 & a_0 & a_{-1} & \dots \\ a_2 & a_1 & a_0 & \dots \\ \dots & \dots & \dots & \dots \end{pmatrix} \tag{45}$$

where $\{a_k\}_{k \in \mathbf{Z}}$ is a sequence of $N \times N$ matrices, $a_k \in \mathbf{C}^{N \times N}$ for all $k \in \mathbf{Z}$. One can show that the matrix (45) induces a bounded operator on l^2 if and only if there exists an essentially bounded matrix function $a : \mathbf{T} \to \mathbf{C}^{N \times N}$ such that

$$a_k = \frac{1}{2\pi} \int_0^{2\pi} a(e^{i\theta}) e^{-ik\theta} \, d\theta \quad (k \in \mathbf{Z}).$$

In that case the matrix (45) is denoted by $T(a)$ and the matrix function a is called the *symbol* of the matrix (45). We remark that the matrix in Example 2.6 is a block Toeplitz matrix; its symbol is the constant function given by

$$a(t) = \begin{pmatrix} 0 & 1 \\ 1 & 0 \end{pmatrix} \quad (t \in \mathbf{T}).$$

Invertibility of block Toeplitz matrices is a difficult problem and results like Theorem 2.4 are not available in the block case. Let $\mathbf{C}^{N \times N}(\mathbf{T})$ denote the collection of all continuous matrix functions $a : \mathbf{T} \to \mathbf{C}^{N \times N}$.

Theorem 8.1 *Let $a \in \mathbf{C}^{N \times N}(\mathbf{T})$. Then $T(a)$ is Fredholm if and only if* $\det a :$ *$\mathbf{T} \to \mathbf{C}$ has no zeros, in which case the index of $T(a)$ is minus the winding number of* $\det a$ *about the origin.*

For a **proof** see e.g. Gohberg and Feldman [1971, Chap. VIII, Theorem 5.1] or Böttcher and Silbermann [1990, Theorem 2.94]. ■

Notice in this connection that formula (25) is also valid in the block case.

Example 2.6 shows that $T(a)$ need not belong to $\Pi\{P_n, P_n\}$ even if a is a matrix polynomial and $T(a)$ is invertible. However, if $T(a)$ is as in Example 2.6, then $P_{2n}T(a)P_{2n}|\text{Im}\, P_{2n}$ is invertible for all $n \geq 1$ and it is clear that $T(a) \in \Pi\{P_{2n}, P_{2n}\}$.

Example 8.2 Is it true that $T(a) \in \Pi\{P_{nN}, P_{nN}\}$ whenever $a \in \mathbf{C}^{N\times N}(\mathbf{T})$ and $T(a)$ is invertible? The answer is no! Indeed, let

$$a(t) = \begin{pmatrix} t & 1 \\ 0 & t^{-1} \end{pmatrix} \quad (t \in \mathbf{T}). \tag{46}$$

Then $T(a)x = y$ may be written in the form

$$\begin{pmatrix} 0 & 1 & 0 & 0 & 0 & 0 & 0 & 0 & \dots \\ 0 & 0 & 0 & 1 & 0 & 0 & 0 & 0 & \dots \\ 1 & 0 & 0 & 1 & 0 & 0 & 0 & 0 & \dots \\ 0 & 0 & 0 & 0 & 0 & 1 & 0 & 0 & \dots \\ 0 & 0 & 1 & 0 & 0 & 1 & 0 & 0 & \dots \\ 0 & 0 & 0 & 0 & 0 & 0 & 0 & 1 & \dots \\ 0 & 0 & 0 & 0 & 1 & 0 & 0 & 1 & \dots \\ 0 & 0 & 0 & 0 & 0 & 0 & 0 & 0 & \dots \\ \dots & & \dots & & \dots & & \dots & & \dots \end{pmatrix} \begin{pmatrix} \xi_1 \\ \eta_1 \\ \xi_2 \\ \eta_2 \\ \xi_3 \\ \eta_3 \\ \xi_4 \\ \eta_4 \\ \dots \end{pmatrix} = \begin{pmatrix} \varphi_1 \\ \psi_1 \\ \varphi_2 \\ \psi_2 \\ \varphi_3 \\ \psi_3 \\ \varphi_4 \\ \psi_4 \\ \dots \end{pmatrix}$$

and the unique solution is

$$\eta_1 = \varphi_1,\ \eta_2 = \psi_1,\ \eta_3 = \psi_2,\ \eta_4 = \psi_3, \dots$$
$$\xi_1 = \psi_2 - \varphi_1,\ \xi_2 = \psi_3 - \varphi_2,\ \xi_3 = \psi_4 - \varphi_3,\ \xi_4 = \psi_5 - \varphi_4, \dots .$$

Hence $T(a)$ is invertible. On the other hand, each finite section of $T(a)$ contains either a zero column or a zero row, so that, in particular, $T(a) \notin \Pi\{P_{2n}, P_{2n}\}$. ■

Given a matrix function $a : \mathbf{T} \to \mathbf{C}^{N\times N}$, we define a new matrix function $\tilde{a} : \mathbf{T} \to \mathbf{C}^{N\times N}$ by $\tilde{a}(t) = a(1/t)$ $(t \in \mathbf{T})$. In the scalar case $(N = 1)$, $T(\tilde{a})$ is the transposed of $T(a)$ and thus $T(\tilde{a})$ is invertible if and only if so is $T(a)$ (recall the proof of Corollary 6.8). The situation is different in the matrix case: if a is the matrix function (46), then

$$\tilde{a}(t) = \begin{pmatrix} t^{-1} & 1 \\ 0 & t \end{pmatrix} \quad (t \in \mathbf{T}),$$

hence

$$T(\tilde{a}) = \begin{pmatrix} 0 & 1 & 1 & 0 & 0 & 0 & \dots \\ 0 & 0 & 0 & 0 & 0 & 0 & \dots \\ & & & & & & \\ 0 & 0 & 0 & 1 & 1 & 0 & \dots \\ 0 & 1 & 0 & 0 & 0 & 0 & \dots \\ & & & & & & \\ 0 & 0 & 0 & 0 & 0 & 1 & \dots \\ 0 & 0 & 0 & 1 & 0 & 0 & \dots \\ & & & & & & \\ \dots & & \dots & & \dots & & \dots \end{pmatrix},$$

and since the second row of $T(\tilde{a})$ contains only zeros, the operator $T(\tilde{a})$ is not surjective.

Theorem 8.3 *Let $a \in \mathbf{C}^{N\times N}(\mathbf{T})$ and let K be a compact operator. Then $T(a)+K \in \Pi\{P_{nN}, P_{nN}\}$ if and only if $T(a)+K$ and $T(\tilde{a})$ are invertible.*

Proof Let $T(a)+K$ and $T(\tilde{a})$ be invertible. Then $a^{-1} \in \mathbf{C}^{N\times N}(\mathbf{T})$ due to Theorem 8.1, and one can show that

$$T(a^{-1}) \text{ is invertible } \Longleftrightarrow T(\tilde{a}) \text{ is invertible.} \tag{47}$$

One way to see this is using our magic wand. Define the block Laurent matrix $M(a)$ by (6). A little thought reveals that $T(a)$ and $T(\tilde{a})$ may be identified with the south-east block $PM(a)P|\text{Im } P$ and the north-west block $QM(a)Q|\text{Im } Q$. So (47) is equivalent to the statement that

$$PM^{-1}(a)P|\text{Im } P \text{ and } QM(a)Q|\text{Im } Q$$

are simultaneously invertible or not, which in turn is an immediate consequence of Proposition 5.2.

The argument of the proof of Theorem 5.3 now shows that $T^{-1}(a^{-1})$ belongs to $\Pi\{P_{nN}, P_{nN}\}$, and because

$$T(a)+K = T^{-1}(a^{-1}) - T^{-1}(a^{-1})H(a)H(\tilde{a}^{-1}) + K,$$

Theorem 4.4 implies that $T(a)+K \in \Pi\{P_{nN}, P_{nN}\}$.

Conversely, suppose $T(a)+K \in \Pi\{P_{nN}, P_{nN}\}$. Let W_n denote the operator on l^2 given by the matrix

$$W_n = \begin{pmatrix} 0 & 0 & \dots & 0 & I & 0 & \dots \\ 0 & 0 & \dots & I & 0 & 0 & \dots \\ \dots & \dots & \dots & \dots & \dots & \dots & \dots \\ I & 0 & \dots & 0 & 0 & 0 & \dots \\ 0 & 0 & \dots & 0 & 0 & 0 & \dots \\ \dots & \dots & \dots & \dots & \dots & \dots & \dots \end{pmatrix} \tag{48}$$

where I is the $N \times N$ identity matrix. We have $W_n^2 = P_{nN}$ and $W_nT(a)W_n = P_{nN}T(\tilde{a})P_{nN}$, whence

$$(P_{nN}T(\tilde{a})P_{nN} + W_nKW_n)^{-1}P_{nN} = (P_{nN}(T(a)+K)P_{nN})^{-1}P_{nN}.$$

Since $\{(P_{nN}(T(a)+K)P_{nN})^{-1}P_{nN}\}$ is a stable sequence, it follows that there is an M such that

$$\|P_{nN}x\| \leq M\|(P_{nN}T(\tilde{a})P_{nN} + W_nKW_n)x\|$$

for all sufficiently large n and all $x \in l^2$. Taking into account that W_nKW_n goes strongly to zero, we find that $\|x\| \leq M\|T(\tilde{a})x\|$ for all $x \in l^2$. Consequently, $T(\tilde{a})$ is injective. Because $T(\tilde{a})$ is Fredholm of index zero by virtue of Theorem 8.1, we conclude that $T(\tilde{a})$ is invertible. ∎

For $K = 0$, the preceding theorem was established in Gohberg and Feldman [1965] and, independently, in Devinatz and Shinbrot [1969]. In the form it is cited here, the theorem was first proved in Silbermann [1982].

The approach developed in Sections 6 and 7 can be extended to the block case in the natural manner: instead of considering sequences $\{A_n\}_{n=1}^{\infty}$ of operators $A_n \in \mathcal{L}(\operatorname{Im} P_n)$, one has to work with sequences $\{A_n\}_{n=1}^{\infty}$ of operators $A_n \in \mathcal{L}(\operatorname{Im} P_{nN})$, Widom's formula (33) remains valid with P_n replaced by P_{nN} and W_n replaced by (48), etc. In this way one can in particular prove that the conclusion of Theorem 8.3 also holds if a is a matrix function with entries in $PC(\mathbf{T})$ (see Gohberg and Feldman [1971], Silbermann [1982], Böttcher and Silbermann [1983b] or [1990].

What we have learned is that if A is a block Toeplitz matrix with a continuous or a piecewise continuous symbol, then $A \in \Pi\{P_{nN}, P_{nN}\}$ if and only if A itself and some other operator $\tilde{A}$ are invertible. The occurence of only the two operators A and $\tilde{A}$ rests on Theorem 6.7 and the circumstance that in the cases considered so far the coset $\{A_n\} + \mathcal{J}$ is automatically invertible in $\mathcal{S}/\mathcal{J}$ if A and $\tilde{A}$ are invertible. However, for more complicated matrices A, the invertibility of $\{A_n\} + \mathcal{J}$ in $\mathcal{S}/\mathcal{J}$ may produce additional conditions.

Example 8.4 Let a_{jk} be a finite collection of functions in $PC(\mathbf{T})$ and put

$$A = \sum_j \prod_k T(a_{jk}).$$

If all the functions a_{jk} are continuous, then

$$A = T\Big(\sum_j \prod_k a_{jk}\Big) + \text{compact operator} \tag{49}$$

due to Proposition 5.4, and consequently, thanks to Theorem 5.5 we have $A \in \Pi\{P_n, P_n\}$ if and only if A is invertible. For piecewise continuous functions the equality (49) is in general not true. In that case we may have recourse to Theorem 6.7 with $A_n = P_nAP_n|\operatorname{Im} P_n$.

It turns out that $\tilde{A}$ is a pure Toeplitz operator:

$$\tilde{A} = T(\sum_j \prod_k \tilde{a}_{jk}). \tag{50}$$

To see this, let, for example, $A = T(a)T(b)$. Then

$$\begin{aligned}
&W_nT(a)T(b)W_n = W_nT(a)P_nT(b)W_n + W_nT(a)Q_nT(b)W_n \\
&= W_nT(a)W_nW_nT(b)W_n + P_nH(\tilde{a})V^{(-n)}V^nH(b)P_n \quad \text{(recall (34))} \\
&= T_n(\tilde{a})T_n(\tilde{b}) + P_nH(\tilde{a})H(b)P_n \\
&\to T(\tilde{a})T(\tilde{b}) + H(\tilde{a})H(b) = T(\tilde{a}\tilde{b}).
\end{aligned}$$

The invertibility of $\{A_n\} + \mathcal{J}$ in $\mathcal{S}/\mathcal{J}$ may be studied with the help of local principles. What results is the following (see e.g. Böttcher and Silbermann [1990, Theorem 7.68]). For each $\tau \in \mathbf{T}$, define a continuous function $\varrho_\tau : [0,1] \to \mathbf{C}$ by

$$\varrho_\tau(\mu) = \sum_j \prod_k (a_{jk}(\tau - 0)(1-\mu) + a_{jk}(\tau + 0)\mu)$$

and put

$$R_\tau(A) = [\varrho_\tau(1), \varrho_\tau(0)] \cup \varrho_\tau([0,1]).$$

Thus, $R_\tau(A)$ is the closed continuous curve consisting of the segment $[\varrho_\tau(1), \varrho_\tau(0)]$ and the curve from $\varrho_\tau(0)$ to $\varrho_\tau(1)$ that is traced out by $\varrho_\tau(\mu)$ as μ goes from 0 to 1. One can show that $R_\tau(A)$ does not contain the origin if both A and $\tilde{A}$ are Fredholm but that $R_\tau(A)$ may have nonzero winding number about the origin even if A and $\tilde{A}$ are invertible. The invertibility of $\{A_n\} + \mathcal{J}$ in $\mathcal{S}/\mathcal{J}$ is equivalent to the condition

$$\text{wind}\, R_\tau(A) = 0 \quad \text{for all} \quad \tau \in \mathbf{T}. \tag{51}$$

To summarize, we have $A \in \Pi\{P_n, P_n\}$ if and only if A itself and the operator $\tilde{A}$ given by (50) are invertible and the condition (51) holds. ■

Until now we have studied operators on $l^2(\mathbf{Z}_+)$ or $l^2(\mathbf{Z})$. The higher-dimensional analogues of such operators are operators on $l^2(K)$ where K is a subset $\mathbf{Z}^N$. For the sake of simplicity, assume $N = 2$.

A *(discrete) convolution equation* on $l^2(K)$ is of the form

$$\sum_{(k,l)\in K} a_{i-k,j-l}x_{kl} = y_{ij} \quad ((i,j) \in K). \tag{52}$$

If there is a function a in L^∞ over the torus $\mathbf{T}^2$ such that

$$a_{m,n} = \frac{1}{(2\pi)^2} \int_0^{2\pi}\int_0^{2\pi} a(e^{i\theta}, e^{i\psi})e^{-im\theta}e^{-in\psi}\, d\theta\, d\psi$$

for all m, n, then the operator defined by the left-hand side of (52) is bounded on $l^2(K)$. This operator is denoted by $T_K(a)$ and a is usually referred to as the *symbol* of $T_K(a)$. The case in which $K = \mathbf{Z}_+ \times \mathbf{Z}_+$ is of particular interest; the operator $T_K(a)$ is then denoted by $T_{++}(a)$ and called a *Toeplitz operator over the quarter-plane*, while the corresponding equation

$$\sum_{k,l\geq 0} a_{i-k,j-l}x_{kl} = y_{ij} \quad (i,j \geq 0) \tag{53}$$

is referred to as a *two-dimensional discrete Wiener–Hopf equation*.

Example 8.5 Let δ_{ij} denote the sequence in $l^2(\mathbf{Z}_+ \times \mathbf{Z}_+)$ which is 1 at the site (i,j) and 0 at all remaining sites. Then $\{\delta_{ij}\}_{i,j=0}^{\infty}$ is an orthonormal basis of $l^2(\mathbf{Z}_+ \times \mathbf{Z}_+)$. If we order the δ_{ij}'s by

$$\delta_{00}, \delta_{01}, \delta_{02}, \ldots, \delta_{10}, \delta_{11}, \delta_{12}, \ldots, \delta_{20}, \delta_{21}, \delta_{22}, \ldots,$$

then the matrix representation of $T_{++}(a)$ becomes

$$T_{++}(a) = \begin{pmatrix} T(b_0) & T(b_{-1}) & T(b_{-2}) & \ldots \\ T(b_1) & T(b_0) & T(b_{-1}) & \ldots \\ T(b_2) & T(b_1) & T(b_0) & \ldots \\ \ldots & \ldots & \ldots & \ldots \end{pmatrix} \tag{54}$$

with certain functions $b_n \in L^\infty(\mathbf{T})$ $(n \in \mathbf{Z})$. Thus, a quarter-plane Toeplitz operator may be interpreted as an infinite Toeplitz matrix whose entries are themselves infinite Toeplitz matrices. ∎

Example 8.6 (Douglas and Howe [1971]). Order the elements of the orthonormal basis $\{\delta_{ij}\}_{i,j=0}^{\infty}$ of $l^2(\mathbf{Z}_+ \times \mathbf{Z}_+)$ as follows:

$$\delta_{00}, \delta_{01}, \delta_{10}, \delta_{02}, \delta_{11}, \delta_{20}, \delta_{03}, \delta_{12}, \delta_{21}, \delta_{30}, \ldots.$$

If $a \in L^\infty(\mathbf{T}^2)$ is of the form $a(\xi,\eta) = b(\xi\eta^{-1})$ with $b \in L^\infty(\mathbf{T})$, then $T_{++}(a)$ has the matrix representation

$$T_{++}(a) = \begin{pmatrix} T_1(b) & 0 & 0 & \ldots \\ 0 & T_2(b) & 0 & \ldots \\ 0 & 0 & T_3(b) & \ldots \\ \ldots & \ldots & \ldots & \ldots \end{pmatrix},$$

where $T_n(b) = P_n T(b) P_n | \mathrm{Im}\, P_n$. Conclusions:

$$T_{++}(a) \text{ is Fredholm} \Longleftrightarrow T(b) \in \Pi\{P_n, P_n\};$$

$$T_{++}(a) \text{ is invertible} \Longleftrightarrow \begin{cases} T(b) \in \Pi\{P_n, P_n\} \text{ and} \\ \det T_n(b) \neq 0 \ \forall n \geq 1. \end{cases}$$ ∎

Simonenko [1968] as well as Douglas and Howe [1971] have shown that if $a \in C(\mathbf{T}^2)$, then $T_{++}(a)$ is Fredholm on $l^2(\mathbf{Z}_+ \times \mathbf{Z}_+)$ if and only if $a(\xi,\eta) \neq 0$ for all $(\xi,\eta) \in \mathbf{T}^2$ and if for each $\tau \in \mathbf{T}$ the two functions $\xi \mapsto a(\xi,\tau)$ and $\eta \mapsto a(\tau,\eta)$ have winding number zero with respect to the origin. In that case the index of $T_{++}(a)$ is zero. Invertibility of quarter-plane Toeplitz operators is a very difficult problem and only a few results pertaining to special symbol classes are available (see e.g. Böttcher and Silbermann [1990]); this should not come as a surprise, since Example 8.6 shows that this problem includes the question of deciding whether **all** finite sections of a Toeplitz matrix are invertible.

A.V. Kozak [1973] was the first to investigate the *finite section method* for the operators $T_K(a)$. Choose a convex polygon Ω in $\mathbf{R}^2$ with nonempty interior whose vertices $v_1, \ldots, v_m$ belong to $\mathbf{Z}^2$. For $n \geq 1$, denote by $l^2(n\Omega)$ the sequences in $l^2(\mathbf{Z}^2)$ that are supported in $n\Omega$ and let $P_{n\Omega}$ stand for the canonical (orthogonal) projection of $l^2(\mathbf{Z}^2)$ onto $l^2(n\Omega)$. Also, denote by $K(v_j)$ the "angular sector" in $\mathbf{Z}^2$ spanned by the vertex v_j:

$$K(v_j) = \mathbf{Z}^2 \cap \{v_j + \lambda p : p \in \Omega, \lambda \geq 0\}.$$

If Ω is as in Figure 8, then $K(v_3)$ consists of the points of $\mathbf{Z}^2$ that are located in the dotted angular sector. In Figure 9, we have plotted $n\Omega$ for the first five values of n; recall that the sequences of $l^2(n\Omega)$ are supported in the lattice points of $\mathbf{Z}^2$ that lie in $n\Omega$. Here is Kozak's result.

Theorem 8.8 *Let $a \in C(\mathbf{T}^2)$. Then $\{P_{n\Omega}T_{\mathbf{Z}^2}(a)P_{n\Omega}|\mathrm{Im}\, P_{n\Omega}\}_{n=1}^{\infty}$ is a stable sequence if and only if the operators*

$$T_{K(v_1)}(a), \dots, T_{K(v_m)}(a)$$

are invertible.

For **proofs** see Kozak [1973] and Böttcher and Silbermann [1990, Theorem 8.57]. ■

For example, in order to solve the equation (53) approximately, we could replace it by the truncated systems

$$\sum_{k,l=0}^{n} a_{i-k,j-l}x_{kl}^{(n)} = y_{ij} \quad (0 \le i, j \le n). \tag{55}$$

Notice that this amounts to replacing the matrix (54) by the finite matrices

$$\begin{pmatrix} T_n(b_0) & T_n(b_{-1}) & \dots & T_n(b_{-n+1}) \\ T_n(b_1) & T_n(b_0) & \dots & T_n(b_{-n+2}) \\ \dots & \dots & \dots & \dots \\ T_n(b_{n-1}) & T_n(b_{n-2}) & \dots & T_n(b_0) \end{pmatrix}.$$

The systems (55) are equivalent to the equations

$$P_{n\Omega}T_{\mathbf{Z}^2}(a)P_{n\Omega}x^{(n)} = P_{n\Omega} \quad (x^{(n)} \in \mathrm{Im}\, P_{n\Omega})$$

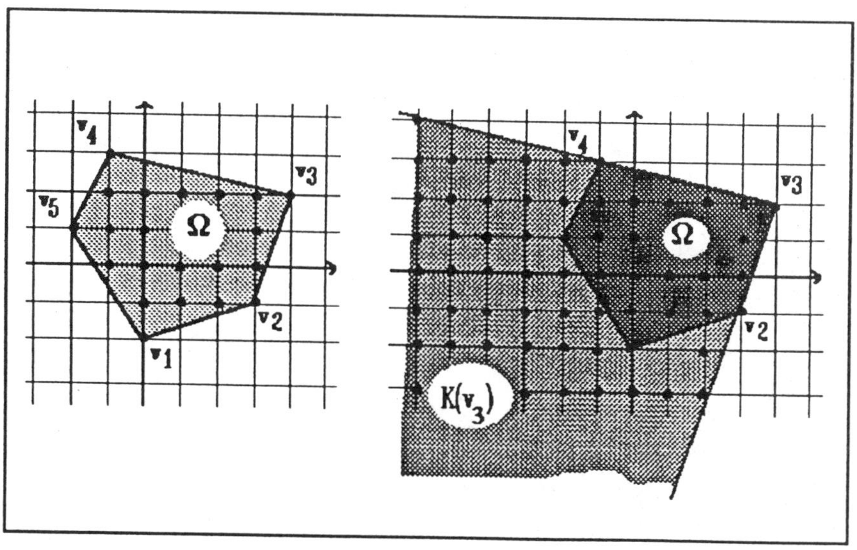

Figure 8

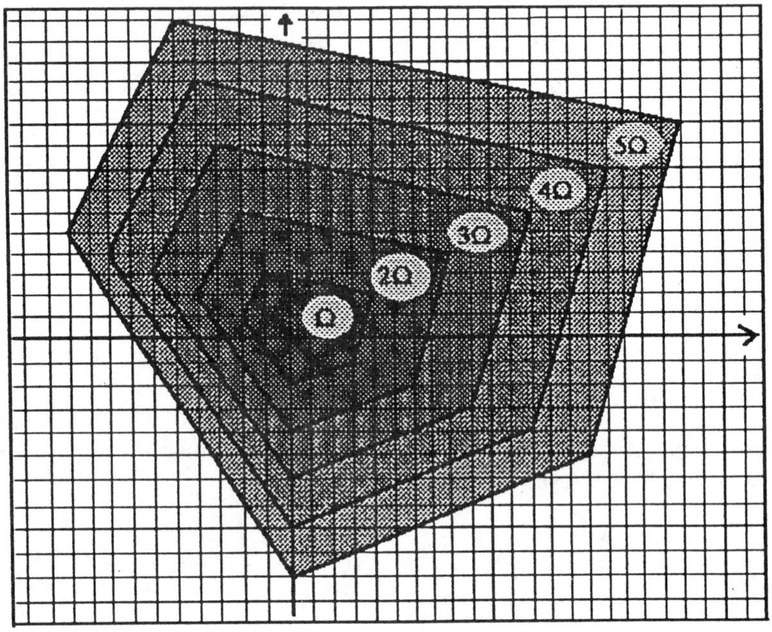

Figure 9

with $\Omega = [0,1] \times [0,1]$. We have

$$v_1 = (0,0), v_2 = (0,1), v_3 = (1,1), v_4 = (1,0)$$

and it is clear that the invertibility of $T_{K(v_1)}(a)$, $T_{K(v_2)}(a)$, $T_{K(v_3)}(a)$, $T_{K(v_4)}(a)$ is equivalent to the invertibility of

$$T_{++}(a),\ T_{-+}(a),\ T_{--}(a),\ T_{+-}(a), \tag{56}$$

respectively, where

$$T_{\pm\pm}(a) := T_{\mathbf{Z}_\pm \times \mathbf{Z}_\pm}(a),\ \mathbf{Z}_+ = \{k \in \mathbf{Z} : k \geq 0\},\ \mathbf{Z}_- = \{k \in \mathbf{Z} : k \leq 0\}.$$

On the other hand, it is not difficult to see that the operators (56) are invertible if and only if the operators

$$T_{++}(a),\ T_{++}(a_1),\ T_{++}(a_2),\ T_{++}(a_{12}) \tag{57}$$

are invertible, where $a_1(\xi,\eta) = a(\xi^{-1},\eta)$; $a_2(\xi,\eta) = a(\xi,\eta^{-1})$, $a_{12}(\xi,\eta) = a(\xi^{-1},\eta^{-1})$. Thus, combining (the $P_{n\Omega}$ analogue of) Proposition 3.2 and Theorem 8.8 we arrive at the following result.

Corollary 8.9 *Let $a \in C(\mathbf{T}^2)$ and $\Omega = [0,1] \times [0,1]$. Then $T_{++}(a)$ belongs to $\Pi\{P_{n\Omega}, P_{n\Omega}\}$ if and only if the four operators (57) are invertible.* ■

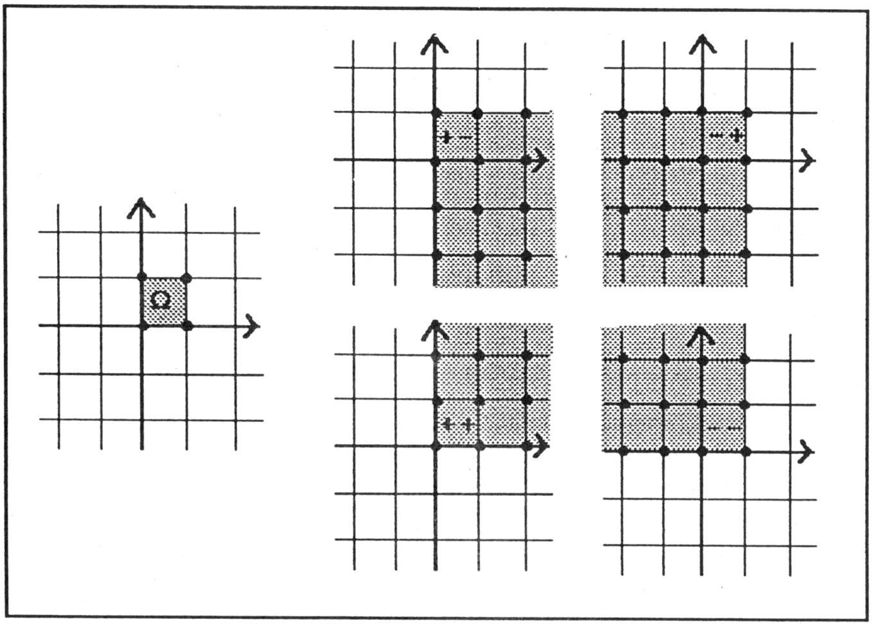

Figure 10

The square $\Omega = [0,1] \times [0,1]$ and the four quarter-planes $K(v_1)$, $K(v_2)$, $K(v_3)$, $K(v_4)$ are shown in Figure 10.

Finally, we remark that $l^2(\mathbf{Z}_+ \times \mathbf{Z}_+)$ may be identified with the Hilbert–space tensor–product $l^2 \otimes l^2$ of two copies of $l^2 = l^2(\mathbf{Z}_+)$. The projections $P_{n\Omega}$ in Corollary 8.9 are equal to $P_n \otimes P_n$ where P_n is given by (17). With W_n defined by (32), we may restate the previous corollary as follows:

If $a \in C(\mathbf{T}^2)$, then $T_{++}(a) \in \Pi\{P_n \otimes P_n, P_n \otimes P_n\}$ if and only if the four operators

$$\begin{aligned}
T_{++}(a) &= \lim_{n\to\infty} (P_n \otimes P_n) T_{++}(a) (P_n \otimes P_n),\\
T_{++}(a_1) &= \lim_{n\to\infty} (W_n \otimes P_n) T_{++}(a) (W_n \otimes P_n),\\
T_{++}(a_2) &= \lim_{n\to\infty} (P_n \otimes W_n) T_{++}(a) (P_n \otimes W_n),\\
T_{++}(a_{12}) &= \lim_{n\to\infty} (W_n \otimes W_n) T_{++}(a) (W_n \otimes W_n)
\end{aligned}$$

(strong limits) are invertible.

9 Banach space phenomena

Until now we have considered operators on Hilbert spaces; after choosing an orthonormal basis in the Hilbert space, these operators become infinite matrices on the space l^2. However, given an infinite matrix, we need not think of it as the matrix representation of some Hilbert space operator, we may also study the properties of this matrix as an operator on sequence spaces different from l^2.

For $1 \le p < \infty$, let l^p denote the Banach space of all sequences $x = \{x_k\}_{k=1}^{\infty}$ such that

$$\|x\|_p := \left(\sum_{k=1}^{\infty} |x_k|^p \right)^{1/p} < \infty.$$

Question: when does the infinite Toeplitz matrix

$$\begin{pmatrix} a_0 & a_{-1} & a_{-2} & \dots \\ a_1 & a_0 & a_{-1} & \dots \\ a_2 & a_1 & a_0 & \dots \\ \dots & \dots & \dots & \dots \end{pmatrix} \tag{58}$$

induce a bounded operator on l^p? We know that if $p = 2$, this is the case if the numbers $\{a_n\}_{n=-\infty}^{\infty}$ are the Fourier coefficients of some function $a \in L^\infty(\mathbf{T})$. The well-known Toeplitz–Brown–Halmos theorem (see Toeplitz [1911], Halmos [1967] or Böttcher and Silbermann [1990]) states that the converse of this is also true: the matrix (58) defines a bounded operator on l^2 if *and only if* the a_n's are the Fourier coefficients of an $L^\infty(\mathbf{T})$ function. The case $p = 1$ is easy: it is not difficult to see that the matrix (58) generates a bounded operator on l^1 if and only if

$$\sum_{n=-\infty}^{\infty} |a_n| < \infty,$$

i.e. if and only if the numbers $\{a_n\}_{n=-\infty}^{\infty}$ are the Fourier coefficients of a function with absolutely convergent Fourier series. The set of these functions is usually denoted by $W(\mathbf{T})$ and referred to as the Wiener algebra. Notice in this connection that it was just Toeplitz operators on l^1 with symbols in $W(\mathbf{T})$ that were considered in Baxter [1963]. The case where $p \notin \{1, 2\}$ is more intricate and leads to what is called a *multiplier problem*. For our purposes it suffices to know that the matrix (58) defines a bounded operator on l^p $(1 < p < \infty)$ whenever the a_n's are the Fourier coefficients of a function $a \in L^\infty(\mathbf{T})$ of bounded variation. In particular, if a is piecewise continuous, if

$$\sum_{\tau \in \mathbf{T}} |a(\tau + 0) - a(\tau - 0)| < \infty,$$

if the continuous pieces of a are C^1 and if

$$\operatorname*{ess\,sup}_{\tau \in \mathbf{T}} |a'(\tau)| < \infty,$$

then a has bounded variation. We designate the set of such functions by PC^1, and for $a \in PC^1$, the operator on l^p $(1 < p < \infty)$ given by the matrix (58) will be denoted by $T(a)$.

Now we are in a position to state one of the most beautiful results of Toeplitz theory, namely to give the answer to the question for which $a \in PC^1$ the operator $T(a)$ is invertible on l^p. Fix $p \in (1, \infty)$ and define $q \in (1, \infty)$ by $1/p + 1/q = 1$. If $a \in PC^1$ has a jump at $\tau \in \mathbf{T}$, denote by $\mathcal{A}_p(a(\tau - 0), a(\tau + 0))$ the circular arc at the points of which the line segment $[a(\tau - 0), a(\tau + 0)]$ is seen at the angle $2\pi / \max\{p, q\}$ and which lies on the right or the left of the line from $a(\tau - 0)$ to $a(\tau + 0)$ depending on whether $1 < p < 2$ or $2 < p < \infty$, respectively. In the case $p = 2$, $\mathcal{A}_2(a(\tau - 0), a(\tau + 0))$ is simply the line segment $[a(\tau - 0), a(\tau + 0)]$ itself.

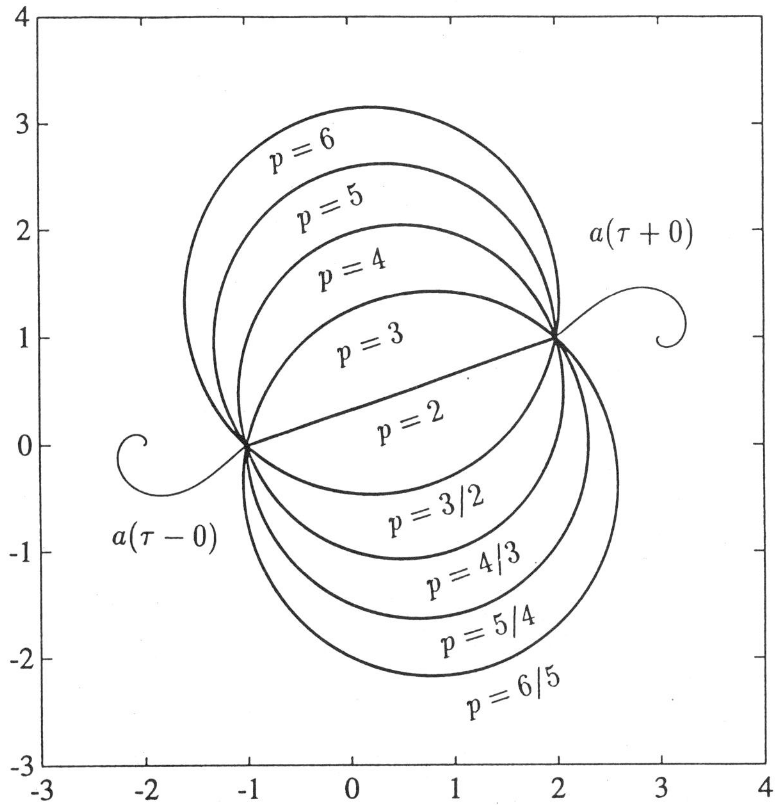

Figure 11

Figure 11 shows $\mathcal{A}_p(a(\tau-0), a(\tau+0))$ for several values of p. Finally, let $a_p^\#$ denote the closed, continuous, and naturally oriented curve resulting from the essential range of a by filling in the arcs $\mathcal{A}_p(a(\tau-0), a((\tau+0))$ between the endpoints of the jumps. Notice that $a_2^\#$ is just the curve $a^\#$ defined in Section 7 and that $a_p^\#$ is nothing but the essential range of a in case a is a continuous function.

Theorem 9.1 *Let $1 < p < \infty$ and $a \in PC^1$. Then the operator $T(a)$ is Fredholm on l^p if and only if $a_p^\#$ does not contain the origin. In that case the index of $T(a)$ equals minus the winding number of the curve $a_p^\#$ about the origin. The operator $T(a)$ is invertible on l^p if and only if it is Fredholm of index zero.*

The earliest paper I know where it is explicitly stated that in order to obtain Fredholm criteria for Toeplitz-like operators on Banach spaces one has to fill in circular arcs between the endpoints of the jumps is the paper Widom [1960], and this phenomen was independently also discovered by Gohberg and Krupnik [1973] in the late sixties. Widom, Gohberg, and Krupnik studied operators on L^p. Theorem 9.1 is concerned with operators on l^p and was established in Duduchava [1972]. It should be mentioned that the bifurcation of $L^2 = l^2$ into L^p and l^p leads to two almost "disjoint theories" whose final results, however, are almost identical. For a proof of the above theorem and for more about this subject the reader is referred to the paper Duduchava [1975], and the books Gohberg and Krupnik [1973], Böttcher and Silbermann [1990]. ■

Example 9.2 Let $T(a)$ be the Hilbert Toeplitz matrix defined in Example 6.2. The curves $a_p^{\#}$ are plotted in Figure 12 for different values of p. Theorem 9.1 implies that $T(a)$ is Fredholm on l^p if and only if $p \neq 2$, that $T(a)$ is invertible on l^p whenever $2 < p < \infty$, and that $T(a)$ is Fredholm of index -1 if $1 < p < 2$. One can show that in the latter case $T(a)$ is injective and that its range has the codimension 1. ■

Let us now turn to the finite section method. We say that the *finite section method converges for* $T(a)$ *on* l^p and write $T(a) \in \Pi_p\{P_n, P_n\}$ in this case if there is an n_0 such that the truncated matrices $T_n(a) = (a_{j-k})_{j,k=1}^n$ are invertible for all $n \geq n_0$ and for every $y \in l^p$ the (unique) solutions $x^{(n)} \in \operatorname{Im} P_n$ of the equations $T_n(a)x^{(n)} = P_n y$ converge in l^p to a solution x of the equation $T(a)x = y$. Here, of course, P_n acts on l^p by the rule (17). The following theorem is an extension of Gohberg's Theorem 1.5.5 to the Banach space case.

Theorem 9.3 *Let* $1 < p < \infty$ *and* $a \in PC^1$. *Then* $T(a) \in \Pi_p\{P_n, P_n\}$ *if and only if both* $T(a)$ *and* $T(\tilde{a})$ *are invertible on* l^p *(where, as usual,* $\tilde{a} \in PC^1$ *is defined by* $\tilde{a}(e^{i\theta}) = a(e^{-i\theta})$*).*

This theorem was established by Verbitsky and Krupnik [1977] for the case where a has at most one jump. The extension of this result to an arbitrary number of jumps turned out to be very difficult and only in 1982, Silbermann and the author were able to dispose of the case of a finite number of jumps (see Böttcher and Silbermann [1982]). A few months after the latter paper was accomplished, Silbermann worked out his "local theory" of the finite section method for Toeplitz operators (i.e. the approach we exhibited in Sections 6 and 7 for the l^2 case) and

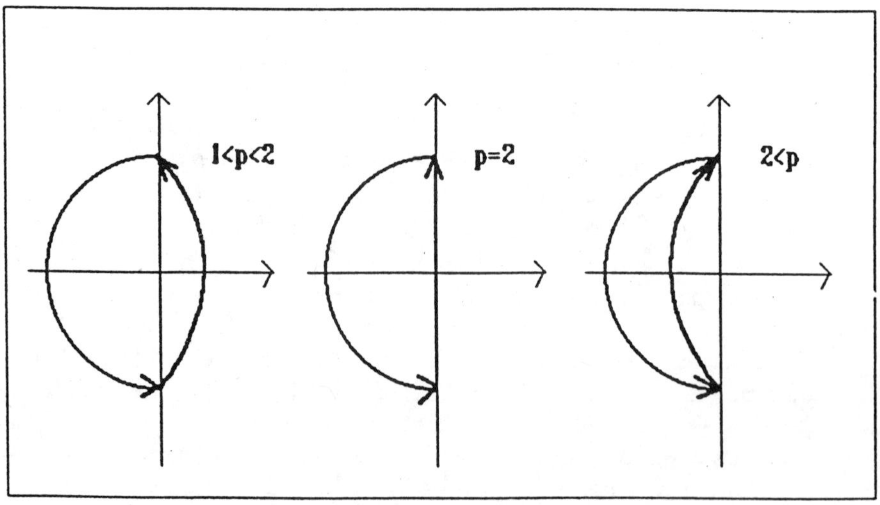

Figure 12

proved Theorem 9.3 for the case of countably many jumps (see Silbermann [1982]). Apart from some technical complications caused by the multiplier problem on l^p, Silbermann's proof is based on algebraizing and localizing as in Sections 6 and 7 and on replacing the "selfadjointness argument" employed in the proof of Theorem 7.5 by the Verbitsky–Krupnik result. For details we refer to Silbermann [1982], Böttcher and Silbermann [1983b] or Böttcher and Silbermann [1990]. ■

Theorem 9.3 admits a nice geometric interpretation. To see this, notice first that, by Theorem 9.1, the operator $T(\tilde{a})$ is invertible on l^p if and only if $T(a)$ itself is invertible on l^q $(1/p + 1/q = 1)$. Secondly, Theorem 9.1 implies that the invertibility of $T(a)$ on both l^p and l^q is equivalent to the invertibility of $T(a)$ on l^r for all r between p and q (inclusively). And once more having recourse to Theorem 9.3, we see that $T(a)$ is invertible on l^r for all $r \in [p,q] := [\min\{p,q\}, \max\{p,q\}]$ if and only if the set

$$a^{\#}_{[p,q]} := \bigcup_{r \in [p,q]} a^{\#}_r$$

does not contain the origin and the winding number of the curve $a^{\#}_2$ (or of $a^{\#}_r$ for any r between p and q) about the origin is zero. We remark that $a^{\#}_{[p,q]}$ results from the essential range of a by filling in the *lentiform domains*

$$\mathcal{O}_p(a(\tau - 0), a(\tau + 0)) := \bigcup_{r \in [p,q]} \mathcal{A}_r(a(\tau - 0), a(\tau + 0))$$

between the endpoints of the jumps (see Figure 13). Thus, we arrive at the conclusion that if $1 < p < \infty$ and $a \in PC^1$, then $T(a) \in \Pi_p\{P_n, P_n\}$ if and only if $0 \notin a^{\#}_{[p,q]}$ and wind $a^{\#}_2 = 0$.

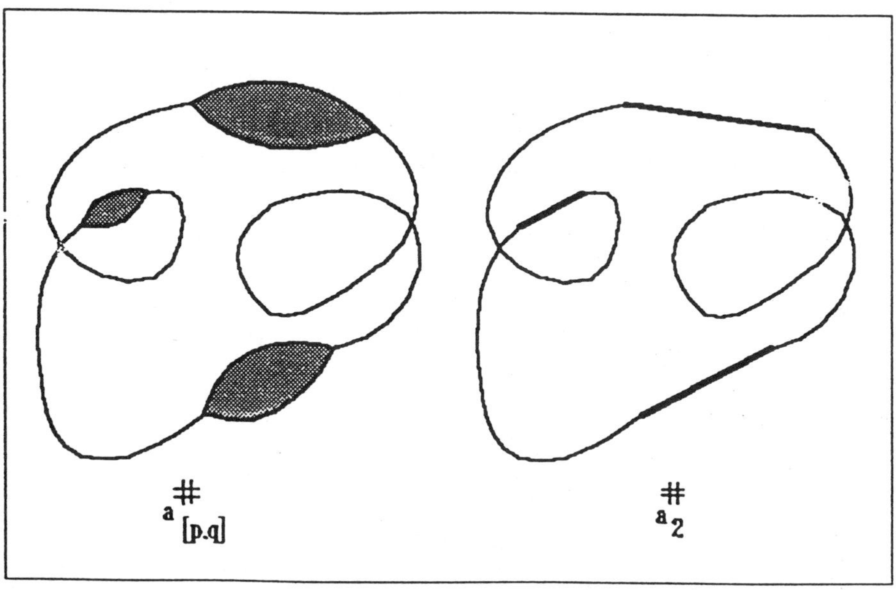

Figure 13

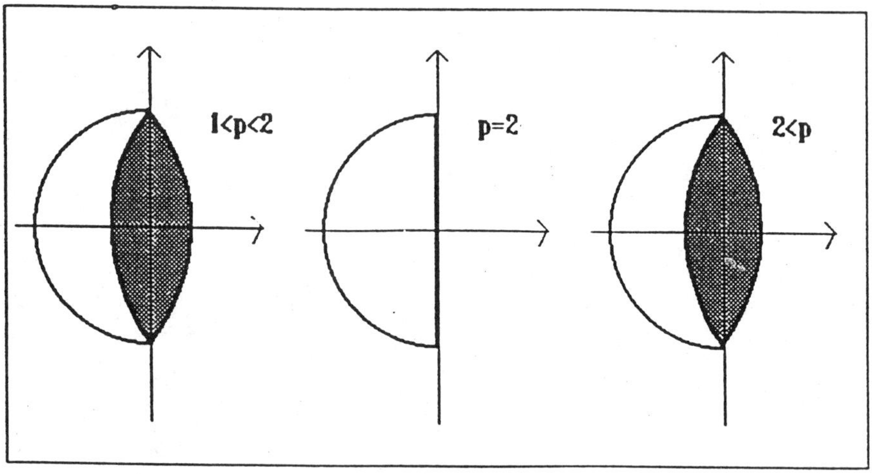

Figure 14

Example 9.4 Let $T(a)$ be the Hilbert Toeplitz matrix. Figure 14 shows the sets $a^{\#}_{[p,q]}$ and consequently, although $T(a)$ is invertible for $p \in (2, \infty)$, there is no p such that the finite section method converges for $T(a)$ on l^p. ■

To conclude this section, we remark that, apart from the multiplier problem, the final results for Toeplitz operators on l^p with continuous symbols do not depend on p. So Theorem 9.1 simply says that if $a \in C^1$, then $T(a)$ is Fredholm on l^p if and only if a has no zeros on $\mathbf{T}$, in which case the index of $T(a)$ is minus the winding number of a about the origin; moreover, $T(a)$ is invertible on l^p if and only if it is Fredholm of index zero. Theorem 9.3 reduces to the assertion that if $a \in C^1$, then $T(a) \in \Pi_p\{P_n, P_n\}$ if and only if $T(a)$ is invertible. Toeplitz operators on l^p with continuous symbols were already considered in Gohberg and Feldman [1971].

10 Norms of inverses and pseudospectra

We now return to the Hilbert space case. So far we have mainly studied convergence of the finite section method, that is, Problem III of Section 2. The purpose of this and the following section is to show how results on the convergence of the finite section method imply nontrivial insights into the asymptotic behavior of the norms $\|A_n^{-1}\|$, the spectral properties of A_n, and the determinants of A_n. As in the previous sections, we will focus our attention on Toeplitz matrices.

Our first concern is Problem II of Section 2. Suppose $A \in \mathcal{L}(l^2)$ and $A \in \Pi\{P_n, P_n\}$. Then, by Proposition 3.1, A is invertible and $A_n^{-1}P_n$ converges strongly to A^{-1} as $n \to \infty$. So the Banach–Steinhaus theorem implies that

$$\liminf_{n\to\infty} \|A_n^{-1}\| \geq \|A^{-1}\|, \tag{59}$$

where $\|A_n^{-1}\|$ stands for the norm of $A_n^{-1}P_n$ on l^2. On the other hand, Proposition 3.2 tells us that

$$\limsup_{n\to\infty} \|A_n^{-1}\| < \infty. \tag{60}$$

The inequalities (59) and (60) are immediate consequences of the fact that $A \in \Pi\{P_n, P_n\}$. Much more can be said in case more is known about the structure of A. For example, if $A = T(a)$ with $a \in PC(\mathbf{T})$, then actually

$$\lim_{n\to\infty} \|T_n^{-1}(a)\| = \|T^{-1}(a)\|, \tag{61}$$

the existence of the limit on the left being part of the conclusion! Equalities like (61) cannot be proved with bare hands. The approach we will use to prove (61) is based upon results on the convergence of the finite section method combined with C^*-algebra arguments. We confine ourselves to pointing out a few central ideas and refer to the papers Silbermann [1993] and Böttcher [1994] for details.

C^*-algebras are especially nice Banach algebras. A map $a \mapsto a^*$ of a (complex) Banach algebra $\mathfrak{A}$ into itself is called an *involution* if

$$a^{**} = a,\ (a+b)^* = a^* + b^*,\ (ab)^* = b^*a^*,\ (\lambda a)^* = \overline{\lambda}a^*$$

for all $a, b \in \mathfrak{A}$ and all $\lambda \in \mathbf{C}$. A *C^*-algebra* is a Banach algebra $\mathfrak{A}$ with an involution such that $\|a\|^2 = \|aa^*\|$ for every $a \in \mathfrak{A}$. For example, if H is a Hilbert space, then $\mathcal{L}(H)$ is a C^*-algebra with passage to the adjoint operator as the involution. Moreover, the algebras $\mathcal{F}, \mathcal{F}/\mathcal{N}, \mathcal{S}$, and $\mathcal{S}/\mathcal{J}$ introduced in Section 6 are C^*-algebras; the involutions in $\mathcal{F}$ and $\mathcal{F}/\mathcal{N}$ are given by

$$\{A_n\}^* := \{A_n^*\} \text{ and } (\{A_n\} + \mathcal{N})^* := \{A_n^*\} + \mathcal{N},$$

respectively.

The spectrum $\operatorname{sp} a$ of an element a of a C^*-algebra $\mathfrak{A}$ with unit element e is defined in the usual way:

$$\operatorname{sp} a = \{\lambda \in \mathbf{C} : a - \lambda e \text{ is not invertible in } \mathfrak{A}\}.$$

C^*-algebras are always *inverse closed.* This means the following: if $\mathfrak{A}$ is a C^*-algebra with unit element and $\mathfrak{A}$ is contained in some larger C^*-algebra $\mathfrak{C}$, then the invertibility of an element $a \in \mathfrak{A}$ in $\mathfrak{C}$ implies that a is invertible in $\mathfrak{A}$.

It is well known that the norm of a normal Hilbert space operator coincides with its spectral radius. The same is true in C^*-algebras with unit element: if $aa^* = a^*a$ then

$$\|a\| = r(a) := \max\{|\lambda| : \lambda \in \operatorname{sp} a\}.$$

The latter inequality has a series of remarkable consequences. Here is the one that forms the foundation of our approach to equalities like (61). Recall that a map $\gamma : \mathfrak{A} \to \mathfrak{B}$ of a C^*-algebra $\mathfrak{A}$ into a C^*-algebra $\mathfrak{B}$ is referred to as a *C^*-algebra homomorphism* if γ is a bounded linear operator of $\mathfrak{A}$ into $\mathfrak{B}$ that preserves products and involutions.

Proposition 10.1 *If a C^*-algebra homomorphism preserves spectra, then it also preserves norms. That is, if $\mathfrak{A}$ and $\mathfrak{B}$ are C^*-algebras with identity elements* *and* $\gamma : \mathfrak{A} \to \mathfrak{B}$ *is a* C^**–algebra homomorphism with the property that* $\operatorname{sp} a = \operatorname{sp} \gamma(a)$ *for every* $a \in \mathfrak{A}$, *then* $\|a\| = \|\gamma(a)\|$ *for all* $a \in \mathfrak{A}$.

Proof Since both aa^* and $\gamma(a)\gamma(a)^*$ are normal (even selfadjoint), we have

$$\begin{aligned}\|a\|^2 &= \|aa^*\| = r(aa^*) = r(\gamma(aa^*)) \\ &= r(\gamma(a)\gamma(a)^*) = \|\gamma(a)\gamma(a)^*\| = \|\gamma(a)\|^2. \blacksquare\end{aligned}$$

Theorem 10.2 *Let* $a \in PC(\mathbf{T})$ *and suppose* $T(a)$ *is invertible on* l^2. *Then the matrices* $T_n(a)$ *are invertible for all sufficiently large* n, *the limit* $\lim_{n\to\infty} \|T_n^{-1}(a)\|$ *exists, and*

$$\lim_{n\to\infty} \|T_n^{-1}(a)\| = \|T^{-1}(a)\|.$$

Proof outline Theorem 7.5 tells us that $T(a) \in \Pi\{P_n, P_n\}$ and hence, due to (59), we are left with showing that

$$\limsup_{n\to\infty} \|T_n^{-1}(a)\| \le \|T^{-1}(a)\|. \tag{62}$$

Define $\mathcal{F}$ and $\mathcal{F}/\mathcal{N}$ as in Section 6. If $\{A_n\} \in \mathcal{F}$, then the norm of the coset $\{A_n\} + \mathcal{N} \in \mathcal{F}/\mathcal{N}$ is given by

$$\|\{A_n\} + \mathcal{N}\| = \limsup_{n\to\infty} \|A_n\|, \tag{63}$$

and thus we have a C^*-algebra whose norm has something to do with the left-hand side of (62).

Unfortunately, the algebra $\mathcal{F}/\mathcal{N}$ is too large to be well understood. However, it contains nice C^*-subalgebras. One of them is the smallest C^*–subalgebra of $\mathcal{F}/\mathcal{N}$ that contains all cosets of the form $\{T_n(b)\} + \mathcal{N}$ with $b \in PC(\mathbf{T})$. We denote this C^*-subalgebra by $\mathfrak{A}$. Notice that $\mathfrak{A}$ is the closure in $\mathcal{F}/\mathcal{N}$ of the collection of all cosets $\{B_n\} + \mathcal{N}$ with

$$B_n = \sum_j \prod_k T_n(b_{jk}), \quad b_{jk} \in PC(\mathbf{T}); \tag{64}$$

the sum and the products are finite. In particular, $\{T_n(a)\} + \mathcal{N} \in \mathfrak{A}$. Moreover, since $\{T_n^{-1}(a)\} + \mathcal{N} \in \mathcal{F}/\mathcal{N}$ by Proposition 6.3, we deduce from the inverse closedness of $\mathfrak{A}$ in $\mathcal{F}/\mathcal{N}$ that also $\{T_n^{-1}(a)\} + \mathcal{N} \in \mathfrak{A}$.

One can show that for every $\{B_n\} + \mathcal{N} \in \mathfrak{A}$ the strong limits

$$B = \lim_{n\to\infty} B_n \quad \text{and} \quad \tilde{B} = \lim_{n\to\infty} W_n B_n W_n$$

exist (here W_n is given by (32)) and that the maps

$$\begin{aligned}\gamma_1 &: \mathfrak{A} \to \mathcal{L}(l^2), \ \{B_n\} + \mathcal{N} \mapsto B \\ \gamma_2 &: \mathfrak{A} \to \mathcal{L}(l^2), \ \{B_n\} + \mathcal{N} \mapsto \tilde{B}\end{aligned}$$

are C^*-algebra homomorphisms. Let $\mathfrak{B}$ denote the direct sum of two copies of $\mathcal{L}(l^2)$ with the norm

$$\|(B, C)\| := \max\{\|B\|, \|C\|\}.$$

Then $\mathfrak{B}$ is a C^*-algebra and what we have constructed is a C^*-algebra homomorphism

$$\gamma : \mathfrak{A} \to \mathfrak{B}\,,\ \{B_n\} + \mathcal{N} \mapsto (B, \tilde{B}). \tag{65}$$

The next step consists in proving that the map (65) preserves spectra. By virtue of Proposition 6.3, this is equivalent to showing that $\{B_n\}$ is stable if and only if both B and $\tilde{B}$ are invertible. The latter equivalence can in turn be established with the help of Theorem 6.7 along with the observation that $\{B_n\} + \mathcal{J}$ is automatically invertible in $\mathcal{S}/\mathcal{J}$ if B is Fredholm (and thus, all the more, if both B and $\tilde{B}$ are invertible).

Now Proposition 10.1 does the rest. Since γ preserves spectra, it also preserves norms, that is,

$$\|\{B_n\} + \mathcal{N}\| = \max\{\|B\|, \|\tilde{B}\|\}$$

for every $\{B_n\} + \mathcal{N} \in \mathfrak{A}$. Letting $B_n = T_n^{-1}(a)$ we arrive in view of (63) at the equality

$$\limsup_{n\to\infty} \|T_n^{-1}(a)\| = \max\{\|T^{-1}(a)\|, \|T^{-1}(\tilde{a})\|\},$$

and because $T^{-1}(\tilde{a})$ is the transposed of $T^{-1}(a)$ and thus the norm equality

$$\|T^{-1}(a)\| = \|T^{-1}(\tilde{a})\|$$

holds, it follows that

$$\limsup_{n\to\infty} \|T_n^{-1}(a)\| = \|T^{-1}(a)\|. \blacksquare$$

We remark that in order to prove Theorem 10.2, which is a theorem for "pure and individual" Toeplitz operators, we had to study the stability of the more complicated sequences of the form (64). This is the price we must pay for letting C^*-algebras do their job. On the other hand, the proof given above almost at once yields several extensions of Theorem 10.2. For example, it shows that if $a \in PC(\mathbf{T})$ and K is a compact operator, then

$$\lim_{n\to\infty} \|(T_n(a) + P_n K P_n)^{-1}\| = \max\{\|(T(a) + K)^{-1}\|, \|T^{-1}(a)\|\},$$

or if $b_{jk} \in PC(\mathbf{T})$, then

$$\lim_{n\to\infty} \left\| \Big(\sum_j \prod_k T_n(b_{jk}) \Big)^{-1} \right\|$$

$$= \max\left\{ \left\| \Big(\sum_j \prod_k T(b_{jk}) \Big)^{-1} \right\|, \left\| \Big(\sum_j \prod_k T(\tilde{b}_{jk}) \Big)^{-1} \right\| \right\},$$

or if $a \in PC^{N\times N}(\mathbf{T})$ (matrix case), then

$$\lim_{n\to\infty} \|T_n^{-1}(a)\| = \max\{\|T^{-1}(a)\|, \|T^{-1}(\tilde{a})\|\}.$$

In the latter two equalities the operators on the right are in general not transposed to one another, so that the "max" cannot be removed.

Only a few years ago, people were happy to have estimates of the form (60), and no one dreamed of proving an equality like the one in Theorem 10.2. That we are now able to prove such equalities is a good occasion to emphasize the usefulness of C^*-algebra techniques in numerical analysis. My experience is that techniques of this kind are usually hesistantly accepted, perhaps since they are too algebraical for numerical analysts and too numerical for C^*-algebraists. I hope that the recent paper Arveson [1994] provides another boost to pay more attention to C^*-algebra methods. Arveson's approach is completely different from the strategy outlined above and aims at asymptotic results for selfadjoint operators arising in the study of one-dimensional quantum systems, but his philosophy coincides with ours: C^*-algebras do a lot of things for us.

Let us now turn to Problem I of Section 2, i.e. to the problem of whether the spectra of $T_n(a)$ for large n mimic the spectrum of $T(a)$. To be more precise, given a sequence $\{E_n\}_{n=1}^{\infty}$ of sets $E_n \subset \mathbf{C}$, we denote by $\lim_{n\to\infty} E_n$ the set of all $\lambda \in \mathbf{C}$ for which there are $n_1, n_2, \dots$ and $\lambda_1, \lambda_2, \dots$ such that

$$n_1 < n_2 < \dots, \quad \lambda_k \in E_{n_k}, \quad \lambda_k \to \lambda,$$

and our question then is whether

$$\lim_{n\to\infty} \operatorname{sp} T_n(a) = \operatorname{sp} T(a). \tag{66}$$

It has been known for about 30 years that (66) holds if a is real-valued and thus $T(a)$ and $T_n(a)$ are selfadjoint (see Widom [1965]). Examples 2.2 and 2.5 tell us that (66) is not true in general, and at the present moment we have not even a sufficiently precise description of the class of symbols for which (66) holds; see Widom [1990], [1994] and Basor and Morrison [1994] for nice discussions of this subject.

A way out of the labyrinth was shown by Reichel and Trefethen [1992]: they considered pseudospectra instead of spectra and were able to show that for a large class of symbols the pseudospectra of truncated Toeplitz matrices mimic **exactly** the pseudospectrum of the original Toeplitz matrix ! For $\varepsilon > 0$, the *ε–pseudospectrum* $\operatorname{sp}_\varepsilon A$ of an operator A is defined as the set

$$\operatorname{sp}_\varepsilon A = \{\lambda \in \mathbf{C} : \lambda \in \operatorname{sp} A \quad \text{or} \quad \|(A - \lambda I)^{-1}\| \geq 1/\varepsilon\}.$$

In the same way the question "Is A invertible?" is in numerical mathematics replaced by the question "What is $\|A^{-1}\|$?", the pseudospectra of "real-world" matrices are, in a sense, of greater import than their usual spectra.

Example 10.3 Define $a : \mathbf{T} \to \mathbf{C}$ by

$$a(t) = -t^{-4} - (3+2i)t^{-3} + it^{-2} + t^{-1} + 10t + (3+i)t^2 + 4t^3 + it^4 \quad (t \in \mathbf{T}).$$

The range $a(\mathbf{T})$ (is it a fish ?) and the eigenvalues of $T_{50}(a)$ are plotted in Figure 15. The theory of Schmidt and Spitzer [1960] tells us that the eigenvalues of $T_n(a)$ cluster along the set plotted in Figure 16 (the skeleton of the fish) as n goes to infinity. Figure 15 beautifully illustrates this result.

But what about pseudospectra? Reichel and Trefethen [1992] observed that for a finite matrix $A \in \mathbf{C}^{N\times N}$ the equality

$$\operatorname{sp}_\varepsilon A = \{\lambda \in \mathbf{C} : \ \lambda \in \operatorname{sp}(A+E) \text{ for some } E \in \mathbf{C}^{N\times N} \text{ with } \|E\| \leq \varepsilon\}$$

holds. Hence, in order to get an idea of the ε–pseudospectrum of a matrix A one may randomly perturb A by matrices E satisfying $\|E\| \leq \varepsilon$ and look at the (usual) spectra of $A + E$. Figure 17 shows the superposition of the (usual) spectra of

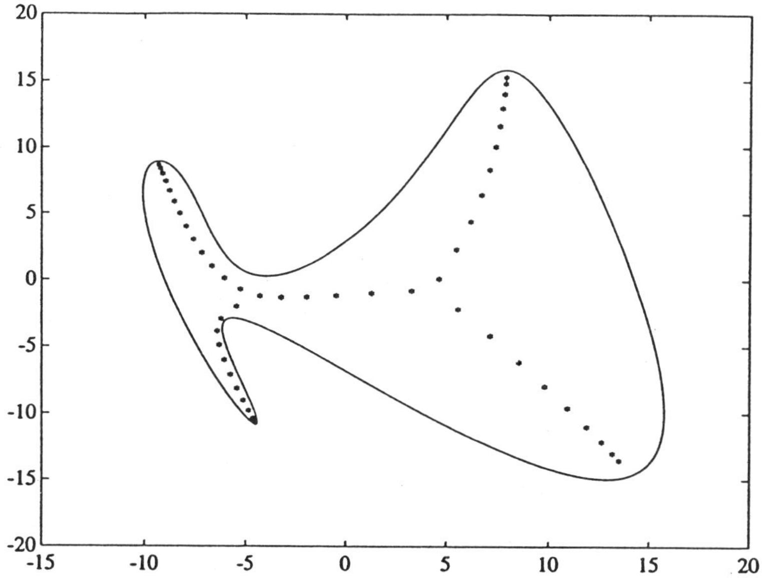

Figure 15

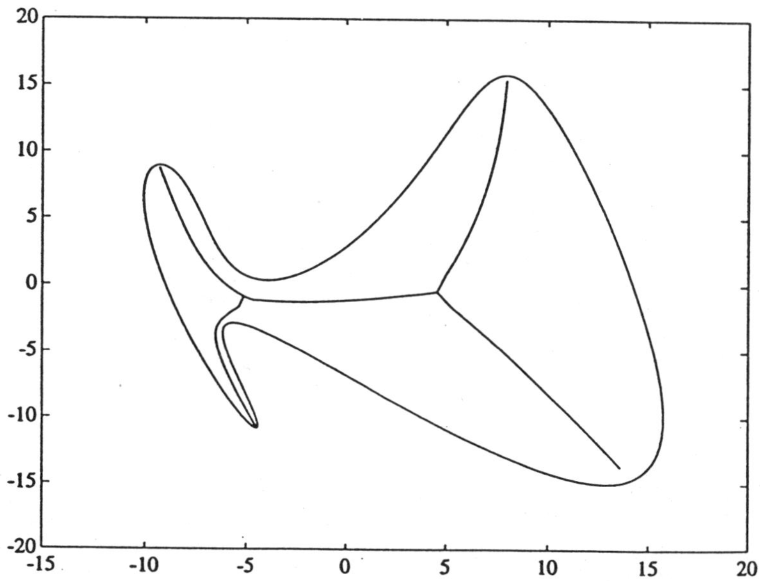

Figure 16

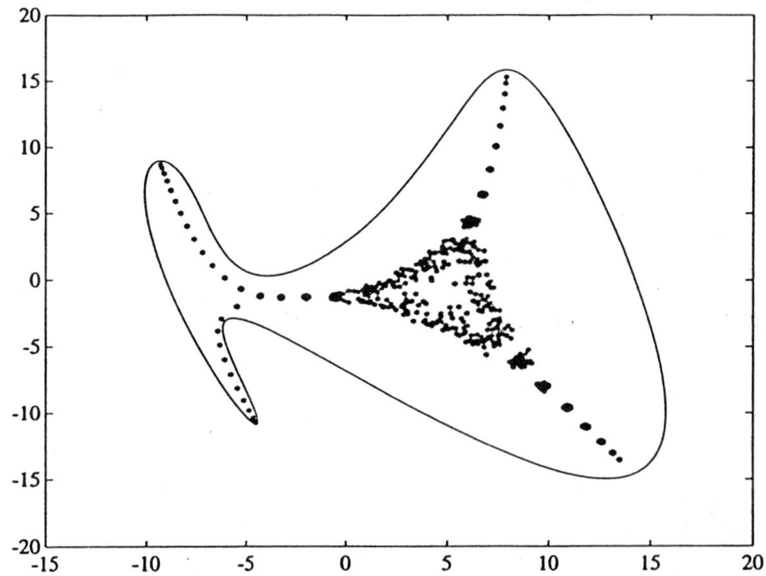

Figure 17

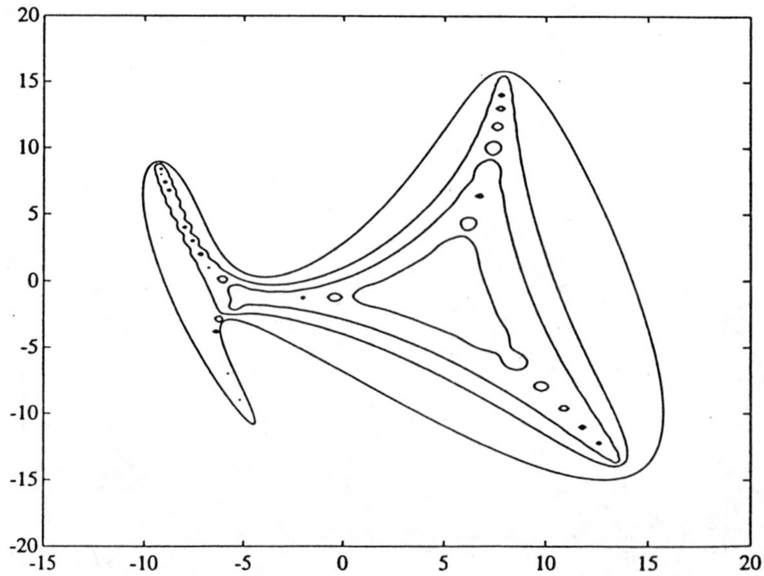

Figure 18

$T_{50}(a)+E$ for 50 randomly chosen matrices E subject to the constraint $\|E\| = 10^{-2}$. Finally, in Figure 18 we have plotted the boundaries of the pseudospectra $\mathrm{sp}_\varepsilon\, T_{50}(a)$ for $\varepsilon = 1/20,\ 1/100,\ 1/1000$.

A whole world of really delightful examples of the previous kind may be found in the book of Trefethen [1995]. ■

Theorem 10.4 *If* $a \in PC(\mathbf{T})$ *and* $\varepsilon > 0$, *then*

$$\lim_{n\to\infty} \mathrm{sp}_{\,\varepsilon} T_n(a) = \mathrm{sp}_\varepsilon\, T(a).$$

Notice how drastically things are simplified by passage from spectra to pseudospectra!

A theorem like this first appeared in Reichel and Trefethen [1992], where it is derived from the results of Widom [1989]. An alternative **proof** is contained in Böttcher [1994]. There it is shown that if A is any operator in $\mathcal{L}(l^2)$ then

$$\lim_{n\to\infty} \mathrm{sp}_\varepsilon\, (P_nAP_n|\mathrm{Im}\, P_n) \supset \mathrm{sp}_\varepsilon\, A$$

for each $\varepsilon > 0$, while the reverse inclusion for Toeplitz operators,

$$\lim_{n\to\infty} \mathrm{sp}_\varepsilon\, T_n(a) \subset \mathrm{sp}_\varepsilon\, T(a) \quad (\varepsilon > 0),$$

can be easily obtained from Theorem 10.2: if $\lambda \notin \mathrm{sp}_\varepsilon\, T(a)$, then $T(a)-\lambda I = T(a-\lambda)$ is invertible and

$$\|T^{-1}(a-\lambda)\| = 1/\varepsilon - 2\delta < 1/\varepsilon,$$

now Theorem 10.2 tells us that there is an n_0 such that

$$\|T_n^{-1}(a-\lambda)\| < 1/\varepsilon - \delta \text{ for all } n > n_0,$$

whence

$$\|T_n^{-1}(a-\mu)\| < 1/\varepsilon$$

if only $n > n_0$ and $|\mu - \lambda| < \varepsilon\delta(1/\varepsilon - \delta)^{-1}$, implying that $\lambda \notin \lim_{n\to\infty} \mathrm{sp}_\varepsilon\, T_n(a)$. ■

May we trust in the computer? Yes, we may, but not blindly. Using *Mathematica*, we numerically computed the eigenvalues of

$$T_{200}(a), T_{300}(a), T_{400}(a), T_{500}(a), T_{700}(a)$$

for a as in Example 10.3. The results are shown in Figures 19, 20, 21, 22, 23, respectively. Thus, if we did not know of the results of Schmidt and Spitzer and blindly trusted in the computer, we would arrive at the guess that the eigenvalues of $T_n(a)$ eventually mimic the range of a. The point is that the computations were done with the standard numerical precision of *Mathematica*, which keeps 15 significant digits in a particular calculation. After increasing the numerical accuracy to 25 digits, we obtained Figure 24 for the eigenvalues of $T_{200}(a)$. Clearly, while Figure 19 does not confirm the forecast of Figure 16, the plot of Figure 24 does it very convincingly.

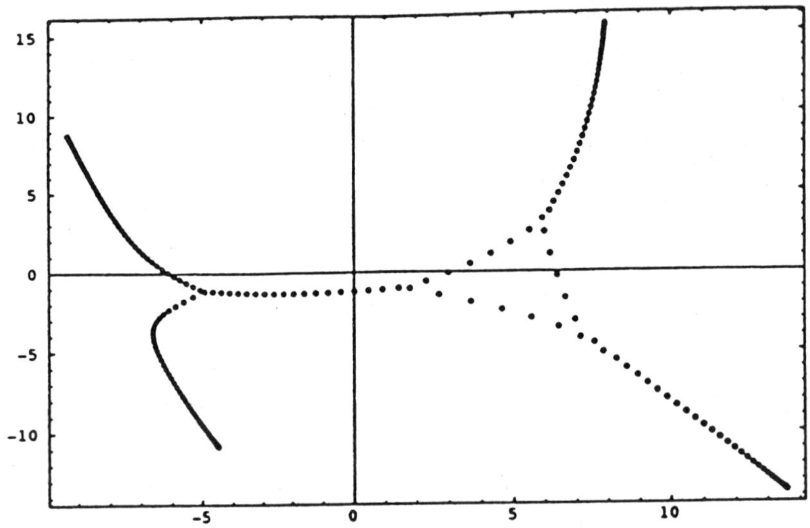

Figure 19 Eigenvalues of $T_{200}a$; numerical precison: 15 digits.

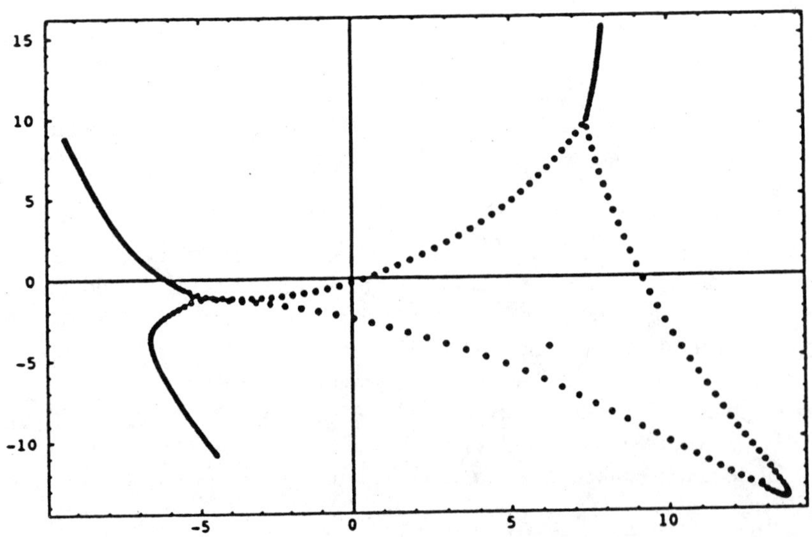

Figure 20 Eigenvalues of $T_{300}a$; numerical precision: 15 digits.

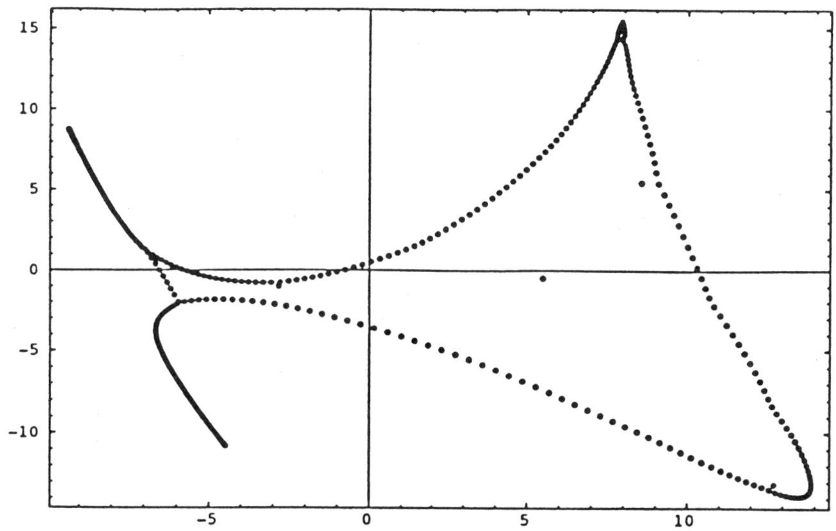

Figure 21 Eigenvalues of $T_{400}a$; numerical precison: 15 digits.

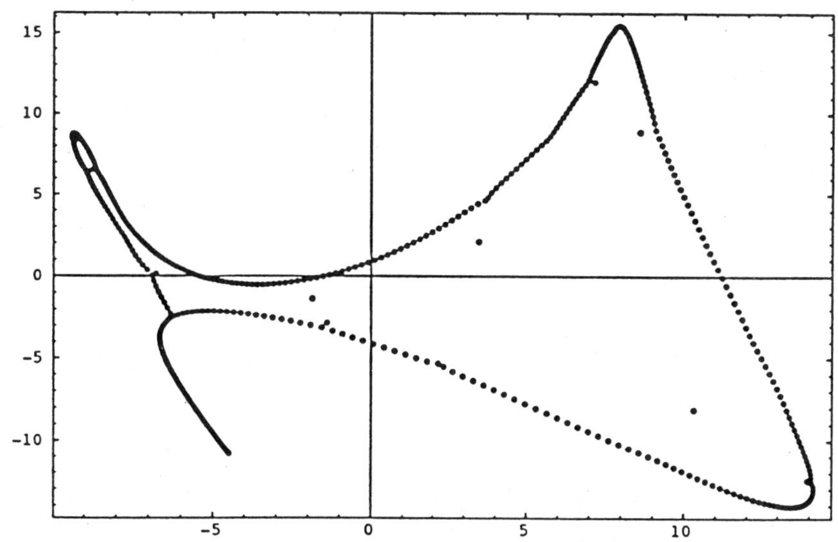

Figure 22 Eigenvalues of $T_{500}a$; numerical precision: 15 digits.

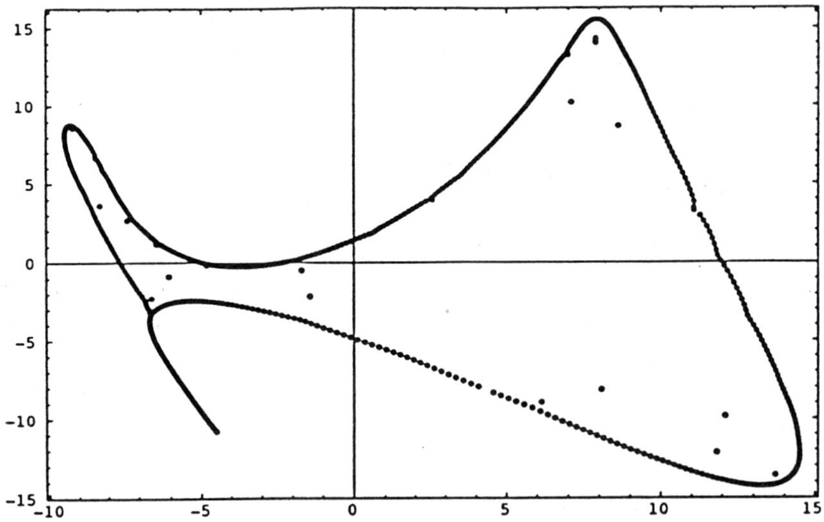

Figure 23 Eigenvalues of $T_{700}a$; numerical precision: 15 digits

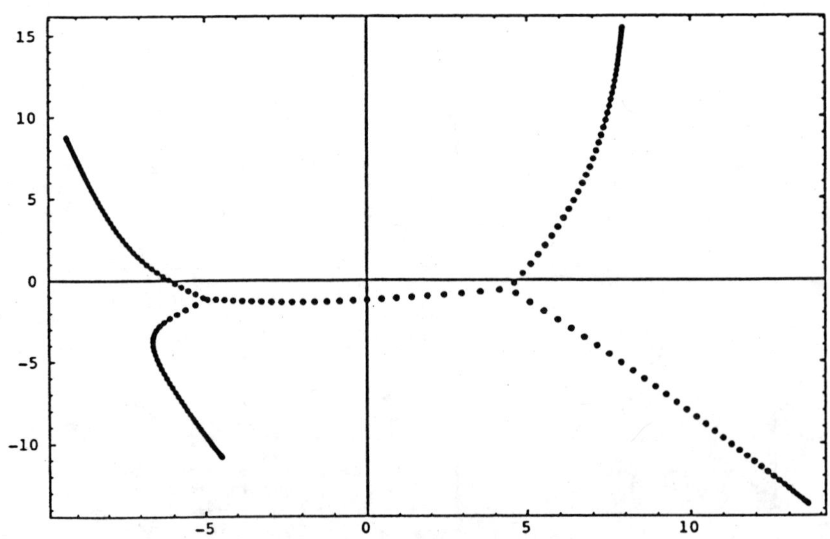

Figure 24 Eigenvalues of $T_{200}a$; numerical precision: 25 digits

11 Toeplitz determinants

We now consider the following problem. Let A be an infinite matrix and denote by $A_n = P_n A P_n | \mathrm{Im}\, P_n$ the finite sections of A. What can be said about the determinants $\det A_n$ as n goes to infinity? One of the most beautiful results in this direction is *Szegö's strong limit theorem*, which concerns the case where $A = T(a)$ is an invertible Toeplitz operator with a sufficiently smooth symbol.

To avoid unnecessary complications, suppose $a \in C(\mathbf{T})$ and

$$\sum_{n=-\infty}^{\infty} |a_n| + \sum_{n=-\infty}^{\infty} (|n|+1)|a_n|^2 < \infty, \tag{67}$$

where, of course, a_n denotes the nth Fourier coefficient of a. Also suppose that a has no zeros on $\mathbf{T}$ and that the winding number of $a(\mathbf{T})$ about the origin is zero. Then a has a logarithm in $C(\mathbf{T})$, i.e. there exists a function $\log a \in C(\mathbf{T})$ such that $a = e^{\log a}$. One can show that $\log a$ also satisfies (67): if we denote by $(\log a)_n$ the nth Fourier coefficient of $\log a$, then

$$\sum_{n=-\infty}^{\infty} |(\log a)_n| + \sum_{n=-\infty}^{\infty} (|n|+1)|(\log a)_n|^2 < \infty. \tag{68}$$

From (68) we infer in particular that the two terms

$$G(a) := \exp(\log a)_0,$$
$$E(a) := \exp \sum_{k=1}^{\infty} k(\log a)_k (\log a)_{-k}$$

are well-defined.

Theorem 11.1 (Szegö's strong limit theorem) *Under the above hypotheses,*

$$\lim_{n\to\infty} \frac{\det T_n(a)}{G(a)^n} = E(a).$$

Thus, $\det T_n(a)$ is asymptotically of the form $G(a)^n E(a)$ with completely identified constants $G(a)$ and $E(a)$. For the history of this theorem the reader is referred to the monographs Grenander and Szegö [1958], Gohberg and Feldman [1971], Böttcher and Silbermann [1983b], [1990], and the paper Widom [1976]. More about this exciting story will also be said in the end of this section. The main purpose of what follows is to indicate how Theorem 11.1 is tied in with the convergence of the finite section method.

An important role in determinant theory is played by *trace class* (or *nuclear*) *operators*. These are especially nice compact operators and they may be defined in various ways. One definition goes as follows: a Hilbert space operator A is called a *Hilbert–Schmidt operator* if there is an orthonormal basis with respect to which A has a matrix representation $(a_{jk})_{j,k=1}^{\infty}$ such that $\sum_{j,k=1}^{\infty} |a_{jk}|^2 < \infty$ (recall Example 4.2(a)), and a trace class operator is an operator that may be represented as the product of two Hilbert–Schmidt operators. One can show that the product of a bounded and a trace class operator is again a trace class operator.

Example 11.2 It is easily seen that the Hankel operator $H(a)$ (Example 4.2(c)) is Hilbert–Schmidt if and only if $\sum_{n=1}^{\infty} n|a_n|^2 < \infty$. In particular, if a is subject to (67), then $H(a)$ is Hilbert-Schmidt. Moreover, one can show that if a satisfies (67) and has no zeros on $\mathbf{T}$, then a^{-1} also satisfies (67). Consequently, $H(a^{-1})$ and $H(\tilde{a}^{-1})$ are also Hilbert–Schmidt in that case. Since, by (26),

$$T(a)T(a^{-1}) - I = H(a)H(\tilde{a}^{-1}) =: L$$

we conclude that $T(a)T(a^{-1}) - I$ as well as $T(a) - T^{-1}(a^{-1}) = LT^{-1}(a^{-1})$ are trace class operators. ■

Let K be a trace class operator and let $\{\lambda_j\}$ denote the sequence of its eigenvalues counted up to algebraic multiplicity (= dimension of the linear space

$$\{x \in H : (K - \lambda I)^n x = 0 \quad \text{for some} \quad n \geq 0\}).$$

It can be shown that $\sum_j |\lambda_j| < \infty$, and hence the *determinant*

$$\det(I + K) := \prod_j (1 + \lambda_j)$$

is well-defined. This determinant enjoys a lot of the properties we know for the determinant of finite matrices. What we need in the following is an approximation property: if B_n converges strongly to B and C_n^* converges strongly to C^*, then $\det(I + B_n K C_n)$ converges to $\det(I + BKC)$ as $n \to \infty$. All the results stated in the previous three paragraphs can be found in Gohberg and Krein [1965] or Simon [1977], for example.

In the end of Section 4 we learned that it might be useful to decompose a given operator into an operator for which the finite section method converges and a compact operator. The following simple result fits in with this philosophy.

Proposition 11.3 *Let $A \in \mathcal{L}(l^2)$, suppose $A \in \Pi\{P_n, P_n\}$, and let K be a trace operator. Then*

$$\lim_{n\to\infty} \frac{\det P_n(A+K)P_n}{\det P_n A P_n} = \det(I + A^{-1}K).$$

Proof Put $A_n = P_n A P_n | \operatorname{Im} P_n$ and $K_n = P_n K P_n | \operatorname{Im} P_n$. Since A_n is invertible for all sufficiently large n and $A_n^{-1} P_n$ converges strongly to A^{-1}, we have

$$\begin{aligned} \frac{\det(A_n + K_n)}{\det A_n} &= \det(I + A_n^{-1} K_n) \\ &= \det(I + A_n^{-1} P_n K P_n) \to \det(I + A^{-1}K). \ ■ \end{aligned}$$

Thus, if we know the asymptotic behavior of $\det P_n A P_n$, then the preceding proposition says something about the asymptotic behavior of $\det P_n(A+K)P_n$. Here is the way Proposition 11.3 can be used to prove Szegö's strong limit theorem. Since $T(a) = T^{-1}(a^{-1}) +$ *a trace class operator* by Example 11.2, we may try our luck by applying Proposition 11.3 with $A = T^{-1}(a^{-1})$ (Theorem 5.3 is very encouraging in this connection).

Proposition 11.4 *Under the same hypothesis as in Theorem 11.1,*

$$\det P_n T^{-1}(a^{-1}) P_n = G(a)^n$$

for all $n \geq 1$.

Proof outline The following may appear mysterious for those who have never heard of Wiener–Hopf factorization and is trivial for those who know what this is. Put

$$a_+(t) = \exp\left(\sum_{k=0}^{\infty} (\log a)_k t^k\right), \quad a_-(t) = \exp\left(\sum_{k=-\infty}^{-1} (\log a)_k t^k\right).$$

Then $a = a_- a_+$ and hence, by (25), $T(a^{-1}) = T(a_-^{-1})T(a_+^{-1})$. It follows that $T^{-1}(a^{-1}) = T(a_+)T(a_-)$, and it is easily seen that

$$P_n T^{-1}(a^{-1}) P_n = P_n T(a_+) T(a_-) P_n = P_n T(a_+) P_n T(a_-) P_n.$$

Hence,

$$\det P_n T^{-1}(a^{-1}) P_n = \det P_n T(a_+) P_n \det P_n T(a_-) P_n$$

and since $P_n T(a_+) P_n$ and $P_n T(a_-) P_n$ are triangular, it follows that the determinant det $P_n T^{-1}(a^{-1}) P_n$ is equal to $C(a)^n$ with some constant $C(a)$. A little computation shows that actually $C(a) = G(a)$. ∎

To prove Theorem 11.1 it remains to combine Propositions 11.3 and 11.4 with

$$A = T^{-1}(a^{-1}) \quad \text{and} \quad K = H(a)H(\tilde{a}^{-1})T^{-1}(a^{-1})$$

(recall Example 11.2). It follows that

$$\begin{aligned} \lim_{n\to\infty} \frac{\det T_n(a)}{G(a)^{n+1}} &= \det(I + (T^{-1}(a^{-1}))^{-1} H(a) H(\tilde{a}^{-1}) T^{-1}(a^{-1})) \\ &= \det(I + T(a^{-1}) H(a) H(\tilde{a}^{-1}) T^{-1}(a^{-1})) \\ &= \det(I + T(a^{-1})(T(a)T(a^{-1}) - I) T^{-1}(a^{-1})) \\ &= \det T(a^{-1}) T(a) \end{aligned}$$

and so we are left with showing that $\det T(a^{-1})T(a)$ equals $E(a)$. As first observed by Widom [1976], this is immediate from writing

$$\begin{aligned} T(a^{-1})T(a) &= T(a_-^{-1})T(a_+^{-1})T(a_-)T(a_+) \\ &= e^{-T(\log a_-)} e^{-T(\log a_+)} e^{T(\log a_-)} e^{t(\log a_+)} \end{aligned}$$

and invoking the following remarkable identity:

$$\det e^A e^B e^{-A} e^{-B} = \exp \text{ trace } (AB - BA),$$

which was shown by Helton, Howe, and Pincus to be valid whenever A and B are bounded operators for which $AB - BA$ is of trace class. ∎

Here now are a few pieces of the story behind Szegö's limit theorem. It turns out that if a ferromagnet is heated, its magnetization deceases. This is no surprise, since we may think of the magnet as consisting of millions and millions of adjusted elementary magnets that are thrown into disorder in the process of heating. However, there is a temperature, the so-called *Curie point* T_c (which equals 1317 Kelvin degrees for iron), at which the magnetization disappears completely and above which it remains zero. If, conversely, a hot ferromagnet is cooled down, magnetization appears at the Curie point and increases as the temperature decreases. In other words, we would understand a behavior as in the left diagram of Figure 25, but in fact we observe a behavior as shown in the right diagram of Figure 25.

What we encounter is a so-called phase transition. Statistical physics provides the apparatus for explaining such macroscopic phenomena by microscopic interactions. In the case at hand, one models a magnet by a large number of elementary magnets located at the sites of the lattice $\mathbf{Z}^N$. Of course, in the real-world case N equals 3. The so-called *Ising model* includes the theory of computing the correlation between two elementary magnets at a large distance (10^{23}, say) via the interaction of neighboring elementary magnets. The presence of such a correlation ("long range order") means that we may observe a global magnetic field, whereas absence of a correlation is interpreted as lack of macroscopic magnetization. The temperature T enters the formula for the correlation as a parameter, and hence the problem is to show that the correlation is positive for $T < T_c$ and zero for $T \geq T_c$.

It turns out, and this is the result of Ernst Ising, that there is no phase transition for a "one-dimensional magnet" ($N = 1$). As far as I know, until now no rigorous and satisfactory results are available in three (and higher) dimensions. However, in 1948, Lars Onsager (who received the Nobel Prize for Chemistry in 1968) surprised the physicists with a beautiful formula for the magnetization of a two-dimensional ferromagnet ($N = 2$). This formula implies that a two-dimensional ferromagnet has a Curie point T_c and that its magnetization deceases as $(T_c - T)^{1/8}$ as T approaches T_c from below.

Onsager has never published a proof of his formula, but from his correspondence with Bruria Kaufman we know that he somehow derived the formula by considering Toeplitz determinants.

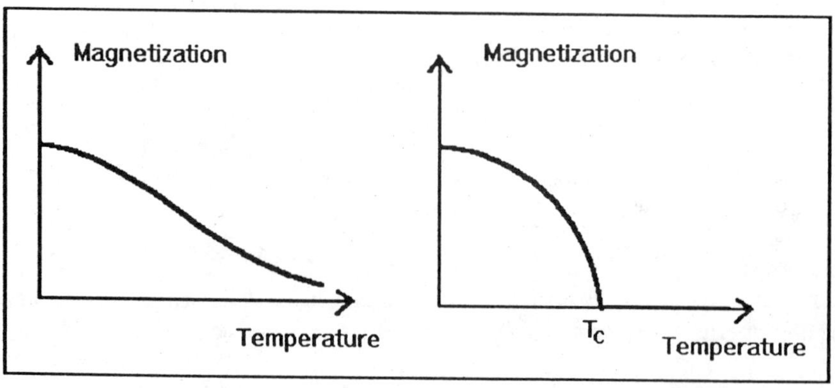

Figure 25

So what has a hot magnet to do with Toeplitz determinants? The correlation between two elementary magnets can be shown to be the sum of certain products. And what is a determinant? It is also a sum of certain products! You guess it ... It was Montroll, Potts, and Ward who were indeed able to rigorously show that the correlation of two elementary magnets in $\mathbf{Z}^2$ may be expressed as the determinant of a Toeplitz (!) matrix.

Now Szegö enters the scene. In 1915, he established an asymptotic formula for Toeplitz determinants with positive (and thus real-valued) symbols. In the early fifties he was told by S. Kakutani that in order to prove Onsager's formula it would be desirable to have an analogous result for certain complex-valued symbols. And Szegö was able to extend his theorem to what is now called the "strong" Szegö limit theorem (Theorem 11.1) and thus to give the missing link for rigorously proving Onsager's formula.

The story does not end at this point. Fisher and Hartwig [1968] took up the matter again. They reminded us of the fact that the correlation function in the two-dimensional Ising model leads to a Toeplitz determinant with a smooth symbol of *nonvanishing* winding number for temperatures $T > T_c$ and with a symbol having a *jump* in case $T = T_c$. They also realized that a number of other problems of statistical physics result in Toeplitz determinants which are generated by symbols with zeros, poles, jumps, or discontinuities of oscillating type.

While the Szegö limit theorem implies an asymptotic behavior of the form $G^n E$ in the case of regular symbols, Fisher and Hartwig conjectured that for a large class of singular symbols the determinants asymptotically behave like $G^n n^q E$ with certain constants G, q, E. Let us consider two concrete examples.

Example 11.5 Given complex numbers $c \neq 0$ and β, define the function $\varphi_{c,\beta}$ by

$$\varphi_{c,\beta}(e^{i\theta}) = ce^{-i\beta\pi}e^{i\beta\theta} \quad (0 \leq \theta < 2\pi).$$

We have

$$\varphi_{c,\beta}(e^{i(0-0)}) = ce^{-i\beta\pi}, \quad \varphi_{c,\beta}(e^{i(0+0)} = ce^{i\beta\pi},$$

and hence $\varphi_{c,\beta}$ has a jump at $e^{i\theta} = 1$ unless β is an integer. Notice that the symbol of the Hilbert Toeplitz matrix (Examples 6.2 and 9.4) is just

$$\varphi_{-\pi,1/2}(e^{i\theta}) = -\pi e^{-i\pi/2}e^{i\theta/2} = \pi i e^{i\theta/2}.$$

Fisher and Hartwig proved that if $\mathrm{Re}\beta \notin \mathbf{Z} \setminus \{0\}$, then

$$\det T_n(\varphi_{c,\beta}) \sim c^n G\left(\frac{3}{2}\right) G\left(\frac{1}{2}\right) n^{-\beta^2} \text{ as } n \to \infty,$$

where $x_n \sim y_n$ means that $x_n/y_n \to 1$ and where $G(z)$ is the so-called Barnes function, which is an entire function satisfying $G(z+1) = G(z)\Gamma(z)$ and is thus a kind of double Gamma function. In case $\mathrm{Re}\beta \in \mathbf{Z} \setminus \{0\}$ we have $\det T_n(\varphi_{c,\beta}) = 0$ for all $n \geq 1$. In particular, for the Hilbert Toeplitz matrix $T(a)$ we obtain

$$\det T_n(a) = (-\pi)^n G\left(\frac{3}{2}\right) G\left(\frac{1}{2}\right) n^{-1/4} \text{ as } n \to \infty$$

with $G\left(\frac{3}{2}\right) G\left(\frac{1}{2}\right) = 0.644....$ ∎

Example 11.6 Now, for complex numbers $c \neq 0$ and α, put

$$\omega_{c,\alpha}(e^{i\theta}) = c(2 - 2\cos\theta)^{\alpha} \quad (0 \leq \theta < 2\pi).$$

If Re$\alpha > -1/2$, then $\omega_{c,\alpha}$ belongs to $L^1(\mathbf{T})$, so it has well-defined Fourier coefficients and induces a Toeplitz matrix $T(\omega_{c,\alpha})$. Clearly, $(2 - 2\cos\theta)^{\alpha}$ has a zero at $e^{i\theta} = 1$ for Re$\alpha > 0$, a pole at $e^{i\theta} = 1$ for Re$\alpha < 0$, and a discontinuity of oscillating type at $e^{i\theta} = 1$ if Re$\alpha = 0$ but Im$\alpha \neq 0$. One can show that

$$\det T_n(\omega_{c,\alpha}) \sim c^n \frac{G(1+\alpha)G(1-\alpha)}{G(1+2\alpha)} n^{\alpha^2} \text{ as } n \to \infty. \blacksquare$$

The *Fisher-Hartwig conjecture* concerns symbols that are composed of symbols of the above two types. It says that if

$$a(e^{i\theta}) = \prod_{j=1}^{N} \varphi_{1,\beta_j}(e^{i(\theta-\theta_j)})\omega_{1,\alpha_j}(e^{i(\theta-\theta_j)})b(e^{i\theta}),$$

where $\theta_1, \dots, \theta_N \in [0, 2\pi)$ and b is a "nice" function (smooth, no zeros, vanishing winding number), then

$$\det T_n(a) \sim G(b)^n n^{\sum(\alpha_j^2 - \beta_j^2)} \tilde{E}(a) \text{ as } n \to \infty$$

with a certain nonzero constant $\tilde{E}(a)$.

During the last 25 years, especially by the work of H. Widom, E. Basor, B. Silbermann, and the author, this conjecture has been shown to be true in important cases and to be not valid in other cases; in the latter cases revised versions of the conjecture have been established and also been partially proved. For more about this topic and for precise results see the papers Basor and Tracy [1991], Basor and Morrison [1994], Böttcher [1995] and the books Böttcher and Silbermann [1983b], [1990].

The proof of Szegö's strong limit theorem given above is heavily based on the finite section method. The proofs of several versions of the Fisher-Hartwig conjecture also make essential use of results on the finite section method. However, in this context it does not suffice to consider only the finite section method for Toeplitz operators on l^2. One has rather to study much more complicated operators and one is even forced to leave Hilbert space and to work in appropriate Banach spaces. For example, the proof of the Fisher-Hartwig conjecture in the case where $|\text{Re}\alpha_j| < 1/2$ and $|\text{Re}\beta_j| < 1/2$ eventually leads to the problem of investigating the finite section method in l^2 for operators of the form $A = T^{-1}(f)T(g)T^{-1}(h)$ with $f, g, h \in PC(\mathbf{T})$. Alternatively, the confirmation of the conjecture in the case of a single but "large" singularity requires studying the finite section method for Toeplitz operators with piecewise continuous symbols on l^p spaces with cleverly chosen weights. Thus, Banach space theory of Toeplitz operators is not at all a purely academical matter.

12 More general projection methods

To conclude these lectures, let us return to the situation we started with: A is a bounded linear operator on some infinite-dimensional separable Hilbert space H and we want to solve the equation $Ax = y$. A *projection method* for approximately solving this equation consists in the following. Choose two sequences of projections $\{R_n\}_{n=1}^{\infty}$ and $\{S_n\}_{n=1}^{\infty}$ on H such that

$$\dim \operatorname{Im} R_n = \dim \operatorname{Im} S_n = n$$

and such that $R_n \to I$ and $S_n \to I$ strongly. Then replace the original equation by the approximate equation

$$R_n A x^{(n)} = R_n y \quad (x^{(n)} \in \operatorname{Im} S_n). \tag{69}$$

The equations (69) are linear algebraic systems with n equations and n unknowns (e.g. the coordinates of $x^{(n)}$ in a basis of Im S_n). For several properties of such more general projection methods we refer the reader to the books Gohberg and Feldman [1971], Prössdorf and Silbermann [1977], Prössdorf and Silbermann [1992].

In practice, projection methods are usually formulated as so-called *Galerkin–Petrov methods*, which means the following. Let

$$\begin{array}{ll} \varphi_1^{(1)} & \psi_1^{(1)} \\ \varphi_1^{(2)},\ \varphi_2^{(2)} & \psi_1^{(2)},\ \psi_2^{(2)} \\ \varphi_1^{(3)},\ \varphi_2^{(3)},\ \varphi_3^{(3)} & \psi_1^{(3)},\ \psi_2^{(3)},\ \psi_3^{(3)} \\ \dots\dots\dots\dots & \dots\dots\dots\dots \end{array} \tag{70}$$

be two systems of elements in H and determine an approximate solution $x^{(n)}$ in the linear span of $\psi_1^{(n)}, \dots, \psi_n^{(n)}$ so that

$$(Ax^{(n)}, \varphi_j^{(n)}) = (y, \varphi_j^{(n)}) \quad \text{for} \quad j = 1, \dots, n. \tag{71}$$

Notice that (71) is again a linear algebraic system with n unknowns and n equations. Since H is often a space of (equivalence classes of) functions, the elements $\psi_j^{(n)}$ are frequently referred to as the *basis functions* and the elements $\varphi_j^{(n)}$ are called the *test functions* of the Galerkin–Petrov method determined by (70). If we denote by R_n and S_n any projections of H onto

$$\operatorname{span}\{\varphi_1^{(n)}, \dots, \varphi_n^{(n)}\} \text{ and } \operatorname{span}\{\psi_1^{(n)}, \dots, \psi_n^{(n)}\},$$

respectively, then (71) may clearly be written in the form (69).

One says that the *projection method* given by (69) *converges* for the operator A and writes $A \in \Pi\{R_n, S_n\}$ in this case if there is an n_0 such that the equations (69) are uniquely solvable for all $n \geq n_0$ and all $y \in H$ and $x^{(n)}$ converges in H to a solution $x \in H$ of the equation $Ax = y$.

The finite section method for operators on l^2 is obviously a Galerkin–Petrov method: the test and basis "functions" (70) are given by

$$\begin{array}{ll} e_1 & e_1 \\ e_1,\ e_2 & e_1,\ e_2 \\ e_1,\ e_2,\ e_3 & e_1,\ e_2,\ e_3 \\ \dots\dots\dots & \dots\dots\dots\ , \end{array}$$

where e_k is the sequence in l^2 whose kth term is 1 and the remaining terms of which are 0. To have a few more concrete examples, let us consider the Toeplitz operator $T^B(a)$ on the Bergman space $A^2(\mathbf{D})$ (Example 1.3). The following material is taken from Böttcher and Wolf [1993].

Example 12.1 Put $e_n(z) = \sqrt{n}z^{n-1}$ $(z \in \mathbf{D})$. Then $\{e_n\}_{n=1}^{\infty}$ is an orthonormal basis in $A^2(\mathbf{D})$. The Galerkin–Petrov method with the test and basis functions

$$\begin{array}{ll} e_1 & e_1 \\ e_1,\ e_2 & e_1,\ e_2 \\ e_1,\ e_2,\ e_3 & e_1,\ e_2,\ e_3 \\ \cdots\cdots\cdots & \cdots\cdots\cdots, \end{array}$$

amounts to the following: we look for an approximate solution of the equation $T^B(a)x = y$ in the form of a polynomial of degree $n-1$ and determine the n unknown coefficients so that the first n Taylor coefficients of $(T^B(a)x)(z)$ and $y(z)$ at $z=0$ coincide. If we represent $T^B(a)$ by its (infinite) matrix with respect to the basis $\{e_n\}_{n=1}^{\infty}$, then the method at hand replaces this matrix by its $n \times n$ truncations and thus turns out to be nothing than the *finite section method.* On defining P_n on $A^2(\mathbf{D})$ as the orthogonal projection of $A^2(\mathbf{D})$ onto span $\{e_1, \dots, e_n\}$, we may write the approximating equations in the form

$$P_n T^B(a) x^{(n)} = P_n y \quad (x^{(n)} \in \operatorname{Im} P_n). \blacksquare$$

Example 12.2 For $\alpha \in \mathbf{D}$, define the function K_α by

$$K_\alpha(z) = \frac{1}{(1-\overline{\alpha}z)^2} \quad (z \in \mathbf{D}).$$

One can show that $K_\alpha \in A^2(\mathbf{D})$ and that

$$x(\alpha) = (x, K_\alpha) := \int_{\mathbf{D}} x(z)\overline{K_\alpha(z)}\, dA(z) \tag{72}$$

for every function $x \in A^2(\mathbf{D})$. Due to (72), K_α is called the *reproducing kernel of* $A^2(\mathbf{D})$ *at the point* α. Now choose a system of points

$$\begin{array}{l} \alpha_1^{(1)} \\ \alpha_1^{(2)},\ \alpha_2^{(2)} \\ \alpha_1^{(3)},\ \alpha_2^{(3)},\ \alpha_3^{(3)} \\ \cdots\cdots\cdots\cdots \end{array} \tag{73}$$

in $\mathbf{D}$ and define the system of test and basis functions (70) by

$$\begin{array}{ll} K_1^{(1)} & e_1 \\ K_1^{(2)},\ K_2^{(2)} & e_1,\ e_2 \\ K_1^{(3)},\ K_2^{(3)},\ K_3^{(3)} & e_1,\ e_2,\ e_3 \\ \cdots\cdots\cdots\cdots & \cdots\cdots\cdots, \end{array}$$

where $K_j^{(n)} := K_{\alpha_j^{(n)}}$ and e_j is as in the preceding example. Because of (72), the approximating equations

$$\left(T^B(a)x^{(n)},\, K_j^{(n)}\right) = \left(y,\, K_j^{(n)}\right) \quad (j = 1, \dots, n) \tag{74}$$

are equivalent to the equations

$$\left(T^B(a)x^{(n)}\right)\left(\alpha_j^{(n)}\right) = y\left(\alpha_j^{(n)}\right) \quad (j = 1, \dots, n). \tag{75}$$

Thus, the Galerkin–Petrov method at hand is a *polynomial collocation method*: the approximate solution $x^{(n)}$ is sought as a polynomial of the degree $n-1$ so that $(T^B(a)x^{(n)})(z)$ and $y(z)$ coincide at the n prescribed collocation points $z = \alpha_1^{(n)}, \dots, \alpha_n^{(n)}$.

Define the orthogonal projections P_n as in Example 12.1 and let L_n denote the Lagrange interpolation projection on $A^2(\mathbf{D})$ associated with the collocation points $\alpha_1^{(n)}, \dots, \alpha_n^{(n)}$. Thus, $L_n x$ is the (uniquely determined) polynomial of degree $n-1$ that coincides with x at the points $\alpha_1^{(n)}, \dots, \alpha_n^{(n)}$. The equations (74) and (75) may now be rewritten in the form

$$L_n T^B(a)x^{(n)} = L_n y \quad (x^{(n)} \in \operatorname{Im} P_n). \tag{76}$$

Since clearly $\operatorname{Im} L_n = \operatorname{Im} P_n$, we might also replace (76) by

$$L_n T^B(a)x^{(n)} = L_n y \quad (x^{(n)} \in \operatorname{Im} L_n). \blacksquare$$

Example 12.3 Once more choose a system of points (73) in $\mathbf{D}$, but now specify the system of test and basis functions (70) to

$$\begin{array}{ll} K_1^{(1)} & K_1^{(1)} \\ K_1^{(2)},\ K_2^{(2)} & K_1^{(2)},\ K_2^{(2)} \\ K_1^{(3)},\ K_2^{(3)},\ K_3^{(3)} & K_1^{(3)},\ K_2^{(3)},\ K_3^{(3)} \\ \dots\dots\dots\dots & \dots\dots\dots\dots\ , \end{array}$$

where, as above, $K_j^{(n)} := K_{\alpha_j^{(n)}}$. The approximating equations are again of the form (74) and (75), but while in Example 12.2 the approximate function $x^{(n)}$ was sought as a polynomial, it is now sought in span $\{K_1^{(n)}, \dots, K_n^{(n)}\}$. Hence, the method under consideration might be called an *analytic element collocation method*: we look for an approximate solution $x^{(n)}$ in the form of a linear combination of $K_1^{(n)}, \dots, K_n^{(n)}$ and determine the n unknown coefficients so that $(T^B(a)x^{(n)})(z)$ and $y(z)$ coincide at the n collocation points $z = \alpha_1^{(n)}, \dots, \alpha_n^{(n)}$.

To rewrite this Galerkin–Petrov method as a projection method, we need a projection of $A^2(\mathbf{D})$ onto span $\{K_1^{(n)}, \dots, K_n^{(n)}\}$. It turns out that L_n^*, the adjoint operator of the Lagrange interpolation projection L_n introduced in the previous example, is such a projection (the proof of this fact, which is based on (72), can be found in Böttcher and Wolf [1993]). Hence, the approximating equations may be written in the form

$$L_n T^B(a)x^{(n)} = L_n y \quad (x^{(n)} \in \operatorname{Im} L_n^*). \blacksquare$$

In Section 5, we showed that $T^B(a) \in \Pi\{P_n, P_n\}$ if a is the harmonic extension into $\mathbf{D}$ of some continuous function on $\mathbf{T}$ and $T^B(a)$ is invertible. Thus, we have a result for the finite section method. Surprisingly, this result has interesting consequences for polynomial collocation (the method $\{L_n, P_n\}$) and analytic element collocation (the method $\{L_n, L_n^*\}$), too. A projection method $\{R_n, S_n\}$ on $A^2(\mathbf{D})$ is called an *asymptotic finite section method* if

$$\|R_n - P_n\| \to 0 \quad \text{and} \quad \|S_n - P_n\| \to 0 \quad \text{as} \quad n \to \infty,$$

where $P_n : A^2(\mathbf{D}) \to \text{span}\,\{e_1, \ldots, e_n\}$ are the orthogonal projections defined in Example 12.1.

Proposition 12.4 *Let $\{R_n, S_n\}$ be an asymptotic finite section method. If $A \in \mathcal{L}(A^2(\mathbf{D}))$, then*

$$A \in \Pi\{R_n, S_n\} \Longleftrightarrow A \in \Pi\{P_n, P_n\}.$$

The simple proof is omitted (see Böttcher and Wolf [1993]). ■

Theorem 12.5 *Fix a number $r \in (0,1)$ and let the points $\alpha_1^{(n)}, \ldots, \alpha_n^{(n)}$ in the nth row of (73) be the zeros of the polynomial $z^n - r^n$. Then*

$$\|L_n - P_n\| = \|L_n^* - P_n\| \to 0 \quad as \quad n \to \infty.$$

In particular, the methods $\{L_n, P_n\}$ and $\{L_n, L_n^*\}$ are asymptotic finite section methods and are thus convergent for $T^B()$ whenever the finite section methods converges for this operator.

A proof is in Böttcher and Wolf [1993]. ■

What we learn from the previous theorem is that the finite section method is of greater import than it might appear at the first glance. It plays a crucial role for establishing the convergence of plenty of approximation methods, not only for Bergman space operators, but also for singular integral operators and their manifold generalizations. The finite section method is something at the heart of numerical analysis and any sort of mathematics that is concerned with questions like the Problems I and II raised in Section 2. Impressive accounts of this "philosophy" are the recent books Prössdorf and Silbermann [1992] and Hagen et al. [1994].

Acknowledgements I wish to thank Israel Gohberg, Bernd Silbermann, and Nick Trefethen for reading an earlier version of this article with great care and making a large number of remarks that led to essential improvements in the presentation. I am also greatly indebted to Harald Heidler and Pedro Santos for their invaluable assistance in producing the Figures 2-7 and 15-24.

Bibliography

Arveson, W. [1994] *C^*-algebras and numerical linear algebra*, J. Funct. Analysis **122**, 333-360.

Bart, H., Gohberg, I. and Kaashoek, M.A. [1984] *The coupling method for solving integral equations*, Operator Theory: Advances and Applications **12**, 39–73; *Addendum:* Integral Equations and Operator Theory **8** (1985), 890–891.

Basor, E. and Morrison, K.E. [1994] *The Fisher–Hartwig conjecture and Toeplitz eigenvalues*, Linear Algebra Appl. **202**, 129-142.

Basor, E. and Tracy, C. [1991] *The Fisher-Hartwig conjecture and generalizations*, Phys. A **177**, 167-173.

Baxter, G. [1963] *A norm inequality for "finite-section" Wiener-Hopf equations*, Illinois J. Math. **7**, 97-103.

Böttcher, A. [1990] *Truncated Toeplitz operators on the polydisk*, Monatshefte f. Math. **110**, 23–32.

Böttcher, A. [1994] *Pseudospectra and singular values of large convolution operators*, J. Integral Equations Appl. **6**, 267–301.

Böttcher, A. [1995] *The Onsager formula, the Fisher-Hartwig conjecture, and their influence on research into Toeplitz operators*, J. Statist. Phys. **78** (Lars Onsager Festschrift), 575–584.

Böttcher, A. and Silbermann, B. [1982] *Über das Reduktionsverfahren für diskrete Wiener–Hopf–Gleichungen mit unstetigem Symbol*, Z. Analysis Anw. **1**, 1–5.

Böttcher, A. and Silbermann, B. [1983a] *The finite section method for Toeplitz operators on the quarter-plane with piecewise continuous symbols*, Math. Nachr. **110**, 279–291.

Böttcher, A. and Silbermann, B. [1983b] *Invertibility and Asymptotics of Toeplitz Matrices*, Akademie–Verlag, Berlin.

Böttcher, A. and Silbermann, B. [1990] *Analysis of Toeplitz Operators*, Springer–Verlag, Berlin, Heidelberg, New York.

Böttcher, A. and Wolf, H. [1993] *Galerkin–Petrov methods for Bergman space Toeplitz operators*, SIAM J. Numer. Analysis **30**, 846-863.

Devinatz, A. and Shinbrot, M. [1969] *General Wiener–Hopf operators*, Trans. Amer. Math. Soc. **145**, 467–494.

Douglas, R.G. [1972] *Banach Algebra Techniques in Operator Theory*, Academic Press, New York.

Douglas, R.G. [1978] *Local Toeplitz operators*, Proc. London Math. Soc., 3rd ser., **36**, 243–272.

Douglas, R.G. and Howe, R. [1971] *On the C^*-algebra of Toeplitz operators on the quarter-plane*, Trans. Amer. Math. Soc. **158**, 203–217.

Duduchava, R.V. [1971] *Discrete Wiener–Hopf equations in l^p spaces with weight*, Soobszh. Akad. Nauk Gruz. SSR **67**, 17–20 (Russian).

Duduchava, R.V. [1975] *On discrete Wiener–Hopf equations*, Trudy Tbilis. Mat. Inst. **50**, 42–59 (Russian).

Fisher, M.E. and Hartwig, R.E. [1968] *Toeplitz determinants: some applications, theorems and conjectures*, Adv. Chem. Phys. **15**, 333–353.

Gaier, D. [1987] *Lectures on Complex Approximation*, Birkhäuser Verlag, Basel, Boston, Stuttgart.

Gohberg, I. [1952] *On an application of the theory of normed rings to singular integral equations*, Uspehi Matem. Nauk **7**, 149-156 (Russian).

Gohberg, I. [1967] *On Toeplitz matrices composed of the Fourier coefficients of piecewise continuous functions*, Funct. Analysis Appl. **1**, 166-167.

Gohberg, I. and Feldman, I.A. [1965] *On the finite section method for systems of the Wiener-Hopf type*, Dokl. Akad. Nauk SSSR **165**, 268-271 (Russian).

Gohberg, I. and Feldman, I.A. [1971] *Convolution Equations and Projection Methods for Their Solution*, Amer. Math. Soc. Transl. of Math. Monographs **41**, Providence, R.I., 1974 (Russian original: Nauka, Moscow 1971).

Gohberg, I. and Krein, M.G. [1965] *Introduction to the Theory of Linear Non-Selfadjoint Operators in Hilbert Space*, Amer. Math. Soc. Transl. of Math. Monographs **18**, Providence, R.I., 1969 (Russian original: Nauka, Moscow 1965).

Gohberg, I. and Krupnik, N. [1973] *One-Dimensional Linear Singular Integral Equations*, Vols. I and II, Birkhäuser Verlag, Basel, Boston, Berlin 1992 (Russian original: Shtiintsa, Kishinev 1973).

Grenander, U. and Szegö, G. [1958] *Toeplitz Forms and Their Applications*, Univ. Calif. Press, Berkeley ans Los Angeles.

Hagen, R., Roch, S. and Silbermann, B. [1994] *Spectral Theory of Approximation Methods for Convolution Equations*, Birkhäuser Verlag, Basel, Boston, Berlin (to appear)

Halmos, P. [1967] *A Hilbert Space Problem Book*, D. Van Nostrand Comp., Inc., Princeton, N.J.

Kozak, A.V. [1973] *On the finite section method for multidimensional discrete convolutions*, Matem. Issled. **8**: 3 (29), 157–160 (Russian).

Power, S.C. [1982] *Hankel Operators on Hilbert Space*, Pitman Research Notes **64**, Pitman, Boston, London, Melbourne.

Prössdorf, S. and Silbermann, B. [1977] *Projektionsverfahren und die näherungsweise Lösung singulärer Gleichungen*, Teubner–Verlag, Leipzig.

Prössdorf, S. and Silbermann, B. [1992] *Numerical Analysis for Integral and Related Operator Equations*, Birkhäuser Verlag, Basel, Boston, Berlin.

Reichel, L. and Trefethen, L.N. [1992] *Eigenvalues and pseudo-eigenvalues of Toeplitz matrices*, Linear Algebra Appl. **162**, 153–185.

Sarason, D. [1977] *Toeplitz operators with piecewise quasicontinuous symbols*, Indiana Univ. Math. J. **26**, 817–838.

Sarason, D. [1978] *Function Theory on the Unit Circle.*, Virginia Polytechnic Inst. and State Univ., Blacksburg.

Schmidt, P. and Spitzer, F. [1960] *The Toeplitz matrices of an arbitrary Laurent polynomial*, Math. Scand. **8**, 15–38.

Silbermann, B. [1982] *Lokale Theorie des Reduktionsverfahrens für Toeplitz-operatoren* Math. Nachr. **104**, 45–56.

Silbermann, B. [1993] *On the limiting set of singular values of Toeplitz matrices*, Linear Algebra Appl. **182**, 35–43.

Simon, B. [1977] *Notes on infinite determinants of Hilbert space operators*, Adv. Math. **24**, 244–273.

Simonenko, I.B. [1968] *On multidimensional discrete convolutions*, Matem. Issled. **3**: 1, 108–122 (Russian).

Speck, F.-O. [1985] *General Wiener–Hopf Factorization Methods*, Pitman Research Notes **119**, Pitman, Boston, London, Melbourne.

Toeplitz, O. [1911] *Zur Theorie der quadratischen Formen von unendlich vielen Veränderlichen*, Math. Annalen **70**, 351–376.

Treil, S.R. [1987] *The invertibility of a Toeplitz operator does not imply its invertibility by the projection method*, Sov. Math. Dokl. **35**, 103–107.

Trefethen, L.N. [1995] *Spectra and Pseudospectra. The Behavior of Non-Normal Matrices and Operators.* Book in preparation.

Verbitsky, I.E. and Krupnik, N. [1977] *On the applicability of the finite section method to discrete Wiener–Hopf equations with piecewise continuous symbol*, Matem. Issled. **45**, 17-28.

Widom, H. [1960] *Singular integral equations on L^p*, Trans. Amer. Math. Soc. **7**, 131–160.

Widom, H. [1965] *Toeplitz matrices*, In: Studies in Real and Complex Analysis (I.I. Hirschman, Jr., ed.) *M.A.A. Studies in Mathematics* **3**, 179–209.

Widom, H. [1976] *Asymptotic behavior of block Toeplitz matrices and determinants II*, Adv. Math. **21**, 1–29.

Widom, H. [1989] *On the singular values of Toeplitz matrices*, Z. Analysis Anw. **8**, 221–229.

Widom, H. [1990] *Eigenvalue distribution of nonselfadjoint Toeplitz matrices and the asymptotics of Toeplitz determinants in the case of nonvanishing index*, Operator Theory: Advances and Applications **48**, 387–421.

Widom, H. [1994] *Eigenvalue distribution for nonselfadjoint Toeplitz matrices (Toeplitz Lecture presented at Tel Aviv University, March 1993)* Operator Theory: Advances and Applications **71**, 1-7.

Zhu, K. [1990] *Operator Theory in Function Spaces.* Marcel Dekker, Inc., New York and Basel.

LECTURE SERIES 2

OPERATOR THEORY AND ORDINARY DIFFERENTIAL OPERATORS

Aad Dijksma
Department of Mathematics
University of Groningen
P.O. Box 800
9700 AV Groningen
The Netherlands

Heinz Langer
Department of Mathematics
Technical University of Vienna
Wiedner Hauptstrasse 8–10
A-1040 Vienna
Austria

Introduction

During the last 10–15 years some "nonclassical" eigenvalue problems have been studied, which are closely related to definitizable operators in Kreĭn spaces. Here by nonclassical we mean that, roughly speaking, the eigenvalue problems are originally not of the form $(\tilde{A} - \lambda\tilde{I})y = 0$ with a selfadjoint operator $\tilde{A}$ in some Hilbert space. More precisely, in the problems we have in mind the basic operator is, for example, a formally symmetric differential operator, but

(i) in the basic equation the term containing the eigenvalue parameter λ contains a weight function r, that is, it is not of the form λy but of the form $\lambda r y$, or

(ii) the boundary conditions also contain the eigenvalue parameter, possibly nonlinearly, or

(iii) besides the boundary conditions there are given interface conditions, which contain the values and derivatives of the unknown function y in an interior point of the considered interval and may also depend on λ; sometimes these conditions can be expressed equivalently by a potential containing delta functions and its derivatives, or

(iv) the basic equation depends nonlinearly (for example, rationally or polynomially) on the eigenvalue parameter, or, which sometimes amounts to the same, the basic operator is given by a matrix whose entries are differential operators of different orders.

What these problems have in common is that often there can be defined a linear operator $\tilde{A}$ in a Pontryagin or Kreĭn space, which is definitizable, hence it has a spectral function with certain critical points. Even if $\tilde{A}$ is not definitizable, it can have a locally definite spectral function on some intervals. Whereas in the problems of type (i) the Kreĭn space is obtained from the classical Hilbert space L^2 just by introducing a new inner product, in the problems (ii) and (iii) the operator is obtained by means of the extension theory of symmetric operators or the theory of nodes or colligations, and in the problems of type (iv) it is obtained by a simple transformation, which can be included into the theory of colligations which we don't do here in order to keep the exposition as simple as possible.

The aim of these notes is to describe how the above problems lead to selfadjoint operators in a Kreĭn, Pontryagin or Hilbert space and to formulate some of the consequences; for the proofs we mainly refer to the literature.

In Section 1 we introduce the notion of a definitizable operator in a Kreĭn space and its spectral function. The special feature of the spectral function of a definitizable operator is the appearance of critical points, and we indicate some criteria for their regularity or singularity.

In Section 2 we first recall some basic facts about Sturm–Liouville operators, in particular, the Titchmarsh–Weyl coefficient is introduced which contains all the spectral information and is of importance also for the nonclasssical problems, in particular, for those with λ–dependent boundary or interface conditions. The main result in Section 2 is Theorem 2.4, which associates with the boundary eigenvalue problem a selfadjoint operator $\tilde{A}$ in some Hilbert, Pontryagin or Kreĭn space which contains the original space L^2. Here $\tilde{A}$ is, for example, definitizable, if the exit space is a π_κ–space. In a natural way the operator $\tilde{A}$ can be characterized by an analytic scalar or matrix function which belongs to a certain class of "generalized Nevanlinna functions."

These connections between holomorphic functions and selfadjoint operators are studied in Section 3 where also some results of the general theory of colligations are used. For more information about colligations we refer to Lecture Series 3 of this volume: *Operators on indefinite inner product spaces* by M.A. Dritschell and J.L. Rovnyak.

The aim of Section 4 is to show that Kreĭn spaces are a natural framework for the study of Sturm–Liouville operators with indefinite weight function. In particular, the question of half range completeness is closely related to the regularity of the critical point ∞: The latter property implies half range completeness, and the known sufficient conditions for half range completeness establish in fact the regularity of the critical point ∞.

In Section 5 we show that λ–dependent interface conditions can, basically, be treated in the same way as λ–dependent boundary conditions. However, Sturm–Liouville problems containing potentials acting at just one point can sometimes be formulated by means of interface conditions, sometimes by means of Pontryagin space extensions of the given minimal operator associated with the differential expression.

In Section 6 about "operator pencils" we have chosen just a few examples in order to show that in many cases a linearization of the problem leads to selfadjoint and even definitizable operator in Kreĭn spaces. The "linearizations" are often matrix operators such that at least one entry is a differential operator. In some cases several entries are unbounded operators, then additional conditions have to be imposed to make sure that the matrix defines a closed operator. With the example of the Klein–Gordon equation, which is in fact a problem which is quadratic in the eigenvalue parameter, it is shown that one and the same problem can be treated in different spaces, and that sometimes ∞ is a regular and sometimes a singular critical point.

We thank Branko Ćurgus for his contribution to Subsection 1.5, the example at the end of Subsection 4.1, and his permission to use some results from his preprint Ćurgus [preprint] in Subsection 4.2.

With pleasure we acknowledge the support and hospitality of the Fields Institute. We also acknowledge the support of the Koninklijke Nederlandse Akademie voor Wetenschappen and the "Fonds zur Förderung der wissenschaftlichen Forschung" of Austria, Project P 09832. It was a privilege to present these lectures and we thank Peter Lancaster for incorporating them in the present volume.

1 Definitizable operators in Kreĭn spaces

1. Basic Concepts (see Andô [1979], Azizov and Iohkvidov [1986], Bognár [1974], and Iokhvidov et al. [1982]). Let $(\mathfrak{K}, [\cdot,\cdot])$ be a **Kreĭn space**, that is, $\mathfrak{K}$ is a complex linear space such that

$$\mathfrak{K} = \mathfrak{K}_+[\dot{+}]\mathfrak{K}_-, \tag{1.1}$$

where $[\dot{+}]$ denotes the direct $[\cdot,\cdot]$-orthogonal sum and $(\mathfrak{K}_\pm, \pm[\cdot,\cdot])$ are Hilbert spaces. The decomposition (1.1) induces, evidently, a Hilbert space inner product on $\mathfrak{K}$, given by

$$(x,y) = [x_+, y_+] - [x_-, y_-], \quad x = x_+ + x_-, \; y = y_+ + y_-, \quad x_\pm, y_\pm \in \mathfrak{K}_\pm. \tag{1.2}$$

Although the decomposition (1.1) is not unique, different such decompositions generate equivalent Hilbert space norms. If not explicitly stated otherwise, topological notions refer to this Hilbert space topology. For example, a subspace of $\mathfrak{K}$ is a linear manifold in $\mathfrak{K}$ which is closed with respect to this Hilbert space topology, continuity of operators means continuity with respect to this Hilbert space topology etc. The set of continuous linear operators from a Kreĭn space $\mathfrak{K}$ to a Kreĭn space $\mathfrak{K}'$ is denoted by $\mathbf{L}(\mathfrak{K}, \mathfrak{K}')$; if $\mathfrak{K} = \mathfrak{K}'$ we write $\mathbf{L}(\mathfrak{K}, \mathfrak{K}) = \mathbf{L}(\mathfrak{K})$.

The operator J: $Jx = x_+ - x_-$ if $x = x_+ + x_-$, $x_\pm \in \mathfrak{K}_\pm$, is called the **fundamental symmetry** corresponding to (1.1). Evidently,

$$(x,y) = [Jx, y], \quad [x,y] = (Jx, y), \quad x, y \in \mathfrak{K},$$

$$J = J^* = J^{-1}, \quad J = P_+ - P_-,$$

where * denotes the adjoint with respect to the Hilbert space inner product (1.2), $P_\pm$ is the $(\cdot,\cdot)$-orthogonal projection onto $\mathfrak{K}_\pm$. The Kreĭn space $(\mathfrak{K}, [\cdot,\cdot])$ is called a **π_κ-space** or **Pontryagin space of index** κ if min $(\dim \mathfrak{K}_+, \dim \mathfrak{K}_-) = \kappa < \infty$. In these lectures we assume that $\kappa = \dim \mathfrak{K}_-$. Here the dimension is either a nonnegative integer or infinity.

If $\mathfrak{L} \subseteq \mathfrak{K}$ the **orthogonal companion** of $\mathfrak{L}$ is defined by the relation

$$\mathfrak{L}^{[\perp]} = \{x : [x, \mathfrak{L}] = \{0\}\}.$$

In general the **isotropic part** $\mathfrak{L}_0 = \mathfrak{L} \cap \mathfrak{L}^{[\perp]}$ of $\mathfrak{L}$ can contain nonzero elements; in this case $\mathfrak{L}$ is called **degenerate**. If $\mathfrak{L}$ is a subspace then

$$\mathfrak{L} \cap \mathfrak{L}^{[\perp]} = \{0\} \iff \left(\mathfrak{L}[\dot{+}]\mathfrak{L}^{[\perp]}\right)^c = \mathfrak{K},$$

where ^{c} stands for closure, in this case in the space $\mathfrak{K}$. The subspace $\mathfrak{L}$ is said to be **projection complete** or **regular** if $\mathfrak{L}[\dot{+}]\mathfrak{L}^{[\perp]} = \mathfrak{K}$. For the subspace $\mathfrak{L}$ the following statements are equivalent:

(a) $\mathfrak{L}$ is regular.
(b) $(\mathfrak{L}, [\cdot,\cdot])$ is a Kreĭn space.
(c) $\mathfrak{L}$ is the range of an operator $E \in \mathbf{L}(\mathfrak{K})$ such that

$$E^2 = E, \quad E = E^+,$$

where $^+$ denotes the adjoint with respect to the Kreĭn space inner product (see below, just before Theorem 1.1).

An element $x \in \mathfrak{K}$ is called **positive, nonnegative** etc., if $[x,x] > 0$, $[x,x] \geq 0$ etc.; a subset $\mathfrak{M} \subseteq \mathfrak{K}$ is called **positive, nonnegative** etc., if all its nonzero elements have this property. An element $x \in \mathfrak{K}$ (or a subset $\mathfrak{M} \subseteq \mathfrak{K}$) is called **neutral** if $[x,x] = 0$ (or $[x,x] = 0$ for all $x \in \mathfrak{M}$). We denote by $\mathcal{P}_+$, $\mathcal{P}_-$, and $\mathcal{P}_0$ the sets

$$\mathcal{P}_+ = \{x \in \mathfrak{K} : [x,x] \geq 0\},\ \mathcal{P}_- = \{x \in \mathfrak{K} : [x,x] \leq 0\},\ \mathcal{P}_0 = \{x \in \mathfrak{K} : [x,x] = 0\}.$$

A subspace $\mathfrak{L}$ of $\mathfrak{K}$ is a **maximal nonnegative subspace** if $\mathfrak{L} \subseteq \mathcal{P}_+$ and if each subspace $\mathfrak{L}'$ with $\mathfrak{L} \subseteq \mathfrak{L}' \subseteq \mathcal{P}_+$ equals $\mathfrak{L}$. **Maximal positive, maximal nonpositive** etc. subspaces are defined accordingly. The subspace $\mathfrak{L} \subseteq \mathfrak{K}$ is called **uniformly positive**, if for some (and hence each) of the Hilbert space norms $\|\cdot\|$ on $\mathfrak{K}$ generated by (1.2) there exists a $\gamma > 0$ such that

$$[x,x] \geq \gamma \|x\|^2, \quad x \in \mathfrak{L}.$$

Maximal semidefinite subspaces can be represented by means of an angular operator. If $\mathfrak{L}$ is, for example, a maximal nonnegative subspace of $\mathfrak{K}$ and a decomposition (1.1) is fixed there exists a contraction $K_{\mathfrak{L}}$ from $(\mathfrak{K}_+, [\cdot,\cdot])$ into $(\mathfrak{K}_-, -[\cdot,\cdot])$ such that

$$\mathfrak{L} = \{x = x_+ + K_{\mathfrak{L}} x_+ : x_+ \in \mathfrak{K}_+\}.$$

If $\mathfrak{L}$ is a positive subspace then

$$\|K_{\mathfrak{L}} x_+\| < \|x_+\|, \quad x_+ \in \mathfrak{K}_+,$$

if $\mathfrak{L}$ is a uniformly positive subspace then $\|K_{\mathfrak{L}}\| < 1$. The operator $K_{\mathfrak{L}}$ is called the **angular operator** for $\mathfrak{L}$.

Kreĭn spaces arise often as follows:

(a) Let σ be a real function of bounded variation on some interval ι, and denote by $|\sigma|$ its total variation. Consider the Hilbert space $L^2_{|\sigma|}$. Equipped with the inner product

$$[f,g] = \int_\iota f(t)\overline{g(t)} d\sigma(t), \quad f,g \in L^2_{|\sigma|},$$

$L^2_{|\sigma|}$ becomes a Kreĭn space which we denote by L^2_σ. Indeed, if σ is decomposed as $\sigma = \sigma_+ - \sigma_-$ with two nondecreasing functions σ_+ and σ_- such that $|\sigma| = \sigma_+ + \sigma_-$ we can choose $(L^2_\sigma)_+ = L^2_{\sigma_+}$ and $(L^2_\sigma)_- = L^2_{\sigma_-}$.

(b) Let $(\mathfrak{H}, (\cdot,\cdot))$ be a Hilbert space and let G be a bounded selfadjoint operator in $\mathfrak{H}$. We consider the space $\mathfrak{H}_G = (\mathfrak{H}, (G\cdot,\cdot))$. If $0 \in \rho(G)$, the resolvent set of G, then $\mathfrak{H}_G$ is a Kreĭn space. If $0 \in \sigma_c(G)$, the continuous spectrum of G, consider the spectral subspaces $\mathfrak{H}_+$ and $\mathfrak{H}_-$ of G corresponding to $(-\infty, 0)$ and $(0,\infty)$, respectively. Completing $\mathfrak{H}_+$ with respect to the positive definite inner product (Gx,y), $x,y \in \mathfrak{H}_+$, and $\mathfrak{H}_-$ with respect to the positive definite inner product $-(Gx,y)$, $x,y \in \mathfrak{H}_-$, and denoting these Hilbert space completions by $\mathfrak{K}_+$ and $\mathfrak{K}_-$, respectively, we form the Kreĭn space $\mathfrak{K} = \mathfrak{K}_+[\dotplus]\mathfrak{K}_-$. This space can be considered as a Kreĭn space completion of $\mathfrak{H}_G$. Observe that if $0 \in \sigma_c(G|_{\mathfrak{H}_+}) \cap \sigma_c(G|_{\mathfrak{H}_-})$ then there are other Kreĭn space completions too; see, for example, Hara [1992]. Here a

Kreĭn space completion of $\mathfrak{H}_G$ is a Kreĭn space which contains $\mathfrak{H}_G$ as a dense subset and is such that

$$[x, y] = (Gx, y), \quad x, y \in \mathfrak{H}_G.$$

(c) A linear space $\mathfrak{L}$ is said to be a π_κ**-lineal**, if on $\mathfrak{L}$ there is defined an inner product $[\cdot, \cdot]$, that is, a sesquilinear hermitian form, such that $\mathfrak{L}$ contains a κ-dimensional subspace $\mathfrak{K}_-$ with $[x, x] < 0$ if $x \in \mathfrak{K}_- \setminus \{0\}$ and no $(\kappa+1)$-dimensional subspace with this property. With a π_κ-lineal $\mathfrak{L}$ a π_κ-space can be associated in a canonical way as follows. Consider the factor space $\mathfrak{L}' = \mathfrak{L}/\mathfrak{L}_0$ with respect to the isotropic part $\mathfrak{L}_0$ of $\mathfrak{L}$. It contains a κ–dimensional negative subspace $\mathfrak{K}'_-$. Write $\mathfrak{L}' = \mathfrak{L}'_+[\dot{+}]\mathfrak{K}'_-$. It is easy to see that $(\mathfrak{L}'_+, [\cdot, \cdot])$ is a pre-Hilbert space. Completing it to a Hilbert space $\mathfrak{K}'_+$ yields a π_κ-space $\mathfrak{K}'_+[\dot{+}]\mathfrak{K}'_-$ which contains $\mathfrak{L}'$ as a dense subset.

(d) Let X be a nonempty set. An $n \times n$ matrix function K on $X \times X$ (called a $\mathbb{C}^n$**–kernel** or simply a **kernel** (in particular if $n = 1$)) is **hermitian** if

$$K(\xi, \eta) = K(\eta, \xi)^*, \quad \xi, \eta \in X.$$

It is said to have κ **negative squares** if it is hermitian and if for each $m \in \mathbb{N}$ and points $\xi_1, \xi_2, \ldots, \xi_m \in X$ the $mn \times mn$ matrix $(K(\xi_i, \xi_j))_{i,j=1}^m$ has at most κ negative eigenvalues (including multiplicities) and for at least one choice of m and $\xi_1, \xi_2, \ldots, \xi_m$ it has exactly κ negative eigenvalues. The $\mathbb{C}^n$–kernel K is called **nonnegative** if $\kappa = 0$. The number of positive squares of a hermitian kernel is defined analogously.

Let K be a $\mathbb{C}^n$–kernel with κ negative squares. Then the set $\mathfrak{L}$ of all formal sums $\sum_{\xi \in X} \varepsilon_\xi a_\xi$ with only a finite number of nonzero $a_\xi \in \mathbb{C}^n$, $\xi \in X$, with componentwise addition and multiplication by complex numbers and with the inner product

$$[\sum_{\xi \in X} \varepsilon_\xi a_\xi, \sum_{\eta \in X} \varepsilon_\eta b_\eta] = \sum_{\xi, \eta \in X} b_\eta^* K(\xi, \eta) a_\xi$$

is a π_κ-lineal. According to (c), the completion of the factor space $\mathfrak{L}/\mathfrak{L}_0$ with respect to the isotropic part $\mathfrak{L}_0$ of $\mathfrak{L}$ is a π_κ-space; see Dijksma et al. [1987].

The linear operators, which we consider in the Kreĭn space $\mathfrak{K}$, will in general be densely defined and closed or closable. If A is such an operator then the **adjoint** A^+ is defined as follows: dom A^+ is the set of all $u \in \mathfrak{K}$ for which there exists a $v \in \mathfrak{K}$ with

$$[Ax, u] = [x, v] \text{ for all } x \in \text{dom } A,$$

and in this case $A^+u = v$, or in graph notation

$$A^+ = \{\{u, v\} \in \mathfrak{K} \oplus \mathfrak{K} : [y, u] = [x, v] \text{ for all } \{x, y\} \in A\}.$$

Then dom A^+ is dense in $\mathfrak{K}$ and A^+ is a closed operator. The operator A in the Kreĭn space $\mathfrak{K}$ is called **selfadjoint** if $A = A^+$, **symmetric** if $A \subseteq A^+$, **unitary** if $A^+A = AA^+ = I$, **isometric** if $A^+A = I$ and **contractive** if $[Ax, Ax] \leq [x, x]$, $x \in \mathfrak{K}$. The following theorem is a basic result for selfadjoint operators in Kreĭn spaces.

Theorem 1.1 *If A is selfadjoint and, for example, bounded in the Kreĭn space $\mathfrak{K}$ and for some fundamental symmetry $J = P_+ - P_-$ the operator P_+AP_- is compact, then A has a maximal nonnegative invariant subspace.*

In particular, a (bounded or unbounded) selfadjoint operator A in a Pontryagin space has a maximal nonnegative and a maximal nonpositive invariant subspace. Since the latter is finite dimensional, A has eigenvalues with corresponding non-positive eigenvectors. This statement can be made more precise by considering multiplicities and the location of the eigenvalues with respect to the real axis.

Here are some properties of the spectrum $\sigma(A)$ of a selfadjoint operator A in a Kreĭn space $\mathfrak{K}$:

(i) If $A = A^+$, then $\sigma(A)$ is symmetric with respect to the real axis.
(ii) If $[Ax, x] \geq 0$ for all $x \in \text{dom } A$ and $\rho(A) \neq \emptyset$, then $A = A^+$ and $\sigma(A) \subseteq \mathbb{R}$.

If $A = A^+$ and $[Ax, x] \geq 0$ for all $x \in \text{dom } A$, then $\sigma(A)$ can be the whole complex plane $\mathbb{C}$. Indeed, consider a Hilbert space $(\mathfrak{H}, (\cdot, \cdot))$ and an unbounded selfadjoint positive operator A_0 in $\mathfrak{H}$. Let $\mathfrak{K}$ be the Kreĭn space which is obtained from $\mathfrak{H} \oplus \mathfrak{H}$ by introducing an indefinite inner product by means of the fundamental symmetry $J = \begin{pmatrix} 0 & I \\ I & 0 \end{pmatrix}$. The operator $A = \begin{pmatrix} 0 & 0 \\ A_0 & 0 \end{pmatrix}$ has all the required properties.

2. Definitizable operators (see Langer [1982]). The selfadjoint operator A in $\mathfrak{K}$ is called **definitizable** if $\rho(A) \neq \emptyset$ and there exists a polynomial p such that

$$[p(A)x, x] \geq 0, \quad x \in \text{dom } A^k, \tag{1.3}$$

where $k = \deg p$, the degree of p; p is called a **definitizing polynomial** of A. Examples of definitizable operators are the following:

(i) Assume dim $\mathfrak{K}_- = \kappa < \infty$ and $A = A^+$. Consider a κ-dimensional nonpositive subspace $\mathfrak{L}$ of $\mathfrak{K}$ which is invariant under A, and let q be the minimal polynomial of $A|_{\mathfrak{L}}$. Then A is definitizable and $p = q\bar{q}$ is a definitizing polynomial: We have $q(A)\mathfrak{L} = \{0\}$, hence if, for example, A is bounded,

$$[\bar{q}(A)\mathfrak{K}, \mathfrak{L}] = [\mathfrak{K}, q(A)\mathfrak{L}] = \{0\},$$

or $\bar{q}(A)\mathfrak{K} \subseteq \mathfrak{L}^{[\perp]}$. As $\mathfrak{L}$ is maximal nonpositive, its orthogonal companion is nonnegative, hence

$$[\bar{q}(A)x, \bar{q}(A)x] \geq 0, \quad x \in \mathfrak{K}.$$

(ii) Assume that $A = A^+$, $\rho(A) \neq \emptyset$ and A is **nonnegative** in $\mathfrak{K}$, that is,

$$[Ax, x] \geq 0, \quad x \in \text{dom } A.$$

Then, obviously, A is definitizable and $p(t) = t$ is a definitizing polynomial of A.
(iii) Assume that $A = A^+$, $\rho(A) \neq \emptyset$ and the form $[Ax, y], x, y \in \text{dom } A$, has a finite number κ of negative squares. Then A is definitizable and a definitizing polynomial of A is of the form $p(t) = tq(t)\bar{q}(t)$ with $\deg q \leq \kappa$. If, additionally, $0 \in \rho(A)$ then A is sometimes called **quasi-uniformly positive**; see Ćurgus and Najman [to appear]. Operators with these properties arise in some applications; see Section 4.

(iv) Let $K(s,t)$ be a continuous hermitian kernel on $[0,1]\times[0,1]$ with κ negative squares. Let σ be a function of bounded variation on $[0,1]$. Then in the Kreĭn space L^2_σ the operator

$$(Af)(s) = \int_0^1 K(s,t)f(t)d\sigma(t)$$

is definitizable. Indeed, it is easy to see that for the (bounded) operator A the form $[Ax,y], x,y \in L^2_\sigma$, has at most κ negative squares.
(v) Let $(\mathfrak{K},[\cdot,\cdot])$ and $(\mathfrak{K}_1,[\cdot,\cdot]_1)$ be Kreĭn spaces and suppose that $\mathfrak{K}_1$ is **contractively** (or **continuously**, respectively) embedded in $\mathfrak{K}$. This means that $\mathfrak{K}_1 \subseteq \mathfrak{K}$, the embedding operator $T : \mathfrak{K}_1 \to \mathfrak{K}$ with $Tx = x,\ x \in \mathfrak{K}_1$, is continuous, and $[x,x] \le [x,x]_1, x \in \mathfrak{K}_1$ (or, for some $\gamma > 0$, $[x,x] \le \gamma[x,x]_1, x \in \mathfrak{K}_1$, respectively). Denote by S the adjoint of T:

$$[Sx,x_1]_1 = [x,Tx_1], \quad x \in \mathfrak{K},\ x_1 \in \mathfrak{K}_1.$$

Then $A = TS$, the "adjoint of the inclusion operator", is a continuous map from $\mathfrak{K}$ into itself and it has the properties $A = A^+$ and

$$A - A^2 \ge 0 \quad (\text{ or } \ \gamma A - A^2 \ge 0, \text{ respectively}).$$

Indeed, we have $[TSx,x] = [Sx,Sx]_1$, which is real for all $x \in \mathfrak{K}$, and

$$[TSx,TSx] \le [Sx,Sx]_1 = [TSx,x], \quad x \in \mathfrak{K},$$

(or

$$[TSx,TSx] \le \gamma[Sx,Sx]_1 = \gamma[TSx,x],\ x \in \mathfrak{K},$$

respectively). Hence A is definitizable with the definitizing polynomial $p(t) = t - t^2$ (or $p(t) = \gamma t - t^2$, respectively).
(vi) Let $(\mathfrak{K},[\cdot,\cdot])$ be a Kreĭn space, which is not a Hilbert or anti–Hilbert space, that is, $\mathcal{P}_0 \ne \{0\}$. Let A be a selfadjoint operator in $\mathfrak{K}$ such that $\rho(A) \ne \emptyset$ and that

$$x \in \text{dom } A,\ [x,x] = 0, \text{ and } \ x \ne 0 \Rightarrow [Ax,x] \ne 0. \tag{1.4}$$

Then A is definitizable with a linear definitizing polynomial. The proof of this fact is an easy consequence of the following result.

Lemma 1.2 *Let $\mathfrak{L}$ be a linear space which is equipped with two hermitian sesquilinear forms $[\cdot,\cdot]$ and $[\cdot,\cdot]_1$, such that $[\cdot,\cdot]$ is indefinite and the assumption*

$$x \in \mathfrak{L} \text{ and } [x,x] = 0 \Rightarrow [x,x]_1 \ge 0 \tag{1.5}$$

is satisfied. Then

$$\text{(a)} \quad \mu = \inf_{[x,x]>0} \frac{[x,x]_1}{[x,x]} > -\infty, \qquad \text{and} \qquad \text{(b)} \quad [x,x]_1 \ge \mu[x,x],\ x \in \mathfrak{L}.$$

Proof. Let $x,y \in \mathfrak{L}$ be such that $[x,x] > 0$ and $[y,y] < 0$. We shall show that

$$\frac{[y,y]_1}{[y,y]} \le \frac{[x,x]_1}{[x,x]}. \tag{1.6}$$

Without loss of generality suppose $[x,x] = 1$ and $[y,y] = -1$. Assume that $-[y,y]_1 > [x,x]_1$. We choose $\lambda \in \mathbb{C}$, $|\lambda| = 1$, such that Re $(\lambda[x,y]) = 0$, and consider the elements $z = x \pm \lambda y$. Then

$$[z,z] = 0, \quad [z,z]_1 = [x,x]_1 \pm 2\text{Re } (\lambda[x,y]_1) + [y,y]_1,$$

and for one of the signs $\pm$ the number $[z,z]_1$ is strictly negative as $[x,x]_1 + [y,y]_1$ is strictly negative. But this is impossible by (1.5), and (1.6) and (a) are proved. The statement (b) follows from (a) if $[x,x] > 0$, from (1.5) if $[x,x] = 0$, and from (1.6) if $[x,x] < 0$. □

The definitizability of A in (vi) follows from the fact that $[Ax,x]$ has to be either positive or negative for all nonzero $x \in$ dom A.

3. The spectral function of a definitizable operator (see Langer [1982], Jonas [1981]). In this subsection let A be a definitizable operator in a Kreĭn space $\mathfrak{K}$ and p a definitizing polynomial of A. Then $p(A)$ is nonnegative, hence $\sigma(p(A)) \subseteq \mathbb{R}$; see Subsection 1.2. On the other hand $\sigma(p(A)) = p(\sigma(A))$, hence p is real on $\sigma(A)$. The investigation of the spectral properties of a bounded definitizable selfadjoint operator A with definitizing polynomial p can be based, for example, on the formula

$$(A-z)^{-1} = \frac{1}{p(z)}\left(p(A)(A-z)^{-1} + Q(z;A)\right), \quad z \in \rho(A), \tag{1.7}$$

where $Q(z;A)$ is a polynomial in z and A. If we observe that the function

$$z \longrightarrow f_x(z) = [p(A)(A-z)^{-1}x, x], \quad x \in \mathfrak{K}, \tag{1.8}$$

is holomorphic on $\mathbb{C}^+ \cap \rho(A)$ and has the property that the kernel

$$N_{f_x}(z,\zeta) = \frac{f_x(z) - \overline{f_x(\zeta)}}{z - \bar{\zeta}}$$

is nonnegative on $\mathbb{C}^+ \cap \rho(A)$, it follows that f_x has a holomorphic continuation onto all of $\mathbb{C}^+$. This holomorphic extension has the property that it maps $\mathbb{C}^+$ into $\mathbb{C}^+ \cup \mathbb{R}$, that is, we have

$$\text{Im } f_x(z) \geq 0, \quad z \in \mathbb{C}^+.$$

Functions which are holomorphic on $\mathbb{C}^+$ and have this property will be called **Nevanlinna functions**; the set of these functions is denoted by $\mathbf{N}_0$. Then, according to (1.7), also $[(A-z)^{-1}x, x]$ has a holomorphic continuation onto all of $\mathbb{C}^+$ with the possible exception of the zeros of p. Thus $\sigma(A)$ is real with the possible exception of isolated (nonreal) points which are zeros of p. These isolated points can easily be studied by means of the Riesz-Dunford calculus: If $\lambda \in \sigma_0(A)$, the nonreal spectrum of A, then the corresponding root subspace $\mathfrak{L}_\lambda$ is neutral, A has finite Riesz index at λ and the restrictions $A|_{\mathfrak{L}_\lambda}$ and $A|_{\mathfrak{L}_{\bar\lambda}}$ are "isomorphic"; see Bognár [1974], Azizov and Iohkvidov [1986]. The Riesz-Dunford projection corresponding to $\sigma_0(A)$ will be denoted by E_0. If we observe that after, possibly, a holomorphic continuation, (1.8) is a Nevanlinna function, inversion formulas of Stieltjes-Livšic type (see, for example, Kac and Kreĭn [1968, Lemma S1.2.1]) can be applied to (1.7) and the Theorem 1.3 below can be proved.

To formulate the theorem we introduce the set $c(A)$ of **critical points** of A as the set of all $\lambda \in \mathbb{R}$ such that $p(\lambda) = 0$ for each definitizing polynomial p of

A, and $\infty \in c(A)$ if one (and hence each) definitizing polynomial is of odd degree and $\sigma(A)$ contains arbitrarily large positive and negative numbers. It can be shown that $c(A) \subseteq \sigma(A)$. We denote by $\mathcal{R}_A$ the Boolean algebra generated by all intervals of $\overline{\mathbb{R}} = \mathbb{R} \cup \{\infty\}$ whose endpoints are not in $c(A)$.

Theorem 1.3 *There exists a unique mapping $E : \mathcal{R}_A \to \mathbf{L}(\mathfrak{K})$ with the following properties (Δ and Δ' are arbitrary elements of $\mathcal{R}_A$):*
(1) $E(\Delta) = E(\Delta)^+$, $E(\emptyset) = 0$.
(2) $E(\Delta \cap \Delta') = E(\Delta)E(\Delta')$ *(in particular, $E(\Delta)$ is a projection).*
(3) $E(\Delta \cup \Delta') = E(\Delta) + E(\Delta') - E(\Delta)E(\Delta')$.
(4) $E(\mathbb{R}) = 1_{\mathfrak{K}} - E_0$, *where E_0 is the Riesz-Dunford projection associated with the nonreal spectrum $\sigma_0(A)$ of A.*
(5) *If $p|_\Delta > 0$ (or $p|_\Delta < 0$) for some definitizing polynomial p of A then $E(\Delta)\mathfrak{K}$ is a positive (negative, respectively) subspace.*
(6) $E(\Delta) \in \{(A-z)^{-1}\}''$, *the double commutant of the resolvent $(A-z)^{-1}$ of A, $z \in \rho(A)$.*
(7) *If Δ is bounded then $E(\Delta)\mathfrak{K} \subseteq \operatorname{dom} A$; if Δ is unbounded then $E(\Delta)\mathfrak{K} \cap \operatorname{dom} A$ is dense in $E(\Delta)\mathfrak{K}$. In both cases $\sigma(A|_{E(\Delta)\cap\mathfrak{K}}) \subset \Delta^c$.*
(8) *If A is bounded then*

$$p(A) = \int_{\mathbb{R}} p(\lambda)E(d\lambda) + N, \tag{1.9}$$

where N is a bounded nonnegative operator in $\mathfrak{K}$ with $N^2 = 0$.

The mapping E of Theorem 1.3 is called the **spectral function (with critical points)** of the definitizable operator A. The statement (5) and the equivalence of (b) and (c) of Subsection 1.1 imply that $\mathfrak{K}_\Delta = E(\Delta)\mathfrak{K}$ is a Hilbert space or anti–Hilbert space if p is positive or negative on $\Delta \cap \sigma(A)$ for some definitizing polynomial p. This space $\mathfrak{K}_\Delta$ reduces A and $A|_{\mathfrak{K}_\Delta}$ is a bounded or unbounded and densely defined Hilbert space selfadjoint operator in $\mathfrak{K}_\Delta$. Thus, with the exception of the (finitely many) points in $c(A)$, the definitizable operator A has locally the same spectral properties as a selfadjoint operator in a Hilbert space.

The critical points $\lambda \in c(A)$ can be characterized as follows:

$$\Delta \in \mathcal{R}_A,\ \lambda \in \Delta \Rightarrow \text{ the inner product } [\cdot,\cdot] \text{ is indefinite on } \mathfrak{K}_\Delta.$$

Critical points will be considered in some detail in the next subsection.

The integral in (1.9) is improper with respect to the points $c(A)$ as at these points the spectral function E is not defined. The statement (8) implies that for a bounded nonnegative operator A

$$Ax = \int_{-\|A\|}^{\|A\|} \lambda dE_\lambda x + Nx, \quad x \in \mathfrak{K},$$

where $\|\cdot\|$ denotes any of the equivalent operator norms on $\mathbf{L}(\mathfrak{K})$.

Assume again that $\rho(A) \neq \emptyset$. The definitizability assumption (1.3) can be strengthened to

$$[p(A)x, x] > 0 \text{ if } x \in \text{dom } A^k, \; x \neq 0, \tag{1.10}$$

or to

$$[p(A)x, x] > \gamma \|x\|^2 \text{ if } x \in \text{dom } A^k, \tag{1.11}$$

with some $\gamma > 0$ and any (of the equivalent) Hilbert space norms $\|\cdot\|$ on $\mathfrak{K}$ given by (1.2). The assumption (1.10) implies that $0 \notin \sigma_p(p(A)) = p(\sigma_p(A))$, hence the zeros of p are not eigenvalues of A. Therefore the critical points of A are not eigenvalues of A, and A has no nonreal point spectrum. The stronger assumption (1.11) implies that $0 \in \rho(p(A))$. Hence $0 \notin \sigma(p(A)) = p(\sigma(A))$ and the zeros of p do not belong to $\sigma(A)$. Therefore A has no finite critical points. If, additionally, A is bounded then it is similar to a selfadjoint operator in Hilbert space. For example, the form on the lefthand side of (1.11) can be introduced as an equivalent Hilbert space norm and A is selfadjoint with respect to this norm.

4. Critical points. Again, let A be a definitizable operator in the Kreĭn space $\mathfrak{K}$ and $\lambda_0 \in c(A)$. Then λ_0 is said to be a **regular** critical point if for some (and hence for all) $\mu \in \mathbb{R} \setminus c(A)$, with, for example, $\mu < \lambda_0$, the limits

$$\lim_{\lambda \uparrow \lambda_0} E([\mu, \lambda)) \quad \text{and} \quad \lim_{\lambda \downarrow \lambda_0} E([\mu, \lambda)) \quad \text{if } \lambda_0 \in \mathbb{R}, \tag{1.12}$$

$$\lim_{\lambda \uparrow \infty} E([\mu, \lambda)) \quad \text{and} \quad \lim_{\lambda \downarrow -\infty} E([\lambda, \mu)) \quad \text{if } \lambda_0 = \infty, \tag{1.13}$$

exist in the strong operator topology; otherwise λ_0 is called a **singular** critical point. The existence of these limits is equivalent to the boundedness of the projections. For example, the first limit exists if and only if $\|E([\mu, \lambda))\| \leq C$ for all λ with $\lambda_0 - \varepsilon < \lambda < \lambda_0 + \varepsilon$, ε sufficiently small, $\lambda \neq \lambda_0$, where, as before, $\|\cdot\|$ denotes any of the equivalent operator norms on $\mathbf{L}(\mathfrak{K})$.

Criteria for the regularity of a critical point can be found, for example, in Jonas [1984]. If the critical point $\lambda_0 \in \mathbb{R}$ is regular then the limits in (1.12), (1.13) define selfadjoint projections

$$E([\mu, \lambda_0)) \quad \text{and} \quad E([\mu, \lambda_0])$$

and the spectral function E can be extended to the Boolean algebra generated by all intervals of $\overline{\mathbb{R}}$ whose endpoints are not in $c(A) \setminus \{\lambda_0\}$. In this case, if $\lambda_0 \in \mathbb{R}$, naturally

$$E(\{\lambda_0\}) = E([\mu, \lambda_0]) - E([\mu, \lambda_0)).$$

If ∞ is a regular critical point then

$$E((-\infty, \mu)) + E([\mu, \infty)) = I - E_0.$$

In the following $\mathfrak{L}_{\lambda_0}$ denotes again the root subspace of A at λ_0.

Proposition 1.4 *If $\lambda_0 \in c(A)$, then $\cap_{\lambda_0 \in \Delta, \Delta \in \mathcal{R}_A} \mathfrak{K}_\Delta = \mathfrak{L}_{\lambda_0}$.*

It follows easily that

(a) if $\lambda_0 \in \mathbb{R}$ and λ_0 is a regular critical point then $\mathfrak{L}_{\lambda_0} = E(\{\lambda_0\})$;

(b) if $\lambda_0 \in c(A)$ and $\mathfrak{L}_{\lambda_0} = \{0\}$ then λ_0 is a zero of odd order of each definitizing polynomial p; λ_0 can be a regular or a singular critical point.

Another classification of the critical points in $\mathbb{R}$ is as follows: $\lambda_0 \in c(A) \cap \mathbb{R}$ is said to be a critical point of **finite type** if for some $\Delta \in \mathcal{R}_A$, $\lambda_0 \in \Delta$, the space $(\mathfrak{K}_\Delta, [\cdot,\cdot])$ or $(\mathfrak{K}_\Delta, -[\cdot,\cdot])$ is a π_κ-space. Then, if Δ is sufficiently small and contains λ_0, the index κ is independent of Δ. In this case $\lambda_0 \in \sigma_p(A)$ and the following simple criterion for the regularity or singularity of λ_0 holds.

Proposition 1.5 *The critical point $\lambda_0 \in c(A) \cap \mathbb{R}$ of finite type is regular if and only if $\mathfrak{L}_{\lambda_0}$ is not degenerate.*

If A is such that $[Ax, y], x, y \in \text{dom } A$, has a finite number of negative squares then all the critical points $\neq 0, \infty$ are of finite type. Criteria for the regularity or singularity of ∞ or the other critical points, which are not of finite type, are more complicated; see, for example, Jonas [1984]. In the next subsection we consider the critical point ∞.

5. The critical point ∞. In this section we suppose that A is a uniformly positive operator in the Kreĭn space $\mathfrak{K}$, that is, $[Ax, x] \geq 0, x \in \text{dom } A$, and $0 \in \rho(A)$. Then ∞ is the only critical point of A. Examples of selfadjoint operators A with singular critical point ∞ can be constructed as infinite orthogonal sums of 2×2 matrices A_n such that

$$\begin{pmatrix} 1 & 0 \\ 0 & -1 \end{pmatrix} A_n$$

is hermitian, each matrix A_n having a pair of eigenvalues λ_n^+, λ_n^- with $\lambda_n^\pm \to \pm\infty$, if $n \to \infty$, and such that the norms of the corresponding projections in the 2-dimensional eigenspace tend to ∞.

There exist also uniformly positive operators A with

$$\sigma(A) = (\lambda_n^+) \cup (\lambda_n^-), \quad \lambda_n^\pm \to \pm\infty, \, n \to \infty,$$

such that the norms of the projections onto the eigenspaces at $\lambda_n^\pm$ remain uniformly bounded, but nevertheless, ∞ is a singular critical point, that is, the norms of the projections onto the spectral subspaces corresponding, for example, to $\{\lambda_1, \ldots, \lambda_n\}$ tend to ∞.

For the following theorem, see Ćurgus [1985].

Theorem 1.6 *The following are equivalent:*
(1) *∞ is a regular critical point of A.*
(2) *A is similar to a selfadjoint operator in a Hilbert space.*
(3) *In $\mathfrak{K}$ there exists a bounded and boundedly invertible uniformly positive operator W such that $W\text{dom } A \subseteq \text{dom } A$.*
(4) *In $\mathfrak{K}$ there exists a fundamental symmetry J such that $J\text{dom } A \subseteq \text{dom } A$.*

The essential implication here is (4) $\Rightarrow$ (1). We shall sketch a proof. If $J\text{dom } A \subseteq \text{dom } A$ then for the positive selfadjoint operator $B = JA$ in the Hilbert space $(\mathfrak{K}, (\cdot,\cdot))$, where $(\cdot,\cdot)$ is defined by (1.2), we have $J\text{dom } B \subseteq \text{dom } B$ and, according to the Heinz inequality $J\text{dom } B^\alpha \subseteq \text{dom } B^\alpha$, $0 < \alpha < 1$, in particular, for $\alpha = 1/2$. Consider the Hilbert space $\mathfrak{K}_A = (\text{dom } B^{1/2}, (\cdot,\cdot)_A)$, where $(x,y)_A = [Ax, y] = (Bx, y)$ first defined on dom A and then via completion extended to dom $B^{1/2}$. Evidently, A^{-1} maps dom A into dom A, hence A^{-1} maps dom $B^{1/2}$ into dom $B^{1/2}$ or $\mathfrak{K}_A$ into $\mathfrak{K}_A$. Define $\hat{A} = A^{-1}|_{\text{dom } B^{1/2}}$ as an operator in $\mathfrak{K}_A$. Then

$$(\hat{A}x, x)_A = [AA^{-1}x, x] = [x, x], \quad x \in \mathfrak{K}_A,$$

hence $\hat{A}$ is a bounded selfadjoint operator in $\mathfrak{K}_A$.

Thus, in $\mathfrak{K}_A$, the operator $|\hat{A}|$ and an inner product

$$\langle x, y\rangle_A = (|\hat{A}|x, y)_A = [A|\hat{A}|x, y] = [\text{sgn } \hat{A}x, y]$$

can be defined. It follows that

$$(x, x) = (A^{-1}Jx, x)_A = (\hat{A}J|_{\mathfrak{K}_A} x, x)_A \leq \|J|_{\mathfrak{K}_A}\|(|\hat{A}|x, x)_A \leq \|J|_{\mathfrak{K}_A}\|\langle x, x\rangle_A,$$

where we have used M.G. Kreĭn 's inequality: If $S = S^*$ and T are bounded in the Hilbert space $\mathfrak{H}$ and satisfy $ST = (ST)^*$, then

$$|(STx, x)| \leq \|T\|(|S|x, x), \quad x \in \mathfrak{H}.$$

It can be shown that also an inequality

$$\langle x, x\rangle_A \leq \text{const. } \|x\|^2, \quad x \in \mathfrak{K},$$

holds; see Ćurgus [1985]. Hence the norms $\langle x, x\rangle_A^{1/2}$ and $\|x\|$ are equivalent and sgn $\hat{A}$ is a fundamental symmetry in $\mathfrak{K}$.

2 Boundary eigenvalue problems for Sturm- Liouville operators and related holomorphic functions

1. The Sturm-Liouville differential operator. We consider the **eigenvalue problem**

$$-f'' + qf - \lambda f = g \tag{2.1}$$

on the open interval (a, b) of $\mathbb{R}$, $-\infty \leq a < b \leq \infty$. In this and the next subsection we assume that

(1) the **potential** $q(x)$ is real valued and locally integrable on (a, b).

In Section 5 we consider the case where the potential also contains a delta function or a derivative of a delta function. The complex number λ is called the **eigenvalue parameter**. The problem (2.1) can be written as $(L - \lambda)f = g$, where L is the second order **Sturm–Liouville** differential expression

$$L = -D^2 + q(x), \quad D = d/dx.$$

Unless stated differently functions are complex valued. A function f is called locally integrable on an open, half open or compact subinterval ι of $\mathbb{R}$ if f is integrable on every compact subinterval of ι, and we use the notation: $f \in L^1_{loc}(\iota)$. A function f is locally absolutely continuous on ι (notation: $f \in AC_{loc}(\iota)$), if there exists a locally integrable function h on ι such that for all compact intervals $[c, d] \subseteq \iota$, $f(d) - f(c) = \int_c^d h(x)dx$; we write f' instead of h. To simplify the notation we write, for example, $L^1_{loc}[a, b)$ instead of $L^1_{loc}([a, b))$ etc.

Let g be a function in $L^1_{loc}(a, b)$. A function f on (a, b) is called a solution of the equation $Lf = g$ if f is continuously differentiable, f' is locally absolutely continuous and the equality $-f''(x) + q(x)f(x) = g(x)$ holds almost everywhere on (a, b). The equation $Lf = g$ is called **homogeneous** if $g = 0$; otherwise it is called **inhomogeneous**. The solutions of the homogeneous equation $Lf = 0$ form a 2-dimensional linear space over $\mathbb{C}$. Evidently, these remarks also apply to the equation $(L - \lambda)f = g$.

We study the equation (2.1) in the Hilbert space $L^2(a, b)$ consisting of all square integrable functions on (a, b) with inner product and norm

$$(f, g) = \int_a^b \overline{g(x)} f(x)dx, \quad \|f\| = \sqrt{(f, f)}, \quad f, g \in L^2(a, b).$$

By the Cauchy-Schwarz-Bunyakowski inequality, all functions in $L^2(a, b)$ are locally integrable on (a, b). We explain the Hilbert space setting and additional assumptions on L in detail. For proofs we refer to Achieser and Glasmann [1981], Coddington and Levinson [1955], Hille 1969] and Weidman [1976].

The **maximal realization** of L in $L^2(a, b)$ is the operator

$$L_{max} = \{\{f, g\} : f \in L^2(a, b) \cap C^1(a, b), f' \in AC_{loc}(a, b), g = Lf \in L^2(a, b)\}.$$

We frequently identify operators with their graphs, so that, for example, $\{f, g\} \in L_{max}$ means that f belongs to dom L_{max} and $L_{max}f = g$. The operator L_{max} is closed and densely defined in $L^2(a,b)$. Using integration by parts and (1), we find that for $f, h \in \text{dom } L_{max}$,

$$\int_a^b \overline{h(x)}Lf(x)dx - \int_a^b \overline{Lh(x)}f(x)dx = \langle f,h\rangle(b) - \langle f,h\rangle(a), \tag{2.2}$$

where with $\langle f,h\rangle(x) = -\overline{h(x)}f'(x) + \overline{h'(x)}f(x), x \in (a,b)$,

$$\langle f,h\rangle(a) = \lim_{x\downarrow a}\langle f,h\rangle(x), \quad \langle f,h\rangle(b) = \lim_{x\uparrow b}\langle f,h\rangle(x).$$

This formula is called the **Lagrange formula** or **Green's identity**.

The **minimal realization** of L in $L^2(a,b)$ is the operator whose graph is given by

$$L_{min} = \{\{f, Lf\} \in L_{max} : f \text{ has compact support in (a,b)}\}^c,$$

where the closure is taken in the orthogonal direct sum $L^2(a,b) \oplus L^2(a,b)$. From the Lagrange formula it follows that $S = L_{min}$ is a densely defined symmetric operator in $L^2(a,b)$ with adjoint $S^* = L_{max}$ and

$$L_{min} = \{\{f, Lf\} \in L_{max} : \langle f,h\rangle(a) = 0, \langle f,h\rangle(b) = 0 \text{ for all } h \in \text{dom } L_{max}\}.$$

We recall some definitions and facts about symmetric operators in Hilbert spaces. Let S be a closed densely defined symmetric operator in a Hilbert space $\mathfrak{H}$. The **von Neumann decomposition formula** for the adjoint S^* of S reads in graph notation: For any complex nonreal number μ,

$$S^* = S + M_\mu(S^*) + M_{\bar\mu}(S^*), \quad \text{direct sum in } \mathcal{H} \oplus \mathcal{H},$$

where $M_\lambda(S^*) = \{\{f,g\} \in S^* : g = \lambda f\}$. As a function of λ the dimension of

$$\text{dom } M_\lambda(S^*) = \ker (S^* - \lambda) = \text{ran } (S - \bar\lambda)^\perp$$

is constant on $\mathbb{C}^+$ and constant on $\mathbb{C}^-$. These constants are denoted by d^+ and d^- and are called the **upper** and **lower defect numbers** of S. Evidently, $\dim S^*/S = \dim (\text{dom } S^*/\text{dom } S) = d^+ + d^-$. If both defect numbers are 0, then S is selfadjoint, that is, $S = S^*$. In the sequel we assume that the defect numbers are finite.

A densely defined selfadjoint operator A in a Kreĭn space $\mathfrak{K}$ is called an **extension** of S in $\mathfrak{H}$ if $\mathfrak{K}$ admits the Kreĭn subspace decomposition $\mathfrak{K} = \mathfrak{H} \oplus (\mathfrak{K} \ominus \mathfrak{H})$ and $S \subseteq A$. When $\mathfrak{K} = \mathfrak{H}$, A is called a **canonical extension**, and in the other case A is said to be an **extension with exit space** $\mathfrak{K} \ominus \mathfrak{H}$ or simply "with exit". Selfadjoint extensions with exit always exist; canonical ones exist if and only if the defect numbers d^+ and d^- of S coincide.

We only consider selfadjoint extensions A of S which are **minimal**, that is, have a nonempty resolvent set $\rho(A)$ and satisfy

$$\mathfrak{K} = \left(\text{span } \{\text{ran } (A-\lambda)^{-1}|_{\mathfrak{H}} : \lambda \in \rho(A)\}\right)^c.$$

Minimal selfadjoint extensions in a Hilbert space always exist; see Achieser and Glasmann [1981]. The compression of the resolvent $(A - \lambda)^{-1}$ of a selfadjoint extension A of S, A acting in $\mathfrak{K}$, to the space $\mathfrak{H}$:

$$R(\lambda) = P_{\mathfrak{H}}(A-\lambda)^{-1}|_{\mathfrak{H}}, \quad \lambda \in \rho(A),$$

where $P_{\mathfrak{H}}$ stands for the projection in $\mathfrak{K}$ onto $\mathfrak{H}$, is called a **generalized resolvent** for S. It is a right inverse for $S^* - \lambda$: $(S^* - \lambda)R(\lambda) = I$.

Applied to the minimal differential operator $S = L_{min}$ in $\mathfrak{H} = L^2(a,b)$, we have that

$$\ker\ (S^* - \lambda) = \{f \in L^2(a,b) : (L-\lambda)f = 0\}.$$

Thus the upper/lower defect number $d^{\pm}$ is the number of $L^2(a,b)$-solutions of the homogenous equation $(L-\lambda)f = 0$ for λ in the upper/lower complex halfplane; it is either $0, 1$ or 2. Since, by assumption (1), q is real valued, L_{min} commutes with complex conjugation and this implies that these numbers are the same: $d^+ = d^-$. Hence L_{min} has canonical selfadjoint extensions. They are called the **selfadjoint realizations** of L in $L^2(a,b)$ and, in particular cases, they will be described below.

The left endpoint a of (a,b) is called **regular** for L, or L is called **regular at** a, if a is finite and the potential q belongs to $L^1_{loc}[a,b)$; otherwise a is called **singular** for L or L is **singular at** a. If a is regular, the Lagrange formula implies that for each function $f \in \text{dom } L_{max}$, the limits

$$f(a) = \lim_{x \downarrow a} f(x), \quad f'(a) = \lim_{x \downarrow a} f'(x)$$

exist and that these point evaluations are continuous with respect to the graph norm of L_{max}:

$$|f(a)| \leq C(\|f\|^2 + \|Lf\|^2)^{1/2}, \quad |f'(a)| \leq C'(\|f\|^2 + \|Lf\|^2)^{1/2}.$$

Moreover, it implies that if f belongs to dom L_{min} then $f(a) = 0$ and $f'(a) = 0$. Thus, if a is a regular endpoint then L_{min} is not selfadjoint in $L^2(a,b)$, and its defect numbers are either 1 or 2.

The regularity and singularity of the right endpoint b are defined similarly. Thus b is singular for L if $b = \infty$ or if $\int_c^b |q(x)|dx = \infty$ for some point $c \in (a,b)$. If a and b are regular endpoints, then

$$L_{min} = \{\{f, Lf\} \in L_{max} : f(a) = f'(a) = f(b) = f'(b) = 0\}$$

and the defect numbers of L_{min} are equal to 2. In this case all selfadjoint realizations A of L are of the form

$$A = A_{C,D} = \left\{\{f, Lf\} \in L_{max} : C\begin{pmatrix} f(a) \\ f'(a) \end{pmatrix} + D\begin{pmatrix} f(b) \\ f'(b) \end{pmatrix} = 0\right\},$$

where C and D are 2×2 matrices such that

$$\text{rank } \begin{pmatrix} C & D \end{pmatrix} = 2, \quad C\mathsf{J}C^* - D\mathsf{J}D^* = 0, \quad \mathsf{J} = i\begin{pmatrix} 0 & -1 \\ 1 & 0 \end{pmatrix}.$$

If a is regular and b is singular the defect numbers can be 1 or 2. Both cases occur. For example, it can be shown that the defect numbers of the minimal realization of the Bessel differential expression $L = -D^2 + (\nu^2 - \frac{1}{4})/x^2$ in $L^2(-1,0)$ are equal to 2 if $0 \leq \nu < 1$ and equal to 1 if $\nu \geq 1$. See Achieser and Glasmann [1981,p. 472].

The following theorem is often called **Weyl's alternative**. It improves on what has been said so far about the number of $L^2(a,b)$–solutions f of $(L-\lambda)f = 0$.

Theorem 2.1 *Suppose that* (1) *is valid and that* b *is a singular endpoint. Then for any point* $c \in (a,b)$, *either:*
(LC) *for all* $\lambda \in \mathbb{C}$, *all solutions* f *of* $(L-\lambda)f = 0$ *belong to* $L^2(c,b)$,
or:
(LP) *for all* $\lambda \in \mathbb{C}$, *there is at least one solution* f *of* $(L-\lambda)f = 0$ *that does not belong to* $L^2(c,b)$.
In the latter case, for all $\lambda \in \mathbb{C} \setminus \mathbb{R}$, *the equation* $(L-\lambda)f = 0$ *has (up to a factor) precisely one nonzero solution* $f \in L^2(c,b)$.

Note that, by (1), any point $c \in (a,b)$ is a regular endpoint for the restriction of L to (c,b). If a is a regular endpoint then any solution of $(L-\lambda)f = 0$ belongs to $L^2(a,c)$. Hence the alternative of Theorem 2.1 also holds for $c = a$ if we assume that (1) holds, a is regular, and b is singular. Weyl's proof of Theorem 2.1 is geometric and involves the construction of a sequence of nested circles. The terminology distinguishes between the two possible limits of this sequence: LC stands for **limit circle** and LP for **limit point**. Here we have considered the alternative at the endpoint b; the limit circle and limit point cases at the endpoint a, when a is singular for L, are defined similarly.

Assume that (1) holds and that L is regular at a and singular at b. In the LC case at b, the defect numbers of L_{min} are equal to 2. In the LP case at b, the Lagrange formula (2.2) becomes

$$\int_a^b \overline{h(x)}Lf(x)dx - \int_a^b \overline{Lh(x)}f(x)dx = -\langle f,h\rangle(a), \quad f,h \in \text{dom } L_{max}, \tag{2.3}$$

that is, $\langle f,h\rangle(b) = 0$, and hence the defect numbers of L_{min} are equal to 1 and

$$L_{min} = \{\{f, Lf\} \in L_{max} : f(a) = f'(a) = 0\}.$$

We study this case in the next subsection.

2. The Titchmarsh–Weyl coefficient. In this subsection we assume that the Sturm–Liouville differential expression $L = -D^2 + q$ considered in Subsection 2.1 satisfies the following conditions:

(1) The potential $q(x)$ is real valued and locally integrable on (a,b).
(2) The endpoint a is regular for L.
(3) The endpoint b is singular for L.
(4) L is limit point at b.

Then all canonical selfadjoint realizations of L in $L^2(a,b)$ are given by

$$A_{\alpha,\beta} = \{\{f, Lf\} \in L_{max} : \alpha f(a) + \beta f'(a) = 0\}, \tag{2.4}$$

where α and β are real numbers not both equal to zero.

Associated with each such realization is the following **boundary eigenvalue problem**: Given $\lambda \in \mathbb{C} \setminus \mathbb{R}$ and $g \in L^2(a,b)$, find $f \in L^2(a,b)$ such that

$$\begin{cases} -f'' + (q-\lambda)f = g, \\ \alpha f(a) + \beta f'(a) = 0. \end{cases} \tag{2.5}$$

The solution is of the form $f = (A_{\alpha,\beta} - \lambda)^{-1}g$ and the resolvent operator $R_{\alpha,\beta}(\lambda) = (A_{\alpha,\beta} - \lambda)^{-1}$ is a bounded integral operator.

For $\lambda \in \mathbb{C}$, let $\varphi(\cdot,\lambda)$ and $\psi(\cdot,\lambda)$ be the unique solutions of $(L-\lambda)f = 0$ on (a,b) that satisfy

$$\varphi(a,\lambda) = 1, \quad \varphi'(a,\lambda) = 0, \quad \psi(a,\lambda) = 0, \quad \psi'(a,\lambda) = -1.$$

Then for each $\lambda \in \mathbb{C}\setminus\mathbb{R}$, there exists a unique complex number $m(\lambda)$ such that the function

$$s(\cdot,\lambda) = m(\lambda)\varphi(\cdot,\lambda) + \psi(\cdot,\lambda) \tag{2.6}$$

is the unique solution of $(L-\lambda)f = 0$ which belongs to $L^2(a,b)$. The function $m(\lambda), \lambda \in \mathbb{C}\setminus\mathbb{R}$, is called the **Titchmarsh–Weyl coefficient** for the differential expression L on (a,b) associated with the boundary condition $f'(a) = 0$. It is locally holomorphic on $\mathbb{C}\setminus\mathbb{R}$ and for all $\lambda,\zeta \in \mathbb{C}\setminus\mathbb{R}$,

$$m(\lambda) = \overline{m(\bar\lambda)}, \quad N_m(\lambda,\zeta) = \frac{m(\lambda) - \overline{m(\zeta)}}{\lambda - \bar\zeta} = (s(\cdot,\lambda), s(\cdot,\zeta)).$$

The last equality follows from the Lagrange formula (2.3):

$$\begin{aligned}(\lambda-\bar\zeta)\,(s(\cdot,\lambda),s(\cdot,\zeta)) &= (Ls(\cdot,\lambda),s(\cdot,\zeta)) - (s(\cdot,\lambda),Ls(\cdot,\zeta)) \\ &= -\langle s(\cdot,\lambda),s(\cdot,\zeta)\rangle(a) = m(\lambda) - \overline{m(\zeta)}.\end{aligned}$$

These properties imply that m belongs to the class $\mathbf{N}_0$ of Nevanlinna functions. Recall from Section 1 that a function Q on $\mathbb{C}\setminus\mathbb{R}$ is a Nevanlinna function if $Q(\lambda)$ is locally holomorphic on $\mathbb{C}\setminus\mathbb{R}$, $Q(\bar\lambda) = \overline{Q(\lambda)}$,

$$\operatorname{Im}\, Q(\lambda) \geq 0, \quad \lambda \in \mathbb{C}^+.$$

The function Q is a Nevanlinna function if and only if it admits the integral representation

$$Q(\lambda) = \mu\lambda + \nu + \int_{\mathbb{R}} \left(\frac{1}{t-\lambda} - \frac{t}{t^2+1}\right) d\sigma(t), \tag{2.7}$$

where μ and ν are real numbers with $\mu \geq 0$ and σ is a nondecreasing function on $\mathbb{R}$ with $\int_{\mathbb{R}}(t^2+1)^{-1}d\sigma(t) < \infty$. In Everitt [1972, formula (3.2)] it is shown that the Titchmarsh–Weyl coefficient $m(\lambda)$ has the asymptotic form

$$m(iy) = \frac{e^{\pi i/4}}{\sqrt{y}} + O(\frac{1}{y}), \quad y \longrightarrow \infty. \tag{2.8}$$

By Kac and Kreĭn [1968, Theorem S1.3.1 and Lemma S1.3.1], this implies that the function σ in the representation (2.7) for m has the property $\int_{\mathbb{R}}(|t|+1)^{-1}d\sigma(t) < \infty$ and that m has the simpler integral representation

$$m(\lambda) = \int_{\mathbb{R}} \frac{d\sigma(t)}{t-\lambda}. \tag{2.9}$$

Here $\sigma(t)$ is called the **spectral measure** for the eigenvalue problem $(L-\lambda)f = 0$ with boundary condition $f'(a) = 0$. To explain this terminology, we consider the selfadjoint realization $A_{0,1}$ of L determined by this boundary condition. For functions $g \in L^2(a,b)$ with compact support, the resolvent operator $R_{0,1}(\lambda) = (A_{0,1} - \lambda)^{-1}$ is of the form

$$(R_{0,1}(\lambda)g)(x) = (G_{0,1}(\lambda)g)(x) + m(\lambda)\varphi(x,\lambda)\int_a^b \varphi(t,\lambda)g(t)dt,$$

where $G_{0,1}(\lambda)$ is the bounded integral operator

$$(G_{0,1}(\lambda)g)(x) = -\varphi(x,\lambda)\int_x^b \psi(t,\lambda)g(t)dt - \psi(x,\lambda)\int_a^x \varphi(t,\lambda)g(t)dt.$$

We write

$$\hat{g}(\lambda) = \int_a^b \varphi(t,\lambda)g(t)dt, \quad \lambda \in \mathbb{C}.$$

Let $E_{0,1}$ be the spectral function of $A_{0,1}$ and set $E_{0,1}(t) = E_{0,1}((-\infty,t))$. Then, on account of (2.9),

$$\int_{\mathbb{R}} \frac{d\,(E_{0,1}(t)g,g)}{t-\lambda} = (R_{0,1}(\lambda)g,g) = (G_{0,1}(\lambda)g,g) + \overline{\hat{g}(\bar{\lambda})}\hat{g}(\lambda)\int_{\mathbb{R}} \frac{d\sigma(t)}{t-\lambda}.$$

The first summand on the righthand side and the term $\overline{\hat{g}(\bar{\lambda})}\hat{g}(\lambda)$ are entire in λ. Applying the Stieltjes-Livšic inversion formula, we obtain the formula

$$(E_{0,1}(\Delta)g,g) = \int_\Delta \overline{\hat{g}(\lambda)}d\sigma(\lambda)\hat{g}(\lambda).$$

The formula holds for all subintervals Δ of $\mathbb{R}$ whose endpoints are continuity points for E and all $g \in L^2(a,b)$ with compact support in (a,b). Letting $\Delta \to \mathbb{R}$, we obtain for functions $g \in L^2(a,b)$ with compact support in (a,b),

$$(g,g) = \int_{-\infty}^{\infty} \overline{\hat{g}(\lambda)}d\sigma(\lambda)\hat{g}(\lambda).$$

This readily leads to the following theorem.

Theorem 2.2 *For $g \in L^2(a,b)$, the integrals*

$$\hat{g}(\lambda) = \int_a^b \varphi(x,\lambda)g(x)dx, \quad \int_{-\infty}^{\infty} \varphi(\cdot,\lambda)d\sigma(\lambda)\hat{g}(\lambda)$$

converge in norm in L^2_σ and in norm in $L^2(a,b)$, respectively, and

$$(g,g) = \int_{-\infty}^{\infty} \overline{\hat{g}(\lambda)}d\sigma(\lambda)\hat{g}(\lambda), \quad g(x) = \int_{-\infty}^{\infty} \varphi(x,\lambda)d\sigma(\lambda)\hat{g}(\lambda). \tag{2.10}$$

The map $g \mapsto \hat{g}$ is a unitary map from $L^2(a,b)$ onto L^2_σ. Under this isomorphism, the operator $A_{0,1}$ on $L^2(a,b)$ becomes the operator of multiplication by the independent variable in L^2_σ.

The first relation in (2.10) is called the **completeness theorem** or **Parseval equality** and the second relation is called the **expansion theorem**; see, for example, Coddington and Levinson [1955, Chapter 9].

A densely defined symmetric operator S in a Hilbert space is called **simple** if S has no invariant subspace on which it acts as a selfadjoint operator. The following result (compare, for example, Dijksma et al. [1988]) will be used in Section 3.

Proposition 2.3 *Assume that the Sturm–Liouville expression L satisfies conditions* (1)–(4). *Then the minimal realization L_{min} of L in $L^2(a,b)$ is simple.*

3. The boundary eigenvalue problem. In this subsection, for the Sturm–Liouville expression $L = -D^2 + q$ we assume that the conditions (1)–(4) of the previous subsection are satisfied. Let $\alpha(\lambda)$ and $\beta(\lambda)$ be functions defined on some subset $\mathcal{D}_{\alpha,\beta}$ of $\mathbb{C}\setminus\mathbb{R}$. We consider the following **boundary eigenvalue problem** where also the boundary condition at a depends on the eigenvalue parameter λ: Given $\lambda \in \mathcal{D}_{\alpha,\beta}$ and $g \in L^2(a,b)$, find $f \in L^2(a,b)$ such that

$$\begin{cases} -f'' + (q-\lambda)f = g, \\ \alpha(\lambda)f(a) + \beta(\lambda)f'(a) = 0. \end{cases} \tag{2.11}$$

Over the years many papers concerning boundary eigenvalue problems of this type have appeared, for instance, Friedman [1956], Kac and Kreĭn [1968], Niessen [1970], Walter [1973], Russakovski [1975], Fulton [1977], Fulton and Pruess [1979], Dijksma [1980], Fulton [1980], Benedek and Panzone [1981], Röh [1981], Schneider [1981], Hinton and Shaw [1987], Dijksma et al. [1987], [1987a], [1988], and [1993], Linden [1991], Binding et al. [1993], Lancaster and Shkalikov [1994], and also Binding and Browne [to appear], Russakovski [to appear].

If $\mathcal{O}$ is a subset of $\mathbb{C}$ we denote by $\mathcal{O}^*$ the set $\mathcal{O}^* = \{\lambda \in \mathbb{C} : \bar\lambda \in \mathcal{O}\}$. Thus $\mathcal{O}$ is symmetric with respect to the real axis if and only if $\mathcal{O} = \mathcal{O}^*$. Concerning the coefficients in the boundary condition in (2.11) we assume that:

(5) $\mathcal{D}_{\alpha,\beta}$ is a nonempty open subset of $\mathbb{C}\setminus\mathbb{R}$ and $\mathcal{D}_{\alpha,\beta} = \mathcal{D}^*_{\alpha,\beta}$,
(6) α and β are locally holomorphic on $\mathcal{D}_{\alpha,\beta}$ and $|\alpha(\lambda)| + |\beta(\lambda)| \neq 0$, $\lambda \in \mathcal{D}_{\alpha,\beta}$,
(7) for all $\lambda \in \mathcal{D}_{\alpha,\beta}$, $\alpha(\lambda)\overline{\beta(\bar\lambda)} - \overline{\alpha(\bar\lambda)}\beta(\lambda) = 0$, and
(8) the set $\mathcal{D}'_{\alpha,\beta} = \{\lambda \in \mathcal{D}_{\alpha,\beta} : \alpha(\lambda)m(\lambda) \neq \beta(\lambda)\}$ is nonempty.

Here $m(\lambda)$ is the Titchmarsh–Weyl coefficient discussed in the previous subsection.

For each $\lambda \in \mathcal{D}_{\alpha,\beta}$, we associate with (2.11) the operator

$$T(\lambda) = \{\{f, Lf\} \in L_{max} : \alpha(\lambda)f(a) + \beta(\lambda)f'(a) = 0\}, \tag{2.12}$$

which is closed and densely defined in $L^2(a,b)$. On account of (4) and (7), we have that for all $\lambda \in \mathcal{D}_{\alpha,\beta}$, $L_{min} \subseteq T(\lambda) \subseteq L_{max}$, dim (dom $T(\lambda)$/dom L_{min}) $= 1$, $T(\lambda) = T(\bar\lambda)^*$, and

$$T(\lambda) = \{\{f, Lf\} \in L_{max} : f(a) = k\overline{\beta(\bar\lambda)},\, f'(a) = -k\overline{\alpha(\bar\lambda)} \text{ for some } k \in \mathbb{C}\}.$$

$T(\lambda)$ is called a **Štraus extension** of L_{min}; in an abstract setting for a symmetric operator S the Štraus extension will be defined in Subsection 3.2. The boundary eigenvalue problem (2.11) is the same as: Given $\lambda \in \mathcal{D}_{\alpha,\beta}$ and $g \in L^2(a,b)$, find $f \in L^2(a,b)$ such that $(T(\lambda) - \lambda)f = g$, or equivalently,

$$\begin{cases} -f'' + (q-\lambda)f = g, \\ f \in \text{dom } T(\lambda). \end{cases} \tag{2.13}$$

If the coefficients in the boundary condition are constant, $\alpha(\lambda) \equiv: \alpha$ and $\beta(\lambda) \equiv: \beta$, then $T(\lambda) \equiv A_{\alpha,\beta}$, the selfadjoint realization of L in $L^2(a,b)$ defined in Subsection 2.2, and $T(\lambda)$ is independent of λ.

In the sequel we encounter the kernel

$$N_{\alpha,\beta}(\lambda,\zeta) = \frac{\alpha(\lambda)\overline{\beta(\zeta)} - \beta(\lambda)\overline{\alpha(\zeta)}}{\lambda - \bar{\zeta}},$$

which, in view of (7), is well defined for all $\lambda, \zeta \in \mathcal{D}_{\alpha,\beta}$. Note that $N_{m,1}(\lambda,\zeta) = N_m(\lambda,\zeta)$. A $k \times k$ matrix function $Q(\lambda)$ defined on an open set $\mathcal{O}$ with $\mathcal{O} = \mathcal{O}^*$ is called a **generalized Nevanlinna function** with κ negative squares if $Q(\lambda)$ is locally holomorphic there, $Q(\lambda) = Q(\bar{\lambda})^*$ and the $\mathbb{C}^k$–kernel

$$N_Q(\lambda,\zeta) = \frac{Q(\lambda) - Q(\zeta)^*}{\lambda - \bar{\zeta}}$$

has κ negative squares. The class of such functions is denoted by $\mathbf{N}_\kappa^{k\times k}$. From the integral representation (2.7) of a Nevanlinna function it follows that the class $\mathbf{N}_0$ of Nevanlinna functions coincides with the class $\mathbf{N}_0^{1\times 1}$. We write $\mathbf{N}_\kappa$ for $\mathbf{N}_\kappa^{1\times 1}$.

If α and β are polynomials, $\beta \not\equiv 0$, and satisfy (7), then α/β belongs to $\mathbf{N}_\kappa$ for some κ. In this case we may assume without loss of generality that the coefficients of α and β are real numbers; if $\beta \equiv 0$ we take $\alpha \equiv 1$.

If α/β belongs to $\mathbf{N}_\kappa$ and $\mathcal{D}_{\alpha,\beta} = \mathbb{C} \setminus \mathbb{R}$, then (8) is automatically satisfied. To see this, assume that (8) does not hold. Then $m \equiv \beta/\alpha$ and, since with α/β also $-\beta/\alpha$ belongs to $\mathbf{N}_\kappa$, the kernel N_m has, besides 0 negative squares, κ positive squares (defined in the obvious way). This implies that m is rational. The asymptotic form (2.8) of $m(\lambda)$ shows that this cannot be true. Hence (8) holds.

4. The linearization of the boundary eigenvalue problem. In this subsection we describe the solution of the boundary eigenvalue problem (2.11). If α and β are constant the solution is given by $f = (A_{\alpha,\beta} - \lambda)^{-1}g$. We show that in the general case the solution can be given in terms of a generalized resolvent of L_{min}.

An operator W from a Kreĭn space $\mathfrak{K}_1$ to a Kreĭn space $\mathfrak{K}_2$ is called a **weak isomorphism** if dom W and ran W are dense linear subsets of $\mathfrak{K}_1$ and $\mathfrak{K}_2$ and W : dom $W \to$ ran W is linear and isometric (and hence bijective). A weak isomorphism is not necesarily continuous. The dimensions of the positive and negative spaces in a fundamental decomposition of a Kreĭn space are invariant under weak isomorphisms. A weak isomorphism W can be extended to an (everywhere defined and continuous) isomorphism (unitary mapping) if dom W contains a maximal uniformly definite subspace; see Shmul'yan [1967]. Two selfadjoint extensions A_1 in a Kreĭn space $\mathfrak{K}_1$ and A_2 in a Kreĭn space $\mathfrak{K}_2$ of a symmetric operator S in a Hilbert space $\mathfrak{H}$ are called **weakly isomorphic** if $\rho(A_1) \cap \rho(A_2) \neq \emptyset$ and there exists a weak isomorphism W from $\mathfrak{K}_1$ to $\mathfrak{K}_2$ which on $\mathfrak{H}$ coincides with the identity operator on $\mathfrak{H}$, such that for some $\mu \in \rho(A_1) \cap \rho(A_2)$,

$$W(A_1 - \mu)^{-1}|_{\text{dom } W} = (A_2 - \mu)^{-1}W.$$

A subset $\mathcal{D}$ of $\mathbb{C}$ will be called a **Jordan region** if it is a simply connected and open set with smooth boundary $\partial\mathcal{D}$.

For the boundary eigenvalue problem (2.11) satisfying (1)–(8) the following result is true.

Theorem 2.4 (1) *For each* $\lambda \in \mathcal{D}'_{\alpha,\beta}$, *the boundary eigenvalue problem* (2.11) *has a unique solution* $f = R(\lambda)g$ *in* $L^2(a,b)$.
(2) *If* $\mathcal{D}$ *is a Jordan region with* $\mathcal{D}^c \subseteq \mathcal{D}_{\alpha,\beta} \cap \mathbb{C}^+$, *then there exists a selfadjoint extension* A *of* L_{min} *in a Kreĭn space* $\mathfrak{K}$ *such that* $(\mathcal{D} \cup \mathcal{D}^*) \setminus \rho(A)$ *is empty or consists of isolated points only and*

$$R(\lambda)g = P_{L^2(a,b)}(A-\lambda)^{-1}g, \quad \lambda \in (\mathcal{D} \cup \mathcal{D}^*) \cap \rho(A), \quad g \in L^2(a,b). \tag{2.14}$$

This extension can be chosen minimal. Then it is unique up to a weak isomorphism and if $\mathfrak{K} \ominus L^2(a,b) = \mathfrak{K}'_+ \oplus \mathfrak{K}_-$ *is a fundamental decomposition of the exit space, the dimensions* $\dim \mathfrak{K}'_+$ *and* $\dim \mathfrak{K}_-$ *are equal to the number of positive and the number of negative squares of the kernel*

$$N_{\alpha,\beta}(\lambda,\zeta) = \frac{\alpha(\lambda)\overline{\beta(\zeta)} - \beta(\lambda)\overline{\alpha(\zeta)}}{\lambda - \bar{\zeta}}, \quad \lambda, \zeta \in \mathcal{D} \cup \mathcal{D}^*.$$

If one of these numbers is finite then the extension is unique up to an isomorphism.
(3) *If, in particular,* α/β *belongs to* $\mathbf{N}_\kappa$ *and* $\mathcal{D}_{\alpha,\beta} = \mathbb{C} \setminus \mathbb{R}$, *then the condition* (8) *is satisfied,* (2.14) *holds with* $\mathcal{D} \cup \mathcal{D}^*$ *replaced by* $\mathbb{C} \setminus \mathbb{R}$, *the minimal selfadoint extension* A *in* $\mathfrak{K}$ *is uniquely determined up to an isomorphism, and the space* $\mathfrak{K}$ *is a* π_κ*-space.*

The proof of Theorem 2.4 will be outlined in Subsection 3.2. A minimal selfadjoint operator A satisfying (2.14) is called a **linearization** of the boundary eigenvalue problem (2.11) or more precisely of the corresponding Štraus extension $T(\lambda)$ in (2.12). The generalized resolvent $R(\lambda)$ in (2.14) is a bounded integral operator. For $g \in L^2(a,b)$ which vanishes near b, it takes the form

$$R(\lambda)g = G(\lambda)g + \begin{pmatrix} \varphi(\cdot,\lambda) & \psi(\cdot,\lambda) \end{pmatrix} \Omega(\lambda) \begin{pmatrix} \int_a^b \varphi(t,\lambda)g(t)dt \\ \int_a^b \psi(t,\lambda)g(t)dt \end{pmatrix}, \tag{2.15}$$

where $G(\lambda)$ is a bounded integral operator on $L^2(a,b)$ and a right inverse for $L_{max} - \lambda$, which for $g \in L^2(a,b)$ vanishing near b is of the form

$$(G(\lambda)g)(x) = -\psi(x,\lambda)\int_x^b \varphi(t,\lambda)g(t)dt - \varphi(x,\lambda)\int_a^x \psi(t,\lambda)g(t)dt,$$

and $\Omega(\lambda)$ is the 2×2 matrix

$$\begin{aligned} \Omega(\lambda) &= \begin{pmatrix} 1 & -m(\lambda) \\ \alpha(\lambda) & -\beta(\lambda) \end{pmatrix}^{-1} \begin{pmatrix} 0 & 1 \\ -\beta(\lambda) & 0 \end{pmatrix} \\ &= (\beta(\lambda) - m(\lambda)\alpha(\lambda))^{-1} \begin{pmatrix} m(\lambda)\beta(\lambda) & \beta(\lambda) \\ \beta(\lambda) & \alpha(\lambda) \end{pmatrix}. \end{aligned}$$

All terms in the formula for $R(\lambda)$, except $\Omega(\lambda)$, are entire in λ and that makes Ω stand out from the rest. If the kernel $N_{\alpha,\beta}(\lambda,\zeta)$ has κ negative squares then $\Omega \in \mathbf{N}_\kappa^{2\times 2}$. This follows readily from the formula

$$-\Omega(\lambda)^{-1} = \begin{pmatrix} \alpha(\lambda)/\beta(\lambda) & -1 \\ -1 & m(\lambda) \end{pmatrix} \tag{2.16}$$

and the fact that Ω belongs to the class $\mathbf{N}_\kappa^{2\times 2}$ if its inverse exists and $-\Omega^{-1}$ belongs to this class. The matrix function Ω is called the **Titchmarsh–Weyl coefficient** corresponding to the boundary condition in (2.11).

5. Special cases. (I) If the boundary condition in (2.11) is such that the kernel $N_{\alpha,\beta}(\lambda,\zeta)$ is nonnegative then Ω belongs to the Nevanlinna class $\mathbf{N}_0^{2\times 2}$, hence has the integral representation

$$\Omega(\lambda) = M\lambda + N + \int_{\mathbb{R}} \left(\frac{1}{t-\lambda} - \frac{t}{t^2+1} \right) d\Sigma(t),$$

where $M = M^* \geq 0$ and $N = N^*$ are constant 2×2 matrices and Σ is a nondecreasing 2×2 matrix function on $\mathbb{R}$ with $\int_{\mathbb{R}} (t^2+1)^{-1} d\Sigma(t) < \infty$. It can be shown that $\frac{1}{y}\Omega(iy) \to 0,\ y \to \infty$, which implies that $M = 0$.

Let E be the spectral function of the linearization A in the Hilbert space $\mathfrak{K}$ of the boundary eigenvalue problem (2.11) and set $F(\Delta) = P_{L^2(a,b)} E(\Delta)|_{L^2(a,b)}$. Then using the Stieltjes-Livšic inversion formula as before, we find that for $g \in L^2(a,b)$ which vanishes near b,

$$(F(\Delta)g, g) = \int_\Delta \hat{g}(\lambda)^* d\Sigma(\lambda) \hat{g}(\lambda),$$

where

$$\hat{g}(\lambda) = \begin{pmatrix} \int_a^b \varphi(t,\lambda) g(t) dt \\ \int_a^b \psi(t,\lambda) g(t) dt \end{pmatrix}.$$

The analogue of Theorem 2.2 now reads as follows.

Theorem 2.5 *For $g \in L^2(a,b)$, the integrals*

$$\hat{g}(\lambda) = \begin{pmatrix} \int_a^b \varphi(t,\lambda) g(t) dt \\ \int_a^b \psi(t,\lambda) g(t) dt \end{pmatrix}, \quad \int_{-\infty}^{\infty} (\, \varphi(\cdot,\lambda) \ \psi(\cdot,\lambda) \,) d\Sigma(\lambda) \hat{g}(\lambda)$$

converge in norm in L^2_Σ and in norm in $L^2(a,b)$, respectively, and

$$g(x) = \int_{-\infty}^{\infty} (\, \varphi(x,\lambda) \ \psi(x,\lambda) \,) d\Sigma(\lambda) \hat{g}(\lambda).$$

The map $g \mapsto \hat{g}$ is an isometry from $L^2(a,b)$ into L^2_Σ, under which the selfadjoint extension A becomes the operator of multiplication by the independent variable in L^2_Σ. The map is surjective if and only if A is canonical.

(II) In certain cases the matrix function $\Omega(\lambda)$ can be replaced by a scalar function $\omega(\lambda)$: For example, let $\alpha(\lambda)$ and $\beta(\lambda)$ be polynomials with real coefficients satisfying the condition (7), and hence also the condition (8) in the previous subsection. Then by the Euclidean algorithm there exist polynomials $\gamma(\lambda)$ and $\delta(\lambda)$ with real coefficients also such that

$$\alpha(\lambda)\gamma(\lambda) + \beta(\lambda)\delta(\lambda) \equiv 1, \quad \deg\, \gamma \leq \deg\, \beta, \quad \deg\, \delta \leq \deg\, \alpha.$$

The 2×2 matrix function $\mathcal{E}$ on $\mathbb{C}$ with

$$\mathcal{E}(\lambda) = \begin{pmatrix} \alpha(\lambda) & -\beta(\lambda) \\ \delta(\lambda) & \gamma(\lambda) \end{pmatrix}$$

is entire in λ, invertible for all λ and satisfies $\mathcal{E}(\lambda)\mathsf{J}\mathcal{E}(\bar{\lambda})^* = \mathsf{J}$. Moreover,

$$\Omega(\lambda) = \mathcal{E}(\lambda)^{-1}\begin{pmatrix} 0 & -1/2 \\ -1/2 & \omega(\lambda) \end{pmatrix}(\mathcal{E}(\bar{\lambda})^*)^{-1},$$

where for $\lambda \in \mathbb{C}\setminus\mathbb{R}$,

$$\omega(\lambda) = -\frac{\delta(\lambda)m(\lambda) + \gamma(\lambda)}{\alpha(\lambda)m(\lambda) - \beta(\lambda)}.$$

We define two new solutions of the homogeneous equation $(L-\lambda)f = 0$, $\lambda \in \mathbb{C}$, namely

$$\eta(\cdot,\lambda) = -\gamma(\lambda)\varphi(\cdot,\lambda) + \delta(\lambda)\psi(\cdot,\lambda), \quad \theta(\cdot,\lambda) = \beta(\lambda)\varphi(\cdot,\lambda) + \alpha(\lambda)\psi(\cdot,\lambda).$$

Replacing Ω in (2.15) by its representation in terms of ω, and regrouping terms in (2.15) we find that the resolvent can be written as

$$R(\lambda)g = H(\lambda)g + \theta(\cdot,\lambda)\omega(\lambda)\int_a^b \theta(t,\lambda)g(t)dt. \tag{2.17}$$

Here $H(\lambda)$ is the bounded integral operator on $L^2(a,b)$, which is entire in λ, a right inverse for $L_{max} - \lambda$, and for $g \in L^2(a,b)$ which vanishes near b given by

$$(H(\lambda)g)(x) = \theta(x,\lambda)\int_x^b \eta(t,\lambda)g(t)dt + \eta(x,\lambda)\int_a^x \theta(t,\lambda)g(t)dt.$$

The function ω is also called the **Titchmarsh–Weyl coefficient** corresponding to the boundary condition in (2.11). It can be shown that if the kernel $N_{\alpha,\beta}(\lambda,\zeta)$ has κ negative squares then $\omega \in \mathbf{N}_\kappa$. If $\kappa = 0$, an analogue of Theorem 2.5 for ω instead of Ω can now be formulated and proved.

(III) We give four examples.
(i) If $\alpha \equiv 0$ and $\beta \equiv 1$, then the extension A mentioned in Theorem 2.4 is the canonical extension $A_{0,1}$ of L_{min} and $\omega = m$.
(ii) The λ-**linear** case. In this case,

$$\alpha(\lambda) = \alpha_0 + \alpha_1\lambda, \quad \beta(\lambda) = \beta_0 + \beta_1\lambda,$$

where the coefficients α_0, α_1, β_0, and β_1 are real numbers not simultaneously zero. Then

$$N_{\alpha,\beta}(\lambda,\zeta) = \alpha_1\beta_0 - \alpha_0\beta_1 =: \tau, \quad \lambda,\zeta \in \mathbb{C}.$$

The minimal selfadjoint linearization A described in the theorem is a canonical extension of L_{min} if $\tau = 0$; otherwise it is an operator in the space $\mathfrak{K} = L^2(a,b)\oplus\mathbb{C}$ with inner product

$$[\begin{pmatrix} f \\ \tilde{f} \end{pmatrix}, \begin{pmatrix} g \\ \tilde{g} \end{pmatrix}] = \int_a^b \overline{g(t)}f(t)dt + \frac{1}{\tau}\bar{\tilde{g}}\tilde{f}, \quad f,g \in L^2(a,b), \tilde{f},\tilde{g} \in \mathbb{C},$$

which is a Hilbert space if $\tau > 0$ and a π_1–space if $\tau < 0$. If $\tau \neq 0$,

$$A = \left\{\left\{\begin{pmatrix} f \\ -\alpha_1 f(a) - \beta_1 f'(a) \end{pmatrix}, \begin{pmatrix} Lf \\ \alpha_0 f(a) + \beta_0 f'(a) \end{pmatrix}\right\} : f \in L_{max}\right\}.$$

In this case we may take $\gamma = -\beta_1/\tau$, $\delta = \alpha_1/\tau$ and

$$\omega(\lambda) = -\frac{1}{\tau}\frac{\alpha_1 m(\lambda) - \beta_1}{\alpha(\lambda)m(\lambda) - \beta(\lambda)}.$$

(iii) Let c be any point in (a,b), let μ and ν be any real numbers not simultaneously equal to zero, and set

$$\alpha(\lambda) = \nu\varphi'(c,\lambda) + \mu\psi'(c,\lambda), \quad \beta(\lambda) = -(\nu\varphi(c,\lambda) + \mu\psi(c,\lambda)). \tag{2.18}$$

Then the selfadoint realization $A_{\mu,\nu}$ of L_{min} in $L^2(a,b)$ is the linearization of the boundary eigenvalue problem on the smaller interval (c,b):

$$\begin{cases} -f'' + (q-\lambda)f = g \text{ in } L^2(c,b), \\ \alpha(\lambda)f(c) + \beta(\lambda)f'(c) = 0. \end{cases}$$

The kernel $N_{\alpha,\beta}$ associated with the functions in (2.18) is given by

$$N_{\alpha,\beta}(\lambda,\zeta) = \int_a^c (\nu\varphi(t,\lambda) + \mu\psi(t,\lambda))\overline{(\nu\varphi(t,\zeta) + \mu\psi(t,\zeta))}dt$$

and is nonnegative. Without loss of generality we may assume that $\mu^2 + \nu^2 = 1$. Then the solution is given by (2.17) with

$$\omega(\lambda) = -\frac{\nu m(\lambda) + \mu}{\mu m(\lambda) - \nu}$$

and

$$\eta(\cdot,\lambda) = -\mu\varphi(\cdot,\lambda) + \nu\psi(\cdot,\lambda), \quad \theta(\cdot,\lambda) = \nu\varphi(\cdot,\lambda) + \mu\psi(\cdot,\lambda).$$

This example shows that a reduction to the scalar case can also be possible even when the functions $\alpha(\lambda)$ and $\beta(\lambda)$ are not polynomials.

(iv) Consider $L = -D^2 + q$ on an interval $(\tilde{a}, b)$ where $\tilde{a} < a < b$. Assume that L satisfies the condition (1) on $(\tilde{a}, b)$ and that the restriction of L to (a,b) still satisfies conditions (1)–(4) of Subsection 2.2. We use the same notation as before except that we denote the $L^2(a,b)$–solution s and the Titchmarsh–Weyl coefficient m by s_+ and m_+. We assume that L is limit point in $\tilde{a}$, or that it is regular at $\tilde{a}$ in which case we only consider functions f which satisfy a selfadjoint boundary condition at $\tilde{a}$. Then there exists a Nevanlinna function $m_-(\lambda)$ such that the function

$$s_-(\cdot,\lambda) = \psi(\cdot,\lambda) - m_-(\lambda)\varphi(\cdot,\lambda)$$

is the only $L^2(\tilde{a}, a)$–solution of $(L-\lambda)f = 0$, or the only solution of this equation that satisfies the boundary condition at $\tilde{a}$, respectively. There is only one realization of L in $L^2(\tilde{a}, b)$ and it is selfadjoint. We denote it by $\tilde{A}$. Then $\tilde{A}$ is the linearization of the problem

$$\begin{cases} -f'' + (q-\lambda)f = g, \\ f(a) - m_-(\lambda)f'(a) = 0. \end{cases}$$

Its compressed resolvent is given by (2.15) with

$$\Omega(\lambda) = -\begin{pmatrix} -1/m_-(\lambda) & -1 \\ -1 & m_+(\lambda) \end{pmatrix}^{-1}. \tag{2.19}$$

3 Operator representations of holomorphic functions

1. Colligations. A collection of operators of the form

$$V = \begin{pmatrix} T & F \\ G & H \end{pmatrix} : \begin{pmatrix} \mathfrak{K} \\ \mathfrak{F} \end{pmatrix} \to \begin{pmatrix} \mathfrak{K} \\ \mathfrak{G} \end{pmatrix},$$

where $\mathfrak{K}, \mathfrak{F}$ and $\mathfrak{G}$ are Kreĭn spaces and T, F, G and H are bounded operators between the corresponding spaces, is called a **colligation**. Alternative notation: $(\mathfrak{K}, \mathfrak{F}, \mathfrak{G}; V)$ or $(\mathfrak{K}, \mathfrak{F}, \mathfrak{G}; T, F, G, H)$. Sometimes $\mathfrak{K}$ is called the **state space**, $\mathfrak{F}$ the **incoming space**, $\mathfrak{G}$ the **outgoing space**, and T the **main operator**. If V is unitary, then the colligation is called **unitary**.

The **characteristic function** of the colligation V is

$$\Theta(z) = \Theta_V(z) = H + zG(I - zT)^{-1}F.$$

Note that $\Theta_V(\bar{z})^* = \Theta_{V^*}(z)$. The characteristic function is an analytic mapping from a neigbourhood in $\mathbb{D}$ of the origin to $\mathbf{L}(\mathfrak{F}, \mathfrak{G})$. Such a function is called a **generalized Schur function**, and the class of such functions is denoted by $\mathbf{S}(\mathfrak{F}, \mathfrak{G})$. For $\Theta \in \mathbf{S}(\mathfrak{F}, \mathfrak{G})$, we denote by $\mathcal{D}_\Theta$ the domain of holomorphy of Θ and by $\mathcal{D}(\Theta)$ the component of $\mathcal{D}_\Theta \cap \mathbb{D}$ which contains the origin. We set $\tilde{\Theta}(z) = \Theta(\bar{z})^*$. Then $\tilde{\Theta} \in \mathbf{S}(\mathfrak{G}, \mathfrak{F})$ and $\mathcal{D}(\tilde{\Theta}) = \mathcal{D}(\Theta)^*$.

The **realization problem** asks for the representation of $\Theta \in \mathbf{S}(\mathfrak{F}, \mathfrak{G})$ as $\Theta(z) = \Theta_V(z)$ in a small neigbourhood of the origin in $\mathbb{D}$ for some colligation V which is unitary and **closely connected**, which means that

$$\mathfrak{K} = (\text{span } \{\text{ran } T^m F, \text{ran } T^{*n} G^* : m, n = 0, 1, \dots\})^c.$$

The closely connectedness condition implies **weak uniqueness**: If $\Theta_{V_1}(z) = \Theta_{V_2}(z)$ on some open disc around the origin and V_1 and V_2 are unitary and closely connected, then $H_1 = H_2$ and there exists a weak isomorphism W from $\mathfrak{K}_1$ to $\mathfrak{K}_2$ such that $T_1 \text{dom } W \subseteq \text{dom } W$, ran $F_1 \subseteq \text{dom } W$, and on dom $W \oplus \mathfrak{F}$

$$\begin{pmatrix} W & 0 \\ 0 & I \end{pmatrix} \begin{pmatrix} T_1 & F_1 \\ G_1 & H_1 \end{pmatrix} = \begin{pmatrix} T_2 & F_2 \\ G_2 & H_2 \end{pmatrix} \begin{pmatrix} W & 0 \\ 0 & I \end{pmatrix}.$$

Let $\mathcal{O}$ be an open set in $\mathbb{C}$ and let $(\mathfrak{K}, [\cdot, \cdot])$ be a Kreĭn space. A function $K : \mathcal{O} \times \mathcal{O} \to \mathbf{L}(\mathfrak{K})$ is called a **$\mathfrak{K}$–kernel** on $\mathcal{O}$. It is **hermitian** if $K(z, w) = K(w, z)^+$, and it is said to have κ **positive/negative squares** if K is hermitian and for every choice of the natural number $n \in \mathbb{N}$, the elements $k_1, k_2, \dots, k_n \in \mathfrak{K}$ and the points $z_1, z_2, \dots, z_n \in \mathcal{O}$, the hermitian $n \times n$ matrices

$$([K(z_i, z_j)k_i, k_j])_{i,j=1}^n$$

have at most κ positive/negative eigenvalues and at least one of them has exactly κ positive/negative eigenvalues counting multiplicities. A $\mathfrak{K}$–kernel is nonnegative/nonpositive if it has zero negative/positive squares. Evidently, this definition is in accordance with and generalizes the corresponding definitions in Subsections 1.1 and 2.3.

Various kernels can be associated with $\Theta \in \mathbf{S}(\mathfrak{F}, \mathfrak{G})$. For example, the $\mathfrak{G}$–kernel

$$\Delta_\Theta(z, w) =: \frac{I - \Theta(z)\Theta(w)^*}{1 - z\bar{w}}, \quad z \in \mathcal{D}(\Theta),$$

the $\mathfrak{F}$–kernel $\Delta_{\tilde{\Theta}}(z,w)$, and the operator matrix $(\mathfrak{G} \oplus \mathfrak{F})$–kernel

$$D_\Theta(z,w) = \begin{pmatrix} \Delta_\Theta(z,w) & \frac{\Theta(z)-\Theta(\bar{w})}{z-\bar{w}} \\ \frac{\tilde{\Theta}(z)-\tilde{\Theta}(\bar{w})}{z-\bar{w}} & \Delta_{\tilde{\Theta}}(z,w) \end{pmatrix}, \quad z,w \in \mathcal{D}(\Theta) \cap \mathcal{D}(\Theta)^*.$$

It can be shown that the matrix kernel has a finite number of positive/negative squares if and only if both kernels on its main diagonal have a finite number of positive/negative squares and if these conditions hold then the number of positive/negative squares of all three kernels are the same; see Alpay et al. [1992].

In the Pontryagin space case the following theorem is valid. We denote by $\text{ind}_- \mathfrak{F}$ the index of the Pontryagin space $\mathfrak{F}$.

Theorem 3.1 *Let $\Theta \in \mathbf{S}(\mathfrak{F}, \mathfrak{G})$, where $\mathfrak{F}$ and $\mathfrak{G}$ are Pontryagin spaces, and let κ be a nonnegative integer. Then the following statements are equivalent:*
(i) $D_\Theta(z,w)$ has κ negative squares.
(ii) $\Delta_\Theta(z,w)$ has κ negative squares and $\text{ind}_- \mathfrak{F} = \text{ind}_- \mathfrak{G}$.
(iii) $\Delta_{\tilde{\Theta}}(z,w)$ has κ negative squares and $\text{ind}_- \mathfrak{F} = \text{ind}_- \mathfrak{G}$.
When these conditions hold, Θ can be extended to a meromorphic function on $\mathbb{D}$ and the number of negative squares of each of the three kernels associated with the extended function is equal to κ.

A solution to the realization problem can be formulated as follows. For a proof we refer to Dijksma et al. [1987a]. The theorem in less detailed form was first stated and proved by T.Ya. Azizov; see Azizov [1982], Azizov [1984].

Theorem 3.2 *Let Θ belong to $\mathbf{S}(\mathfrak{F}, \mathfrak{G})$ and let $\mathcal{D}$ be a Jordan region with $0 \in \mathcal{D}$ and $\mathcal{D}^c \subseteq \mathcal{D}(\Theta)$. Then Θ coincides on $\mathcal{D}$ with the characteristic function of a unitary colligation $(\mathfrak{K}, \mathfrak{F}, \mathfrak{G}; V)$. This colligation can be chosen closely connected. Then it is unique up to a weak isomorphism, and if $\mathfrak{K} = \mathfrak{K}_+ \oplus \mathfrak{K}_-$ is any fundamental decomposition of the state space,* $\dim \mathfrak{K}_+$ *and* $\dim \mathfrak{K}_-$ *are equal to the number of positive and the number of negative squares of the kernel $D_\Theta(z,w)$, respectively. If one of these numbers is finite then the colligation is unique up to an isomorphism.*

As a first application of Theorem 3.2 we describe how a closed densely defined symmetric operator S in a Hilbert space $\mathfrak{H}$ can be characterized by a holomorphic operator function. We assume that S is simple. We choose a point $\mu \in \mathbb{C}^+$ and define the function $z(\lambda) = z_\mu(\lambda)$ by

$$z(\lambda) = \frac{\lambda - \mu}{\lambda - \bar{\mu}}.$$

Observe that it maps the real axis and the upper/lower half plane onto the unit circle and its interior/exterior, respectively. The **Cayley transform** $C_\tau(T)$ of an operator T, $\tau \in \mathbb{C}$, is formally defined by $C_\tau(T) = (T - \bar{\tau})(T - \tau)^{-1}$, and in graph notation by

$$C_\tau(T) = \{\{y - \tau x, y - \bar{\tau} x\} : \{x, y\} \in T\}.$$

We consider the Cayley transform $V = C_\mu(S)$ of S. Then V is an isometry on $\mathfrak{H}$ with $\text{dom } V = \text{ran } (S - \mu)$, $\text{ran } V = \text{ran } (S - \bar{\mu})$. We extend V to a partial isometry $\tilde{V}$ on $\mathfrak{H}$ by setting $\tilde{V}x = Vx$ if $x \in \text{dom } V$ and $\tilde{V}x = 0$ if $x \in \mathfrak{H} \ominus \text{dom } V$. We set $\mathfrak{F} = \mathfrak{H} \ominus \text{ran } V = \ker (S^* - \mu)$ and $\mathfrak{G} = \mathfrak{H} \ominus \text{dom } V = \ker (S^* - \bar{\mu})$. The colligation $(\mathfrak{H}, \mathfrak{F}, \mathfrak{G}; \tilde{V}, I_{\mathfrak{F}}, P_{\mathfrak{G}}, 0)$ is unitary and closely connected. We denote its characteristic function by $X(z)$:

$$X(z) = zP_{\mathfrak{G}}(I - z\tilde{V})^{-1}|_{\mathfrak{F}}.$$

Then $X \in \mathbf{S}(\mathfrak{F}, \mathfrak{G})$, $X(0) = 0$ and the kernel $\Delta_X(z, w)$ is nonnegative. We call the function $K(\lambda) = X(z(\lambda)), \lambda \in \mathbb{C}^+$, the **characteristic function** of S. It can be shown that, since S is simple, Theorem 3.2 implies that $K(\lambda)$ determines S uniquely up to an isomorphism. It can more conveniently be described as:

$$K(\lambda) = z(\lambda)P|_{\ker\ (S^* - \mu)},$$

where P stands for the projection in $\mathfrak{H}$ onto $\ker\ (S^* - \bar{\mu})$ parallel to $\mathrm{ran}\ (S - \lambda)$. Note that for $\lambda \in \mathbb{C}^+$, the space $\mathfrak{H}$ can be written as the direct sum

$$\mathfrak{H} = \ker\ (S^* - \bar{\mu}) + \mathrm{ran}\ (S - \lambda).$$

We calculate the characteristic function of the differential operator L_{min} in $L^2(a, b)$.

Proposition 3.3 *Assume that the Sturm–Liouville expression L satisfies the conditions* (1)–(4) *of Subsection 2.2. Then the characteristic function of the minimal realization L_{min} of L in $L^2(a, b)$ is given by*

$$(K(\lambda)s(\cdot, \mu))(x) = \frac{m(\lambda) - m(\mu)}{m(\lambda) - m(\bar{\mu})}s(x, \bar{\mu}), \quad \lambda \in \mathbb{C}^+,$$

where $s(\cdot, \lambda)$ is the $L^2(a, b)$–function in (2.6).

Indeed, we have that $K(\lambda)s(\cdot, \mu) = k(\lambda)z(\lambda)s(\cdot, \bar{\mu})$, where $k(\lambda) \in \mathbb{C}$ is such that

$$s(\cdot, \mu) - k(\lambda)s(\cdot, \bar{\mu}) \perp \ker\ (L_{max} - \bar{\lambda}) = \mathrm{span}\ \{s(\cdot, \bar{\lambda})\},$$

that is,

$$k(\lambda) = \frac{(s(\cdot, \mu), s(\cdot, \bar{\lambda}))}{(s(\cdot, \bar{\mu}), s(\cdot, \bar{\lambda}))} = \frac{1}{z(\lambda)}\frac{m(\lambda) - m(\mu)}{m(\lambda) - m(\bar{\mu})},$$

from which the desired formula follows.

2. Linearizations and Štraus extensions. Let S be a closed densely defined symmetric operator in a Hilbert space $\mathfrak{H}$. It is possible to describe a selfadjoint extension of S in a Kreĭn space in terms of certain objects in the smaller space $\mathfrak{H}$. These objects are the Štraus extension and the generalized resolvent of S. A family of operators $T(\lambda)$ is called a **Štraus extension** of S if

$$T(\lambda) = \{\{P_{\mathfrak{H}}f, P_{\mathfrak{H}}g\} : \{f, g\} \in A,\ g - \lambda f \in \mathfrak{H}\}, \quad \lambda \in \mathbb{C},$$

for some selfadjoint extension A of S; any such A is called a **linearization** of $T(\lambda)$.

Note that the generalized resolvent of S corresponding to A:

$$R(\lambda) = P_{\mathfrak{H}}(A - \lambda)^{-1}|_{\mathfrak{H}}, \quad \lambda \in \rho(A),$$

also defines a family of operators with parameter λ. The generalized resolvent $R(\lambda)$ and the Štraus extension $T(\lambda)$ corresponding to the same selfadjoint extension A of S are related as follows:

$$R(\lambda) = (T(\lambda) - \lambda)^{-1}.$$

In this section we describe the Štraus extensions $T(\lambda)$ of S. For $\lambda \in \rho(A)$ the following relations hold:

$$S \subseteq T(\lambda) \subseteq S^*, \quad T(\lambda) = T(\bar{\lambda})^*.$$

In connection with differential operators the requirement $f \in \text{dom } T(\lambda)$ in (2.13) means that f must satisfy certain boundary conditions which may depend on the eigenvalue parameter λ. To make this more transparent in this abstract setting we introduce (abstract) boundary operators.

We simplify by assuming that S has equal and finite defect numbers: $d = d^+ = d^- < \infty$. This is the case for the differential operators we consider here. We have dim (dom S^*/dom S) $= 2d$. A **boundary operator** is any surjective linear operator $\mathsf{b} : \text{dom } S^* \to \mathbb{C}^{2d}$ with the property that $\mathsf{b}\{f\} = 0$ for all $f \in \text{dom } S$. So b is constant on the cosets of dom S^*/dom S and induces a bijection from dom S^*/dom S onto $\mathbb{C}^{2d}$.

Let Q be the uniquely determined $2d \times 2d$ matrix such that for all $f, h \in$ dom S^*,

$$(1/i)\,((S^*f, h) - (f, S^*h)) = \mathsf{b}\{h\}^*\mathsf{Q}\mathsf{b}\{f\}. \tag{3.1}$$

In the theory of differential operators, this formula is the Lagrange formula or Green's identity; compare (2.2). The matrix Q is invertible and hermitian, and the numbers of positive and negative eigenvalues of Q are equal to d, counting multiplicities.

For $\lambda \in \mathbb{C}\backslash\mathbb{R}$, let $s_1(\lambda), s_2(\lambda), \ldots, s_d(\lambda)$ be a basis for ker $(S^* - \lambda)$, and denote by $S(\lambda)$ the $2d \times d$ matrix

$$S(\lambda) = \begin{pmatrix} \mathsf{b}\{s_1(\lambda)\} & \mathsf{b}\{s_2(\lambda)\} & \cdots & \mathsf{b}\{s_d(\lambda)\} \end{pmatrix}.$$

The basis elements can be chosen locally holomorphic in $\lambda \in \mathbb{C} \setminus \mathbb{R}$, and then the matrix $S(\lambda)$ is locally holomorphic also.

Theorem 3.4 (i) *Let A in $\mathfrak{K}$ be a selfadjoint extension of S in $\mathfrak{H}$ with $\mu \in \rho(A) \cap \mathbb{C}^+$ and let $\mathcal{O}_\mu$ be an open subset of $\mathbb{C} \setminus \mathbb{R}$ such that $\mu \in \mathcal{O}_\mu = \mathcal{O}_\mu^*$. Then for $\lambda \in \mathcal{O}_\mu$, the Štraus extension $T(\lambda)$ of S corresponding to A admits the representation:*

$$T(\lambda) = \{\{f, g\} \in S^* : U(\lambda)\mathsf{b}\{f\} = 0\},$$

where $U(\lambda)$ is a locally holomorphic $d \times 2d$ matrix on $\mathcal{O}_\mu \cap \mathbb{C} \setminus \mathbb{R}$ such that for $\lambda \in \mathcal{O}_\mu \cap \mathbb{C}^+$, rank $U(\lambda) = d$, rank $U(\bar{\lambda}) = d$, *the $d \times d$ matrices $U(\lambda)S(\mu)$ and $U(\bar{\lambda})S(\bar{\mu})$ are invertible, and*

$$U(\lambda)\mathsf{Q}^{-1}U(\bar{\lambda})^* = 0.$$

(ii) *Conversely, let $U(\lambda)$ with these properties be given on some open connected subset $\mathcal{O} = \mathcal{O}^*$ of $\mathbb{C}\backslash\mathbb{R}$ with $\mu \in \mathcal{O} \cap \mathbb{C}^+$ and let $\mathcal{D}$ be a Jordan region with $0 \in \mathcal{D}$ and $\mathcal{D}^c \subseteq \mathbb{D}$ such that $z^{-1}(\mathcal{D}^c) \subseteq \mathcal{O}$. Then $T(\lambda)$ defined above for $\lambda \in z^{-1}(\mathcal{D}) \cup z^{-1}(\mathcal{D})^*$ is a Štraus extension of S whose linearization is a selfadjoint extension A_{lin} in a*

Kreĭn space $\mathfrak{K}$ *with* $\mu \in \rho(A_{lin})$.
(iii) *The extension* A_{lin} *in* (ii) *can be chosen minimal. Then it is unique up to a weak isomorphism and if* $\mathfrak{K} \ominus \mathfrak{H} = \mathfrak{K}'_+ \oplus \mathfrak{K}_-$ *is any fundamental decomposition of the exit space,* $\dim \mathfrak{K}'_+$ *and* $\dim \mathfrak{K}_-$ *are equal to the number of positive and the number of negative squares of the* $d \times d$ *matrix kernel*

$$k(\lambda, \zeta) = \frac{1}{i}\frac{U(\bar{\zeta})\mathsf{Q}^{-1}U(\bar{\lambda})^*}{\lambda - \bar{\zeta}}, \quad \lambda, \zeta \in \mathcal{D} \cup \mathcal{D}^*.$$

If one of these numbers is finite then the extension is unique up to an isomorphism.
(iv) *If the negative index of* $k(\lambda, \zeta)$ *is zero or finite, then in* (i) *and* (ii) *the condition on the invertibility of the matrices* $U(\lambda)S(\mu)$ *and* $U(\bar{\lambda})S(\bar{\mu})$ *is always satisfied.*
(v) *The Štraus extension* $T(\lambda)$ *coincides with a canonical selfadjoint extension of* S *if and only if* $U(\lambda)$ *can be chosen independent of* λ, *that is,*

$$T(\lambda) = \{\{f, g\} \in S^* : U\mathsf{b}\{f\} = 0\},$$

where U *is a constant* $d \times 2d$ *matrix with* $\operatorname{rank} U = d$ *and* $U\mathsf{Q}^{-1}U^* = 0$.

The proof of the theorem is based on the theory of colligations and the Cayley transforms. For S and A in part (i) of the theorem we have that $C_\mu(S)$ is an isometric operator in $\mathfrak{H}$ with domain $= \operatorname{ran}\ (S - \mu)$ and range $= \operatorname{ran}\ (S - \bar{\mu})$, and $C_\mu(A) \in \mathbf{L}(\mathfrak{K})$ is a unitary extension of $C_\mu(S)$. It is easy to see that

$$C_\mu(A) = C_\mu(S) \oplus V_\mu, \quad C_{\bar{\mu}}(A) = C_{\bar{\mu}}(S) \oplus V_\mu^*, \tag{3.2}$$

where $V_\mu = C_\mu(A)\,|_{\mathfrak{K} \ominus \operatorname{ran}\ (S - \mu)}$ is a unitary colligation of the form

$$V_\mu = (\mathcal{K} \ominus \mathcal{H}, \ker\ (S^* - \bar{\mu}), \ker\ (S^* - \mu); T, F, G, H). \tag{3.3}$$

Denote by Θ its characteristic function; then $\Theta \in \mathbf{S}(\ker\ (S^* - \bar{\mu}), \ker\ (S^* - \mu))$.

Applying the inverse of the Cayley transform to both sides of the first equality in (3.2), we get in graph notation

$$\begin{aligned} A \;=\; & \left\{\left\{\begin{pmatrix} 0 \\ f \end{pmatrix}, \begin{pmatrix} 0 \\ g \end{pmatrix}\right\} : \{f, g\} \in S\right\} \\ & + \left\{\left\{\begin{pmatrix} k \\ \psi \end{pmatrix}, \begin{pmatrix} Tk + F\psi \\ Gk + H\psi \end{pmatrix}\right\} : k \in \mathcal{K} \ominus \mathcal{H},\ \psi \in \ker\ (S^* - \bar{\mu})\right\}, \end{aligned}$$

direct sum in $\begin{pmatrix} \mathcal{K} \ominus \mathcal{H} \\ \mathcal{H} \end{pmatrix} \oplus \begin{pmatrix} \mathcal{K} \ominus \mathcal{H} \\ \mathcal{H} \end{pmatrix}$. For $\lambda \in \mathbb{C}^+$, the corresponding Štraus extension is given by

$$T(\lambda) = S + \{\{\Theta(z(\lambda))\psi - \psi, \mu\Theta(z(\lambda))\psi - \bar{\mu}\psi\} : \psi \in \ker\ (S^* - \bar{\mu})\},$$

direct sum in $\mathcal{H} \oplus \mathcal{H}$. Taking the adjoints of the operators on both sides we find that for λ in a neighborhood of $\bar{\mu}$,

$$T(\lambda) = \{\{f, g\} \in S^* : U(\lambda)\,\mathsf{b}\{f\} = 0\},$$

where

$$U(\lambda) = \left(\mathsf{b}\{\Theta(z(\bar{\lambda}))s_1(\bar{\mu}) - s_1(\bar{\mu})\} \cdots \mathsf{b}\{\Theta(z(\bar{\lambda}))s_d(\bar{\mu}) - s_d(\bar{\mu})\}\right)^* \mathsf{Q}. \tag{3.4}$$

From similar calculations starting with the second equality in (3.2) we obtain that for $\lambda \in \mathbb{C}^-$,

$$T(\lambda) = S + \{\{\Theta(z(\bar{\lambda}))^*\varphi - \varphi, \bar{\mu}\Theta(z(\bar{\lambda}))^*\varphi - \mu\varphi\} : \varphi \in \ker\ (S^* - \mu)\},$$

direct sum in $\mathcal{H} \oplus \mathcal{H}$. Hence for λ in a neigborhood of μ,

$$T(\lambda) = \{\{f, g\} \in S^* : U(\lambda)\mathsf{b}\{f\} = 0\}$$

with

$$U(\lambda) = (\mathsf{b}\{\Theta(z(\lambda))^* s_1(\mu) - s_1(\mu)\} \cdots \mathsf{b}\{\Theta(z(\lambda))^* s_d(\mu) - s_d(\mu)\})^* \,\mathsf{Q}. \quad (3.5)$$

From the defining relation (3.1) of Q the indicated properties of $U(\lambda)$ can be verified.

The converse in (ii) can be proved by first showing that $U(\lambda)$, apart from an invertible $d \times d$ matrix function as a left factor, can be written in the form (3.4) and (3.5) and then by applying the realization theorem of the previous section to the function $\Theta(z)$.

Part (iii) can be proved by relating the number of positive/negative squares of the kernel $k(\lambda, \zeta)$ to those of the kernel $D_\Theta(z(\lambda), z(\zeta))$. The observation in part (iv) of the theorem is related to the fact that not all selfadjoint extensions A of S in a Kreĭn space have a nonempty resolvent set. The invertibility of the indicated matrices assures that the point μ belongs to $\rho(A)$. When the exit space is a Hilbert or Pontryagin space then all selfadjoint operators do have a nonempty resolvent set and hence the invertibility condition is always satisfied.

The above formulas show that the essentially one–to–one correspondence between the Štraus extensions of S and the minimal selfadjoint extensions A of S stems from the one–to–one correspondence of each of these families of extensions with the characteristic functions $\Theta \in \mathbf{S}(\ker\ (S^* - \bar{\mu}), \ker\ (S^* - \mu))$ of the colligations (3.3).

We illustrate the outline of the proof of Theorem 3.4 (ii) given above by considering the Štraus extension (2.12) defined by the boundary eigenvalue problem (2.11) satisfying conditions (1)–(8). In that case we have $S = L_{min}$ in $\mathfrak{H} = L^2(a, b)$, $d = 1$, and $s_1(\lambda) = s(\cdot, \lambda)$ given by (2.6). The boundary operator is necessarily of the form

$$\mathsf{b}\{f\} = \begin{pmatrix} f(a) & f'(a) \end{pmatrix}^T,$$

where ^{T} stands for transpose, and (3.1) holds with $\mathsf{Q} = \mathsf{J}$. The 2×1 matrix function $S(\lambda) = \mathsf{b}\{s_1(\lambda)\}$ can now be calculated:

$$S(\lambda) = \begin{pmatrix} m(\lambda) & -1 \end{pmatrix}^T.$$

We set $U(\lambda) = \begin{pmatrix} \alpha(\lambda) & \beta(\lambda) \end{pmatrix}$, so that

$$U(\lambda)S(\lambda) = \alpha(\lambda)m(\lambda) - \beta(\lambda), \quad U(\lambda)\mathsf{Q}^{-1}U(\bar{\lambda})^* = -i\left(\alpha(\lambda)\overline{\beta(\bar{\lambda})} - \overline{\alpha(\bar{\lambda})}\beta(\lambda)\right).$$

Then $T(\lambda)$ given by (2.12) is of the form described in Theorem 3.4. We find that

$$U(\lambda) = p(\lambda)\,(\mathsf{b}\{\Theta(z(\lambda))^* s(\cdot, \mu) - s(\cdot, \mu)\})^* \,\mathsf{J},$$

for some factor $p(\lambda)$ if

$$(\Theta(z(\lambda))s(\cdot,\bar{\mu}))(x) = \frac{\alpha(\lambda)m(\bar{\mu})-\beta(\lambda)}{\alpha(\lambda)m(\mu)-\beta(\lambda)}s(x,\mu).$$

The colligation of which Θ is the characteristic function leads to a linearization A of $T(\lambda)$. These observations also provide the essential ingredients of the proof of Theorem 2.4. Note that if $\beta(\lambda)\equiv 0$, then the kernels $k(\lambda,\zeta)$ and $N_{\alpha,\beta}(\lambda,\zeta)$ are everywhere zero also; otherwise,

$$k(\lambda,\zeta) = \overline{q(\zeta)}N_{\alpha,\beta}(\lambda,\zeta)q(\lambda),$$

where $q(\lambda) = \overline{\beta(\bar{\lambda})}/\beta(\lambda)$. Thus, the number of positive/negative squares of the two kernels coincide. For later reference we record part of the foregoing remarks concerning the boundary eigenvalue problem (2.11) satisfying conditions (1)–(8) of Subsections 2.2 and 2.3.

Proposition 3.5 *Assume that $N_{\alpha,\beta}(\lambda,\zeta)$ has κ negative squares. Then the linearization of the Štraus extension* (2.12) *of L_{min} in $L^2(a,b)$ is a selfadjoint operator in a π_κ- space and is uniquely determined up to an isomorphism by the function*

$$(\Theta(z(\lambda))s(\cdot,\bar{\mu}))(x) = \frac{\alpha(\lambda)m(\bar{\mu})-\beta(\lambda)}{\alpha(\lambda)m(\mu)-\beta(\lambda)}s(x,\mu),$$

where $s(\cdot,\lambda)$ is the $L^2(a,b)$-function in (2.6). *The kernels $D_\Theta(z,w)$ and $\Delta_\Theta(z,w)$ have κ negative squares.*

The Propositions 3.3 and 3.5 show that a simple symmetric operator S and a minimal selfadjoint extension of S in a π_κ-space are uniquely characterized by two functions. This will be used in the next subsection.

We conclude this subsection with one more example related to Theorem 3.4. Assume that the Sturm-Liouville expression L on the interval (a,b) considered in Subsection 2.1 is regular in a as well as in b. Then $S = L_{min}$ is a closed symmetric operator in $\mathfrak{H} = L^2(a,b)$ with defect numbers $d = d^+ = d^- = 2$. All selfadjoint realizations of L in $L^2(a,b)$ are of the form

$$A_{C,D} = \left\{\{f,Lf\} : C\begin{pmatrix} f(a) \\ f'(a)\end{pmatrix} + D\begin{pmatrix} f(b) \\ f'(b)\end{pmatrix} = 0\right\},$$

where C and D are 2×2 matrices with

$$\text{rank }(C\ D) = 2, \quad CJC^* - DJD^* = 0.$$

It can be shown that all Štraus extensions of L whose linearizations act in π_κ-spaces are of the form

$$T(\lambda) = \left\{\{f,Lf\} : C(\lambda)\begin{pmatrix} f(a) \\ f'(a)\end{pmatrix} + D(\lambda)\begin{pmatrix} f(b) \\ f'(b)\end{pmatrix} = 0\right\},$$

where C and D are 2×2 locally meromorphic matrix functions on $\mathbb{C}\setminus\mathbb{R}$ with

$$\text{rank }(C(\lambda)\ D(\lambda)) = 2, \quad C(\lambda)JC(\bar{\lambda})^* - D(\lambda)JD(\bar{\lambda})^* = 0$$

and the kernel

$$\frac{C(\bar{\zeta})JC(\bar{\lambda})^* - D(\bar{\zeta})JD(\bar{\lambda})^*}{\lambda-\bar{\zeta}}$$

has κ negative squares. If the kernel condition is replaced by the condition that suitably chosen matrices are invertible then these formulas describe all Štraus extensions whose linearizations have a nonempty resolvent set. The boundary operator of L_{min} is given by

$$\mathsf{b}\{f\} = \begin{pmatrix} f(a) & f'(a) & f(b) & f'(b) \end{pmatrix}^T,$$

formula (3.1) holds with Q = diagonal $(\mathsf{J}, -\mathsf{J})$, and the characterization of the Štraus extensions can easily be obtained by writing $U(\lambda) = (\,C(\lambda)\ D(\lambda)\,)$.

3. Models for the Titchmarsh–Weyl coefficients. See, for example, Kreĭn and Langer [1973] and Dijksma et al. [1987], Dijksma et al. [1988], Dijksma et al. [unpublished]. A generalized Nevanlinna function $Q \in \mathbf{N}_\kappa^{k\times k}$ with the property that $\lim_{y\to\infty} \frac{1}{y} Q(iy) = 0$ admits a represention of the form

$$Q(\lambda) = Q(\bar{\mu}) + (\lambda - \bar{\mu})\Gamma^*(I + (\lambda - \mu)(A - \lambda)^{-1})\Gamma, \quad \lambda \in \mathcal{D}_Q, \tag{3.6}$$

where A is a selfadjoint operator in some Pontryagin space Π with index κ and $\mu \in \rho(A) \cap \mathbb{C}^+$, and Γ is a linear mapping from $\mathbb{C}^k$ to Π. It follows that $\emptyset \neq \rho(A) \subseteq \mathcal{D}_Q$, the domain of holomorphy of Q. The operator A and the space Π in which it acts can be chosen Γ–**minimal** which means that

$$\Pi = \overline{\operatorname{span}}\,\{(I + (\lambda - \mu)(A - \lambda)^{-1})\Gamma f : f \in \mathbb{C}^k, \lambda \in \rho(A)\},$$

and then the representation is unique up to an isomorphism. Below we outline the construction of this model for Q using the ε–method from Kreĭn and Langer [1973]. The construction also gives a symmetric operator S of which A is a canonical extension. We apply the results to the Titchmarsh–Weyl coeffients m, Ω and ω considered in Section 2.

Let $Q \in \mathbf{N}_\kappa^{k\times k}$ be given and assume that $\lim_{y\to\infty} \frac{1}{y} Q(iy) = 0$. Consider the linear space $\mathfrak{L}$ of all finite formal sums $\sum_\lambda \varepsilon_\lambda f_\lambda$, where $\lambda \in \mathcal{D}_Q$, $f_\lambda \in \mathbb{C}^k$ and ε_λ is a symbol associated with each $\lambda \in \mathcal{D}_Q$, and provide $\mathfrak{L}$ with the (possibly degenerate) indefinite inner product

$$[\sum_\lambda \varepsilon_\lambda f_\lambda, \sum_\zeta \varepsilon_\zeta g_\zeta] = (\sum_{\lambda,\zeta} N_Q(\lambda,\zeta) f_\lambda, g_\zeta)_{\mathbb{C}^k}.$$

Define in $\mathfrak{L}$ the linear operators A_0 and S_0 by (in graph notation)

$$\begin{aligned} A_0 &= \{\{\textstyle\sum_\lambda \varepsilon_\lambda f_\lambda, \sum_\lambda \varepsilon_\lambda \lambda f_\lambda\} : \sum_\lambda f_\lambda = 0\}, \\ S_0 &= \{\{\textstyle\sum_\lambda \varepsilon_\lambda f_\lambda, \sum_\lambda \varepsilon_\lambda \lambda f_\lambda\} : \sum_\lambda f_\lambda = 0, \sum_\lambda Q(\lambda) f_\lambda = 0\} \\ &= \{\{\textstyle\sum_\lambda \varepsilon_\lambda f_\lambda, \sum_\lambda \varepsilon_\lambda \lambda f_\lambda\} \in A_0 : \sum_\lambda \varepsilon_\lambda (\lambda - \zeta) f_\lambda \,[\perp]\, \varepsilon_{\bar\zeta} g \text{ for all } g \in \mathbb{C}^k\}, \end{aligned}$$

where ζ is any point in $\mathcal{D}_Q$. Finally define the mapping $\Gamma_0 : \mathbb{C}^k \to \mathfrak{L}$ by $\Gamma f = \varepsilon_\mu f$, $f \in \mathbb{C}^k$, where μ is a fixed point in $\mathcal{D}_Q \cap \mathbb{C}^+$. Then one can show that A_0 and S_0 are symmetric and the relations

$$\begin{aligned} (A_0 - \lambda)^{-1}\varepsilon_\zeta &= (\lambda - \zeta)^{-1}(\varepsilon_\lambda - \varepsilon_\zeta) = (A_0 - \zeta)^{-1}\varepsilon_\lambda, \\ Q(\lambda) &= Q(\bar\mu) + (\lambda - \bar\mu)\Gamma_0^*(I + (\lambda - \mu)(A_0 - \lambda)^{-1})\Gamma_0 \end{aligned}$$

hold.

The completion of the quotient space $\mathfrak{L}/(\mathfrak{L}\cap\mathfrak{L}^{[\perp]})$ with respect to the inner product yields the π_κ-space Π, and S_0, A_0 and Γ_0 give rise to a closed densely defined symmetric operator S, a canonical selfadjoint extension A of S with $\mu \in \rho(A)$, and a linear mapping $\Gamma : \mathbb{C}^k \to \Pi$ such that (3.6) is valid. In this model A is Γ-minimal, so that $\mathfrak{L}$, the inner product $[\cdot,\cdot]$, Π, A and S are unique up to an isomorphism. We denote the dependence of these objects on Q by writing, for example, $\Pi(Q)$, $A(Q)$ etc. The minimality of the model implies that $S(Q)$ is simple.

We follow the same procedure and construct a Pontryagin space $\Pi(\alpha,\beta)$ associated with the functions α and β satisfying the conditions (4)–(8) in Subsection 2.2 and the kernel $N_{\alpha,\beta}(\lambda,\zeta)$ defined there. We assume that the kernel has κ negative squares. We start from the linear space $\mathfrak{L}(\alpha,\beta)$ of all finite sums $\sum_\lambda \delta_\lambda f_\lambda$, where $f_\lambda \in \mathbb{C}$, $\lambda \in \mathcal{D}_{\alpha,\beta}$, equipped with the inner product

$$[\sum_\lambda \delta_\lambda f_\lambda, \sum_\zeta \delta_\zeta g_\zeta] = \sum_{\lambda,\zeta} \bar{g}_\zeta N_{\alpha,\beta}(\lambda,\zeta) f_\lambda,$$

and then we define the space $\Pi(\alpha,\beta)$ as the completion of the factor space $\mathfrak{L}(\alpha,\beta)/\mathfrak{L}(\alpha,\beta)\cap\mathfrak{L}(\alpha,\beta)^{[\perp]}$.

We apply the models to the Titchmarsh–Weyl coefficients considered in Section 2. Recall that $S = L_{min}$, the minimal realization in $\mathfrak{H} = L^2(a,b)$ of the Sturm–Liouville differential expression L which satisfies the conditions (1)–(4) in Subsection 2.2. The Titchmarsh–Weyl coefficient m appearing there belongs to $\mathbf{N}_0$ and $\lim_{y\to\infty}\frac{1}{y}m(iy) = 0$. Thus the space $\Pi(m)$ is a Hilbert space. We denote by A_{lin} in $\mathfrak{K}$ the linearization of the Štraus extension (2.12) associated with the boundary eigenvalue problem (2.11) in which the coefficients α and β satisfy the conditions (4)–(8) in Subsection 2.2. We assume that the kernel $N_{\alpha,\beta}(\lambda,\zeta)$ has κ negative squares. Then the Titchmarsh–Weyl coefficient

$$\Omega(\lambda) = (\beta(\lambda) - m(\lambda)\alpha(\lambda))^{-1}\begin{pmatrix} m(\lambda)\beta(\lambda) & \beta(\lambda) \\ \beta(\lambda) & \alpha(\lambda)\end{pmatrix}$$

belongs to $\mathbf{N}_\kappa^{2\times 2}$ and satisfies $\lim_{y\to\infty}\frac{1}{y}\Omega(iy) = 0$. Thus the spaces $\mathfrak{K}$ and $\Pi(\Omega)$ are Pontryagin spaces with negative index κ.

Theorem 3.6 *The space $\Pi(\Omega)$ can be written as the orthogonal direct sum $\Pi(\Omega) = \Pi(m)\oplus\Pi(\alpha,\beta)$ and there exists an isomorphism $W : \mathfrak{K} \to \Pi(\Omega)$ such that $W\mathfrak{H} = \Pi(m)$ and*

$$S(m)W = WS, \quad A(\Omega)W = WA_{lin}.$$

The orthogonal decomposition of the space $\Pi(\Omega)$ follows from the formula

$$-\Omega(\lambda)^{-1} = \begin{pmatrix} \alpha(\lambda)/\beta(\lambda) & -1 \\ -1 & m(\lambda)\end{pmatrix}.$$

One shows in the model for m that ker $(S(m)^* - \lambda) =$ span $\{\varepsilon_\lambda\}$ and that the characteristic function of $S(m)$ is given by

$$K(\lambda)\varepsilon_\mu = \frac{m(\lambda) - m(\mu)}{m(\lambda) - m(\bar{\mu})}\varepsilon_{\bar{\mu}}.$$

Then one proves that $A(\Omega)$ is a minimal extension of $S(m)$ determined by the function

$$\Theta(z(\lambda))\varepsilon_{\bar{\mu}} = \frac{\alpha(\lambda)m(\bar{\mu}) - \beta(\lambda)}{\alpha(\lambda)m(\mu) - \beta(\lambda)}\varepsilon_{\mu}.$$

The theorem now follows from Propositions 3.3 and 3.5 and the fact that S and A_{lin} are uniquely determined by these functions. Here we use the fact that $S(m)$ and, according to Proposition 2.3, S are simple symmetric operators.

Theorem 3.6 also holds for ω instead of Ω, that is, when, for example, the functions $\alpha(\lambda)$ and $\beta(\lambda)$ are polynomials with real coefficients. The proof of the theorem for ω is based on the representation (3.6) for ω and the identity

$$[(A_{lin} - \lambda)^{-1}f, g] = \hat{\bar{g}}(\lambda)\omega(\lambda)\hat{f}(\lambda) + \text{ an entire function,} \tag{3.7}$$

where (see Subsection 2.5(II) and formula (2.17)),

$$\hat{f}(\lambda) = \int_a^b \theta(t,\lambda)f(t)dt, \quad \theta(\cdot,\lambda) = \beta(\lambda)\varphi(\cdot,\lambda) + \alpha(\lambda)\psi(\cdot,\lambda).$$

Since the operators A_{lin} and $A(\omega)$ are definitizable, they have spectral functions which we denote by E and E_{ω}, respectively. For bounded subintervals Δ of $\mathbb{R}$ whose endpoints do not coincide with the critical points of A_{lin} and $A(\omega)$, the above identity yields

$$[(A_{lin} - \lambda)^{-1}E(\Delta)f, E(\Delta)g] = \hat{\bar{g}}(\lambda)[(A(\omega)^{\Delta} - \lambda)^{-1}u, u]\hat{f}(\lambda) + \text{ an entire function,}$$

where

$$A(\omega)^{\Delta} = A(\omega)\,|_{E_{\omega}(\Delta)\Pi(\omega)}$$

and u is some generating element in $E_{\omega}(\Delta)\Pi(\omega)$. Contour integration and functional calculus then imply that

$$A_{lin}\,|_{E(\Delta)\mathfrak{K}} \quad \text{and} \quad A(\omega)^{\Delta}$$

are isomorphic. As Δ can be made arbitrarily large we see that

$$A_{lin}\,|_{E(\mathbb{R})\mathfrak{K}} \quad \text{and} \quad A(\omega)\,|_{E_{\omega}(\mathbb{R})\Pi(\omega)}$$

are isomorphic. It follows from the representation (3.6) for ω and (3.7) that the nonreal spectra of A_{lin} and $A(\omega)$ coincide, and one can show that the corresponding algebraic eigenspaces are isomorphic. Hence A_{lin} and $A(\omega)$ are isomorphic.

For more details we refer to Dijksma et al. [1987]. There one can also find an analysis of the nonreal and real eigenvalues of A_{lin} and the corresponding chains. These eigenvalues occur as zeros and, for example, as generalized zeros of negative type of the function $-\omega(\lambda)^{-1}$, which also belongs to $\mathbf{N}_{\kappa}$.

4 Sturm-Liouville operators with indefinite weight

1. The problem (see Ćurgus and Langer [1989], Daho and Langer [1977]). The basic facts about Sturm-Liouville operators mentioned in Subection 2.1 remain true for the more general problems with a weight function r at the eigenvalue parameter λ:

$$Lf - \lambda r f \equiv -f'' + qf - \lambda r f = 0 \text{ on } (0, b), \tag{4.1}$$

if r is positive. Then the endpoint 0 is regular if $q, r \in L^1_{loc}[0, b)$, and singular otherwise; similarly, the endpoint b is regular if $b < \infty$ and $q, r \in L^1_{loc}(0, b]$, and singular otherwise. The Hilbert space $L^2(0, b)$ in Subsection 2.1 has to be replaced by the Hilbert space L^2_r with inner product

$$[f, g] = \int_0^b \overline{g(x)} f(x) r(x) dx, \tag{4.2}$$

and the symmetric operator S in L^2_r is now formally defined as $S = (1/r)L$. Note that we write L^2_r instead of L^2_σ with σ defined by $d\sigma(x) = r(x)dx$; hopefully this inconsistent notation does not lead to confusion.

In this section we consider the case where the weight function r in (4.1) changes its sign, that is, where both sets

$$\Delta_+ = \{x : r(x) > 0\}, \quad \Delta_- = \{x : r(x) < 0\}$$

have positive Lebesgue measure. Then the space L^2_r with inner product (4.2) is a Kreĭn space. For simplicity we assume that r is nonzero almost everywhere.

In addition to (4.1) we introduce the problem

$$Lf - \lambda |r| f = 0 \text{ on } (0, b), \tag{4.3}$$

in the Hilbert space $L^2_{|r|}$, to which the results of Subsection 2.1 apply. The inner product in $L^2_{|r|}$ is denoted by $(\cdot, \cdot)$:

$$(f, g) = \int_0^b \overline{g(x)} f(x) |r(x)| dx.$$

As in Subection 2.1, the maximal operator $M_{\max}$ and the minimal operator M_{min} are associated with the problem (4.3) as follows:

$$M_{max} = \{\{f, g\} : f, g \in L^2_{|r|}, Lf = |r| g\}$$

and M_{min} is the closure of the restriction of M_{max} to those functions $f \in \operatorname{dom} M_{max}$ which vanish near $x = 0$ and $x = b$. For $u \in \operatorname{dom} M_{min}$ and $v \in \operatorname{dom} M_{max}$, it follows that

$$(M_{min} u, v) = \int_0^b (u'(x)\overline{v'(x)} + q(x) u(x) \overline{v(x)}) dx = (u, M_{min} v). \tag{4.4}$$

Hence M_{min} is symmetric in $L^2_{|r|}$. Observe that this form is independent of the weight function r.

In the following we always suppose that q and r in (4.1) are such that

(α) the point 0 is a regular point for (4.3), that is, $q, r \in L^1_{loc}[0, b)$;
(β) the limit point case prevails at b for (4.3).

The assumption (β) can be replaced by the assumption that the limit circle case prevails at b or that b is regular; then a selfadjoint boundary condition has to be imposed at b. Then the operator M_{min} has defect numbers $d_+ = d_- = 1$. For the problem (4.1) or (4.3) we impose a boundary condition at $x = 0$:

$$f(0)\cos\alpha - f'(0)\sin\alpha = 0 \tag{4.5}$$

for some $\alpha \in [0, \pi)$. For the problem (4.3), the restriction of M_{max} by this boundary condition is denoted by M_α. As in Subsection 2.1, M_α is a selfadjoint operator in $L^2_{|r|}$ and for $f, g \in \text{dom } M_\alpha$, we have

$$(M_\alpha f, g) = \cot\alpha f(0)\overline{g(0)} + \int_0^b (f'(x)\overline{g'(x)} + q(x)f(x)\overline{g(x)})dx.$$

Proposition 4.1 (a) *If $q \geq 0$ then there exists an $\alpha_0 \in [\pi/2, \pi)$ such that $M_\alpha \geq 0$ if $\alpha \in [0, \alpha_0]$ and such that M_α has one negative eigenvalue if $\alpha \in (\alpha_0, \pi)$.*
(b) *If b is regular then M_α has a finite number of negative eigenvalues.*
(c) *If $q(x) \geq 0$ for all x sufficiently close to b then M_α has a finite number of negative eigenvalues.*

Now we return to the original problem (4.1). Let $\mathfrak{K}$ be the Kreĭn space L^2_r, that is, the Hilbert space $L^2_{|r|}$ equipped with the inner product (4.2):

$$[f, g] = \int_0^b \overline{g(x)}f(x)r(x)dx.$$

A fundamental symmetry in $\mathfrak{K}$ is the operator J:

$$Jf = (\text{sgn } r)f, \quad f \in L^2_{|r|}.$$

In L^2_r we consider the operator

$$A_\alpha = JM_\alpha.$$

Then

$$A_\alpha f = g \Longleftrightarrow \begin{cases} -f'' + qf = rg, \\ f(0)\cos\alpha - f'(0)\sin\alpha = 0, \end{cases}$$

and

$$[A_\alpha f, f] = (M_\alpha f, f), \quad f \in \mathfrak{D}_\alpha = \text{dom } A_\alpha = \text{dom } M_\alpha. \tag{4.6}$$

Therefore the numbers of negative squares of the two forms $[A_\alpha f, f]$ and $(M_\alpha f, f)$ on $\mathfrak{D}_\alpha$ coincide. Although always $\rho(M_\alpha) \neq \emptyset$, if b is singular we have to impose an additional assumption to assure that $\rho(A_\alpha) \neq \emptyset$.

Proposition 4.2 *If b is singular, the following statements are equivalent:*
(1) *There exists an $\alpha \in [0, \pi)$ such that $\rho(A_\alpha) \neq \emptyset$.*
(2) *There exists an $\alpha \in [0, \pi)$ and a $\lambda_0 \in \mathbb{C}$ such that ran $(A_\alpha - \lambda_0)$ is closed.*
(3) *$\rho(A_\alpha) \neq \emptyset$ for all $\alpha \in [0, \pi)$.*
Each of the following conditions is sufficient for (1)–(3) to hold:
(a) *$q \geq \gamma|r|$ for some $\gamma > 0$.*

(b) Δ_+ *or* Δ_- *is bounded.*
(c) M_α^{-1} *is compact.*

It follows immediately from Subsection 1.2 (iii) that if $\rho(A_\alpha) \neq \emptyset$ and the form $[A_\alpha f, g], f, g \in \mathfrak{D}_\alpha$, has a finite number of negative squares then A_α is definitizable on L_r^2. Sufficient conditions for $\rho(A_\alpha) \neq \emptyset$ are given in Proposition 4.2. Sufficient conditions for $[A_\alpha f, g], f, g \in \mathfrak{D}_\alpha$, having a finite number of negative squares follow immediately from Proposition 4.1 and (4.6).

If $\rho(A_\alpha) \neq \emptyset$ and $[A_\alpha f, g], f, g \in \mathfrak{D}_\alpha$, has a finite number κ of negative squares, the results of Section 1 imply immediately: The critical points in $(0, \infty)$ ($(-\infty, 0)$, respectively) are of finite negative (positive, respectively) index and in $(0, \infty)$ ($(-\infty, 0)$, respectively) the spectrum of A_α of negative (positive, respectively) type consists only of isolated eigenvalues of finite total multiplicity and the sum of all these indices and multiplicities does not exceed κ. However, if an interval $\Delta \in \mathcal{R}_{A_\alpha}$, $\Delta \subseteq (0, \infty)$ ($\Delta \subseteq (-\infty, 0)$, respectively), does not contain a point of $c(A_\alpha)$ and an eigenvalue of negative (positive, respectively) type, then $(\mathfrak{K}_\Delta, [\cdot, \cdot])$ is a Hilbert space (anti-Hilbert space, respectively).

We consider the following example:

$$\begin{cases} -f''(x) - \lambda \operatorname{sgn}(x) f(x) = 0 \text{ on } [-1, 1], \\ \quad f(-1) = 0, \qquad f(+1) = 0. \end{cases}$$

Then both endpoints $+1$ and -1 are regular, the corresponding Kreĭn space $\mathcal{K}$ is the space $L^2(-1, 1)$ equipped with the inner product

$$[f, g] = \int_{-1}^{1} f(x)\overline{g(x)}dx,$$

and the corresponding selfadjoint operator A is nonnegative in $\mathcal{K}$:

$$[Af, f] = \int_{-1}^{1} |f'(x)|^2 \, dx \geq 0, \qquad f \in \operatorname{dom} A.$$

Evidently, $\lambda = 0$ is not an eigenvalue of A. If λ is an eigenvalue of A with a corresponding eigenfunction φ, then also $-\lambda$ is an eigenvalue of A with an eigenfunction $\varphi_1 : \varphi_1(x) = \varphi(-x)$. Therefore it is sufficient to consider the positive eigenvalues $\lambda_1^+ < \lambda_2^+ < \cdots$ of A. In order to find them, we solve for $\lambda > 0$ the initial problems

$$\begin{aligned} -f_-'' &= -\lambda f_- \text{ on } [-1, 0], \qquad & f_-(-1) = 0, \\ -f_+'' &= \lambda f_+ \text{ on } [0, +1], \qquad & f_+(+1) = 0. \end{aligned}$$

The solutions are

$$f_-(x) = C_- \sinh(\sqrt{\lambda}(1 + x)), \quad f_+(x) = C_+ \sinh(\sqrt{\lambda}(1 - x)).$$

The numbers C_{pm} have to be chosen such that

$$f_-(0) = f_+(0), \quad f_-'(0) = f_+'(0),$$

which yields the equation

$$-\tanh(\sqrt{\lambda}) = \tan(\sqrt{\lambda}).$$

It follows easily that the positive solutions $0 < \lambda_1^+ < \lambda_2^+ < \cdots$ of this equation behave asymptotically as follows:

$$\lambda_n^+ \sim (n\pi - \frac{\pi}{4})^2.$$

Here are the first six eigenvalues and their asymptotic values:

n	$\lambda_n^+ \approx$	$(n\pi - \arctan(1))^2 \approx$
1	5.5933213620143	5.5516524756128
2	30.2258479317809	30.2256634783362
3	74.6388838245438	74.6388832832383
4	138.7913118916898	138.7913118903191
5	222.6829492995819	222.6829492995786
6	326.3137955107455	326.3137955110169

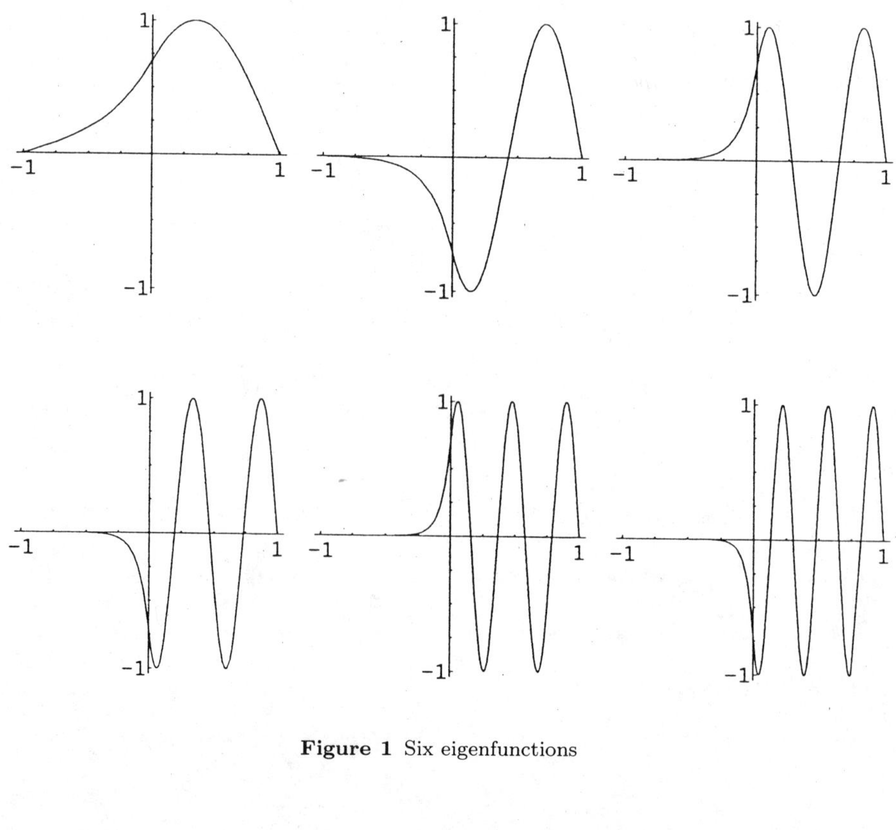

Figure 1 Six eigenfunctions

The corresponding eigenfunctions are given by

$$\varphi_n^+(x) = \begin{cases} \dfrac{\sin(\sqrt{\lambda_n})}{\sinh(\sqrt{\lambda_n})} \sinh\left(\sqrt{\lambda_n}(1+x)\right), & -1 \le x < 0 \\ \sin\left((\sqrt{\lambda_n}(1-x)\right), & 0 \le x \le 1 \end{cases}, \quad n = 1, 2, \dots .$$

The first six eigenfunctions are sketched in Figure 1. Observe that they are oscillating on $\Delta_+ = (0,1]$, the n-th eigenfunction φ_n^+ having n zeros there, whereas on $[1,0)$ they are isotonic or antitonic.

2. Full range and half range completeness (see Beals [1985], Ćurgus and Langer [1989]). We suppose in this section that A_α is nonnegative in L_r^2 and that $0 \in \rho(A_\alpha)$. Then ∞ is the only critical point of A_α. Additionally, in Theorem 4.3 below the spectrum $\sigma(A_\alpha)$ is assumed to be discrete. Then $\sigma(A_\alpha)$ consists of a sequence (λ_j^+) of positive simple eigenvalues and a sequence (λ_j^-) of negative simple eigenvalues with eigenvectors φ_j^+, φ_j^-, respectively, $j = 1, 2, \cdots$. Let $r_\pm(x) = \max(\pm r(x), 0), 0 \le x < b$, consider the fundamental decomposition

$$L_r^2 = L_{r_+}^2 [\dot{+}] L_{r_-}^2, \tag{4.7}$$

and denote the orthogonal projections onto $L_{r_+}^2$ and $L_{r_-}^2$ by P_+ and P_-, respectively. Theorem 1.2 implies that the closed linear span of all the eigenvectors φ_j^+, φ_j^-, $j = 1, 2, \cdots$, is L_r^2, that is, the eigenvectors of A_α form a complete system in L_r^2, which is sometimes called **full range completeness**. The so called **half range completeness** for A_α, which means that the vectors $P_+\varphi_j^+$ ($P_-\varphi_j^-$, respectively), $j = 1, 2, \dots$, form a Riesz basis of $L_{r_+}^2$ ($L_{r_-}^2$, respectively), is more involved.

Theorem 4.3 *Suppose that A_α is nonnegative in L_r^2, $0 \in \rho(A_\alpha)$ and that $\sigma(A_\alpha)$ is discrete. The following statements are equivalent:*
(i) *∞ is a regular critical point of A_α.*
(ii) *In $L_{|r|}^2$ there exists a Riesz basis of eigenvectors $\varphi_j^\pm, j = 1, 2, \dots$, of A_α; they can be chosen such that*

$$[\varphi_j^\pm, \varphi_k^\pm] = \pm\delta_{jk}, \quad [\varphi_j^+, \varphi_k^-] = 0, \quad j, k = 1, 2, \dots .$$

If (i) *or* (ii) *hold, then half range completeness holds for A_α.*

Theorem 4.3 could be formulated in the general situation of Subsection 1.5. The equivalence of (i) and (ii) follows from Theorem 1.6.

In order to prove the last claim, suppose that (i) holds. Then the space decomposes as $L_r^2 = \mathfrak{L}_+ [\dot{+}] \mathfrak{L}_-$, where $\mathfrak{L}_+$ ($\mathfrak{L}_-$, respectively) is a maximal uniformly positive (uniformly negative, respectively) subspace, namely the spectral subspace of A corresponding to $(0, \infty)$ ($(-\infty, 0)$, respectively). Let K denote the angular operator of $\mathfrak{L}_+$ with respect to the decomposition (4.7), $K : L_{r_+}^2 \to L_{r_-}^2$. Then $\|K\| < 1$ and

$$\varphi_j^+ = \begin{pmatrix} P_+\varphi_j^+ \\ KP_+\varphi_j^+ \end{pmatrix}.$$

The relation $[\varphi_j^+, \varphi_k^+] = \delta_{jk}$ is equivalent to

$$(TP_+\varphi_j^+, P_+\varphi_k^+) = \delta_{jk}$$

with $T = I - K^*K$ and, as $0 \ll T \ll I$, the elements $P_+\varphi_j^+$, $j = 1, 2, \ldots$, form a Riesz basis in L_r^2.

According to Theorem 4.3, an element $f_+ \in L^2_{r_+}$ has an expansion in terms of $P_+\varphi_j^+$, $j = 1, 2, \ldots$, of the form

$$f_+ = \sum_{j=1}^{\infty} (Tf_+, P_+\varphi_j^+) P_+\varphi_j^+. \tag{4.8}$$

If we retain the assumptions that A_α is nonnegative, $0 \in \rho(A_\alpha)$, and ∞ is a regular critical point, but the spectrum of A_α is not necessarily discrete, we consider again the spectral subspace $\mathfrak{L}_+$ of $(0, \infty)$ and its angular operator K: $\|K\| < 1$. Then for each interval $\Delta \subseteq [0, \infty)$ there exists an orthogonal projection $P(\Delta) \in L^2_{r_+}$ such that

$$E(\Delta) = \begin{pmatrix} P(\Delta)(P(\Delta)TP(\Delta))^{-1}P(\Delta) & -P(\Delta)(P(\Delta)TP(\Delta))^{-1}P(\Delta)K^* \\ KP(\Delta)(P(\Delta)TP(\Delta))^{-1}P(\Delta) & -KP(\Delta)(P(\Delta)TP(\Delta))^{-1}P(\Delta)K^* \end{pmatrix}.$$

Then $F(\Delta) = P(\Delta)(P(\Delta)TP(\Delta))^{-1}P(\Delta)T$ is symmetric with respect to the T–inner product and nondecreasing in $L^2_{r_+}$ as a function of Δ. As $P((0, \infty)) = I$, we have $F((0, \infty)) = I$, and $f_+ \in L^2_{r_+}$ now admits the expansion

$$f_+ = \int_0^\infty F(d\lambda) f_+,$$

which is the continuous analogue of (4.8).

In order to give sufficient conditions for ∞ to be a regular critical point of A_α, we suppose for simplicity that the weight function r has just one turning point $x_0 \in (0, b)$. Then it can be shown that ∞ is a regular critical point of A_α if in a one–sided neighbourhood of x_0, for example, for $x_0 < x < x_0 + \delta$, the weight function r is of the form

$$r(x) = (x - x_0)^\tau \rho(x)$$

with $\tau > -1$ and $\rho \in C^1([x_0, x_0 + \delta])$. The proof in Ćurgus and Langer [1989] and Fleige [1995] uses a construction which goes back to Baouendi and Grisvard [1968], Beals [1985] in order to show the existence of an operator W as in Theorem 1.6.

Remarks (1) Half range completeness is of interest in connection with boundary value problems arising in transport theory (Greenberg et al. [1987]) where, for a diffusion type equation, "initial" values at $t = 0$ and at $t = 1$ are given. Consider, for example, the problem

$$xu_t = u_{xx} \quad \text{on} \quad -1 \leq x \leq 1, \tag{4.9}$$

with boundary conditions $u(-1, t) = u(1, t) = 0$, $u(x, 0) = g_+(x)$ for $x \in [0, 1]$ and $u(x, 1) = g_-(x)$ for $x \in [-1, 0]$. Looking for functions of the form $u(x, t) = \mathrm{e}^{-\lambda t} v(x)$, the equation (4.9) becomes

$$-\lambda x v(x) = v''(x), \quad -1 \leq x \leq 1,$$

where the weight function $r(x) = x$ has one turning point $x_0 = 0$.
(2) The solution of the problem in Remark (1) can be obtained according to the following scheme. Consider

$$\frac{du}{dt} = -Au, \quad 0 \le t \le 1, \tag{4.10}$$

where u is function on $[0,1]$ with values in a Kreĭn space $\mathfrak{K}$ and A is a nonnegative operator on $\mathfrak{K}$ which generates a C_0–semigroup. There is given an (initial) fundamental decomposition of $\mathfrak{K}$: $\mathfrak{K} = \mathfrak{K}_+[\dot{+}]\mathfrak{K}_-$ with corresponding projections P_+ and P_- and the "spectral decomposition" corresponding to A: $\mathfrak{K} = \mathfrak{L}_+[\dot{+}]\mathfrak{L}_-$ with corresponding projections Q_+ and Q_-. Here ∞ is supposed to be a regular critical point of A. Further, an "initial condition" is given for u:

$$P_+u(0) + P_-u(1) = g. \tag{4.11}$$

Looking for a solution of (4.10) and (4.11) in the form

$$u(t) = (\mathrm{e}^{-tA}Q_+ + \mathrm{e}^{(1-t)A}Q_-)h$$

we obtain

$$g = P_+u(0) + P_-u(1) = [P_+Q_+ + P_-Q_- + (P_+Q_-\mathrm{e}^{A} + P_-Q_+\mathrm{e}^{-A})]h = Vh$$

or

$$u(t) = (\mathrm{e}^{-tA}Q_+ + \mathrm{e}^{(1-t)A}Q_-)V^{-1}g.$$

(3) The assumption "∞ is a regular critical point of A " is fulfilled if a fundamental symmetry J_0 maps the domain dom A^α of some power A^α of A, $0 < \alpha < 1$, into itself; see Ćurgus [1985]. For some elliptic differential operators these domains are known to contain piecewise continuous functions and using this fact the existence of such an operator J_0 can be shown. This is closely related to the considerations of Pyatkov [1989]; see also Ćurgus and Najman [1994]) where, in fact, the regularity of the critical point ∞ is formulated in terms of an interpolation condition. For some elliptic differential operators the construction in Beals [1985] was generalized in Faierman [in preparation].

3. The Titchmarsh-Weyl coefficient (see Daho and Langer [1977]). Consider again the problem (4.1), (4.5) under the assumption that it is in the limit point case at b and that the operator A_α is definitizable. Then, with the possible exception of finitely many eigenvalues, the spectrum of A_α is real. Let φ_α, ψ_α be a pair of solutions of the equation (4.1), satisfying the initial conditions

$$\begin{aligned} \varphi_\alpha(0,\lambda) &= \sin\alpha, & \psi_\alpha(0,\lambda) &= \cos\alpha, \\ \varphi_\alpha'(0,\lambda) &= \cos\alpha, & \psi_\alpha'(0,\lambda) &= -\sin\alpha. \end{aligned}$$

As $\rho(A_\alpha)$ is nonempty, it can be shown that there exists a unique function m_α of λ such that the solution s_α of (4.1):

$$s_\alpha = m_\alpha\varphi_\alpha + \psi_\alpha$$

belongs to the Kreĭn space L^2_r, that is, it is square integrable with respect to $|r|$. If the problem (4.1) is in the limit circle case at b or if b is regular, a "boundary condition" has to be imposed at b, and for each selfadjoint boundary condition such a function m_α exists.

If the weight function r is nonnegative the **Titchmarsh-Weyl coefficient** m_α belongs to the Nevanlinna class $\mathbf{N}_0$ and determines the operator A_α up to unitary equivalence. In the situation of this subsection the following result holds.

Theorem 4.4 *Suppose that $\rho(A_\alpha) \neq \emptyset$ and that the form $[A_\alpha f, g]$, $f, g \in \text{dom } A_\alpha$, has a finite number κ_α negative squares. Then the function*

$$\hat{m}_\alpha = \begin{cases} m_0(\lambda)/\lambda, & \text{if } \alpha = 0, \\ \lambda(m_\alpha(\lambda) + \cot\alpha), & \text{if } \alpha \in (0, \pi), \end{cases}$$

belongs to the class $\mathbf{N}_{\kappa_\alpha}$

With the function m_α a Kreĭn space and a selfadjoint operator $\tilde{A}_\alpha$ in this Kreĭn space can be associated such that $\tilde{A}_\alpha$ is weakly isomorphic to A_α. In particular, if an interval Δ does not contain 0 or any critical points of A_α in its closure, for the spectral projection $E(\Delta)$ of A_α and for functions $f, g \in L^2_r$, which vanish identically at b, we have:

$$[E(\Delta)f, g] = \int_\Delta f_\alpha(\lambda)\overline{g_\alpha(\lambda)}\frac{d\tau_\alpha(\lambda)}{\lambda},$$

where the function $\hat{f}_\alpha(\lambda)$ is given by

$$f_\alpha(\lambda) = \int_0^b f(x)\varphi_\alpha(x, \lambda)r(x)dx$$

and τ_α is the spectral function of the function $\hat{m}_\alpha$; see Kreĭn and Langer [1977]. Observe that the measure $d\tau_\alpha(\lambda)/\lambda$ is negative if $\Delta \subseteq (-\infty, 0)$ and positive if $\Delta \subseteq (0, \infty)$. Also, the nonreal spectrum of A_α coincides with the set of nonreal singularities of m_α.

5 Interface conditions and singular potentials

1. Interface conditions. In the following we could have considered Sturm–Liouville expressions on an interval (a, b) with any potential q satisfying condition (1) in Subsection 2.1, and then studied the differential expression of the form $-D^2 + q + \hat{q}$, where $\hat{q}$ is some function with a singular behavior at an interior point such as a delta function or one of its derivatives. Such functions sometimes lead to interface conditions, in which q plays no role. Therefore we simply take $q \equiv 0$ and we consider the expression $L = -D^2$ on the interval $(-1, 1)$.

We shall study differential operators with interface conditions at the origin. We eliminate the influence of the behavior of the functions at the endpoints ± 1 in these conditions by imposing separated selfadjoint boundary conditions in ± 1. Clearly, the endpoints of $(-1, 1)$ are regular for L.

Let $\widehat{A}$ be the (canonical) selfadjoint realization of L in $L^2(-1, 1)$ determined by the separated boundary conditions $f(-1) = 0,\ f(1) = 0$ (see Subsection 2.2):

$$\widehat{A} = \{\{f, -f''\} \in L_{max} : f(-1) = 0,\ f(1) = 0\}.$$

The boundary eigenvalue problem

$$\begin{cases} -f'' - \lambda f = g, \\ f \in \operatorname{dom} \widehat{A}, \end{cases} \tag{5.1}$$

with $\lambda \in \mathbb{C} \setminus \mathbb{R}$ has the unique solution $f = \widehat{R}(\lambda) g$, where $\widehat{R}(\lambda) = (\widehat{A} - \lambda)^{-1}$. This resolvent operator is an integral operator:

$$(\widehat{R}(\lambda) g)(x) = \int_{-1}^{1} G(x, y, \lambda) g(y) dy.$$

Here the kernel $G(x, y, \lambda)$ is the Green's function. It is given by

$$G(x, y, \lambda) = u(x, \lambda) v(y, \lambda) 1_{[-1,x)}(y) + u(y, \lambda) v(x, \lambda) 1_{(x,1]}(y),$$

where 1_ι denotes the indicator function of the interval ι and $u,\ v$ are solutions of the equation $-D^2 y = \lambda y$ such that $u(1, \lambda) = 0,\ v(-1, \lambda) = 0$ and their Wronskian $uv' - u'v \equiv 1$ on $(-1, 1)$.

For example,

$$u(x, \lambda) = \frac{\sin(\sqrt{\lambda}(1 - x))}{\sin 2\sqrt{\lambda}}, \quad v(x, \lambda) = \frac{\sin(\sqrt{\lambda}(x + 1))}{\sqrt{\lambda}}.$$

Note that $G(0, y, \lambda)$ is continuous at $y = 0$ and that

$$\frac{\partial}{\partial y} G(0, y, \lambda) \mid_{y=0+} - \frac{\partial}{\partial y} G(0, y, \lambda) \mid_{y=0-} = -1.$$

The operator $\widehat{R}(\lambda)$ is bounded on $L^2(-1, 1)$, $\widehat{R}(\lambda) = \widehat{R}(\bar{\lambda})^*$, and

$$\widehat{R}(\lambda) - \widehat{R}(\zeta) = (\lambda - \zeta) \widehat{R}(\lambda) \widehat{R}(\zeta),$$

for all nonreal numbers λ and ζ. Moreover, $\widehat{A}$ can be obtained from $\widehat{R}(\lambda)$ by

$$\widehat{A} = \{\{\widehat{R}(\lambda) g, (I + \lambda \widehat{R}(\lambda)) g\} : g \in L^2(-1, 1)\}.$$

Thus the righthand side is independent of λ and $\operatorname{dom} \widehat{A} = \operatorname{ran} \widehat{R}(\lambda)$.

For future reference, we define ext dom $\widehat{A}$ as the set of all $f \in L^2(-1,1)$ such that $f \in C^1(-1,0), f' \in AC_{loc}(-1,0), f \in C^1(0,1), f' \in AC_{loc}(0,1), f'' \in L^2(-1,1)$, and $f(\pm 1) = 0$. Note that all endpoints of $(-1,0)$ and $(0,1)$ are regular for $-D^2$. Hence for $f \in$ ext dom $\widehat{A}$, the four limits $f(0-), f(0+), f'(0-)$, and $f'(0+)$ also exist.

Interface conditions at 0 occur if we "eliminate" the origin from the interval $(-1,1)$ by considering the operator S which is the restriction of $\widehat{A}$ to all those functions $f \in$ dom $\widehat{A}$ for which $f(0) = f'(0) = 0$. Then S is a closed densely defined symmetric operator in $L^2(-1,1)$ with defect numbers $d^+ = d^- = 2$. It is the direct sum of two symmetric differential operators associated with L, one in $L^2(-1,0)$ and one in $L^2(0,1)$, and S^* is the direct sum of the adjoints of these two operators. We have dom $S^* =$ ext dom $\widehat{A}$. The boundary operator for S is given by

$$\mathsf{b}\{f\} = \begin{pmatrix} f(0-) & f'(0-) & f(0+) & f'(0+) \end{pmatrix}^T,$$

and the matrix Q in (3.1) is the 4×4 matrix

$$\mathsf{Q} = \text{diagonal } (-\mathsf{J}, \mathsf{J}), \quad \mathsf{J} = i \begin{pmatrix} 0 & -1 \\ 1 & 0 \end{pmatrix}.$$

From Theorem 3.4 we obtain the following result.

Theorem 5.1 *Each Štraus extension $T(\lambda)$ of S in $L^2(-1,1)$ whose linearization acts in a π_κ-space $\mathfrak{K}$ is of the form*

$$T(\lambda) = \left\{ \{f, Lf\} : f \in \text{ext dom } \widehat{A},\ C(\lambda) \begin{pmatrix} f(0-) \\ f'(0-) \end{pmatrix} + D(\lambda) \begin{pmatrix} f(0+) \\ f'(0+) \end{pmatrix} = 0 \right\},$$

where C and D are locally meromorphic 2×2 matrix functions on $\mathbb{C} \setminus \mathbb{R}$ such that

$$\text{rank } \begin{pmatrix} C(\lambda) & D(\lambda) \end{pmatrix} = 2, \quad C(\lambda)\mathsf{J}C(\bar{\lambda})^* - D(\lambda)\mathsf{J}D(\bar{\lambda})^* = 0,$$

and the kernel

$$N_{C,D}(\lambda, \zeta) = \frac{1}{i} \frac{D(\bar{\zeta})\mathsf{J}D(\bar{\lambda})^* - C(\bar{\zeta})\mathsf{J}C(\bar{\lambda})^*}{\lambda - \bar{\zeta}}$$

has κ negative squares. The (finite or infinite) number of positive squares of this kernel is equal to $\dim (\mathfrak{K} \ominus L^2(-1,1)) - \kappa$.

Similar formulas can be found at the end of Subsection 3.2. Boundary conditions of the type appearing in the description of the Štraus extensions in Theorem 5.1 are called **interface conditions** at the origin. All canonical selfadjoint extensions of S are obtained by taking C and D constant 2×2 matrices satisfying the above conditions; these extensions are denoted by $A_{C,D}$.

As an example consider the eigenvalue problem: Given $g \in L^2(-1,1)$ and $\lambda \in \mathbb{C}$, find $h \in L^2(-1,1)$ such that

$$\left\{ \begin{array}{l} -h'' - \lambda h + \lambda n(\lambda) h(0) w = g, \\ h \in \text{dom } \widehat{A}, \end{array} \right. \tag{5.2}$$

where $w(x) = |x| - 1, x \in (-1,1)$, and $n(\lambda)$ is a generalized Nevanlinna function with κ negative squares. For example, $n(\lambda) = a/(b - \lambda)$, $a, b \in \mathbb{R}$, then $\kappa = 0$ if

$a > 0$ and $\kappa = 1$ if $a < 0$. The third term on the lefthand side of the differential equation in (5.2) can formally be written as

$$\lambda n(\lambda)(h, \delta_0)w = \frac{1}{2}\lambda n(\lambda)(h, D^2 w)w,$$

hence it is a λ–dependent potential involving δ_0.

The problem (5.2) is equivalent to the interface eigenvalue problem

$$\begin{cases} -f'' - \lambda f = g \text{ on } (-1,0) \cup (0,1), \\ f \in \text{ext dom } \widehat{A}, \\ C(\lambda) \begin{pmatrix} f(0-) \\ f'(0-) \end{pmatrix} + D(\lambda) \begin{pmatrix} f(0+) \\ f'(0+) \end{pmatrix} = 0, \end{cases} \tag{5.3}$$

in which

$$C(\lambda) = \begin{pmatrix} 1 & 0 \\ 0 & 1 + n(\lambda) \end{pmatrix}, \quad D(\lambda) = -\begin{pmatrix} 1 & 0 \\ 2n(\lambda) & 1 + n(\lambda) \end{pmatrix}.$$

To verify this, simply make the substitution $f = h - n(\lambda)h(0)w$. Then, since h belongs to dom $\widehat{A}$, we see that the differential equation in (5.2) becomes

$$-f'' - \lambda f = g \text{ on } (-1,0) \cup (0,1),$$

f belongs to ext dom $\widehat{A}$, f is continuous at the origin, $f(0) = (1 + n(\lambda))h(0)$, and that

$$f'(0+) - f'(0-) = -2n(\lambda)h(0) = -\frac{2n(\lambda)}{1 + n(\lambda)} f(0).$$

This implies the equivalence of the two problems (5.2) and (5.3).

The selfadjoint linearization of the interface eigenvalue problem (5.3) necessarily acts in a π_κ–space with κ equal to the number of negative squares of the kernel N_n. This follows from Theorem 3.4 and the identity

$$N_{C,D}(\lambda, \zeta) = 2 \text{ diagonal } (0, \ N_n(\lambda, \zeta)).$$

The problems involving interface conditions considered here are one–dimensional versions of quantum mechanical models with point interactions in higher dimensions; compare Caspers [1992], Karpeshina [1992], Pavlov [1987], Fewster [1993], and Grossmann et al. [1980]. Canonical selfadjoint differential operators with interface conditions in one or finitely many points have been characterized in, for example, Everitt and Zettl [1986], Krall [1968], Krall and Zettl [1988], Zettl [1968], Zettl [1988]; operators with interface conditions in infinitely many points have been studied in Everitt and Zettl [1992] and Gesztesy and Kirsch [1985].

In Atkinson et al. [1988] the eigenvalue problem

$$\begin{cases} -f''(x) - \frac{1}{x} f(x) - \lambda f(x) = 0 \text{ on } [a, b], \\ f(a) = f(b) = 0, \end{cases}$$

where $-\infty < a < 0 < b < \infty$, is studied in the context of interface conditions; see Everitt et al. [1988] and also Atkinson [1988], where the problem is reduced to an eigenvalue problem for a system of differential equations.

Interface conditions are obtained by "splitting the interval at an interior point." A generalization of this point of view is the notion of a graph of operators studied in Exner and Šeba [1989] and Gerasimenko and Pavlov [1988].

2. A singular potential continuous in the graph norm of $\hat{A}$. In this subsection we consider an eigenvalue problem associated with the Sturm–Liouville expression $-D^2+q$ where the potential is concentrated at 0:

$$q = s\left(\cdot, \bar{a}\delta_0 + \bar{b}\delta_0{}'\right)(\bar{a}\delta_0 + \bar{b}\delta_0{}'),$$

here s is a real number and a and b are complex numbers with $a\bar{b} \in \mathbb{R}$. We show that this problem can be formulated equivalently with interface conditions.

The calculations will be presented in the context of **rigged Hilbert spaces**

$$\mathfrak{H}_+ \subset L^2(-1,1) \subset \mathfrak{H}_-,$$

where $\mathfrak{H}_+ = \text{dom } \widehat{A}$ provided with the graph norm of $\widehat{A}$, called a **space with positive norm**, and $\mathfrak{H}_-$ is its dual space with the duality defined by the inner product of $L^2(-1,1)$. The space $\mathfrak{H}_-$ is called a **space with negative norm**. For details, see Berezanskiĭ [1968], Berezanskiĭ [1986] and Tsekanovskii and Shmul'yan [1977].

Observe that $\mathfrak{H}_-$ contains the delta distribution δ_0 and its derivative δ_0':

$$(f,\delta_0) = f(0) = \overline{(\delta_0, f)}, \quad (f, \delta_0{}') = -f'(0) = \overline{(\delta_0{}', f)}.$$

These point evaluations are continuous with respect to the graph norm of $\widehat{A}$. We consider the operator $H_s : \mathfrak{H}_+ \to \mathfrak{H}_-$ defined by

$$H_s = -D^2 + s\left(\cdot, \bar{a}\delta_0 + \bar{b}\delta_0{}'\right)(\bar{a}\delta_0 + \bar{b}\delta_0{}').$$

Note that in the duality $(\cdot,\cdot)$ between the spaces with positive and negative norms H_s is "symmetric", that is, for all $f,\ g \in \mathfrak{H}_+$,

$$(H_s f, g) = (f, H_s g)\,.$$

In the following theorem S is the symmetric operator in $L^2(-1,1)$ which is the restriction of $\widehat{A}$ to the functions $f \in \text{dom } \widehat{A}$ for which $f(0) = f'(0) = 0$.

Theorem 5.2 *The operator H_s admits an interpretation as the canonical self-adjoint extension A_s of S in $L^2(-1,1)$ given by*

$$A_s = A_{C_s,D_s} = \left\{\{f, Lf\} : f \in \text{ext dom } \widehat{A},\ C_s\begin{pmatrix} f(0-) \\ f'(0-)\end{pmatrix} + D_s\begin{pmatrix} f(0+) \\ f'(0+)\end{pmatrix} = 0\right\},$$

where

$$C_s = I = \begin{pmatrix} 1 & 0 \\ 0 & 1\end{pmatrix}, \quad D_s = \begin{pmatrix} a\bar{b}s - 1 & -|b|^2 s \\ |a|^2 s & -\bar{a}bs - 1\end{pmatrix}.$$

To construct A_s, we first extend H_s to the space $\mathfrak{D}$ of all functions $f \in \text{ext dom } \widehat{A}$ for which $af - bf'$ is continuous at 0, that is,

$$af(0-) - bf'(0-) = af(0+) - bf'(0+). \tag{5.4}$$

The distribution $\bar{a}\delta_0 + \bar{b}\delta_0{}'$ can be applied to a function $f \in \mathfrak{D}$ and

$$\left(f, \bar{a}\delta_0 + \bar{b}\delta_0{}'\right) = (af - bf')(0).$$

Secondly, we extend the resolvent operator $\widehat{R}(\lambda)$ to $\mathfrak{H}_-$ by putting

$$\left(\widehat{R}(\lambda)f, g\right) = \left(f, \widehat{R}(\bar{\lambda})g\right), \quad f \in \mathfrak{H}_-, \quad g \in \mathfrak{H}_+.$$

It follows that

$$\widehat{R}(\lambda)\delta_0 = G(0,\cdot,\lambda)$$

and

$$\widehat{R}(\lambda)\delta_0{}' = -u'(0,\lambda)v(\cdot,\lambda)1_{[-1,0)}(\cdot) - v'(0,\lambda)u(\cdot,\lambda)1_{(0,1]}(\cdot).$$

Then we have for $f \in \mathfrak{D}$,

$$(H_s - \lambda)f = (-D^2 - \lambda)\{f + s(af - bf')(0)\widehat{R}(\lambda)(\bar{a}\delta_0 + \bar{b}\delta_0{}')\}.$$

We define dom A_s to be set of all $f \in \mathfrak{D}$ for which the expression between $\{\quad\}$ is C^1 in a neighborhood of the origin, that is,

$$\begin{cases} f(0-) &= f(0+) - s\bar{b}(af - bf')(0), \\ f'(0-) &= f'(0+) - s\bar{a}(af - bf')(0). \end{cases} \tag{5.5}$$

Note that each of the three conditions on $f \in \text{ext dom } \widehat{A}$ in (5.4) and (5.5) can be deduced from the other two. Finally we define

$$A_s = \{\{f, Lf\} : f \in \text{dom } A_s\},$$

and we find that A_s is selfadjoint and that $A_s = A_{C_s,D_s}$. Observe that A_0 coincides with $\widehat{A}$.

Remark An alternative way of obtaining the interface conditions which define the operator A_s is as follows. Choose a sequence of smooth functions u_n with compact support in $(-1,1)$ such that : $u_n \to \bar{a}\delta_0 + \bar{b}\delta_0{}'$ and $(\widehat{R}(\lambda)u_n, u_n)$ is convergent as $n \to \infty$. Consider the problem

$$\begin{cases} -D^2 f - \lambda f + s\,(f, u_n)\,u_n = g, \\ \qquad f \in \text{dom } \widehat{A}, \end{cases} \tag{5.6}$$

which is an approximation of the problem

$$-D^2 f - \lambda f + s\left(f, \bar{a}\delta_0 + \bar{b}\delta_0{}'\right)\left(\bar{a}\delta_0 + \bar{b}\delta_0{}'\right) = g. \tag{5.7}$$

The solutions of (5.6) are of the form

$$f_n = \widehat{R}(\lambda)g - s\,(f_n, u_n)\,\widehat{R}(\lambda)u_n.$$

Taking the L^2 inner product of the functions on both sides of this equality with u_n we get that

$$(f_n, u_n) = s\frac{(\widehat{R}g, u_n)}{1 + s(\widehat{R}(\lambda)u_n, u_n)}.$$

Hence (f_n, u_n) is a convergent sequence with limit λ_g, say, and

$$f_n \to f = \widehat{R}(\lambda)g - s\lambda_g\widehat{R}(\lambda)(\bar{a}\delta_0 + \bar{b}\delta_0{}').$$

The function f has the properties

$$\begin{cases} f(0-) - f(0+) &= -s\bar{b}\lambda_g, \\ f'(0-) - f'(0+) &= -s\bar{a}\lambda_g, \end{cases}$$

which shows that $af - bf'$ is continuous at 0 and $\lambda_g = \left(f, \bar{a}\delta_0 + \bar{b}\delta_0{}'\right)$. Thus we see that

$$f_n \to f = (A_s - \lambda)^{-1}g,$$

which can be considered as the solution of the "limit" problem (5.7).

3. A singular potential discontinuous in the graph norm of $\hat{A}$. The papers of Yu.G. Shondin (Shondin [1988] and Shondin [1991]) and of J.F. van Diejen and A. Tip (Diejen and Tip [1991]) contain a wealth of material pertaining to interface conditions and more general conditions, such as integral conditions. In what follows we only consider a very special case of their theory. The method developed in Subsection 5.2 breaks down if in the "potential" $(\cdot, \bar{a}\delta_0 + \bar{b}\delta_0{}')\,(\bar{a}\delta_0 + \bar{b}\delta_0{}')$ we replace δ_0 or $\delta_0{}'$ by a higher order derivative, for example, $\delta_0{}''$, as it does not belong to the space $\mathfrak{H}_-$. To include it in a theory, one extends the rigging above to a longer chain:

$$\mathfrak{H}_2 \subseteq \mathfrak{H}_1 \subseteq \mathfrak{H}_0 = L^2(-1,1) \subseteq \mathfrak{H}_{-1} \subseteq \mathfrak{H}_{-2},$$

where $\mathfrak{H}_{\pm 1} = \mathfrak{H}_\pm$, $\mathfrak{H}_2 = \operatorname{dom}\ \widehat{A}^2$ provided with the graph norm of $\widehat{A}^2$ and $\mathfrak{H}_{-2}$ is the dual space of $\mathfrak{H}_2$. It can be verified that $\delta_0{}''$ belongs to $\mathfrak{H}_{-2}$.

As an example, we consider the potential $s\,(\cdot,\chi)\,\chi$, where $\chi = -\delta_0{}'' - \lambda_0\delta_0$ in which λ_0 is a point in the resolvent set of $\widehat{A}$. We show how Shondin and Van Diejen and Tip associate with the family of eigenvalue problems

$$-f'' - \lambda f - s\,(f,\chi)\,\chi = g$$

parametrized over $s \in \mathbb{R}$, the family of interface eigenvalue problems parametrized over $t \in \mathbb{R}$ and of the form

$$\begin{cases} -f'' - \lambda f = g \quad \text{on} \quad (-1,0)\cup(0,1), \\ f \in \operatorname{ext}\operatorname{dom}\ \widehat{A}, \\ C_t(\lambda)\begin{pmatrix} f(0-) \\ f'(0-) \end{pmatrix} + D_t(\lambda)\begin{pmatrix} f(0+) \\ f'(0+) \end{pmatrix} = 0, \end{cases} \tag{5.8}$$

where, with $\Delta \in \mathbb{R}$ and $p(\lambda) = (\lambda - \lambda_0)(\lambda - \overline{\lambda_0})$,

$$C_t(\lambda) = \begin{pmatrix} 1 & 0 \\ 0 & \lambda\Delta - t \end{pmatrix}, \qquad D_t(\lambda) = -\begin{pmatrix} 1 & 0 \\ p(\lambda) & \lambda\Delta - t \end{pmatrix}.$$

The kernel associated with $C_t(\lambda)$ and $D_t(\lambda)$ is of the form

$$N_{C_t,D_t}(\lambda,\zeta) = \text{ diagonal } \ (0, K(\lambda,\zeta)),$$

where

$$K(\lambda,\zeta) = -\lambda\overline{\zeta} + t(\lambda + \overline{\zeta})|\lambda_0|^2\Delta - 2t\mathrm{Re}\ \lambda_0.$$

We have that for $\lambda_1, \lambda_2, \ldots, \lambda_n \in \mathbb{C}$ and $\xi_1, \xi_2, \ldots, \xi_n \in \mathbb{C}$,

$$\sum_{i,j} \overline{\xi_j} K(\lambda_i,\lambda_j)\xi_i = \begin{pmatrix} \sum_i \lambda_i\xi_i \\ \sum_i \xi_i \end{pmatrix}^* \begin{pmatrix} -\Delta & t \\ t & |\lambda_0|^2\Delta - 2t\mathrm{Re}\ \lambda_0 \end{pmatrix} \begin{pmatrix} \sum_i \lambda_i\xi_i \\ \sum_i \xi_i \end{pmatrix}.$$

Since the determinant of the 2×2 matrix on the righthand side equals $-|\lambda_0\Delta - t|^2 < 0$, the kernels K and N_{C_t,D_t} have exactly one positive and one negative square. This implies that the selfadjoint linearization of the problem (5.8) acts in a Pontryagin space; the exit space is 2-dimensional and has negative index 1.

The relation between the two eigenvalue problems is obtained in the following way: Observe that χ belongs to $\mathfrak{H}_{-2} \setminus \mathfrak{H}_{-1}$, and that

$$\chi_{-1} = \widehat{R}(\lambda_0)\chi = \delta_0 \in \mathfrak{H}_{-1} \setminus \mathfrak{H}_0, \quad \chi_0 = \widehat{R}(\lambda_0)\chi_{-1} = G(0,\cdot,\lambda_0) \in \mathfrak{H}_0 \setminus \mathfrak{H}_1.$$

Hence the space $\mathfrak{H}_1$ and the elements χ_0 and χ_{-1} are linearly independent. The various dualities $(\cdot,\cdot)$ between the spaces in the rigging can be extended so that

$$\begin{aligned}(\chi_0,\chi_0) &= \int_{-1}^{1} G(0,y,\lambda_0)\overline{G(0,y,\lambda_0)}\,dy,\\ (\chi_0,\chi_{-1}) &= G(0,0,\lambda_0).\end{aligned}$$

By the resolvent identity for $\widehat{R}(\lambda)$ we have that

$$(\chi_0,\chi_{-1})-(\chi_{-1},\chi_0)=(\lambda_0-\overline{\lambda_0})\,(\chi_0,\chi_0)\,.$$

Below we assign the real value Δ, say, to the inner product $(\chi_{-1},\chi_{-1})\in\mathbb{R}$.

We consider two inner product spaces. Denote by $\mathfrak{L}$ the linear space of all elements Φ of the form

$$\Phi=\varphi_1+c_0\chi_0+c_{-1}\chi_{-1},\quad \varphi_1\in\mathfrak{H}_1,\quad c_0,c_{-1}\in\mathbb{C}, \tag{5.9}$$

(such a decomposition is unique) equipped with the indefinite inner product

$$\begin{aligned}[\Phi,\widetilde{\Phi}] \;=\; &(\varphi_1+c_0\chi_0,\widetilde{\varphi}_1+\tilde{c}_0\chi_0)\\ &+c_{-1}\,(\chi_{-1},\widetilde{\varphi}_1+\tilde{c}_0\chi_0)+\overline{\tilde{c}_{-1}}\,(\varphi_1+c_0\chi_0,\widetilde{\chi}_{-1})+\overline{\tilde{c}_{-1}}\Delta c_{-1}.\end{aligned}$$

Let $\mathfrak{K}$ be the Pontryagin space $\mathfrak{H}_0\oplus\mathbb{C}^2$, where the space $\mathbb{C}^2$ is provided with inner product

$$[\begin{pmatrix}c_1\\c_2\end{pmatrix},\begin{pmatrix}d_1\\d_2\end{pmatrix}]=\begin{pmatrix}d_1\\d_2\end{pmatrix}^*\begin{pmatrix}0&1\\1&\Delta\end{pmatrix}\begin{pmatrix}c_1\\c_2\end{pmatrix}.$$

Then the map

$$\Phi\stackrel{\tau}{\longrightarrow}\begin{pmatrix}\varphi_1+c_0\chi_0\\(\varphi_1+c_0\chi_0,\chi_{-1})\\c_{-1}\end{pmatrix}$$

defines an isometry from $\mathfrak{L}$ into $\mathfrak{K}$. It has a dense range: Evidently, the space $\{0\}\oplus\{0\}\oplus\mathbb{C}$ is contained in ran τ. Since the map

$$\mathfrak{H}_0\supseteq C[-1,1]\ni f\to(f,\chi_{-1})=f(0)$$

(which plays a role in the second entry of $\tau\Phi$) is discontinuous, there exists a sequence φ_1^n in $\mathfrak{H}_{-1}$ such that $\varphi_1^n\to0$ in $\mathfrak{H}_0$ and $\varphi_1^n(0)\to1$ in $\mathbb{C}$ as $n\to\infty$. This implies that the space $\{0\}\oplus\mathbb{C}\oplus\{0\}$ is also contained in ran τ. Finally, since $\mathfrak{H}_1=\operatorname{dom}\,\widehat{A}$ is dense in $\mathfrak{H}_0$, we see that ran τ is dense. Thus $\mathfrak{L}$ is a π_1–lineal and $\mathfrak{K}$ is the (unique) completion of the space $\mathfrak{L}$.

It is easily verified that, on account of the resolvent identity,

$$\widehat{R}(\lambda)\Phi=\widehat{R}(\lambda)\varphi_1+c_0\widehat{R}(\lambda)\chi_0+c_{-1}(\chi_0+(\lambda-\overline{\lambda_0})\widehat{R}(\lambda)\chi_0)$$

belongs to $\mathfrak{H}_0$, and hence

$$\tau(\widehat{R}(\lambda)\Phi)=\begin{pmatrix}\widehat{R}(\lambda)\Phi\\\left(\widehat{R}(\lambda)\Phi,\chi_{-1}\right)\\0\end{pmatrix}.$$

For each $\lambda \in \rho(\widehat{A})$ the operator $\tau \widehat{R}(\lambda)\tau^{-1}$ defined on ran τ admits a bounded closure on $\mathfrak{K}$. We denote this operator by $R^\infty(\lambda)$; it has the matrix representation

$$R^\infty(\lambda)\begin{pmatrix} f \\ d_1 \\ d_2 \end{pmatrix} = \begin{pmatrix} \widehat{R}(\lambda) & 0 & \widehat{R}(\lambda)\chi_{-1} \\ \left(\cdot, \widehat{R}(\overline{\lambda})\chi_{-1}\right) & 0 & \left(\widehat{R}(\lambda)\chi_{-1}, \chi_{-1}\right) \\ 0 & 0 & 0 \end{pmatrix}\begin{pmatrix} f \\ d_1 \\ d_2 \end{pmatrix},$$

where $f \in \mathfrak{H}_0$ and $d_1, d_2 \in \mathbb{C}$. From this representation it can be shown that $R^\infty(\lambda)$ has the properties:

$$\begin{aligned} &R^\infty(\lambda) - R^\infty(\zeta) = (\lambda - \zeta)R^\infty(\lambda)R^\infty(\zeta), \\ &R^\infty(\lambda)^+ = R^\infty(\overline{\lambda}), \\ &\ker\ R^\infty(\lambda) = \{0\} \oplus \mathbb{C} \oplus \{0\}. \end{aligned}$$

In the second equality the adjoint is taken with respect to the indefinite inner product of the Pontryagin space $\mathfrak{K}$.

It follows that $R^\infty(\lambda)$ is the resolvent of the selfadjoint relation

$$\begin{aligned} H^\infty &= \left\{\left\{R^\infty(\lambda_0)\begin{pmatrix} f \\ d_1 \\ d_2 \end{pmatrix}, (I + \lambda_0 R^\infty(\lambda_0))\begin{pmatrix} f \\ d_1 \\ d_2 \end{pmatrix}\right\} : f \in \mathfrak{H}_0, d_1, d_2 \in \mathbb{C}\right\} \\ &= \left\{\left\{\begin{pmatrix} h + d_2\chi_0 \\ (h + d_2\chi_0, \chi_{-1}) \\ 0 \end{pmatrix}, \begin{pmatrix} \widehat{A}h + d_2\lambda_0\chi_0 \\ d_1 \\ d_2 \end{pmatrix}\right\} : h \in \operatorname{dom}\ \widehat{A}, d_1, d_2 \in \mathbb{C}\right\}. \end{aligned}$$

We now choose a restriction S of H^∞ (so S is necessarily symmetric in $\mathfrak{K}$) such that the defect space $\ker\ (S^+ - \overline{\lambda_0}) = \operatorname{ran}\ (S - \lambda_0)^{[\perp]}$ of S is of dimension 1 and spanned by the vector $\tau\chi_{-1} = (0, 0, 1)^T$. Since

$$\mathfrak{K} \ominus \operatorname{span}\left\{\begin{pmatrix} 0 \\ 0 \\ 1 \end{pmatrix}\right\} = \left\{\begin{pmatrix} f \\ d_1 \\ d_2 \end{pmatrix} : f \in \mathfrak{H}_0, d_2 \in \mathbb{C}, d_1 = -\Delta d_2\right\},$$

we find that

$$\begin{aligned} S &= \left\{\{u, v\} \in H^\infty : [v - \lambda_0 u, \begin{pmatrix} 0 \\ 0 \\ 1 \end{pmatrix}]_{\mathfrak{K}} = 0\right\} \\ &= \left\{\left\{\begin{pmatrix} h + d\chi_0 \\ (h + d\chi_0, \chi_{-1}) \\ 0 \end{pmatrix}, \begin{pmatrix} \widehat{A}h + d\lambda_0\chi_0 \\ -d\Delta + \lambda_0 (h + d\chi_0, \chi_{-1}) \\ d \end{pmatrix}\right\} : \right. \\ &\qquad\qquad \left. h \in \operatorname{dom}\ \widehat{A}, d \in \mathbb{C} \right\}. \end{aligned}$$

Clearly, S is symmetric, and straightforward calculations yield

$$S^+ = \left\{\left\{\begin{pmatrix} h + c\chi_0 \\ (h + c\chi_0, \chi_{-1}) \\ a \end{pmatrix}, \begin{pmatrix} \widehat{A}h + c\lambda_0\chi_0 \\ b \\ c + \lambda_0 a \end{pmatrix}\right\} : h \in \operatorname{dom}\ \widehat{A}, a, b, c \in \mathbb{C}\right\}.$$

All canonical selfadjoint extensions H^t of S in the Pontryagin space $\mathfrak{K}$ can be determined and parametrized over $t \in \mathbb{R} \cup \{\infty\}$:

$$H^t = \left\{ \left\{ \begin{pmatrix} h + c\chi_0 \\ (h + c\chi_0, \chi_{-1}) \\ a \end{pmatrix}, \begin{pmatrix} \widehat{A}h + c\lambda_0\chi_0 \\ d_1 \\ c + a\lambda_0 \end{pmatrix} \right\} : h \in \operatorname{dom} \widehat{A}\,, \right.$$

$$\left. a, c \in \mathbb{C}, d_1 = a(t - \lambda_0\Delta) - c\Delta + \overline{\lambda_0}\,(h + c\chi_0, \chi_{-1}) \right\}.$$

One can obtain H^∞ from this formula by replacing a by a/t and letting t tend to infinity. To rewrite H^t, we put $f = h + c\chi_0$ and we recall that $\chi_0 = G(0, \cdot, \lambda_0)$. Then f is continuous at 0, and

$$f'(0-) - f'(0+) = h'(0-) - h'(0+) + c(\chi_0'(0-) - \chi_0'(0+)) = c.$$

Hence, since $\chi_{-1} = \delta_0$,

$$H^t = \left\{ \left\{ \begin{pmatrix} f \\ f(0) \\ a \end{pmatrix}, \begin{pmatrix} -D''f \\ d_1 \\ c + a\lambda_0 \end{pmatrix} \right\} : f \in \operatorname{ext\,dom}\, \widehat{A},\ a \in \mathbb{C}, \right.$$

$$\left. f(0-) = f(0+), c = f'(0-) - f'(0+), d_1 = a(t - \lambda_0\Delta) - c\Delta + \overline{\lambda_0} f(0) \right\}.$$

Finally, after some straightforward computations we find that the eigenvalue problem

$$(H^t - \lambda) \begin{pmatrix} f \\ f(0) \\ a \end{pmatrix} = \begin{pmatrix} g \\ 0 \\ 0 \end{pmatrix}$$

coincides with the interface eigenvalue problem (5.8).

4. Strings with point masses and dipoles (see Kac and Kreĭn [1968], Langer [1972]). In this subsection we show that certain λ–linear interface conditions arise in the consideration of vibrating strings containing a point mass or a dipole. Consider the boundary eigenvalue problem

$$\begin{cases} -f'' - \lambda f = g \text{ on } (-1,0) \cup (0,1), \\ f \in \operatorname{ext\,dom}\, \widehat{A} \text{ is continuous at the origin}, \\ m_0\lambda f(0) = f'(0-) - f'(0+). \end{cases}$$

If $m_0 = 0$ then the problem is equivalent to $(\widehat{A} - \lambda)f = g$ and the solution is $g = \widehat{R}(\lambda)f$. If $m_0 \neq 0$ then the linearization of the problem is the selfadjoint operator

$$A_{lin} = \left\{ \left\{ \begin{pmatrix} f \\ f(0) \end{pmatrix}, \begin{pmatrix} -f'' \\ \frac{f'(0-) - f'(0+)}{m_0} \end{pmatrix} \right\} : f \in \operatorname{ext\,dom}\, \widehat{A}, f(0+) = f(0-) \right\}$$

acting in the space $\mathfrak{K} = L^2(-1,1) \oplus \mathbb{C}$, where $\mathbb{C}$ is equipped with the inner product $[x, y] = \bar{y} m_0 x$. The space $\mathfrak{K}$ is a Hilbert space if $m_0 > 0$ and a π_1–space if $m_0 < 0$.

The operator A_{lin} coincides with the operator

$$l_M = -\frac{d}{dM(x)}\frac{d}{dx}$$

acting in the space L^2_M, where

$$M(x) = \begin{cases} x, & \text{if } -1 \le x < 0, \\ x + m_0, & \text{if } 0 \le x \le 1. \end{cases}$$

The operator l_M is the generalized second order derivative with mass distribution function M, which was introduced independently by M.G. Kreĭn and W. Feller; see, for example, Kac and Kreĭn [1968], where more general strings are considered. By definition, the relation $l_M f = g$ with $g \in L^2_M$ is equivalent to the following: there exist numbers $a, b \in \mathbb{C}$ such that for all $x \in [-1, 1]$,

$$\begin{aligned} f(x) &= a + bx - \textstyle\int_{-1}^{x}(x-t)g(t)dM(t) \\ &= \begin{cases} a + bx - \int_{-1}^{x}(x-t)g(t)dt, & -1 \le x \le 0, \\ a + (b - m_0 g(0))x - \int_{-1}^{x}(x-t)g(t)dt, & 0 < x \le 1, \end{cases} \end{aligned}$$

and this means that $f \in \operatorname{ext}\operatorname{dom} \widehat{A}$, $g = -f''$, and

$$g(0) = \frac{f'(0-) - f'(0+)}{m_0}.$$

It now follows easily that A_{lin} and l_M are isomorphic selfadjoint operators under the isomorphism

$$f \mapsto \begin{pmatrix} f \\ f(0) \end{pmatrix}$$

from L^2_M onto $\mathfrak{K}$.

From the definition of the operator l_M we see that if we integrate both sides of the equality

$$-df' - \lambda f dM = g dM$$

over the interval $[-1, x]$ we obtain the relation $l_M f - \lambda f = g$.

Further, a dipole at the point $x = 0$ is mathematically described by a differential expression of the form

$$-df' - \lambda f dM - \lambda^2 dD,$$

where $D(x) = d_0 1_{[0,1]}(x), x \in [-1, 1]$, for some $d_0 \in \mathbb{R}$. Integration of both sides of the equality

$$-df' - \lambda f dM - \lambda^2 dD = g dM$$

yields the interface eigenvalue problem

$$\begin{cases} -f'' - \lambda f = g \ \text{ on } \ (-1, 0) \cup (0, 1), \\ C(\lambda)\begin{pmatrix} f(0-) \\ f'(0-) \end{pmatrix} + D(\lambda)\begin{pmatrix} f(0+) \\ f'(0+) \end{pmatrix} = 0, \end{cases} \tag{5.10}$$

in which

$$C(\lambda) = \begin{pmatrix} 1 & 0 \\ 0 & 1 \end{pmatrix}, \qquad D(\lambda) = -\begin{pmatrix} 1 & 0 \\ m_0\lambda + d_0\lambda^2 & 1 \end{pmatrix}.$$

From the relation

$$N_{C,D}(\lambda, \lambda) = \text{diagonal} \quad (0, m_0 + d_0(\lambda + \overline{\lambda}))$$

it follows that if $|m_0| + |d_0| \neq 0$ the selfadjoint linearization of this problem acts in a Pontryagin space unless $d_0 = 0$ and $m_0 > 0$, in which case it acts in a Hilbert space.

6 Operator pencils

1. Basic notations. Let L be a function which is defined and holomorphic on some domain $\mathcal{D}_L \subseteq \mathbb{C}$ and with values in the set of bounded or closed operators in a Hilbert space $\mathfrak{H}$ (an "operator pencil"). In the case of closed operators $L(\lambda)$ the definition of holomorphy involves some technicalities, therefore we restrict ourselves mainly to bounded operators; we refer to Lancaster et al. [1993] and the literature quoted there.

The point $\lambda \in \mathcal{D}_L$ is said to belong to $\rho(L)$ ($\sigma(L)$, $\sigma_p(L)$ etc.) if $0 \in \rho(L(\lambda))$ ($\sigma(L(\lambda))$, $\sigma_p(L(\lambda))$ etc.). If $\lambda_0 \in \mathcal{D}_L$, a Jordan chain of L at λ_0: $x_0, x_1, x_2, \dots, x_n$ is defined by the relations

$$\begin{gathered} L(\lambda_0)x_0 = 0, \quad L(\lambda_0)x_1 + \tfrac{1}{1!}L'(\lambda_0)x_0 = 0, \quad \dots, \\ L(\lambda_0)x_n + \tfrac{1}{1!}L'(\lambda_0)x_{n-1} + \dots + \tfrac{1}{n!}L^{(n)}(\lambda_0)x_0 = 0. \end{gathered}$$

Here, we shall be concerned mainly with selfadjoint pencils, that is, $\mathcal{D}_L$ is symmetric with respect to the real axis and $L(\bar{\lambda}) = L(\lambda)^*$. With the pencils L considered in this section a "linearization" $\tilde{A}$ in some Hilbert or Kreĭn space $\tilde{\mathfrak{K}}$ containing $\mathfrak{H}$ will be associated such that the relation

$$L(\lambda)^{-1} = Q(\tilde{A} - \lambda)^{-1}P \tag{6.1}$$

holds, where P is a mapping from $\mathfrak{H}$ into $\tilde{\mathfrak{K}}$ and Q maps $\tilde{\mathfrak{K}}$ into $\mathfrak{H}$. In particular, if $\mathfrak{H} \subseteq \tilde{\mathfrak{K}}$, if Q is the projection from $\tilde{\mathfrak{K}}$ onto $\mathfrak{H}$ and P is the embedding from $\mathfrak{H}$ into $\tilde{\mathfrak{K}}$, then $L(\lambda)^{-1}$ is a compressed resolvent of $\tilde{A}$, compare Section 2. Under some minimality assumptions the spectral properties of L will be reflected by those of $\tilde{A}$; the symmetry property $L(\bar{\lambda}) = L(\lambda)^*$ implies in the examples that $\tilde{A}$ is symmetric with respect to a Hilbert or a Kreĭn space inner product. In particular, an indefinite inner product yields a classification of the eigenvalues of $\tilde{A}$, and hence also of L, into those of positive, negative or mixed type.

Often the operator $\tilde{A}$ is definitizable, or locally definite; see Langer et al. [in preparation]. In the sequel we shall give some examples. We mention that also linear pencils $L(\lambda) = \lambda F + G$ in a Hilbert space can lead to definitizable operators, for example, if one of the operators F or G has a finite number of negative eigenvalues. Problems of this type will not be considered here; see Lancaster et al. [1993].

2. Operator polynomials (see Lancaster [1964], Gohberg et al. [1982], Rodman [1989], Markus [1988]). Let

$$L(\lambda) = \lambda^n I + \lambda^{n-1}A_{n-1} + \dots + \lambda A_1 + A_0,$$

where A_j, $j = 0, 1, \dots, n-1$ are bounded selfadjoint operators in $\mathfrak{H}$. Here $\mathcal{D}_L = \mathbb{C}$ and the relation (6.1) holds with $\tilde{\mathfrak{K}} = \mathfrak{H}^n$ and

$$\tilde{A} = \begin{pmatrix} -A_{n-1} & \dots & -A_1 & -A_0 \\ 1 & & 0 & 0 \\ \vdots & & \vdots & \vdots \\ 0 & & 1 & 0 \end{pmatrix}, \quad P = \begin{pmatrix} I \\ 0 \\ \vdots \\ 0 \end{pmatrix}, \quad Q = \begin{pmatrix} 0 & \dots & 0 & I \end{pmatrix}.$$

The operator $\tilde{A}$ is selfadjoint with respect to the $\tilde{G}$–inner product $[\cdot,\cdot]$ on $\tilde{\mathfrak{K}}$, where

$$\tilde{G} = \begin{pmatrix} 0 & 0 & \dots & 0 & 1 \\ 0 & 0 & & 1 & A_{n-1} \\ \vdots & \vdots & & & \vdots \\ 0 & 1 & & & A_2 \\ 1 & A_{n-1} & \dots & A_2 & A_1 \end{pmatrix}.$$

Then $\sigma_p(L) = \sigma_p(\tilde{A})$ etc., and for an eigenvector x_0 of L at the eigenvalue λ_0 and the corresponding eigenvector $\tilde{x}_0$ of $\tilde{A}$ we have

$$\tilde{x}_0 = \begin{pmatrix} \lambda_0^{n-1} x_0 & \dots & \lambda_0 x_0 & x_0 \end{pmatrix}^T \quad \text{and} \qquad [\tilde{x}_0, \tilde{x}_0] = (L'(\lambda_0)x_0, x_0).$$

There is also a bijective correspondence between the Jordan chains of L and $\tilde{A}$ at λ_0. In examples the operators $A_0, \dots, A_{n-1}$ are often unbounded. Sometimes the linearization $\tilde{A}$ can also be considered in this situation, sometimes by a simple transformation the given pencil can be transformed into one with bounded operators.

A large number of linear vibration problems leads to polynomial pencils of operators. We mention the following ones (see, for example, Kreĭn and Langer [1978], Lancaster and Skhalikov [1994], Barkwell et al. [1992] and papers quoted there):

$$-T_0\frac{\partial^2 u}{\partial x^2} + \gamma\frac{\partial u}{\partial t} + \rho\frac{\partial^2 u}{\partial t^2} = 0, \qquad \text{damped vibrating string,}$$
$$-T_0\Delta u + \gamma\frac{\partial u}{\partial t} + \rho\frac{\partial^2 u}{\partial t^2} = 0, \qquad \text{damped membrame,}$$
$$\frac{\partial^4 u}{\partial x^4} + \alpha\frac{\partial^5 u}{\partial x^4 \partial t} + \beta(x)\frac{\partial u}{\partial t} + \rho(x)\frac{\partial^2 u}{\partial t^2} = 0, \quad \text{damped beam.}$$

Pencils arise if we look for solutions of the form $u(t,x) = \mathrm{e}^{i\mu t}v(x)$. Formally, these problems are of the form

$$Fv + \mu Gv + \mu^2 Hv = 0$$

with unbounded operators F, G, H, where F and H are positive. Introducing $\lambda = 1/\mu$ and $w = F^{1/2}v$ we can rewrite this equation as

$$(\lambda^2 I + \lambda A_1 + A_0)w = 0,$$

where A_0 is positive and compact.

Here is a more specific assumption which implies definitizability of $\tilde{A}$: The pencil

$$L(\lambda) = \lambda^2 I + \lambda A_1 + A_0 \tag{6.2}$$

is called **weakly hyperbolic**, if for each $x \in \mathfrak{H}, x \neq 0$, the quadratic equation $(L(\lambda)x, x) = 0$ has two real solutions $\lambda_1(x)$, $\lambda_2(x)$, say, $\lambda_1(x) \le \lambda_2(x)$.

Theorem 6.1 *If the pencil* (6.2) *is weakly hyperbolic, then* $\lambda_1(x) \le \lambda_2(y)$, $x, y \in \mathfrak{H} \setminus \{0\}$, *and if* α *satisfies*

$$\sup_{0\neq x\in\mathfrak{H}} \lambda_1(x) \le \alpha \le \inf_{0\neq x\in\mathfrak{H}} \lambda_2(x)$$

then

$$[(\tilde{A} - \alpha)\tilde{x}, \tilde{x}] \ge 0, \quad \tilde{x} \in \tilde{\mathfrak{K}}.$$

That is, in this case $\tilde{A}$ *is definitizable with definitizing polynomial* $p(t) = t - \alpha$.

Evidently the pencil (6.2) is weakly hyperbolic if and only if

$$(A_1x, x)^2 \geq 4(A_0x, x)\|x\|^2, \quad x \in \mathfrak{H}.$$

For vibrating systems this assumption means that the system is strongly damped. For weakly hyperbolic pencils of degree n, see Markus [1988]. Vibrating systems which lead to definitizable operators with a definitizing polynomial of higher than second degree are considered in Barkwell et al. [1992].

Consider again the pencil (6.2). Then

$$\tilde{G}\tilde{A} = \begin{pmatrix} I & 0 \\ 0 & -A_0 \end{pmatrix},$$

therefore the number of negative squares of $[\tilde{A}\tilde{x}, \tilde{y}], \tilde{x}, \tilde{y} \in \tilde{\mathfrak{K}}$, coincides with the number of negative eigenvalues of $-A_0$. Thus, if this number is finite, the operator $\tilde{A}$ is definitizable. As an example to which these considerations apply (at least formally, as the operator A_0 is unbounded), consider the problem (see Pivovarcik [1988]):

$$f'' + (\lambda^2 + u - \lambda q)f = 0 \text{ on } [0, \infty), \quad f(0) = 0, \tag{6.3}$$

where the functions u and q are, for example, real and essentially bounded. Then

$$\tilde{G}\tilde{A} = \begin{pmatrix} I & 0 \\ 0 & -D^2 - u \end{pmatrix},$$

where D^2 is the operator $\frac{d^2}{dx^2}$ with initial condition $f(0) = 0$. Hence the number κ of negative squares of $[\tilde{A}\tilde{x}, \tilde{y}], \tilde{x}, \tilde{y} \in \tilde{\mathfrak{K}}$, equals the number of negative eigenvalues of $-D^2 - u$ and is independent of q. Moreover, if it is known that the problem (6.3) has no real eigenvalues, then κ is the number of eigenvalues of (6.3) in the open upper (or lower) half plane.

Finally, we mention that if for the pencil (6.2) we have $A_0 \geq 0$, then a natural linearization which is similar to $\tilde{A}$, is

$$\tilde{A}_0 = \begin{pmatrix} -A_1 & -A_0^{1/2} \\ +A_0^{1/2} & 0 \end{pmatrix}.$$

If, additionally, A_0 is compact then $\tilde{A}_0$ has a spectral function with finite type citical points outside 0; see Langer et al. [in preparation].

3. A λ–rational pencil. Let $\mathfrak{H}$, $\mathfrak{H}_1$ be Hilbert spaces, $A = A^*$ in $\mathfrak{H}$, $B \in \mathbf{L}(\mathfrak{H}, \mathfrak{H}_1)$, $D = D^* \in \mathbf{L}(\mathfrak{H}_1)$ and

$$L(\lambda) = \lambda - A - B^*(D - \lambda)^{-1}B.$$

A linearization of L is

$$\tilde{A} = \begin{pmatrix} A & B^* \\ -B & D \end{pmatrix} \quad \text{in } \tilde{\mathfrak{K}} = \mathfrak{H} \oplus \mathfrak{H} \tag{6.4}$$

which is symmetric with respect to the $\tilde{G}$–inner product, where

$$\tilde{G} = \text{diagonal } (I, -I).$$

Theorem 6.2 *If, additionally to the assumptions of this subsection, the resolvent of A is compact and A is bounded from below, then $\tilde{A}$ in (6.4) is definitizable.*

Indeed, choose β such that $D \ll \beta$ and consider the operator

$$\tilde{G}(\tilde{A}-\beta) = \begin{pmatrix} A-\beta & B^* \\ B & -D+\beta \end{pmatrix}.$$

Replace A by

$$A_0 = A - \sum_{j=1}^{k} \lambda_j P_j + \lambda_{k+1} \sum_{j=1}^{k} P_j,$$

where $\lambda_1, \lambda_2 \ldots, \lambda_{k+1}$ are the smallest eigenvalues of A, $\lambda_1 < \lambda_2 < \ldots < \lambda_{k+1}$, and $P_1, P_2 \ldots, P_{k+1}$ are the orthogonal projections onto the corresponding eigenspaces. Then for sufficiently large k the operator

$$\begin{pmatrix} A_0-\beta & B^* \\ B & -D+\beta \end{pmatrix}$$

is, evidently, nonnegative, hence with

$$\tilde{A}_0 = \begin{pmatrix} A_0 & B^* \\ -B & D \end{pmatrix}$$

also $\tilde{G}(\tilde{A}_0-\beta)$ is nonnegative and $\tilde{A}_0$ is definitizable with the definitizing polynomial $p(t) = t-\beta$. As $\tilde{A}$ is a finite dimensional perturbation of $\tilde{A}_0$ it is also definitizable.

As an example we consider the problem

$$-f'' + (\lambda + \frac{q}{u-\lambda})f = 0 \text{ on } [0,1], \quad f(0) = f(1) = 0,$$

with continuous functions q, u on $[0,1]$, $q < 0$. In this case

$$\begin{pmatrix} -D_0^2 & |q|^{1/2} \\ -|q|^{1/2} & u \end{pmatrix},$$

where D_0^2 is the operator $\frac{d^2}{dx^2}$ with boundary conditions $f(0) = f(1) = 0$, and the above assumptions are satisfied. We mention that in this example also a Titchmarsh-Weyl coefficient m can be introduced and it turns out that $\lambda m(\lambda) - \beta$ belongs to some class $\mathbf{N}_\kappa$; compare Langer et al. [1990].

4. The Klein-Gordon equation. (see Langer and Najman [unpublished manuscript]). The equation

$$[(\frac{\partial}{\partial t} - ieq)^2 + m^2 - \Delta]u = 0$$

describes the motion of a relativistic particle of mass m and charge e in an electrostatic field with potential q. Formally it can be written as

$$(\frac{d}{dt} - iV)^2 u + H_0 u = 0 \tag{6.5}$$

with $H_0 = m^2 - \Delta$ and $V = eq$, or as

$$\frac{d\tilde{v}}{dt} = iA\tilde{v} \tag{6.6}$$

with

$$A = \begin{pmatrix} V & I \\ H_0 & V \end{pmatrix}, \quad \tilde{v} = \begin{pmatrix} u \\ -i\frac{du}{dt} - Vu \end{pmatrix}. \tag{6.7}$$

We consider the problem for a symmetric operator V and a positive selfadjoint operator H_0 in a Hilbert space $\mathfrak{H}$ with $H_0 \geq m^2 > 0$. Additionally we suppose:

(c_1) V is $H_0^{1/2}$-compact, that is, dom $H_0^{1/2} \subseteq$ dom V and $S = VH_0^{-1/2}$ is compact.

Then A from (6.7) with its natural domain can be closed in $\tilde{\mathfrak{H}}_1 = \mathfrak{H} \oplus \mathfrak{H}$; denote the closure by $\tilde{A}_1$, and by $\tilde{\mathfrak{K}}_1$ the Kreĭn space which is obtained if $\tilde{\mathfrak{H}}_1$ is equipped with the $\begin{pmatrix} 0 & I \\ I & 0 \end{pmatrix}$–inner product.

There are other possibilities to associate a Kreĭn space and a selfadjoint operator with A. For example, let $\mathfrak{G}_\alpha$ be the Hilbert space which is the completion of dom H_0 with respect to the norm $\|x\|_\alpha = \|H_0^\alpha\|$ and equipped with this norm. Suppose instead of (c_1):

(c_2) dom $H_0^{3/4} \subseteq$ dom V and V is bounded as a mapping from $\mathfrak{G}_{3/4}$ into $\mathfrak{G}_{1/4}$.

Then the matrix A defines an operator A_2 in $\tilde{\mathfrak{H}}_2 = \mathfrak{G}_{3/4} \oplus \mathfrak{G}_{1/4}$ with domain dom $A_2 =$ dom $H_0^{3/4} \oplus$ dom $H_0^{1/4}$. Its closure exists and is denoted by $\tilde{A}_2$. With the $\begin{pmatrix} 0 & H_0^{-1/2} \\ H_0^{1/2} & 0 \end{pmatrix}$–inner product the space $\tilde{\mathfrak{H}}_2$ is a Kreĭn space which we denote by $\tilde{\mathfrak{K}}_2$.

Theorem 6.3 *If the conditions* (c_1) *and* (c_2) *are satisfied, for* $j = 1, 2$ *then:*
(i) $\tilde{A}_j$ *is a definitizable selfadjoint operator in* $\tilde{\mathfrak{K}}_j$.
(ii) *The number of negative squares of the form* $[\tilde{A}_j\tilde{x}, \tilde{y}], \tilde{x}, \tilde{y} \in$ dom $\tilde{A}_j$, *coincides with the total multiplicity of the negative eigenvalues of* $I - S^*S$.
(iii) $\sigma(\tilde{A}_1) = \sigma(\tilde{A}_2)$, $\sigma_{ess}(\tilde{A}_1) = \sigma_{ess}(\tilde{A}_2) = \{\lambda : \lambda^2 \in \sigma_{ess}(\tilde{H}_0)\}$.
(iv) *In the algebraic eigenspaces corresponding to isolated eigenvalues of finite algebraic multiplicity, the operators* $\tilde{A}_1$ *and* $\tilde{A}_2$ *are isomorphic.*

Although $\tilde{A}_1$ and $\tilde{A}_2$ have the same spectral properties in the finite complex domain, there is an essential difference in the behavior of the spectral functions of $\tilde{A}_1$ and $\tilde{A}_2$ at ∞. Namely, if $V = 0$ and H_0 is unbounded, then ∞ is a singular critical point of $\tilde{A}_1$ and a regular critical point of $\tilde{A}_2$. Under certain boundedness conditions on V, ∞ is also a regular critical point of $\tilde{A}_2$ if $V \neq 0$. Thus, in this case, a C_0-semigroup can be associated with the differential equations (6.5) and (6.6) .

Bibliography

Achieser, N.I. and Glasmann, I.M. [1981] *Theorie der Linearen Operatoren im Hilbertraum*, Akademie–Verlag, Berlin.

Alpay, D., Dijksma, A., Ploeg, J. van der and Snoo, H.S.V. de [1992] *Holomorphic operators between Kreĭn spaces and the number of squares of associated kernels*, Operator Theory: Adv. Appl. 59, Birkhäuser Verlag, Basel, 11–29.

Andô, T. [1979] *Linear Operators on Kreĭn Spaces*, Hokkaido University, Research Institute of Applied Electricity, Division of Applied Mathematics, Sapporo.

Atkinson, F.V. [1988] *Asymptotics of an eigenvalue problem involving an interior singularity*, in Proceedings of the Focused Research Program on Spectral Theory and Boundary Value Problems, Singular Differential Equations, Argonne National Laboratory ANL-87-26, vol. 2, 1-18.

Atkinson, F.V., Everitt, W.N. and Zettl, A. [1988] *Regularization of a Sturm–Liouville problem with an interior singularity using quasi–derivatives*, Differential and Integral Equations 1, 213–221.

Azizov, T.Ya. [1982] *On the theory of extensions of isometric and symmetric operators in spaces with an indefinite metric*, Preprint Voronesh University. Deposited paper 3420-82 (Russian).

Azizov, T.Ya. [1984] *Extensions of J–isometric and J–unitary operators*, Functional Anal. i Priložen 18, 57–58; English translation: Functional. Anal. Appl., 18 (1984), 46–48.

Azizov, T.Ya. and Iohkvidov, I.S. [1986] *Foundations of the Theory of Linear Operators in Spaces with Indefinite Metric*, "Nauka", Moscow, 1986; English translation: *Linear Operators in Spaces with an Indefinite Metric*, John Wiley, New York, 1990.

Baouendi, N.S. and Grisvard, P. [1968] *Sur une équation d'évolution changeant de type*, J. Functional Analysis 2, 352-367.

Barkwell, L., Lancaster, P. and Markus, A.S. [1992] *Gyroscopically stabilized systems: A class of quadratic eigenvalue problems with real spectrum*, Canadian J. of Math. 44, 42-53.

Beals, R. [1985] *Indefinite Sturm–Liouville problems and half range completeness*, J. Differential Equations 56, 391-407.

Benedek, A.I. and Panzone, R. [1981] *On Sturm-Liouville problems with the square-root of the eigenvalue parameter contained in the boundary conditions*, Notas de Algebra y Analysis 10, Universidad Nacional del Sur, Bahia Blanca.

Berezanskiĭ, Yu.M. [1968] *Expansions in Eigenfunctions of Selfadjoint Operators*, Transl. Amer. Math. Soc. 17, Providence, Rhode Island.

Berezanskiĭ, Yu.M. [1986] *Selfadjoint Operators in Spaces of Functions of Infinitely Many Variables*, Transl. Amer. Math. Soc. 63, Providence, Rhode Island.

Binding, P.A., Browne, P.J. and Seddighi, K. [1993] *Sturm–Liouville problems with eigenparameter dependent boundary conditions*, Proc. Edinburgh Math. Soc. 37, 57–72.

Binding, P.A. and Browne, P.J., *Application of two parameter eigencurves to Sturm–Liouville problems with eigenparameter dependent boundary conditions*, to appear in Proc. Roy. Soc. Edin.

Bognár, J. [1974] *Indefinite Inner Product Spaces*, Springer Verlag, Berlin.

Caspers, W. [1992] *On Point Interactions*, Thesis, Technical University Delft.

Coddington, E.A. and Levinson, N. [1955] *Theory of Ordinary Differential Equations*, McGraw-Hill, New York.

Ćurgus, B. [1985] *On the regularity of the critical point infinity of definitizable operators*, Integral Equations and Operator Theory 8, 462–488.

Ćurgus, B. *Abstract boundary value problems in Kreĭn spaces*, preprint.

Ćurgus, B. and Langer, H. [1989] *A Kreĭn space approach to symmetric ordinary differential operators with an indefinite weight function*, J. Differential Equations 79, 31-61.

Ćurgus B. and Najman, B. [1994] *A Kreĭn space approach to elliptic eigenvalue problems with an indefinite weight*, Differential and Integral Equations 7, 1241-1252.

Ćurgus B. and Najman, B., *Quasi uniformly positive operators in Kreĭn spaces*, to appear.

Daho, K. and Langer, H. [1977] *Sturm–Liouville operators with an indefinite weight function*, Proc. Roy. Soc. Edinburgh 87A, 161-191.

Diejen, J.F. van and Tip, A. [1991] *Scattering from generalized point interactions using selfadjoint extensions in Pontryagin spaces*, J. Math. Phys. 32 (3), 630-641.

Dijksma, A. [1980] *Eigenfunction expansions for a class of J-selfadjoint ordinary differential operators with boundary conditions containing the eigenvalue parameter*, Proc. Roy. Soc. Edinburgh 86A, 1–27.

Dijksma, A., Langer, H. and Snoo, H.S.V. de [1987] *Symmetric Sturm–Liouville operators with eigenvalue depending boundary conditions*, Canadian Math. Soc., Conference Proceedings 8, 87-116.

Dijksma, A., Langer, H. and Snoo, H.S.V. de [1987a] *Unitary colligations in Kreĭn spaces and their role in the extension theory of isometries and symmetric linear relations in Hilbert spaces*, Functional Analysis II, Proceedings Dubrovnik 1985, Lecture Notes in Mathematics 1242, Springer Verlag, Berlin, 1–42.

Dijksma, A., Langer, H. and Snoo, H.S.V. de [1988] *Hamiltonian systems with eigenvalue depending boundary conditons*, Operator Theory: Adv. Appl. 35, Birkhäuser Verlag, Basel, 37–83.

Dijksma, A., Langer, H. and Snoo, H.S.V. de [1993] *Eigenvalues and pole functions of Hamiltonian systems with eigenvalue depending boundary conditons*, Math. Nachr. 161, 107–154.

Dijksma, A., Langer, H. and Snoo, H.S.V. de, *Selfadjoint Kreĭn and Pontryagin space extensions and their parametrization*, unpublished manuscript.

Everitt, W.N. [1972] *On a property of the m–coefficient of a second order linear differential equation*, J. London Math. Soc. (2) 4, 443-457.

Everitt, W.N., Gunson, J. and Zettl, A. [1988] *Some comments on Sturm–Liouville eigenvalue problems with interior singularities*, J. Appl. Math. and Physics (ZAMP) 38, 813–838.

Everitt, W.N. and Zettl, A. [1986] *Sturm-Liouville differential operators on direct sum spaces*, Rocky Mountain J. of Math. 16(3), 497-516.

Everitt, W.N. and Zettl, A. [1992] *Differential operators generated by a countable number of quasi-differential expressions on the real line*, Proc. London Math. Soc. (3) 64, 524–544.

Exner, P. and Šeba, P. [1989] *Free quantum motion on a branching graph*, Reports on Mathematical Physics 28, 7-26.

Faierman, M., *On an elliptic eigenvalue problem involving an indefinite weight*, in preparation.

Fewster, C.J. [1993] *Resonance point interactions*, Dept. Appl. Math. Theor. Physics, Cambridge, preprint.

Fleige, A., [1995] *The turning point condition of R. Beals for indefinite Sturm-Liouville problems*, Math. Nachr., 172, 109–112.

Friedman, B. [1956] *Principles and Techniques of Applied Mathematics*, Wiley, New York.

Fulton, C.T. [1977] *Two–point boundary value problems with eigenvalue parameter contained in the boundary conditions*, Proc. Roy. Soc. Edinburgh 77A, 293–308.

Fulton, C.T. [1980] *Singular eigenvalue problems with eigenvalue parameter contained in the boundary conditions*, Proc. Roy. Soc. Edinburgh 87A, 1–34.

Fulton, C.T. and Pruess, S. [1979] *Numerical methods for a singular eigenvalue problem with eigenparameter in the boundary conditions*, J. Math. Anal. Appl. 71, 431–462.

Gerasimenko, N.I. and Pavlov, B.S. [1988] *Scattering problems on noncompact graphs*, Theor. Math. Phys. 74, 230-240.

Gesztesy, F. and Kirsch, W. [1985] *One–dimensional Schrödinger operators with interactions on a discrete set*, J. für Mathematik 362, 28-50.

Gohberg, I., Lancaster, P. and Rodman, L.[1982] *Matrix Polynomials*, Academic Press, New York.

Greenberg, W., Mee, C. van der and Protopopescu, V. [1987] *Boundary Value Problems in Abstract Kinetic Theory*, Operator Theory: Adv. Appl. 23, Birkhäuser Verlag, Basel.

Grossmann, A., Hoegh–Krohn, R. and Mebkhout, M. [1980] *A class of explicitly soluble, local, many–center Hamiltonians for one–particle quantum mechanics in two and three dimensions. I*, J. Math. Phys. 21 (9), 2376–2385.

Hara, T. [1992] *Operator inequalities and construction of Kreĭn spaces*, Integral Equations and Operator Theory 15, 551–567.

Hille, E. [1969] *Lectures on Ordinary Differential Equations*, Addison–Wesley, Reading.

Hinton, D.B. and Shaw, J.K. [1987] *Spectrum of a Hamiltonian system with spectral parameter in a boundary condition*, Can. Math. Soc. Conference Proc. 8, 171–186.

Iokhvidov, I.S., Kreĭn, M.G. and Langer, H. [1982] *Introduction to the Spectral Theory of Operators in Spaces with an Indefinite Metric*, Mathematical Research 9, Akademie–Verlag, Berlin.

Jonas, P. [1981] *On the functional calculus and the spectral function for definitizable operators in Kreĭn spaces*, Beiträge Anal. 16, 121-135.

Jonas, P. [1984] *Regularity criteria for critical points of definitizable operators*, Operator Theory: Adv. Appl. 14, Birkhäuser Verlag, Basel, 179-195.

Kac, I.S. and Kreĭn, M.G. [1968] *R–functions–analytic functions mapping the upper halfplane into itself*, Supplement I to the Russian edition of F.V. Atkinson, *Discrete and continuous boundary problems*, Mir, Moscow, 1968 (English translation: Amer. Math. Soc. Transl., (2) 103 (1974), 1-18).

Kac, I.S. and Kreĭn, M.G. [1968] *On the spectral functions of a string*, Supplement II to the Russian edition of F.V. Atkinson, *Discrete and continuous boundary problems*, Mir, Moscow, 1968 (English translation: Amer. Math. Soc. Transl., (2) 103 (1974), 19 -102).

Karpeshina, Yu.E. [1992] *Zero–range potential model of p–scattering by a potential well*, Forschungsinstitut für Mathematik, ETH, Zürich, and Univ. of St. Petersburg, preprint

Krall, A.M. [1968] *Differential operators and their adjoints under integral and multipoint boundary conditions*, J. Differential Equations 4, 327–336.

Krall, A.M. and Zettl, A. [1988] *Singular selfadjoint Sturm-Liouville problems II: Interior points*, Siam J. Math. Anal. 19(5), 1135-1141.

Kreĭn, M.G. and Langer, H. [1973] *Über die Q–Funktion eines π-hermiteschen Operators im Raume Π_κ*, Acta Sci. Math. (Szeged) 34, 191–230.

Kreĭin, M.G. and Langer, H. [1977] *Über einige Fortsetzungsprobleme, die eng mit der Theorie hermitescher Operatoren im Raume Π_κ zusammenhängen. I. Einige Funktionenklassen und ihre Darstellungen*, Math. Nachr., 77 (1977), 187-236; II. *Verallgemeinerte Resolventen, u–Resolventen und ganze Operatoren*, J. Funct. Anal., 30 (1978), 390–447; III. *Indefinite analogues of the Hamburger and Stieltjes moment problems. Part I*, Beiträge Anal., 14 (1979), 25–40; III. *Indefinite analogues of the Hamburger and Stieltjes moment problems. Part II*, Beiträge Anal., 15 (1980), 27–45; IV. *Continuous analogues of orthogonal polynomials on the unit circle with respect to an indefinite weight* *and related continuation problems for some classes of functions*, J. Operator Theory, 13 (1985), 299–417.

Kreĭn, M.G. and Langer, H. [1978] *On some mathematical principles in the linear theory of damped oscillations of continua*, Integral Equations and Operator Theory 1, 364-399 and 539-566.

Lancaster, P. [1964] *Lambda Matrices and Vibrating Systems*, Pergamon Press.

Lancaster, P. and Skhalikov, A. [1994] *Damped vibrations of beams and related spectral problems*, Canadian Applied Mathematics Quaterly 2, 1, 45-90.

Lancaster, P., Skhalikov, A. and Ye, Q. [1993] *Strongly definitizable linear pencils in Hilbert space*, Integral Equations and Operator Theory 17, 338-360.

Langer, H. [1972] *Zur Spektraltheorie verallgemeinerter gewöhnlicher Differentialoperatoren zweiter Ordnung mit einer nichtmonotonen Gewichtsfunktion*, Report 14, University of Jyväskylä.

Langer, H. [1982] *Spectral Functions of Definitizable Operators in Kreĭn Spaces*, Lecture Notes in Mathematics 948, Springer Verlag, Berlin, 1-46.

Langer, H., Markus, A. and Matsaev, V. *Locally definite operators in indefinite inner product spaces*, in preparation.

Langer H., Mennicken, R. and Möller, M. [1990] *A second order differential operator depending nonlinearly on the eigenvalue parameter*, Operator Theory: Adv. Appl. 48, Birkhäuser Verlag, Basel, 319-332.

Langer, H. and Najman, B. [ ] *A Kreĭn space approach to the Klein-Gordon equations*, unpublished manuscript.

Langer H. and Textorius, B. [1977] *On generalized resolvents and Q-functions of symmetric linear relations (subspaces) in a Hilbert space*, Pacific J. Math. 72 (1), 135–165.

Linden, H. [1991] *Leighton's bounds for Sturm–Liouville eigenvalues with eigenvalue parameter in the boundary conditions*, J. Math. Anal. Appl. 156, 444–456.

Markus, A.S. [1988] *Introduction to the spectral theory of operator polynomials*, Transl. of Math. Monographs 71, Amer. Math. Soc., Providence.

Naimark, M.A. [1967] *Linear Differential Operators* I, II, Harrap, London.

Niessen, H.D. [1970] *Singuläre S–hermitesche Rand–Eigenwertprobleme*, Manuscripta Math. 3, 35–68.

Pavlov, B.S. [1987] *The theory of extensions and explicitly–soluble models*, Uspekhi Mat. Nauk 42 (6), 99–131; English translation: Russian Math Surveys 42 (6) (1987), 127–168.

Pivovarcik, V.N. [1988] *On the number of eigenvalues of a Sturm–Liouville on the semi–axis with a potential which is linear in the parameter*, Diff. Uravnenija 24, 4, 705–708.

Pyatkov, S.G. [1989] *Some properties of eigenfunctions of linear pencils*, Sibirskiĭ Mat. Z. 30, 4, 111-124.

Rodman, L. [1989] *An Introduction to Operator Polynomials*, Operator Theory: Adv. Appl. 38, Birkhäuser Verlag, Basel.

Röh, H. [1981] *Self–adjoint subspace extensions satisfying* λ*-linear boundary conditions*, Proc. Royal Soc. Edinburgh 90A, 107-124.

Russakovskii, E.M. [1975] *Operator treatment of boundary problems with spectral parameters entering via polynomials in the boundary conditions*, Functional Anal. Appl. 9, 358–359.

Russakovskii, E.M., *The theory of V–Bezoutians and its applications*, to appear.

Schneider, A. [1981] *On spectral theory for the linear selfadjoint equation* $Fy = \lambda Gy$, in Ordinary and Partial Differential Equations, Proceedings, Dundee Scotland, 1980, Lecture Notes in Mathematics 846, Springer Verlag, Berlin, 306–332.

Shmul'yan, Yu.L. [1967] *Division in the class of J–expansive operators*, Mat. Sb. (N.S.)74 (116), 516–525; Math. USSR–Sbornik, 3 (1967), 471–479.

Shondin, Yu.G. [1988] *Quantum-mechanical models in* R_n *associated with extensions of the energy operator in Pontryagin space*, Theor. Math. Phys. 74, 220-230.

Shondin, Yu.G. [1991] *Perturbation of differential operators on high-codimensional manifold and the extension theory for symmetric linear relations in an indefinite metric space*, Lecture given at the workshop "Mathematical aspects of scattering theory and applications", Leningrad, May 1991.

Tsekanovskii, E.R. and Shmul'yan, Yu.L. [1977] *The theory of bi–extensions of operators on rigged Hilbert spaces, unbounded operator colligations and characteristic functions*, Russian Math. Surveys 32(5), 73-131.

Walter, J. [1973] *Regular eigenvalue problems with eigenvalue parameter in the boundary conditions*, Math. Z. 133, 301–312.

Weidman, J. [1976] *Lineare Operatoren in Hilberträumen*, B.G. Teubner, Stuttgart.

Zettl, A. [1968] *Adjoint and self–adjoint boundary value problems with interface conditions*, SIAM J. Appl. Math. 16, 851–859.

Zettl, A. [1988] *The Legendre equation on the whole line*, Proceedings of the International Conference on Theory and Applications of Differential Equations, Ohio University, 502–508.

Symbols Used in Lecture Series 2

$[\dot{+}]$ 77
$\mathbf{L}(\mathfrak{K}, \mathfrak{K}')$ 77
$\mathbf{L}(\mathfrak{K})$ 77
J 77
* 77
$\mathfrak{L}_0$ 77
$[\perp]$ 77
^{c} 77
$\mathcal{P}_+, \mathcal{P}_-, \mathcal{P}_0$ 78
$K_{\mathfrak{L}}$ 78
$L^2_{|\sigma|}$ 78
L^2_σ 78
$\mathfrak{H}_G$ 78
$^+$ 79
$\mathbf{N}_0$ 82
$c(A)$ 82
$\mathcal{R}_A$ 83
$L^1_{loc}(\iota)$ 87
$AC_{loc}(\iota)$ 87
$L^2(a,b)$ 87
L_{max} 87
L_{min} 88
J 89
$\mathcal{O}^*$ 93
$\mathcal{D}_{\alpha,\beta}$ 93
$\mathcal{D}'_{\alpha,\beta}$ 93
$N_{\alpha,\beta}(\lambda, \zeta)$ 94
$N_Q(\lambda, \zeta)$ 94
$\mathbf{N}_\kappa^{k\times k}$ 94
$\mathbf{N}_\kappa$ 94
$\mathbf{S}(\mathfrak{F}, \mathfrak{G})$ 99
$\mathcal{D}_\Theta$ 99
$\mathcal{D}(\Theta)$ 99
$\tilde{\Theta}(z)$ 99
$\Delta_\Theta(z,w)$ 99
$\Delta_{\tilde{\Theta}}(z,w)$ 100
$D_\Theta(z,w)$ 100
$\mathrm{ind}_-\,\mathfrak{F}$ 100
$C_\tau(T)$ 100
$K(\lambda)$ 101
b 102
Q 102
^{T} 104
L^2_r 109
Δ_+ 109
Δ_- 109
$L^2_{|r|}$ 109
$M_{\max}$ 109
M_{min} 109
M_α 110
A_α 110
l_M 126

LECTURE SERIES 3

OPERATORS ON INDEFINITE INNER PRODUCT SPACES

Michael A. Dritschel and James Rovnyak
Department of Mathematics
University of Virginia
Charlottesville, Virginia 22903-3199

Preface

Kreĭn space operator theory has long been a principal area of research in eastern mathematics, and it is now enjoying a surge of interest in western countries as well. The present lectures survey some recent methods and results which center around factorization properties of operators. Topics include the geometry of Kreĭn spaces, Pontryagin spaces, isometric operators, contractions, dilation and extension problems, reproducing kernel spaces, operator colligations, definitizable operators, and invariant subspaces.

Factorization ideas lead to notions of defect and Julia operators, which are tools for reducing the study of general operators to the isometric and unitary cases, where Hilbert space methods can often be used. Connections with interpolation theory and reproducing kernel spaces are explored. Factorization of selfadjoint operators also has applications in spectral theory and invariant subspace problems.

It is not possible to represent the field in six lectures, and the selection of topics is based on a particular viewpoint. Roughly, the viewpoint is that Kreĭn spaces provide a natural setting for operator theory. We assume Hilbert spaces or Pontryagin spaces only when there is some reason for it. Insights into the Hilbert space case come out of this approach, which separates the effects of positivity of vectors and positivity of operators. Connections between operator and spatial positivity then arise in the form of index formulas. Hilbert space methods find new applications in this approach, and operator theory is enriched by concepts which originate from the indefiniteness of inner products.

The list of omitted topics is long. For example, we do not discuss mappings of operator spheres and domains of holomorphy, connections with univalent functions, commutant lifting, interpolation theory, or complementation theory for Kreĭn spaces. Our bibliography is likewise limited, and we regret omissions. On any topic, the bibliography needs to be supplemented from the usual sources. Authoritative literature notes may be found in the books of Azizov and Iokhvidov [1989], Bognár [1974], and Iokhvidov, Kreĭn and Langer [1982].

Two other lecture series in this work complement our account. Systems aspects are treated in the lectures of M. A. Kaashoek on *State-Space Theory of Rational Matrix Functions and Applications.* Applications to differential operators and related topics such as operator polynomials are covered in the lectures of Aad Dijksma and Heinz Langer on *Operator Theory and Ordinary Differential Operators.*

We thank Peter Lancaster and the Program Committee for inviting us to contribute this series of lectures. We also thank The Fields Institute for their support, efficient organization, and hospitality. Both authors are supported by the National Science Foundation.

1991 *Mathematics Subject Classification.* Primary 47-02, 47B50, 47A20; Secondary 46C20, 46E22, 47A15, 47A48, 47A57.

Introduction: Preliminaries and Notation

A. Prerequisites

A **Hilbert space** is a strictly positive inner product space $\mathfrak{H}$ over the complex numbers which is complete in its norm metric. We also include separability in the axioms for a Hilbert space. Some familiarity with Hilbert spaces and operator theory is presumed. Otherwise the material covered in a first year graduate course of analysis is sufficient background. For example, see Halmos [1951] and Rudin [1974].

B. Inner product spaces

As we use the term, an **inner product space** is a complex vector space $\mathfrak{H}$ together with a complex-valued function $\langle\cdot,\cdot\rangle = \langle\cdot,\cdot\rangle_{\mathfrak{H}}$ on $\mathfrak{H} \times \mathfrak{H}$ which satisfies these axioms:

(1) (linearity) $\langle af + bg, h\rangle = a\langle f, h\rangle + b\langle g, h\rangle$ for all $f, g, h \in \mathfrak{H}$ and $a, b \in \mathbf{C}$;

(2) (symmetry) $\langle f, g\rangle = \overline{\langle g, f\rangle}$ for all $f, g \in \mathfrak{H}$.

The function $\langle\cdot,\cdot\rangle$ is called an **inner product**. By the **antispace** of an inner product space $(\mathfrak{H}, \langle\cdot,\cdot\rangle)$ we mean the inner product space $(\mathfrak{H}, -\langle\cdot,\cdot\rangle)$. Suppressing inner products, we sometimes write $-\mathfrak{H}$ for the antispace of an inner product space $\mathfrak{H}$. Thus $\langle\cdot,\cdot\rangle_{-\mathfrak{H}} = -\langle\cdot,\cdot\rangle_{\mathfrak{H}}$.

In any inner product space $\mathfrak{H}$, we use the notions of subspace, orthogonality, and direct sum in the same manner as for Hilbert spaces. For example, vectors f and g in an inner product space $\mathfrak{H}$ are called **orthogonal** if $\langle f, g\rangle = 0$. Elementary properties which follow as in the Hilbert space case are taken for granted.

An inner product $\langle\cdot,\cdot\rangle$ on $\mathfrak{H}$ is said to be **nondegenerate** if the only vector f in $\mathfrak{H}$ such that $\langle f, g\rangle = 0$ for all vectors $g \in \mathfrak{H}$ is $f = 0$. In applications, this is usually a consequence of stronger assumptions. For example, Hilbert spaces, Pontryagin spaces, and Kreĭn spaces (defined below) are nondegenerate inner product spaces. An inner product $\langle\cdot,\cdot\rangle$ on $\mathfrak{H}$ is said to be

(1) **positive** if $\langle f, f\rangle \geq 0$ for all $f \in \mathfrak{H}$,

(2) **negative** if $\langle f, f\rangle \leq 0$ for all $f \in \mathfrak{H}$,

(3) **definite** if it is either positive or negative,

(4) **indefinite** if it is neither positive nor negative, and

(5) **neutral** if it is both positive and negative (and hence the inner product is identically zero by the polarization identity).

A positive (negative) inner product is called **strictly positive (strictly negative)** if $\langle f, f\rangle = 0$ only for $f = 0$. These terms are applied to the space $\mathfrak{H}$ as well as the inner product. They are also applied to subspaces $\mathfrak{M}$ of $\mathfrak{H}$ if the restriction of the inner product has the corresponding property.

The term **operator** always refers to a linear transformation. A complex number λ is identified with the operator of multiplication by λ; the identity operator on a space $\mathfrak{H}$ is denoted 1 or $1_{\mathfrak{H}}$. We write dom A and ran A for the domain and range of an operator A.

Two inner product spaces $(\mathfrak{H}, \langle\cdot,\cdot\rangle_{\mathfrak{H}})$ and $(\mathfrak{K}, \langle\cdot,\cdot\rangle_{\mathfrak{K}})$ are called **isomorphic** if there is a one-to-one operator A from $\mathfrak{H}$ onto $\mathfrak{K}$ such that $\langle Af, Ag\rangle_{\mathfrak{K}} = \langle f, g\rangle_{\mathfrak{H}}$ for all $f, g \in \mathfrak{H}$. Such an operator A is said to be an **isomorphism**.

C. Kreĭn spaces and fundamental decompositions

The definition of a Kreĭn space is made in such a way that the class of Kreĭn spaces includes all Hilbert spaces, antispaces of Hilbert spaces, and is closed under the formation of orthogonal direct sums. It is the smallest such class of inner product spaces.

Definition. (1) *By a* **Kreĭn space** *we mean an inner product space $\mathfrak{H}$ which can be expressed as an orthogonal direct sum*

$$\mathfrak{H} = \mathfrak{H}_+ \oplus \mathfrak{H}_-, \tag{0.1}$$

where $\mathfrak{H}_+$ is a Hilbert space and $\mathfrak{H}_-$ is the antispace of a Hilbert space. Any such representation is called a **fundamental decomposition** *of the space.*

(2) *Given a Kreĭn space $\mathfrak{H}$ and fundamental decomposition* (0.1), *by $|\mathfrak{H}|$ we mean the Hilbert space obtained from* (0.1) *by replacing $\mathfrak{H}_-$ by its antispace $|\mathfrak{H}_-| = -\mathfrak{H}_-$:*

$$|\mathfrak{H}| = \mathfrak{H}_+ \oplus |\mathfrak{H}_-|. \tag{0.2}$$

The Hilbert space norm $\|\cdot\| = \|\cdot\|_{|\mathfrak{H}|}$ is called a **norm** *for the Kreĭn space $\mathfrak{H}$.*

In this situation, the operator $J_{\mathfrak{H}}$ defined on $\mathfrak{H}$ by

$$J_{\mathfrak{H}} f = f_+ - f_-$$

whenever $f = f_+ + f_-$ with $f_\pm \in \mathfrak{H}_\pm$ is called a **fundamental symmetry** or **signature operator** for the space. The relationship between the Kreĭn space

$$(\mathfrak{H}, \langle\cdot,\cdot\rangle_{\mathfrak{H}})$$

and the Hilbert space

$$(|\mathfrak{H}|, \langle\cdot,\cdot\rangle_{|\mathfrak{H}|})$$

is that $\mathfrak{H}$ and $|\mathfrak{H}|$ coincide as vector spaces and

$$\langle f, g\rangle_{|\mathfrak{H}|} = \langle J_{\mathfrak{H}} f, g\rangle_{\mathfrak{H}} \qquad \text{and} \qquad \langle f, g\rangle_{\mathfrak{H}} = \langle J_{\mathfrak{H}} f, g\rangle_{|\mathfrak{H}|}$$

for any vectors f and g.

Hilbert spaces and antispaces of Hilbert spaces are Kreĭn spaces. Excluding these cases, a Kreĭn space has infinitely many fundamental decompositions and therefore infinitely many associated Hilbert spaces $|\mathfrak{H}|$ and norms $\|\cdot\| = \|\cdot\|_{|\mathfrak{H}|}$. We compare these structures for two fundamental decompositions.

Theorem. *Let $\mathfrak{H}$ be a Kreĭn space with two fundamental decompositions*

$$\mathfrak{H} = \mathfrak{H}_+ \oplus \mathfrak{H}_- \qquad \text{and} \qquad \mathfrak{H} = \mathfrak{H}'_+ \oplus \mathfrak{H}'_-. \tag{0.3}$$

Define $X \in \mathfrak{L}(\mathfrak{H}_+, \mathfrak{H}'_+)$ by

$$Xf_+ = f'_+, \qquad f_+ \in \mathfrak{H}_+,$$

if $f_+ = f'_+ + f'_-$, $f'_\pm \in \mathfrak{H}'_\pm$. Define $Y \in \mathfrak{L}(|\mathfrak{H}_-|, |\mathfrak{H}'_-|)$ by

$$Yf_- = f'_-, \qquad f_- \in |\mathfrak{H}_-|,$$

if $f_- = f'_+ + f'_-$, $f'_\pm \in \mathfrak{H}'_\pm$. Then X and Y are one-to-one and onto.

Proof. Let us show that X has a closed graph. For any $z \in \mathfrak{H}$, the mapping $f \to \langle f, z\rangle_{\mathfrak{H}}$ is a continuous linear functional on $\mathfrak{H}_+$ and on $\mathfrak{H}'_+$. If $f_n = g_n + h_n$, where $f_n \in \mathfrak{H}_+$, $g_n \in \mathfrak{H}'_+$, and $h_n \in \mathfrak{H}'_-$ for all $n \geq 1$, and if $f_n \to f$ in $\mathfrak{H}_+$ and $g_n \to g$ in $\mathfrak{H}'_+$, then $f_n - g_n \perp \mathfrak{H}'_+$ in the inner product of $\mathfrak{H}$ for all $n \geq 1$, and so $f - g \perp \mathfrak{H}'_+$ in the inner product of $\mathfrak{H}$. In other words, $f = g + h$ with $h = f - g \in \mathfrak{H}'_-$. It follows that the graph of X is closed, and so X is continuous by the closed graph theorem.

Suppose that $f_+ \in \mathfrak{H}_+$ and $f_+ = f'_+ + f'_-$ with $f'_\pm \in \mathfrak{H}'_\pm$. Since

$$\langle f_+, f_+\rangle_{\mathfrak{H}_+} = \langle f'_+, f'_+\rangle_{\mathfrak{H}'_+} + \langle f'_-, f'_-\rangle_{\mathfrak{H}'_-} \leq \langle f'_+, f'_+\rangle_{\mathfrak{H}'_+},$$

$\|Xf_+\| \geq \|f_+\|$. Therefore X is one-to-one and has closed range.

The range of X is all of $\mathfrak{H}'_+$. For suppose that h belongs to $\mathfrak{H}'_+$ and is orthogonal to the range of X. Let $f_+ \in \mathfrak{H}_+$ and $f_+ = f'_+ + f'_-$ with $f'_\pm \in \mathfrak{H}'_\pm$. Then h is orthogonal to both $f'_+ = Xf_+$ and f'_-, and so h is orthogonal to f_+. Therefore $h \in \mathfrak{H}_-$ and $\langle h, h\rangle_{\mathfrak{H}'_+} = \langle h, h\rangle_{\mathfrak{H}_-} \leq 0$, and so $h = 0$. We have now shown that X is one-to-one and onto.

The assertions for Y may be deduced similarly or by applying what we just proved to the antispace of $\mathfrak{H}$. □

Corollary 1. *If $\mathfrak{H}$ is a Kreĭn space, the numbers*

$$\text{ind}_\pm \mathfrak{H} = \dim \mathfrak{H}_\pm$$

do not depend on the choice of decomposition (0.1).

We call $\text{ind}_\pm \mathfrak{H}$ the **positive** and **negative** indices of the space. A **Pontryagin space** is a Kreĭn space $\mathfrak{H}$ with $\text{ind}_- \mathfrak{H} < \infty$.

Corollary 2. *Let $\mathfrak{H}$ be a Kreĭn space with two fundamental decompositions* (0.3) *and associated Hilbert spaces*

$$|\mathfrak{H}| = \mathfrak{H}_+ \oplus |\mathfrak{H}_-| \qquad \text{and} \qquad |\mathfrak{H}|' = \mathfrak{H}'_+ \oplus |\mathfrak{H}'_-|. \tag{0.4}$$

Then $|\mathfrak{H}|$ and $|\mathfrak{H}|'$ have equivalent norms, that is, there exist $m > 0$ and $M > 0$ such that

$$m\,\|f\|_{|\mathfrak{H}|} \leq \|f\|_{|\mathfrak{H}|'} \leq M\,\|f\|_{|\mathfrak{H}|} \tag{0.5}$$

for all $f \in \mathfrak{H}$. Hence the norm topologies of the Hilbert spaces (0.4) *are identical.*

Proof. Represent an arbitrary vector f in $\mathfrak{H}$ in the form $f = f_+ + f_- = f'_+ + f'_-$, where $f_\pm \in \mathfrak{H}_\pm$ and $f'_\pm \in \mathfrak{H}'_\pm$. If X, Y are the operators of the theorem, then

$$\begin{aligned} f_+ &= Xf_+ + (1-X)f_+, \\ f_- &= (1-Y)f_- + Yf_-, \end{aligned}$$

where $Xf_+, (1-Y)f_- \in \mathfrak{H}'_+$ and $(1-X)f_+, Yf_- \in \mathfrak{H}'_-$. It follows that

$$\begin{aligned} f'_+ &= Xf_+ + (1-Y)f_-, \\ f'_- &= (1-X)f_+ + Yf_-. \end{aligned}$$

Similar relations hold with the roles of the fundamental decompositions reversed, and we obtain (0.5) for suitable constants m and M. □

Definition. *The* **strong topology** *of a Kreĭn space $\mathfrak{H}$ is the norm topology of any Hilbert space* (0.2) *associated with a fundamental decomposition* (0.1) *for $\mathfrak{H}$.*

Notions of convergence and continuity are understood to be with respect to the strong topology unless otherwise stated.

The closed graph theorem holds for Kreĭn spaces: an everywhere defined linear transformation on a Kreĭn space $\mathfrak{H}$ into a Kreĭn space $\mathfrak{K}$ is continuous if it has a closed graph. The proof is simply the observation that the strong topology of a Kreĭn space is a Hilbert space topology, and the closed graph theorem is true for Hilbert spaces.

The Riesz representation theorem holds for any Kreĭn space $\mathfrak{H}$: every continuous linear functional L on $\mathfrak{H}$ has the form

$$Lf = \langle f, g \rangle, \qquad f \in \mathfrak{H},$$

for a unique $g \in \mathfrak{H}$.

D. Example

Define the **Minkowski space $\mathbf{M}^{n+1}$** as the set of $(n+1)$-dimensional column vectors

$$c = \begin{pmatrix} c_1 \\ \vdots \\ c_n \\ \gamma \end{pmatrix}$$

with complex entries. Typographically, it is easier to write $c = (c_1, \dots, c_n, \gamma)^t$, using a superscript t to indicate a matrix transpose. Given two elements $a = (a_1, \dots, a_n, \alpha)^t$ and $b = (b_1, \dots, b_n, \beta)^t$ of the space, define their inner product by

$$\langle a, b \rangle_{\mathbf{M}^{n+1}} = a_1 \bar{b}_1 + \cdots + a_n \bar{b}_n - \alpha \bar{\beta}.$$

Then $\mathbf{M}^{n+1}$ is a Kreĭn space with $\operatorname{ind}_+ \mathbf{M}^{n+1} = n$ and $\operatorname{ind}_- \mathbf{M}^{n+1} = 1$. If we represent operators on the space as matrices in the usual way, then a fundamental symmetry for the space is given by

$$J_{\mathbf{M}^{n+1}} = \begin{pmatrix} I_n & 0 \\ 0 & -1 \end{pmatrix},$$

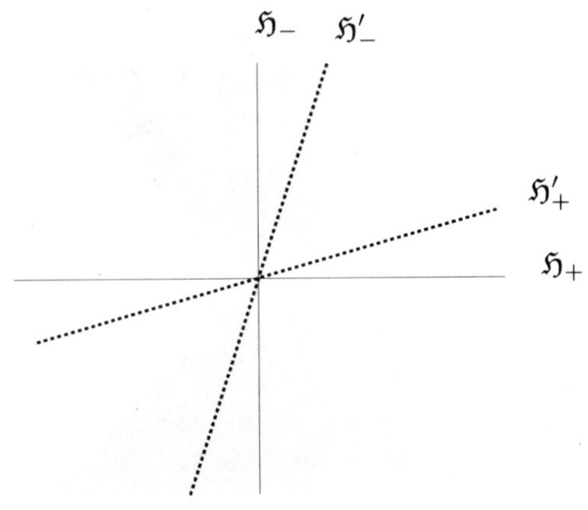

Figure 1

where I_n is the identity matrix of order n. Then $|\mathbf{M}^{n+1}| = \mathbf{C}^{n+1}$ in the Euclidean metric.

One fundamental decomposition $\mathfrak{H} = \mathfrak{H}_+ \oplus \mathfrak{H}_-$ of the space $\mathbf{M}^2$ is obtained with $\mathfrak{H}_+$ equal to the span of $(1,0)^t$ and $\mathfrak{H}_-$ equal to the span of $(0,1)^t$. A second can be chosen, for example, with the spans of $(4,1)^t$ and $(1,4)^t$, as shown in figure 1. In figure 1, the shaded region represents all vectors f such that $\langle f, f \rangle \leq 0$ (negative cone), the light region all vectors f such that $\langle f, f \rangle \geq 0$ (positive cone).

Figure 1 is actually a 2-dimensional slice of a 4-dimensional space. A similar figure for $\mathbf{M}^3$ can also be visualized. In this case, the "negative" part of the space is a solid cone. This method of visualization is common in approaches that feature applications, such as Hassibi, Sayed, and Kailath [1994], and relativity texts.

E. Books and expositions

Standard sources on indefinite inner product spaces are Andô [1979], Azizov and Iokhvidov [1989], Bognár [1974], and Iokhvidov, Kreĭn, and Langer [1982]. The finite-dimensional case is treated in Gohberg, Lancaster, and Rodman [1983]. Dritschel and Rovnyak [1990] also include some introductory material.

Lecture 1: Kreĭn Spaces and Operators

Key ideas:

- Kreĭn spaces are treated like Hilbert spaces, with the same notation used for geometric concepts and classes of operators, including selfadjoint operators, projections, and isometries.
- The focus is on unique objects (strong topology, inner product, operators), and we de-emphasize nonunique ones (fundamental decompositions, norms).

The geometry of Kreĭn spaces has many similarities with the Hilbert space case but also some important differences such as the presence of negative subspaces. It is convenient to study Kreĭn space geometry concurrently with basic classes of operators, including selfadjoint operators, projections, and isometries. In this lecture we introduce the main classes of operators and show their relationship to the geometry of Kreĭn spaces.

Kreĭn spaces and their strong topologies have already been defined in §C of the Introduction. Underlying spaces throughout are assumed to be Kreĭn spaces if nothing further is stated. By $\mathfrak{L}(\mathfrak{H}, \mathfrak{K})$ we mean the space of continuous operators on $\mathfrak{H}$ to $\mathfrak{K}$. Set $\mathfrak{L}(\mathfrak{H}) = \mathfrak{L}(\mathfrak{H}, \mathfrak{H})$. For every $A \in \mathfrak{L}(\mathfrak{H}, \mathfrak{K})$ there is a unique $A^* \in \mathfrak{L}(\mathfrak{K}, \mathfrak{H})$ such that

$$\langle Af, g \rangle_{\mathfrak{K}} = \langle f, A^* g \rangle_{\mathfrak{H}}, \qquad f \in \mathfrak{H},\ g \in \mathfrak{K}.$$

We call A^* the **adjoint** of A. The existence of the adjoint follows from the Riesz representation theorem, and elementary properties are immediate as in the Hilbert space case. For example, when the operations are defined,

$$(A + B)^* = A^* + B^*, \qquad (AB)^* = B^* A^*.$$

The space $\mathfrak{L}(\mathfrak{H}, \mathfrak{K})$ has the structure of a Banach space depending on choices of fundamental decompositions $\mathfrak{H} = \mathfrak{H}_+ \oplus \mathfrak{H}_-$ and $\mathfrak{K} = \mathfrak{K}_+ \oplus \mathfrak{K}_-$ and associated fundamental symmetries $J_{\mathfrak{H}}$ and $J_{\mathfrak{K}}$. Let $\|\cdot\|_{|\mathfrak{H}|}$ and $\|\cdot\|_{|\mathfrak{K}|}$ be the norms for the corresponding Hilbert spaces $|\mathfrak{H}| = \mathfrak{H}_+ \oplus |\mathfrak{H}_-|$ and $|\mathfrak{K}| = \mathfrak{K}_+ \oplus |\mathfrak{K}_-|$. Our notation for fundamental symmetries and associated Hilbert spaces follows §C of the Introduction. Then $\mathfrak{L}(\mathfrak{H}, \mathfrak{K})$ coincides with $\mathfrak{L}(|\mathfrak{H}|, |\mathfrak{K}|)$. We call the norm $\|\cdot\|$ of $\mathfrak{L}(|\mathfrak{H}|, |\mathfrak{K}|)$ an **operator norm** for $\mathfrak{L}(\mathfrak{H}, \mathfrak{K})$. For any $A \in \mathfrak{L}(\mathfrak{H}, \mathfrak{K})$,

$$|\langle Af, g \rangle_{\mathfrak{K}}| \leq \|A\| \, \|f\|_{|\mathfrak{H}|} \, \|g\|_{|\mathfrak{K}|}, \qquad f \in \mathfrak{H},\ g \in \mathfrak{K}.$$

Any two operator norms for $\mathfrak{L}(\mathfrak{H}, \mathfrak{K})$ are equivalent, and we thus obtain a unique **uniform topology** for $\mathfrak{L}(\mathfrak{H}, \mathfrak{K})$. **Strong** and **weak operator topologies** for $\mathfrak{L}(\mathfrak{H}, \mathfrak{K})$ are similarly defined from the corresponding Hilbert space notions.

The Kreĭn space and Hilbert space concepts of adjoint are related. Viewing $A \in \mathfrak{L}(\mathfrak{H}, \mathfrak{K})$ as an element of $\mathfrak{L}(|\mathfrak{H}|, |\mathfrak{K}|)$ as above, we write $A^\times \in \mathfrak{L}(|\mathfrak{K}|, |\mathfrak{H}|)$ for the Hilbert space adjoint. Then $A^* = J_\mathfrak{H} A^\times J_\mathfrak{K}$. For if $f \in \mathfrak{H}$ and $g \in \mathfrak{K}$, then

$$\langle Af, g\rangle_\mathfrak{K} = \langle J_\mathfrak{K} Af, g\rangle_{|\mathfrak{K}|} = \langle f, A^\times J_\mathfrak{K} g\rangle_{|\mathfrak{H}|} = \langle f, J_\mathfrak{H} A^\times J_\mathfrak{K} g\rangle_\mathfrak{H}.$$

In particular, $\|A^*\| = \|A\|$ for any $A \in \mathfrak{L}(\mathfrak{H}, \mathfrak{K})$ and any operator norm on $\mathfrak{L}(\mathfrak{H}, \mathfrak{K})$.

We say that

(1) $A \in \mathfrak{L}(\mathfrak{H})$ is **selfadjoint** if $A^* = A$;
(2) $A \in \mathfrak{L}(\mathfrak{H})$ is a **projection** if A is selfadjoint and $A^2 = A$;
(3) $A \in \mathfrak{L}(\mathfrak{H}, \mathfrak{K})$ is **isometric** if $A^*A = 1_\mathfrak{H}$;
(4) $A \in \mathfrak{L}(\mathfrak{H}, \mathfrak{K})$ is **partially isometric** if $AA^*A = A$;
(5) $A \in \mathfrak{L}(\mathfrak{H}, \mathfrak{K})$ is **unitary** if both A and A^* are isometric;
(6) $A \in \mathfrak{L}(\mathfrak{H}, \mathfrak{K})$ is a **contraction** if $A^*A \le 1_\mathfrak{H}$;
(7) $A \in \mathfrak{L}(\mathfrak{H}, \mathfrak{K})$ is a **bicontraction** if both A and A^* are contractions.

The definition of a contraction in part (6) uses the partial ordering of selfadjoint operators. The selfadjoint operators on $\mathfrak{H}$ are partially ordered by writing $A \ge 0$ if

$$\langle Af, f\rangle \ge 0, \qquad f \in \mathfrak{H}.$$

Then $A \le B$ means that $B - A \ge 0$. A selfadjoint operator A is said to be **positive** or **negative** according as $A \ge 0$ or $A \le 0$.

An operator $A \in \mathfrak{L}(\mathfrak{H}, \mathfrak{K})$ is a contraction if and only if

$$\langle Af, Af\rangle_\mathfrak{K} \le \langle f, f\rangle_\mathfrak{H}, \qquad f \in \mathfrak{H}.$$

An operator $A \in \mathfrak{L}(\mathfrak{H}, \mathfrak{K})$ is unitary if and only if it is an isomorphism between the Kreĭn spaces $\mathfrak{H}$ and $\mathfrak{K}$.

Theorem 1.1 (Bognár-Krámli Factorization) *Let $\mathfrak{H}$ be a Kreĭn space. Any selfadjoint operator $A \in \mathfrak{L}(\mathfrak{H})$ can be written in the form*

$$A = DD^*,$$

where $D \in \mathfrak{L}(\mathfrak{D}, \mathfrak{H})$ for some Kreĭn space $\mathfrak{D}$ and $\ker\ D = \{0\}$.

Proof. Let $J_\mathfrak{H}$ be a fundamental symmetry for $\mathfrak{H}$ and $|\mathfrak{H}|$ the associated Hilbert space. Since $A = A^* = J_\mathfrak{H} A^\times J_\mathfrak{H}$, $AJ_\mathfrak{H}$ is selfadjoint as an element of $\mathfrak{L}(|\mathfrak{H}|)$. Its polar decomposition has the form

$$AJ_\mathfrak{H} = RU = UR,$$

where R is a nonnegative operator in $\mathfrak{L}(|\mathfrak{H}|)$ and U is a partial isometry in $\mathfrak{L}(|\mathfrak{H}|)$ with initial and final space equal to $\overline{\text{ran } R}$. In fact, U is also selfadjoint, and

$$\mathfrak{D} = \overline{\text{ran } R}$$

is a Kreĭn space with signature operator $J_\mathfrak{D} = U\big|_\mathfrak{D}$. Define $D \in \mathfrak{L}(\mathfrak{D}, \mathfrak{H})$ by

$$Df = R^{1/2} f, \qquad f \in \mathfrak{D}.$$

A short calculation shows that $D^* = UR^{1/2} J_\mathfrak{H}$, and so

$$DD^* = R^{1/2} U R^{1/2} J_\mathfrak{H} = AJ_\mathfrak{H} J_\mathfrak{H} = A.$$

By construction, D has zero kernel. $\square$

If $\mathfrak{M}$ is a subspace of a Kreĭn space $\mathfrak{H}$, its **orthogonal companion** is defined by

$$\mathfrak{M}^{\perp} = \{g \;:\; g \in \mathfrak{H} \text{ and } \langle f, g\rangle = 0 \text{ for all } f \in \mathfrak{M}\}.$$

For any subspaces $\mathfrak{M}$ and $\mathfrak{N}$ of $\mathfrak{H}$,

(1) $\mathfrak{M}^{\perp\perp} = \overline{\mathfrak{M}}$,

(2) $(\mathfrak{M} + \mathfrak{N})^{\perp} = \mathfrak{M}^{\perp} \cap \mathfrak{N}^{\perp}$,

(3) if $\mathfrak{M}$ and $\mathfrak{N}$ are closed, then $(\mathfrak{M} \cap \mathfrak{N})^{\perp} = \overline{\mathfrak{M}^{\perp} + \mathfrak{N}^{\perp}}$,

(4) $\overline{\mathfrak{M}} = \mathfrak{H}$ if and only if $\mathfrak{M}^{\perp} = \{0\}$.

The bars indicate closure. The assertions can be deduced from the Hilbert space case by choosing a fundamental symmetry $J_{\mathfrak{H}}$, and then noting that $\mathfrak{M}^{\perp}$ coincides with the orthogonal complement of $J_{\mathfrak{H}}\mathfrak{M}$ in the corresponding Hilbert space $|\mathfrak{H}|$.

The term "orthogonal complement" for $\mathfrak{M}^{\perp}$ is avoided, because the projection theorem fails in general for Kreĭn spaces. For a Hilbert space $\mathfrak{H}$, the projection theorem asserts that the relation

$$\mathfrak{H} = \mathfrak{M} \oplus \mathfrak{M}^{\perp} \tag{1.1}$$

holds for any closed subspace $\mathfrak{M}$ of $\mathfrak{H}$. An example where (1.1) fails is the span $\mathfrak{M}$ of the vector

$$\begin{pmatrix} 1 \\ 1 \end{pmatrix}$$

in the Minkowski space $\mathbf{M}^2$ (Introduction, §D). If (1.1) holds for a closed subspace $\mathfrak{M}$ of a Kreĭn space $\mathfrak{H}$, we also call $\mathfrak{M}^{\perp}$ the **orthogonal complement** of $\mathfrak{M}$.

Every closed subspace $\mathfrak{M}$ of a Kreĭn space $\mathfrak{H}$ has a **closed complement**: by this we mean a closed subspace $\mathfrak{N}$ of $\mathfrak{H}$ such that

$$\mathfrak{H} = \mathfrak{M} + \mathfrak{N} \qquad \text{and} \qquad \mathfrak{M} \cap \mathfrak{N} = \{0\}.$$

This is proved by choosing a fundamental decomposition for $\mathfrak{H}$ and taking $\mathfrak{N}$ to be the Hilbert space orthogonal complement of $\mathfrak{M}$ in the associated space $|\mathfrak{H}|$. Then $\mathfrak{M}$ and $\mathfrak{N}$ are orthogonal relative to the Hilbert space inner product but generally not in the Kreĭn space inner product. A closed subspace $\mathfrak{M}$ of a Kreĭn space $\mathfrak{H}$ typically has infinitely many closed complements. It is an important problem to identify those subspaces which have orthogonal complements, that is, for which (1.1) holds.

Definition 1.2 *By a* **Kreĭn subspace** *of a Kreĭn space $\mathfrak{H}$ we mean a subspace $\mathfrak{M}$ of $\mathfrak{H}$ which is itself a Kreĭn space in the inner product of $\mathfrak{H}$, that is,*

$$\langle f, g\rangle_{\mathfrak{M}} = \langle f, g\rangle_{\mathfrak{H}}, \qquad f, g \in \mathfrak{M}.$$

In the literature, these are called regular subspaces. For an example, let $\mathfrak{M}$ be a finite-dimensional subspace of $\mathfrak{H}$ spanned by the linearly independent vectors $e_1, \dots, e_n$. Then $\mathfrak{M}$ is a Kreĭn subspace of $\mathfrak{H}$ if and only if the **Gram matrix**

$$G(e_1, \dots, e_n) = \left(\langle e_j, e_k\rangle_{\mathfrak{H}}\right)_{j,k=1}^{n}$$

is invertible. Hint for the proof: put the Gram matrix in diagonal form.

The next result shows that Kreĭn space projections are similar to Hilbert space projections, but the geometry applies only to Kreĭn subspaces.

Theorem 1.3 *If $\mathfrak{M}$ is a subspace of a Kreĭn space $\mathfrak{H}$, the following assertions are equivalent:*

(1) *$\mathfrak{M}$ is a Kreĭn subspace;*

(2) *$\mathfrak{H} = \mathfrak{M} \oplus \mathfrak{M}^{\perp}$;*

(3) *$\mathfrak{M}$ is the range of a projection operator P.*

In this case $\mathfrak{M}$ is closed and the strong topology of $\mathfrak{M}$ viewed as a Kreĭn space in the inner product of $\mathfrak{H}$ coincides with the restriction of the strong topology of $\mathfrak{H}$ to $\mathfrak{M}$.

We often denote the projection of a Kreĭn space $\mathfrak{H}$ onto a Kreĭn subspace $\mathfrak{M}$ by $\mathrm{Pr}_{\mathfrak{M}}$.

Proof of Theorem 1.3. (3) $\Rightarrow$ (1) Let $\mathfrak{M} = \operatorname{ran} P$, where $P \in \mathfrak{L}(\mathfrak{H})$ is a projection. Use the Bognár-Krámli factorization to write $P = DD^*$, where $D \in \mathfrak{L}(\mathfrak{D}, \mathfrak{H})$ and $\ker\ D = \{0\}$. Since $P^2 = P$,

$$DD^*DD^* = DD^*.$$

Since D has zero kernel, D^* has dense range, and therefore $DD^*D = D$ and $D^*D = 1$. It follows that D preserves inner products and $\mathfrak{M} = \operatorname{ran} DD^* = \operatorname{ran} D$. Now (1) follows because by what we have shown, D is an isomorphism from the Kreĭn space $\mathfrak{D}$ onto $\mathfrak{M}$ in the inner product of $\mathfrak{H}$. Notice that $\mathfrak{M} = \operatorname{ran} D$ is a closed subspace of $\mathfrak{H}$ because D has a left inverse.

(2) $\Rightarrow$ (3) If (2) holds, then every $h \in \mathfrak{H}$ has a unique representation $h = f + g$ with $f \in \mathfrak{M}$ and $g \in \mathfrak{M}^{\perp}$. The relation $Ph = f$ defines an operator $P \in \mathfrak{L}(\mathfrak{H})$ by the closed graph theorem. Using (2), we verify that P is a projection. Clearly $\mathfrak{M} = \operatorname{ran} P$, and so (3) follows.

The proof will be completed later. □

Recall the classification of subspaces in §B of the Introduction: A subspace $\mathfrak{M}$ of a Kreĭn space $\mathfrak{H}$ is positive, negative, definite, indefinite, neutral, strictly positive, or strictly negative if the restriction of the inner product of $\mathfrak{H}$ to $\mathfrak{M}$ has the corresponding property. These notions are illustrated for the space $\mathbf{M}^2$ in figure 2. Subspaces $\mathfrak{M}_0$ in the boundary of the shaded region in figure 2 are neutral. Subspaces $\mathfrak{M}_-$ in the shaded region are negative (strictly negative when they are not on the boundary). Subspaces $\mathfrak{M}_+$ in the unshaded region are positive (strictly positive when they are not on the boundary).

Definition 1.4 *Let $\mathfrak{H}$ be a Kreĭn space with norm $\|\cdot\|$. A subspace $\mathfrak{M}$ of $\mathfrak{H}$ is called* **uniformly positive** *if there is a number $\delta > 0$ such that*

$$\langle f, f \rangle \geq \delta \|f\|^2, \qquad f \in \mathfrak{M}.$$

We say that $\mathfrak{M}$ is **uniformly negative** *if there is a number $\delta > 0$ such that*

$$\langle f, f \rangle \leq -\delta \|f\|^2, \qquad f \in \mathfrak{M}.$$

These notions are independent of the choice of norm.

A closed subspace $\mathfrak{M}$ of a Kreĭn space $\mathfrak{H}$ is uniformly positive if and only if it is a Hilbert space in the inner product of $\mathfrak{H}$. The necessity part of this assertion is clear. Conversely, suppose that $\mathfrak{M}$ is a Hilbert space in the inner product of $\mathfrak{H}$. Denote this Hilbert space $\mathfrak{M}_1$. Let $\mathfrak{M}_2$ be $\mathfrak{M}$ in any norm of $\mathfrak{H}$. The identity mapping from

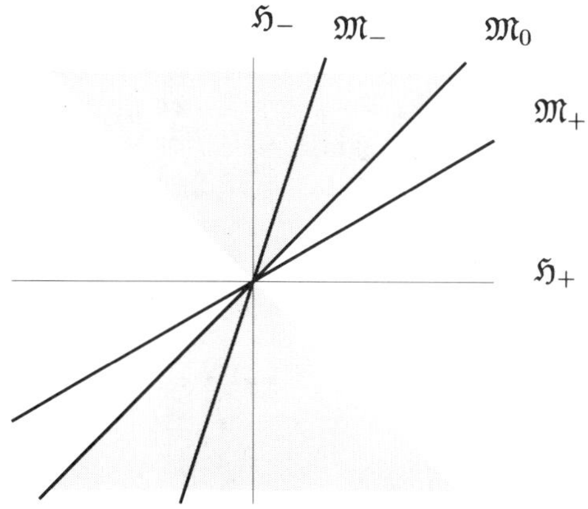

Figure 2

$\mathfrak{M}_2$ to $\mathfrak{M}_1$ is bounded and hence continuous. By the open mapping theorem, the inverse mapping is also bounded, and therefore $\mathfrak{M}$ is uniformly positive. Similar remarks apply to uniformly negative subspaces.

A subspace is said to be **maximal** with respect to some property if it has the property and is not properly contained in another subspace which has the property. For example, a maximal positive subspace is a positive subspace $\mathfrak{M}$ of $\mathfrak{H}$ which is not properly contained in another positive subspace of $\mathfrak{H}$. For "positive" we can substitute "negative," "uniformly positive," and so forth.

Definition 1.5 (Graph representation of a definite subspace) *Let $\mathfrak{H}$ be a Kreĭn space with fundamental decomposition $\mathfrak{H} = \mathfrak{H}_+ \oplus \mathfrak{H}_-$, and let $\mathfrak{M}$ be a negative subspace of $\mathfrak{H}$. We identify $\mathfrak{H}$ as a vector space with $\mathfrak{H}_+ \times |\mathfrak{H}_-|$, so $\mathfrak{H}$ may alternatively be viewed as a space of pairs*

$$(g, f)^t = \begin{pmatrix} g \\ f \end{pmatrix}$$

with $g \in \mathfrak{H}_+$ and $f \in |\mathfrak{H}_-|$. The **angle operator** *for $\mathfrak{M}$ is the Hilbert space contraction operator K with* dom $K \subseteq |\mathfrak{H}_-|$ *and* ran $K \subseteq \mathfrak{H}_+$ *such that as a subset of $\mathfrak{H}_+ \times |\mathfrak{H}_-|$,*

$$\mathfrak{M} = \text{Graph } (K) = \left\{ \begin{pmatrix} Kf \\ f \end{pmatrix} \ : \ f \in \text{dom } K \right\}.$$

The graph representation of a positive subspace is defined analogously.

To see that such an operator K exists, observe that $\mathfrak{M}$ can contain no nonzero element of the form $(g, 0)^t$ because $\mathfrak{M}$ is negative. This is the condition that $\mathfrak{M} = \text{Graph } (K)$ for some operator K with dom $K \subseteq |\mathfrak{H}_-|$ and ran $K \subseteq \mathfrak{H}_+$. Since $\mathfrak{M}$ is negative, for each $f \in \text{dom } K$,

$$\langle Kf + f, Kf + f \rangle_{\mathfrak{H}} \leq 0$$

and so

$$\|Kf\|_{\mathfrak{H}_+} \leq \|f\|_{|\mathfrak{H}_-|} .$$

Thus $\|K\| \leq 1$. This proves the existence of the angle operator.

Let $\mathfrak{M}$ be a negative subspace of a Kreĭn space $\mathfrak{H}$ represented as a graph as in Definition 1.5. The following assertions are straightforward:

(1) $\mathfrak{M}$ is closed if and only if dom K is closed in $|\mathfrak{H}_-|$;

(2) $\mathfrak{M}$ is uniformly negative if and only if $\|K\| < 1$;

(3) $\mathfrak{M}$ is maximal negative if and only if dom $K = |\mathfrak{H}_-|$, that is, the projection of $\mathfrak{M}$ onto $\mathfrak{H}_-$ is all of $\mathfrak{H}_-$;

(4) $\mathfrak{M}$ is maximal uniformly negative if and only if dom $K = |\mathfrak{H}_-|$ and $\|K\| < 1$.

It follows that a maximal uniformly negative subspace is maximal negative. Every maximal negative subspace is closed.

Property (3) has these frequently used consequences. Let $\mathfrak{K} = \mathfrak{H} \oplus \mathfrak{G}$, where $\mathfrak{H}$ is a Kreĭn space and $\mathfrak{G}$ is a Hilbert space. Every maximal negative subspace of $\mathfrak{H}$ is a maximal negative subspace of $\mathfrak{K}$. The projection of any maximal negative subspace of $\mathfrak{K}$ into $\mathfrak{H}$ is maximal negative in $\mathfrak{H}$.

Theorem 1.6 *Let $\mathfrak{H}$ be a Kreĭn space.*

(1) *A subspace $\mathfrak{M}$ of $\mathfrak{H}$ is maximal negative if and only if $\mathfrak{M}^\perp$ is maximal positive in $\mathfrak{H}$.*

(2) *If $\mathfrak{M}_+$ is a maximal uniformly positive subspace of $\mathfrak{H}$, then $\mathfrak{M}_- = \mathfrak{M}_+^\perp$ is maximal uniformly negative. Moreover $\mathfrak{H} = \mathfrak{M}_+ \oplus \mathfrak{M}_-$ is a fundamental decomposition.*

(3) *Every negative subspace of $\mathfrak{H}$ is contained in a maximal negative subspace of $\mathfrak{H}$.*

(4) *Every uniformly negative subspace of $\mathfrak{H}$ is contained in a maximal uniformly negative subspace of $\mathfrak{H}$.*

(5) *Assume that $\mathfrak{M}_\pm$ are uniformly positive/negative subspaces of $\mathfrak{H}$ such that $\mathfrak{M}_+ \perp \mathfrak{M}_-$. Then there exists a fundamental decomposition $\mathfrak{H} = \mathfrak{K}_+ \oplus \mathfrak{K}_-$ such that $\mathfrak{M}_\pm \subseteq \mathfrak{K}_\pm$.*

A similar result holds with the words "positive" and "negative" interchanged. Part (5) is related to Phillips' theorem in Lecture 6. We prove parts (2), (4), and (5). See, for example, Andô [1979] and Azizov and Iokhvidov [1989] for further details and historical notes on the graph method.

Proof of Theorem 1.6, parts (2), (4) and (5). We use the notation for fundamental decompositions in Definition 1.5.

(2) By the discussion above, $\mathfrak{M}_+$ has the form

$$\mathfrak{M}_+ = \left\{ \begin{pmatrix} u \\ Ku \end{pmatrix} : u \in \mathfrak{H}_+ \right\},$$

where $K \in \mathfrak{L}(\mathfrak{H}_+, |\mathfrak{H}_-|)$ is a uniform contraction: $\|K\| < 1$. Then

$$\mathfrak{M}_- = \left\{ \begin{pmatrix} K^\times v \\ v \end{pmatrix} : v \in |\mathfrak{H}_-| \right\}.$$

Since $K^\times \in \mathfrak{L}(|\mathfrak{H}_-|, \mathfrak{H}_+)$ and $\|K^\times\| < 1$, $\mathfrak{M}_-$ is maximal uniformly negative. By the remarks following Definition 1.4, $\mathfrak{M}_+$ and $-\mathfrak{M}_-$ are Hilbert spaces in the inner product of $\mathfrak{H}$. By elementary algebra, every element of $\mathfrak{H}_+ \oplus |\mathfrak{H}_-|$ has a unique representation

$$\begin{pmatrix} f \\ g \end{pmatrix} = \begin{pmatrix} u \\ Ku \end{pmatrix} + \begin{pmatrix} K^\times v \\ v \end{pmatrix},$$

with $u \in \mathfrak{H}_+$ and $v \in |\mathfrak{H}_-|$. It follows that $\mathfrak{H} = \mathfrak{M}_+ \oplus \mathfrak{M}_-$ is a fundamental decomposition.

(4) Let $\mathfrak{M}$ be a uniformly negative subspace of $\mathfrak{H}$, and represent $\mathfrak{M}$ as a graph as in Definition 1.5. Then $\|K\| < 1$. Extend K in any way to an operator in $\mathfrak{L}(|\mathfrak{H}_-|, \mathfrak{H}_+)$ with norm less than 1. The graph of the extension is a maximal uniformly negative subspace of $\mathfrak{H}$ which contains $\mathfrak{M}$.

(5) By (4), we can extend $\mathfrak{M}_+$ to a maximal uniformly positive subspace $\tilde{\mathfrak{M}}_+$ of $\mathfrak{H}$ (strictly speaking, we use the version of (4) for positive subspaces). By the remarks following Definition 1.4, $\mathfrak{M}_+$ and $\tilde{\mathfrak{M}}_+$ are Hilbert spaces in the inner product of $\mathfrak{H}$. By the projection theorem for Hilbert spaces,

$$\tilde{\mathfrak{M}}_+ = \mathfrak{M}_+ \oplus \mathfrak{N}_+,$$

where $\mathfrak{N}_+$ is a subspace of $\mathfrak{H}$ which is a Hilbert space in the inner product of $\mathfrak{H}$. By (2), $\tilde{\mathfrak{M}}_- = \tilde{\mathfrak{M}}_+^{\perp}$ is maximal uniformly negative and $\mathfrak{H} = \tilde{\mathfrak{M}}_+ \oplus \tilde{\mathfrak{M}}_-$ is a fundamental decomposition. Therefore

$$\mathfrak{H} = (\mathfrak{M}_+ \oplus \mathfrak{N}_+) \oplus \tilde{\mathfrak{M}}_- = \mathfrak{M}_+ \oplus (\mathfrak{N}_+ \oplus \tilde{\mathfrak{M}}_-).$$

One sees easily out of this that $\mathfrak{N}_+ \oplus \tilde{\mathfrak{M}}_- = \mathfrak{M}_+^{\perp}$ is a Kreĭn space in the inner product of $\mathfrak{H}$. Since $\mathfrak{M}_-$ is now a uniformly negative subspace of $\mathfrak{M}_+^{\perp}$, we can apply a similar argument to show that

$$\mathfrak{M}_+^{\perp} = \mathfrak{M}_- \oplus (\mathfrak{P}_+ \oplus \mathfrak{P}_-),$$

where $\mathfrak{P}_+$ and $-\mathfrak{P}_-$ are subspaces of $\mathfrak{H}$ which are Hilbert spaces in the inner product of $\mathfrak{H}$. Therefore

$$\mathfrak{H} = \mathfrak{M}_+ \oplus \mathfrak{M}_+^{\perp} = (\mathfrak{M}_+ \oplus \mathfrak{P}_+) \oplus (\mathfrak{M}_- \oplus \mathfrak{P}_-).$$

The required fundamental decomposition is obtained with $\mathfrak{K}_\pm = \mathfrak{M}_\pm \oplus \mathfrak{P}_\pm$. □

The proof of Theorem 1.3 can now be completed.

Proof of Theorem 1.3, remaining parts. (1) ⇒ (2) Assume that $\mathfrak{M}$ is a Kreĭn space in the inner product of $\mathfrak{H}$. Then $\mathfrak{M} = \mathfrak{M}_+ \oplus \mathfrak{M}_-$, where $\mathfrak{M}_+$ is a Hilbert space and $\mathfrak{M}_-$ is the antispace of a Hilbert space in the inner product of $\mathfrak{H}$. By Theorem 1.6(5), $\mathfrak{M}_\pm \subseteq \mathfrak{H}_\pm$ for some fundamental decomposition $\mathfrak{H} = \mathfrak{H}_+ \oplus \mathfrak{H}_-$ of $\mathfrak{H}$. We easily check that $\mathfrak{M}^{\perp}$ is the direct sum of $\mathfrak{H}_+ \ominus \mathfrak{M}_+$ and $\mathfrak{H}_- \ominus \mathfrak{M}_-$, and so (2) holds.

The equivalence of the three conditions has been established. Recall that we showed $\mathfrak{M}$ closed in the implication (3) ⇒ (1). The remaining topological properties at the end of the theorem are clear from the relationship between the fundamental decompositions of $\mathfrak{M}$ and $\mathfrak{H}$ in the implication (1) ⇒ (2). □

Partial isometries on Kreĭn spaces have much the same behavior as in the Hilbert space case. The schematic diagram in figure 3 shows the behavior of a partial isometry $A \in \mathfrak{L}(\mathfrak{H}, \mathfrak{K})$ with respect to its initial and final subspaces $\mathfrak{M}$ and $\mathfrak{N}$. Details are in Theorems 1.7 and 1.8.

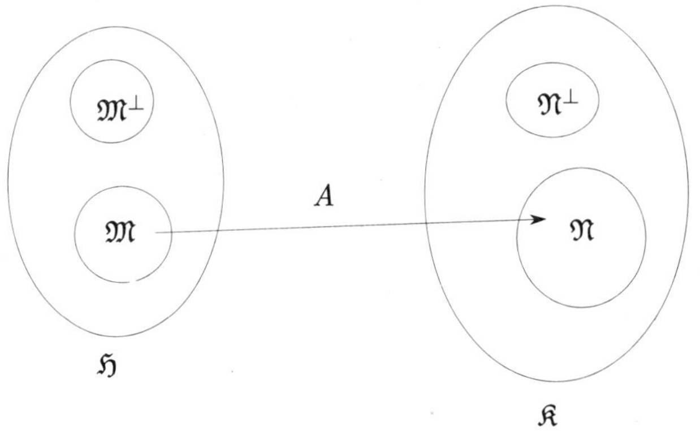

Figure 3

Theorem 1.7 *If $A \in \mathfrak{L}(\mathfrak{H}, \mathfrak{K})$ for some Kreĭn spaces $\mathfrak{H}$ and $\mathfrak{K}$, then A is a partial isometry if and only if there exist Kreĭn subspaces $\mathfrak{M}$ of $\mathfrak{H}$ and $\mathfrak{N}$ of $\mathfrak{K}$ such that*

(1) *A maps $\mathfrak{M}$ in a one-to-one way onto $\mathfrak{N}$, and*

$$\langle Af, Ag\rangle_{\mathfrak{K}} = \langle f, g\rangle_{\mathfrak{H}}, \qquad f, g \in \mathfrak{M},$$

(2) *$Af = 0$ for all $f \in \mathfrak{M}^\perp$.*

In this case, we call $\mathfrak{M}$ and $\mathfrak{N}$ the **initial** *and* **final subspaces** *for A. Moreover A^* is a partial isometry with initial space $\mathfrak{N}$ and final space $\mathfrak{M}$, and*

$$\begin{aligned} A^*A &= \mathrm{Pr}_{\mathfrak{M}}, \qquad & \ker\ A &= \ker\ \mathrm{Pr}_{\mathfrak{M}}, \\ AA^* &= \mathrm{Pr}_{\mathfrak{N}}, \qquad & \ker\ A^* &= \ker\ \mathrm{Pr}_{\mathfrak{N}}. \end{aligned}$$

Theorem 1.8 *Let $A \in \mathfrak{L}(\mathfrak{H}, \mathfrak{K})$, where $\mathfrak{H}$ and $\mathfrak{K}$ are Kreĭn spaces. The following assertions are equivalent:*

(1) *A is a partial isometry;*

(2) *A^*A is a projection operator and $\ker\ A^*A = \ker\ A$;*

(3) *AA^* is a projection operator and $\ker\ AA^* = \ker\ A^*$.*

Proofs of Theorems 1.7 and 1.8 follow basically Hilbert space methods. Where positivity of inner products is apparently required, the conditions on kernels provide what is needed. See Dritschel and Rovnyak [1990], pp. 8–10.

Notice that in Theorem 1.8, the condition $\ker\ A^*A = \ker\ A$ in (2) holds automatically if $\mathfrak{K}$ is a Hilbert space, and the condition $\ker\ AA^* = \ker\ A^*$ in (3) holds automatically if $\mathfrak{H}$ is a Hilbert space. An example of an operator A such that A^*A is a projection but $\ker\ A^*A \neq \ker\ A$ is any nonzero operator whose range is a neutral subspace, since then $A^*A = 0$. Such operators appear later in Lecture 3. We can even construct an example such that both A^*A and AA^* are projections but A is not a partial isometry. To do this, let A be a Hilbert space projection onto a nonzero closed neutral subspace of a Kreĭn space $\mathfrak{H}$ relative to some fundamental symmetry J. As before, $A^*A = 0$. This means that $JAJA = 0$, and hence $AA^* = AJAJ = 0$. The operator A is not a partial isometry because $AA^*A \neq A$.

The results on partial isometries apply in particular to isometries. Using this information and some elementary reasoning, we deduce:

Corollary 1.9 *An isometry* $A \in \mathfrak{L}(\mathfrak{H}, \mathfrak{K})$ *is a one-to-one operator whose range is a Kreĭn subspace of* $\mathfrak{K}$. *Therefore* A *is a Kreĭn-space isomorphism from* $\mathfrak{H}$ *onto* ran A *viewed as a Kreĭn space in the inner product of* $\mathfrak{K}$.

In particular, an isometry $A \in \mathfrak{L}(\mathfrak{H}, \mathfrak{K})$ maps closed subspaces of $\mathfrak{H}$ onto closed subspaces of $\mathfrak{K}$ and negative (positive) subspaces of $\mathfrak{H}$ onto negative (positive) subspaces of $\mathfrak{K}$. It also maps Kreĭn subspaces onto Kreĭn subspaces.

The structure of Kreĭn space isometries is more complicated than in the Hilbert space case. The Wold decomposition has no completely satisfactory generalization for Kreĭn spaces. Even when one exists, the shift component may not have the same behavior as in the Hilbert space situation. Those Kreĭn space isometries which have a structure analogous to Hilbert space shifts have been characterized by McEnnis [1990].

Lecture 2:
Julia Operators and Contractions

Key ideas:

- Basic tools of Kreĭn space operator theory include defect and Julia operators, Shmul′yan's theorem on extending densely defined operators, and Pontryagin's theorem on dense sets in a Pontryagin space.
- Contractions are an important special case, and they are simpler because their defect spaces are Hilbert spaces.

It will be seen that an arbitrary Kreĭn space operator $A \in \mathfrak{L}(\mathfrak{H}, \mathfrak{K})$ has extensions to isometric, coisometric, and unitary operators. The construction is an application of the Bognár-Krámli factorization. The extensions provide a very useful tool for the study of general operators.

A sharper form of the Bognár-Krámli factorization is needed. It uses a notion of index for selfadjoint operators. If $A \in \mathfrak{L}(\mathfrak{H})$ is a selfadjoint operator, let $\mathrm{ind}_-(A)$ be the supremum of all positive integers r such that there exists a nonpositive and invertible matrix of the form $\left(\langle Af_j, f_k\rangle_{\mathfrak{H}}\right)_{j,k=1}^{r}$; if no such r exists, set $\mathrm{ind}_-(A) = 0$. Put $\mathrm{ind}_+(A) = \mathrm{ind}_-(-A)$. We call $\mathrm{ind}_\pm(A)$ the **indices** of A. They measure how much A behaves like a positive or negative operator. For example, $A \geq 0$ if and only if $\mathrm{ind}_-(A) = 0$.

Theorem 2.1 *Let $\mathfrak{H}$ be a Kreĭn space. Any selfadjoint operator $A \in \mathfrak{L}(\mathfrak{H})$ can be written in the form*

$$A = DD^*,$$

where $D \in \mathfrak{L}(\mathfrak{D}, \mathfrak{H})$ for some Kreĭn space $\mathfrak{D}$ and ker $D = \{0\}$. For any such factorization, $\mathrm{ind}_\pm \mathfrak{D} = \mathrm{ind}_\pm(A)$.

The first statement in Theorem 2.1 is Theorem 1.1. The second is proved in Dritschel and Rovnyak [1990], Th. 1.2.1.

Definition 2.2 *Let $T \in \mathfrak{L}(\mathfrak{H}, \mathfrak{K})$.*

(i) *By a* **defect operator** *for T we mean any operator $\tilde{D} \in \mathfrak{L}(\tilde{\mathfrak{D}}, \mathfrak{H})$ having zero kernel such that*

$$\begin{pmatrix} T \\ \tilde{D}^* \end{pmatrix} \in \mathfrak{L}(\mathfrak{H}, \mathfrak{K} \oplus \tilde{\mathfrak{D}}) \tag{2.1}$$

is an isometry, that is, $T^*T + \tilde{D}\tilde{D}^* = 1_{\mathfrak{H}}$. We then call $\tilde{\mathfrak{D}}$ a **defect space** *for T.*

(ii) *By a* **Julia operator** *for T we mean any unitary operator*

$$\begin{pmatrix} T & D \\ \tilde{D}^* & -L^* \end{pmatrix} \in \mathfrak{L}(\mathfrak{H} \oplus \mathfrak{D}, \mathfrak{K} \oplus \tilde{\mathfrak{D}}) \tag{2.2}$$

such that $D \in \mathfrak{L}(\mathfrak{D}, \mathfrak{K})$ and $\tilde{D} \in \mathfrak{L}(\tilde{\mathfrak{D}}, \mathfrak{H})$ have zero kernels.

The notation in (2.2) is a change from Dritschel and Rovnyak [1990], [1991]: the 22-entry is now written $-L^*$ instead of L. This is closer to the notation in Sz.-Nagy and Foiaş [1970] for the Hilbert space case, and it makes some later formulas more symmetric.

If U is the matrix (2.2), the relations

$$U^*U = 1_{\mathfrak{H}\oplus\mathfrak{D}} \qquad \text{and} \qquad UU^* = 1_{\mathfrak{K}\oplus\tilde{\mathfrak{D}}}$$

are equivalent to the six identities

$$\left.\begin{aligned} T^*T + \tilde{D}\tilde{D}^* &= 1_{\mathfrak{H}}, \\ D^*D + LL^* &= 1_{\mathfrak{D}}, \\ TT^* + DD^* &= 1_{\mathfrak{K}}, \\ \tilde{D}^*\tilde{D} + L^*L &= 1_{\tilde{\mathfrak{D}}}, \end{aligned}\right\} \tag{2.3a, b, c, d}$$

and

$$\left.\begin{aligned} T^*D &= \tilde{D}L^*, \\ T\tilde{D} &= DL, \end{aligned}\right\} \tag{2.4a, b}$$

In particular, $\tilde{D} \in \mathfrak{L}(\tilde{\mathfrak{D}}, \mathfrak{H})$ is a defect operator for T and $D \in \mathfrak{L}(\mathfrak{D}, \mathfrak{K})$ is a defect operator for T^*. The operator U^* is a Julia operator for T^*.

We use the Bognár-Krámli factorization to show that Julia operators exist. This argument is similar to the method used by Davis [1970] to construct unitary dilations of general operators on a Hilbert space: the method of Davis works with minor adaptations for Kreĭn space operators. An existence proof based on a lemma of Kreĭn was given by Arsene, Constantinescu, and Gheondea [1987]. A generalization is given in Theorem 6.2.

Theorem 2.3 *Let $T \in \mathfrak{L}(\mathfrak{H}, \mathfrak{K})$.*

(i) *A defect operator $\tilde{D} \in \mathfrak{L}(\tilde{\mathfrak{D}}, \mathfrak{H})$ for T exists, and for any such operator,*

$$\operatorname{ind}_{\pm} \tilde{\mathfrak{D}} = \operatorname{ind}_{\pm} (1 - T^*T). \tag{2.5}$$

(ii) *If $\tilde{D} \in \mathfrak{L}(\tilde{\mathfrak{D}}, \mathfrak{H})$ is any defect operator for T, there exists a Julia operator for T of the form*

$$\begin{pmatrix} T & D \\ \tilde{D}^* & -L^* \end{pmatrix} \in \mathfrak{L}(\mathfrak{H} \oplus \mathfrak{D}, \mathfrak{K} \oplus \tilde{\mathfrak{D}}). \tag{2.6}$$

For any such operator,

$$\left.\begin{aligned} \operatorname{ind}_{\pm} \mathfrak{D} &= \operatorname{ind}_{\pm} (1 - TT^*), \\ \operatorname{ind}_{\pm} \tilde{\mathfrak{D}} &= \operatorname{ind}_{\pm} (1 - T^*T). \end{aligned}\right\} \tag{2.7}$$

Proof. (i) Apply Theorem 2.1 to the selfadjoint operator $1 - T^*T$.

(ii) If $\tilde{D} \in \mathfrak{L}(\tilde{\mathfrak{D}}, \mathfrak{H})$ is a defect operator for T, then

$$V = \begin{pmatrix} T \\ \tilde{D}^* \end{pmatrix} \in \mathfrak{L}(\mathfrak{H}, \mathfrak{K} \oplus \tilde{\mathfrak{D}})$$

is an isometry. By Theorem 1.7, VV^* is the projection onto ran V, and therefore $1 - VV^*$ is the projection onto $(\operatorname{ran} V)^{\perp} = \ker\ V^*$. Factor

$$1 - VV^* = BB^*,$$

where $B \in \mathfrak{L}(\mathfrak{D}, \mathfrak{K} \oplus \tilde{\mathfrak{D}})$ has zero kernel. Then B has the form

$$B = \begin{pmatrix} D \\ -L^* \end{pmatrix},$$

where $D \in \mathfrak{L}(\mathfrak{D}, \mathfrak{K})$ and $L \in \mathfrak{L}(\tilde{\mathfrak{D}}, \mathfrak{D})$. We verify that with this choice of D and L, (2.6) is a Julia operator.

By construction, BB^* is a projection and ker BB^* = ker B^* (because B has zero kernel). It follows from Theorem 1.8 that B is a partial isometry. Thus B is an isometry with ran B = ker V^*. Therefore $V^*B = 0$ and $B^*B = 1$. These identities imply (2.3b) and (2.4a). The relation (2.3a) follows from the assumption that $\tilde{D}$ is a defect operator for T. The identities (2.3c,d) and (2.4b) follow from the relation $1 - VV^* = BB^*$.

The operator $\tilde{D}$ has zero kernel because it is a defect operator by hypothesis. It remains to show that D has zero kernel. If $Df = 0$ for some $f \in \mathfrak{D}$, then $L^*f = 0$ by (2.4a). Hence $Bf = 0$, and so $f = 0$. Therefore D has zero kernel, and we have shown that (2.6) is a Julia operator.

The relations (2.7) follow from part (i) and the fact that $\tilde{D} \in \mathfrak{L}(\tilde{\mathfrak{D}}, \mathfrak{H})$ is a defect operator for T and $D \in \mathfrak{L}(\mathfrak{D}, \mathfrak{K})$ is a defect operator for T^*. □

In the proof of Theorem 2.3(ii), when $T \in \mathfrak{L}(\mathfrak{H}, \mathfrak{K})$ is an isometry, $\tilde{\mathfrak{D}}$ is the zero space and so $T = V$. It follows that an isometry $V \in \mathfrak{L}(\mathfrak{H}, \mathfrak{K})$ has Julia operator

$$\begin{pmatrix} V & D \end{pmatrix} \in \mathfrak{L}(\mathfrak{H} \oplus \mathfrak{D}, \mathfrak{K}),$$

where $D \in \mathfrak{L}(\mathfrak{D}, \mathfrak{K})$ is a defect operator for V^*.

A characterization of contractions and bicontractions follows from Theorem 2.3. Let $T \in \mathfrak{L}(\mathfrak{H}, \mathfrak{K})$ have Julia operator (2.6). By (2.7), T is a contraction if and only if $\tilde{\mathfrak{D}}$ is a Hilbert space, and T is a bicontraction if and only if both $\mathfrak{D}$ and $\tilde{\mathfrak{D}}$ are Hilbert spaces.

Theorem 2.4 *If $T \in \mathfrak{L}(\mathfrak{H}, \mathfrak{K})$, then*

$$\text{ind}_\pm \mathfrak{H} + \text{ind}_\pm (1 - TT^*) = \text{ind}_\pm \mathfrak{K} + \text{ind}_\pm (1 - T^*T). \tag{2.8}$$

In particular, if $\mathfrak{H}$ and $\mathfrak{K}$ are Pontryagin spaces with the same negative index, then

$$\text{ind}_- (1 - TT^*) = \text{ind}_- (1 - T^*T). \tag{2.9}$$

Proof. Let (2.6) be a Julia operator for T. This establishes an isomorphism between the Kreĭn spaces $\mathfrak{H} \oplus \mathfrak{D}$ and $\mathfrak{K} \oplus \tilde{\mathfrak{D}}$. Hence by (2.7),

$$\begin{aligned} \text{ind}_\pm \mathfrak{H} + \text{ind}_\pm (1 - TT^*) &= \text{ind}_\pm \mathfrak{H} + \text{ind}_\pm \mathfrak{D} \\ &= \text{ind}_\pm \mathfrak{K} + \text{ind}_\pm \tilde{\mathfrak{D}} \\ &= \text{ind}_\pm \mathfrak{K} + \text{ind}_\pm (1 - T^*T). \end{aligned}$$

The second statement follows immediately from the first. □

Corollary 2.5 *If $\mathfrak{H}$ and $\mathfrak{K}$ are Pontryagin spaces with the same negative index, then every contraction $T \in \mathfrak{L}(\mathfrak{H}, \mathfrak{K})$ is a bicontraction.*

Proof. By (2.9), $1 - TT^* \geq 0$ if and only if $1 - T^*T \geq 0$. □

Two Julia operators

$$\begin{pmatrix} T & D \\ \tilde{D}^* & -L^* \end{pmatrix} \in \mathfrak{L}(\mathfrak{H} \oplus \mathfrak{D}, \mathfrak{K} \oplus \tilde{\mathfrak{D}})$$

and

$$\begin{pmatrix} T & D' \\ \tilde{D}'^* & -L'^* \end{pmatrix} \in \mathfrak{L}(\mathfrak{H} \oplus \mathfrak{D}', \mathfrak{K} \oplus \tilde{\mathfrak{D}}')$$

for a given $T \in \mathfrak{L}(\mathfrak{H}, \mathfrak{K})$ are abstractly indistinguishable if there exist Kreĭn space isomorphisms $\varphi : \mathfrak{D} \to \mathfrak{D}'$ and $\psi : \tilde{\mathfrak{D}} \to \tilde{\mathfrak{D}}'$ such that

$$D = D'\varphi, \qquad \tilde{D} = \tilde{D}'\psi, \qquad \varphi L = L'\psi,$$

since this has the effect merely to replace the defect spaces $\mathfrak{D}$ and $\tilde{\mathfrak{D}}$ by isomorphic copies. We say that T has **essentially unique Julia operator** if any two Julia operators for T are so related.

Theorem 2.6 *An operator $T \in \mathfrak{L}(\mathfrak{H}, \mathfrak{K})$ has essentially unique Julia operator if either* $\mathrm{ind}_-(1_{\mathfrak{H}} - T^*T) < \infty$ *or* $\mathrm{ind}_-(1_{\mathfrak{K}} - TT^*) < \infty$. *In particular, T has essentially unique Julia operator if either T or T^* is a contraction.*

For a proof, see Dritschel and Rovnyak [1990], Th. B4.

The basic properties of contraction operators on a Kreĭn space were worked out by Potapov, Ginzburg, Kreĭn, and Shmul'yan in the 1950's and 60's (references are given in Dritschel and Rovnyak [1990], p. 33). The key to these properties is the behavior of a contraction operator on negative subspaces.

Theorem 2.7 *If $T \in \mathfrak{L}(\mathfrak{H}, \mathfrak{K})$ is a contraction, then*

(1) *T maps every closed negative subspace $\mathfrak{M}$ of $\mathfrak{H}$ in a one-to-one way onto a closed negative subspace of $\mathfrak{K}$;*

(2) *T maps every closed uniformly negative subspace $\mathfrak{M}$ of $\mathfrak{H}$ in a one-to-one way onto a closed uniformly negative subspace of $\mathfrak{K}$;*

(3) *the kernel of T is a Hilbert space in the inner product of $\mathfrak{H}$.*

Proof. Fix norms for $\mathfrak{H}$ and $\mathfrak{K}$. Choose a defect operator $\tilde{D} \in \mathfrak{L}(\tilde{\mathfrak{D}}, \mathfrak{H})$ for T, so

$$V = \begin{pmatrix} T \\ \tilde{D}^* \end{pmatrix}$$

is an isometry.

(1) Since T is a contraction, $\tilde{\mathfrak{D}}$ is a Hilbert space, and

$$\begin{aligned} \|\tilde{D}^* f\|^2 &= \langle (1 - T^*T)f, f\rangle_{\mathfrak{H}} \\ &= \langle f, f\rangle_{\mathfrak{H}} - \langle Tf, Tf\rangle_{\mathfrak{K}} \\ &\le -\langle Tf, Tf\rangle_{\mathfrak{K}} \\ &\le \|Tf\|^2 \end{aligned} \tag{2.10}$$

whenever $f \in \mathfrak{H}$ and $\langle f, f\rangle_{\mathfrak{H}} \le 0$. In particular, if $\langle f, f\rangle_{\mathfrak{H}} \le 0$ and $Tf = 0$, then $Vf = 0$ and $f = 0$ by Corollary 1.9. Thus T is one-to-one on any negative subspace.

Suppose that $\mathfrak{M}$ is a closed negative subspace of $\mathfrak{H}$. Clearly $T\mathfrak{M}$ is negative, because T is a contraction. Suppose that $\{f_n\}_1^\infty$ is a sequence in $\mathfrak{M}$ and $Tf_n \to g$

in the strong topology of $\mathfrak{K}$. By (2.10), $\tilde{D}^* f_n \to h$ in the strong topology of $\tilde{\mathfrak{D}}$ for some $h \in \tilde{\mathfrak{D}}$. Then

$$Vf_n \to \begin{pmatrix} g \\ h \end{pmatrix}.$$

Since $V\mathfrak{M}$ is closed,

$$\begin{pmatrix} g \\ h \end{pmatrix} = Vf$$

for some $f \in \mathfrak{M}$, so $g = Tf \in T\mathfrak{M}$. This shows that $T\mathfrak{M}$ is closed.

(2) Let $\mathfrak{M}$ be a closed uniformly negative subspace of $\mathfrak{H}$. Since $\tilde{\mathfrak{D}}$ is a Hilbert space and $V\mathfrak{M}$ is uniformly negative, there exists a $\delta > 0$ such that for all $f \in \mathfrak{M}$,

$$\begin{aligned} \langle Tf, Tf \rangle_{\mathfrak{K}} &\leq \langle Vf, Vf \rangle_{\mathfrak{K} \oplus \tilde{\mathfrak{D}}} \\ &\leq -\delta \|Vf\|^2 \\ &= -\delta \left(\|Tf\|^2 + \|\tilde{D}^* f\|^2 \right) \\ &\leq -\delta \|Tf\|^2. \end{aligned} \tag{2.11}$$

It follows that $T\mathfrak{M}$ is uniformly negative. Since $T\mathfrak{M}$ is closed by (1), (2) follows.

(3) Let $\mathfrak{N} = \ker T$. Then $V\mathfrak{N} = \{0\} \oplus \tilde{D}^*\mathfrak{N}$ is closed and uniformly positive. Since V^* acts as an isometry on ran V into $\mathfrak{H}$, V^* maps closed uniformly positive subspaces of ran V onto closed uniformly positive subspaces of $\mathfrak{H}$. Thus $V^*V\mathfrak{N} = \mathfrak{N}$ is a Hilbert space in the inner product of $\mathfrak{H}$. □

By Theorem 2.7(2), a contraction $T \in \mathfrak{L}(\mathfrak{H}, \mathfrak{K})$ maps a closed negative subspace $\mathfrak{M}$ of $\mathfrak{H}$ onto a closed negative subspace $T\mathfrak{M}$ of $\mathfrak{K}$. If $\mathfrak{M}$ is maximal negative in $\mathfrak{H}$, it is closed and $T\mathfrak{M}$ is therefore closed and negative in $\mathfrak{K}$; however, $T\mathfrak{M}$ is not necessarily maximal negative in $\mathfrak{K}$. For an example, consider the operator $T = 0$ on a Hilbert space $\mathfrak{H}$ into a Kreĭn space $\mathfrak{K}$ which is not a Hilbert space. Notice that in this example, T is a contraction but not a bicontraction.

Theorem 2.8 (Ginzburg) *If $T \in \mathfrak{L}(\mathfrak{H}, \mathfrak{K})$ is a contraction, the following are equivalent:*

(1) *T is a bicontraction;*

(2) *T maps some maximal negative subspace of $\mathfrak{H}$ onto a maximal negative subspace of $\mathfrak{K}$;*

(3) *T maps every maximal negative subspace of $\mathfrak{H}$ onto a maximal negative subspace of $\mathfrak{K}$.*

In this case, T maps every maximal uniformly negative subspace of $\mathfrak{H}$ onto a maximal uniformly negative subspace of $\mathfrak{K}$.

Other equivalent conditions for a contraction to be a bicontraction are given in an addendum at the end of this lecture.

Proof. (1) $\Rightarrow$ (3) Assume that T is a bicontraction. Then in any Julia operator

$$U = \begin{pmatrix} T & D \\ \tilde{D}^* & -L^* \end{pmatrix} \in \mathfrak{L}(\mathfrak{H} \oplus \mathfrak{D}, \mathfrak{K} \oplus \tilde{\mathfrak{D}})$$

for T, the defect spaces $\mathfrak{D}$ and $\tilde{\mathfrak{D}}$ are Hilbert spaces. Let $\mathfrak{M}$ be any maximal negative subspace of $\mathfrak{H}$. Then $\mathfrak{M}$ is a maximal negative subspace of $\mathfrak{H} \oplus \mathfrak{D}$. Therefore $U\mathfrak{M}$

is a maximal negative subspace of $\mathfrak{K} \oplus \tilde{\mathfrak{D}}$. It follows that

$$T\,\mathfrak{M} = \mathrm{Pr}_{\mathfrak{K}}\ U\,\mathfrak{M}$$

is a maximal negative subspace of $\mathfrak{K}$, by remarks preceding Theorem 1.6. Here $\mathrm{Pr}_{\mathfrak{K}}$ denotes projection onto $\mathfrak{K}$. Thus (3) holds.

(3) $\Rightarrow$ (2) Trivial.

(2) $\Rightarrow$ (1) Assume that T maps the maximal negative subspace $\mathfrak{M}$ of $\mathfrak{H}$ onto the maximal negative subspace $\mathfrak{N}$ of $\mathfrak{K}$. Since T is a contraction by hypothesis, $\tilde{\mathfrak{D}}$ is a Hilbert space. Therefore the operator

$$T' = \mathrm{Pr}_{\mathfrak{K}}\ U \in \mathfrak{L}(\mathfrak{H} \oplus \mathfrak{D}, \mathfrak{K})$$

is a contraction: for any $f \in \mathfrak{H}$ and $g \in \mathfrak{D}$,

$$\begin{aligned}\left\langle T' \begin{pmatrix} f \\ g \end{pmatrix}, T' \begin{pmatrix} f \\ g \end{pmatrix} \right\rangle_{\mathfrak{K}} &= \left\langle \mathrm{Pr}_{\mathfrak{K}}\ U \begin{pmatrix} f \\ g \end{pmatrix}, \mathrm{Pr}_{\mathfrak{K}}\ U \begin{pmatrix} f \\ g \end{pmatrix} \right\rangle_{\mathfrak{K}} \\ &\le \left\langle U \begin{pmatrix} f \\ g \end{pmatrix}, U \begin{pmatrix} f \\ g \end{pmatrix} \right\rangle_{\mathfrak{K}\oplus\tilde{\mathfrak{D}}} = \left\langle \begin{pmatrix} f \\ g \end{pmatrix}, \begin{pmatrix} f \\ g \end{pmatrix} \right\rangle_{\mathfrak{H}\oplus\mathfrak{D}}.\end{aligned}$$

If $\mathfrak{D}_-$ is any closed negative subspace of $\mathfrak{D}$, then $\mathfrak{M}' = \mathfrak{M} \oplus \mathfrak{D}_-$ is a closed negative subspace of $\mathfrak{H} \oplus \mathfrak{D}$. By Theorem 2.7(1), T' maps $\mathfrak{M}'$ in a one-to-one way onto a negative subspace $\mathfrak{N}'$ of $\mathfrak{K}$. Since

$$T'\,\mathfrak{M} = \mathrm{Pr}_{\mathfrak{K}}\ U\,\mathfrak{M} = T\,\mathfrak{M} = \mathfrak{N},$$

$\mathfrak{N}' \supseteq \mathfrak{N}$, and the inclusion is proper if $\mathfrak{D}_- \neq \{0\}$. By the maximality of $\mathfrak{N}$, $\mathfrak{D}_- = \{0\}$ and so $\mathfrak{D}$ is a Hilbert space. Therefore T^* is a contraction, which proves (1).

The last assertion follows from Theorem 2.7(2) and the graph representation of negative subspaces. □

In Hilbert space operator theory, a densely defined contraction can automatically be extended by continuity to an everywhere defined contraction. In contrast, there exist discontinuous densely defined operators T_0 on a Kreĭn space $\mathfrak{H}$ into a Kreĭn space $\mathfrak{K}$ such that

$$\langle T_0 f, T_0 g\rangle_{\mathfrak{K}} = \langle f, g\rangle_{\mathfrak{H}}, \qquad f, g \in \mathrm{dom}\, T_0. \tag{2.12}$$

Example:

There exists a densely defined operator V_0 on a Hilbert space $\mathfrak{H}$ into a Kreĭn space $\mathfrak{K}$ such that

$$\langle V_0 f, V_0 g\rangle_{\mathfrak{K}} = \langle f, g\rangle_{\mathfrak{H}}, \qquad f, g \in \mathrm{dom}\, V_0,$$

but V_0 has no extension to an operator $V \in \mathfrak{L}(\mathfrak{H}, \mathfrak{K})$. Let $\mathfrak{H}$ have orthonormal basis $\varphi_1, \varphi_2, \ldots$. Let $\mathfrak{K} = \mathfrak{K}_+ \oplus \mathfrak{K}_-$ be a fundamental decomposition, and assume that $\mathfrak{K}_\pm$ has orthonormal basis $e_1^\pm, e_2^\pm, \ldots$. Let dom V_0 be the span of $\varphi_1, \varphi_2, \ldots$, and define V_0 on this domain so as to be linear and satisfy

$$V_0 \varphi_n = a_n e_n^+ + b_n e_n^-,$$

where the constants are chosen such that $|a_n|^2 - |b_n|^2 = 1$ for all $n \ge 1$. In figure 4 we take $\mathfrak{K}$ to be the direct sum of a sequence of copies of $\mathbf{M}^2$, and we choose $V_0 \varphi_n$

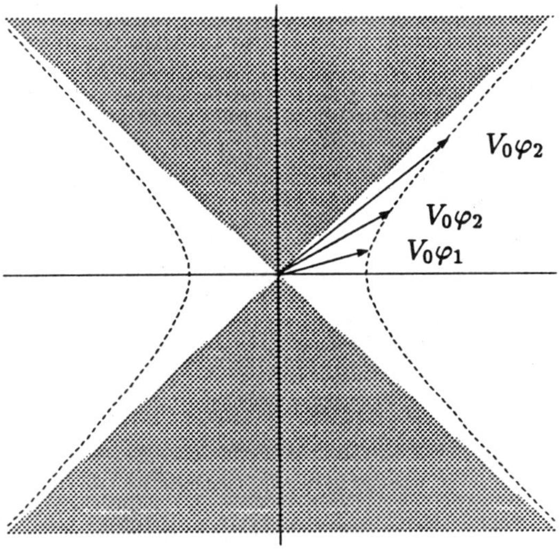

Figure 4

in the n-th copy. The figure is obtained from superimposing the copies. It is easy to see that

$$\langle V_0\varphi_m, V_0\varphi_n\rangle_{\mathfrak{K}} = \delta_{mn}, \qquad n \geq 1,$$

and so V_0 preserves inner products on its domain. If the constants are unbounded, V_0 can have no extension to an operator $V \in \mathfrak{L}(\mathfrak{H}, \mathfrak{K})$.

A theorem of Shmul′yan [1967] shows how such behavior may be excluded.

Theorem 2.9 (Shmul′yan) *Let T_0 be a densely defined operator on a Kreĭn space $\mathfrak{H}$ into a Kreĭn space $\mathfrak{K}$ satisfying*

$$\langle T_0 f, T_0 f\rangle_{\mathfrak{K}} \leq \langle f, f\rangle_{\mathfrak{H}}, \qquad f \in \operatorname{dom} T_0. \tag{2.13}$$

Assume that dom T_0 *contains a maximal uniformly negative subspace $\mathfrak{M}$ and that $T_0\mathfrak{M}$ is maximal uniformly negative in $\mathfrak{K}$. Then T_0 has an extension to a bicontraction $T \in \mathfrak{L}(\mathfrak{H}, \mathfrak{K})$.*

The proof in Shmul′yan [1967] is attributed to M. L. Brodskiĭ. Special cases were apparently known earlier, but we do not know references. See also Dritschel and Rovnyak [1990], Theorem 1.4.4.

In the case of Pontryagin spaces, the following result is helpful in checking the hypotheses on maximal uniformly negative subspaces in Theorem 2.9. For different proofs, see Iokhvidov, Kreĭn, and Langer [1982], Lemma 2.1 on p. 18, or Bognár [1974], Theorem 1.4 on p. 185.

Pontryagin's Theorem. *If $\mathfrak{M}$ is a dense subspace of a Pontryagin space $\mathfrak{H}$, there is a fundamental decomposition $\mathfrak{H} = \mathfrak{H}_+ \oplus \mathfrak{H}_-$ such that $\mathfrak{H}_- \subseteq \mathfrak{M}$.*

Proof. The argument is based on this evident property: If a dense subspace of a Kreĭn space $\mathfrak{K}$ contains no nonzero vector f such that $\langle f, f\rangle < 0$, then $\mathfrak{K}$ is a Hilbert space.

Suppose that $\mathfrak{H}$ has negative index n and that $\mathfrak{M}$ is a dense subspace of $\mathfrak{H}$. If $n = 0$, the conclusion is trivial. If $n > 0$, by the property, there is a vector f_1 in $\mathfrak{M}$ such that $\langle f_1, f_1 \rangle < 0$. Let $\mathfrak{N}_1$ be the span of f_1. Then $\mathfrak{N}_1$ is a one-dimensional Kreĭn subspace of $\mathfrak{H}$. It is easy to see that

$$\mathfrak{M} = \mathfrak{N}_1 \oplus (\mathfrak{M} \ominus \mathfrak{N}_1)$$

and that $\mathfrak{M}_1 = \mathfrak{M} \ominus \mathfrak{N}_1$ is dense in $\mathfrak{H}_1 = \mathfrak{H} \ominus \mathfrak{N}_1$. Here $\mathfrak{H}_1$ is a Pontryagin space with negative index $n - 1$. If $n > 1$, we can repeat the argument and find a vector f_2 in $\mathfrak{M}_1$ such that $\langle f_2, f_2 \rangle < 0$. The span $\mathfrak{N}_2$ of f_2 is a Kreĭn subspace of $\mathfrak{H}_1$, and $\mathfrak{M}_2 = \mathfrak{M}_1 \ominus \mathfrak{N}_2$ is a dense subspace of the Pontryagin space $\mathfrak{H}_2 = \mathfrak{H} \ominus \mathfrak{N}_2$. The process will terminate after n stages, and then $\mathfrak{H}_- = \mathfrak{N}_1 \oplus \cdots \oplus \mathfrak{N}_n$ is a subspace of $\mathfrak{H}$ with the required properties. □

Using Pontryagin's theorem, it is not hard to deduce the following version of Shmul′yan's theorem.

Theorem 2.10 *Let $\mathfrak{H}$ and $\mathfrak{K}$ be Pontryagin spaces with the same negative index, and let T_0 be a densely defined linear operator from $\mathfrak{H}$ into $\mathfrak{K}$ satisfying* (2.13). *Then T_0 has a continuous extension to a bicontraction $T \in \mathfrak{L}(\mathfrak{H}, \mathfrak{K})$.*

The version for Pontryagin spaces is simpler to prove than Theorem 2.9. A nice proof by T. Ya. Azizov appears in Alpay, Dijksma, and de Snoo [1994]. The following version, also taken from the same source, is particularly useful in applications. Details will appear in Alpay, Dijksma, Rovnyak, and de Snoo [in preparation].

Theorem 2.10 (Alternative Form) *Let $\mathfrak{H}$ and $\mathfrak{K}$ be Pontryagin spaces with the same negative index. Let $\mathbf{R}$ be a linear subspace of $\mathfrak{H} \times \mathfrak{K}$ such that*

(1) *$\langle g, g \rangle_{\mathfrak{K}} \leq \langle f, f \rangle_{\mathfrak{H}}$ for all $(f, g) \in \mathbf{R}$, and*

(2) *the set of all $f \in \mathfrak{H}$ such that $(f, g) \in \mathbf{R}$ for some $g \in \mathfrak{K}$ is dense in $\mathfrak{H}$.*

Then there is a bicontraction $T \in \mathfrak{L}(\mathfrak{H}, \mathfrak{K})$ such that $g = Tf$ for all $(f, g) \in \mathbf{R}$.

We conclude this lecture with some formal properties of Julia operators which are very useful in the study of extension properties of operators. Assumptions are only made on the form of the extensions, and otherwise the theorems hold for arbitrary operators $T \in \mathfrak{L}(\mathfrak{H}, \mathfrak{K})$. All underlying spaces in Theorems 2.11, 2.12, and 2.13 are Kreĭn spaces.

Theorem 2.11 *Assume $T \in \mathfrak{L}(\mathfrak{H}, \mathfrak{K})$, and let*

$$\begin{pmatrix} T & D_T \\ \tilde{D}_T^* & -L_T^* \end{pmatrix} \in \mathfrak{L}(\mathfrak{H} \oplus \mathfrak{D}_T, \mathfrak{K} \oplus \tilde{\mathfrak{D}}_T)$$

be a Julia operator. Assume $X \in \mathfrak{L}(\mathfrak{F}, \mathfrak{D}_T)$, and let

$$\begin{pmatrix} X & D_X \\ \tilde{D}_X^* & -L_X^* \end{pmatrix} \in \mathfrak{L}(\mathfrak{F} \oplus \mathfrak{D}_X, \mathfrak{D}_T \oplus \tilde{\mathfrak{D}}_X)$$

be a Julia operator. Then a Julia operator for

$$R = \begin{pmatrix} T & D_T X \end{pmatrix} \in \mathfrak{L}(\mathfrak{H} \oplus \mathfrak{F}, \mathfrak{K})$$

is given by

$$\begin{pmatrix} R & D_R \\ \tilde{D}_R^* & -L_R^* \end{pmatrix} = \begin{pmatrix} (T \quad D_T X) & (D_T D_X) \\ \begin{pmatrix} \tilde{D}_T^* & -L_T^* X \\ 0 & \tilde{D}_X^* \end{pmatrix} & -\begin{pmatrix} L_T^* D_X \\ L_X^* \end{pmatrix} \end{pmatrix}$$

$$= \begin{pmatrix} T & D_T & 0 \\ \tilde{D}_T^* & -L_T^* & 0 \\ 0 & 0 & 1 \end{pmatrix} \begin{pmatrix} 1 & 0 & 0 \\ 0 & X & D_X \\ 0 & \tilde{D}_X^* & -L_X^* \end{pmatrix} \tag{2.14}$$

as an operator on $\mathfrak{L}(\mathfrak{H} \oplus \mathfrak{F} \oplus \mathfrak{D}_X, \mathfrak{K} \oplus \tilde{\mathfrak{D}}_T \oplus \tilde{\mathfrak{D}}_X)$. *Thus defect spaces for* R *and* R^* *are given by*

$$\tilde{\mathfrak{D}}_R = \tilde{\mathfrak{D}}_T \oplus \tilde{\mathfrak{D}}_X \qquad \text{and} \qquad \mathfrak{D}_R = \mathfrak{D}_X.$$

As in Theorem 2.4, the unitarity of the various operators in Theorem 2.11 gives a sequence of useful index formulas. Repeating (2.8), these read:

$$\left.\begin{aligned} \operatorname{ind}_\pm (1 - R^*R) &= \operatorname{ind}_\pm (1 - T^*T) + \operatorname{ind}_\pm (1 - X^*X), \\ \operatorname{ind}_\pm (1 - RR^*) &= \operatorname{ind}_\pm (1 - XX^*), \\ \operatorname{ind}_\pm \mathfrak{H} + \operatorname{ind}_\pm (1 - TT^*) &= \operatorname{ind}_\pm \mathfrak{K} + \operatorname{ind}_\pm (1 - T^*T), \\ \operatorname{ind}_\pm \mathfrak{F} + \operatorname{ind}_\pm (1 - XX^*) &= \operatorname{ind}_\pm (1 - TT^*) + \operatorname{ind}_\pm (1 - X^*X). \end{aligned}\right\} \tag{2.15}$$

Theorem 2.12 *Assume* $T \in \mathfrak{L}(\mathfrak{H}, \mathfrak{K})$, *and let*

$$\begin{pmatrix} T & D_T \\ \tilde{D}_T^* & -L_T^* \end{pmatrix} \in \mathfrak{L}(\mathfrak{H} \oplus \mathfrak{D}_T, \mathfrak{K} \oplus \tilde{\mathfrak{D}}_T)$$

be a Julia operator. Assume $Y \in \mathfrak{L}(\mathfrak{G}, \tilde{\mathfrak{D}}_T)$, *and let*

$$\begin{pmatrix} Y & D_Y \\ \tilde{D}_Y^* & -L_Y^* \end{pmatrix} \in \mathfrak{L}(\mathfrak{G} \oplus \mathfrak{D}_Y, \tilde{\mathfrak{D}}_T \oplus \tilde{\mathfrak{D}}_Y)$$

be a Julia operator. Then a Julia operator for

$$C = \begin{pmatrix} T \\ Y^* \tilde{D}_T^* \end{pmatrix} \in \mathfrak{L}(\mathfrak{H}, \mathfrak{K} \oplus \mathfrak{G})$$

is given by

$$\begin{pmatrix} C & D_C \\ \tilde{D}_C^* & -L_C^* \end{pmatrix} = \begin{pmatrix} \begin{pmatrix} T \\ Y^* \tilde{D}_T^* \end{pmatrix} & \begin{pmatrix} D_T & 0 \\ -Y^* L_T^* & \tilde{D}_Y \end{pmatrix} \\ (D_Y^* \tilde{D}_T^*) & -(D_Y^* L_T^* \quad L_Y) \end{pmatrix}$$

$$= \begin{pmatrix} 1 & 0 & 0 \\ 0 & Y^* & \tilde{D}_Y \\ 0 & D_Y^* & -L_Y \end{pmatrix} \begin{pmatrix} T & D_T & 0 \\ \tilde{D}_T^* & -L_T^* & 0 \\ 0 & 0 & 1 \end{pmatrix} \tag{2.16}$$

as an operator on $\mathfrak{L}(\mathfrak{H} \oplus \mathfrak{D}_T \oplus \tilde{\mathfrak{D}}_Y, \mathfrak{K} \oplus \mathfrak{G} \oplus \mathfrak{D}_Y)$. *Thus defect spaces for* C *and* C^* *are given by*

$$\tilde{\mathfrak{D}}_C = \mathfrak{D}_Y \qquad \text{and} \qquad \mathfrak{D}_C = \mathfrak{D}_T \oplus \tilde{\mathfrak{D}}_Y.$$

The corresponding index formulas are:

$$\left.\begin{aligned} \operatorname{ind}_\pm (1 - C^*C) &= \operatorname{ind}_\pm (1 - YY^*), \\ \operatorname{ind}_\pm (1 - CC^*) &= \operatorname{ind}_\pm (1 - TT^*) + \operatorname{ind}_\pm (1 - Y^*Y), \\ \operatorname{ind}_\pm \mathfrak{H} + \operatorname{ind}_\pm (1 - TT^*) &= \operatorname{ind}_\pm \mathfrak{K} + \operatorname{ind}_\pm (1 - T^*T), \\ \operatorname{ind}_\pm \mathfrak{G} + \operatorname{ind}_\pm (1 - YY^*) &= \operatorname{ind}_\pm (1 - T^*T) + \operatorname{ind}_\pm (1 - Y^*Y). \end{aligned}\right\} \tag{2.17}$$

Theorem 2.13 *Assume $T \in \mathfrak{L}(\mathfrak{H}, \mathfrak{K})$, and let*

$$\begin{pmatrix} T & D_T \\ \tilde{D}_T^* & -L_T^* \end{pmatrix} \in \mathfrak{L}(\mathfrak{H} \oplus \mathfrak{D}_T, \mathfrak{K} \oplus \tilde{\mathfrak{D}}_T)$$

be a Julia operator. Assume $X \in \mathfrak{L}(\mathfrak{F}, \mathfrak{D}_T)$, $Y \in \mathfrak{L}(\mathfrak{G}, \tilde{\mathfrak{D}}_T)$, $Z \in \mathfrak{L}(\tilde{\mathfrak{D}}_X, \tilde{\mathfrak{D}}_Y)$, and let

$$\begin{pmatrix} X & D_X \\ \tilde{D}_X^* & -L_X^* \end{pmatrix} \in \mathfrak{L}(\mathfrak{F} \oplus \mathfrak{D}_X, \mathfrak{D}_T \oplus \tilde{\mathfrak{D}}_X),$$
$$\begin{pmatrix} Y & D_Y \\ \tilde{D}_Y^* & -L_Y^* \end{pmatrix} \in \mathfrak{L}(\mathfrak{G} \oplus \mathfrak{D}_Y, \tilde{\mathfrak{D}}_T \oplus \tilde{\mathfrak{D}}_Y),$$
$$\begin{pmatrix} Z & D_Z \\ \tilde{D}_Z^* & -L_Z^* \end{pmatrix} \in \mathfrak{L}(\tilde{\mathfrak{D}}_X \oplus \mathfrak{D}_Z, \tilde{\mathfrak{D}}_Y \oplus \tilde{\mathfrak{D}}_Z),$$

be Julia operators. Then a Julia operator for

$$A = \begin{pmatrix} T & D_T X \\ Y^* \tilde{D}_T^* & -Y^* L_T^* X + \tilde{D}_Y Z \tilde{D}_X^* \end{pmatrix} \in \mathfrak{L}(\mathfrak{H} \oplus \mathfrak{F}, \mathfrak{K} \oplus \mathfrak{G})$$

is given by

$$\begin{pmatrix} A & D_A \\ \tilde{D}_A^* & -L_A^* \end{pmatrix}$$

$$= \begin{pmatrix} \begin{pmatrix} T & D_T X \\ Y^* \tilde{D}_T^* & -Y^* L_T^* X + \tilde{D}_Y Z \tilde{D}_X^* \end{pmatrix} & \begin{pmatrix} D_T D_X & 0 \\ -Y^* L_T^* D_X - \tilde{D}_Y Z L_X^* & \tilde{D}_Y D_Z \end{pmatrix} \\ \begin{pmatrix} D_Y^* \tilde{D}_T^* & -D_Y^* L_T^* X - L_Y Z \tilde{D}_X^* \\ 0 & \tilde{D}_Z^* \tilde{D}_X^* \end{pmatrix} & -\begin{pmatrix} D_Y^* L_T^* D_X - L_Y Z L_X^* & L_Y D_Z \\ \tilde{D}_Z^* L_X^* & L_Z^* \end{pmatrix} \end{pmatrix}$$

$$= \begin{pmatrix} 1 & 0 & 0 & 0 \\ 0 & Y^* & \tilde{D}_Y & 0 \\ 0 & D_Y^* & -L_Y & 0 \\ 0 & 0 & 0 & 1 \end{pmatrix} \begin{pmatrix} T & D_T & 0 & 0 \\ \tilde{D}_T^* & -L_T^* & 0 & 0 \\ 0 & 0 & Z & D_Z \\ 0 & 0 & \tilde{D}_Z^* & -L_Z^* \end{pmatrix} \begin{pmatrix} 1 & 0 & 0 & 0 \\ 0 & X & D_X & 0 \\ 0 & \tilde{D}_X^* & -L_X^* & 0 \\ 0 & 0 & 0 & 1 \end{pmatrix} \tag{2.18}$$

as an operator on $\mathfrak{L}(\mathfrak{H} \oplus \mathfrak{F} \oplus \mathfrak{D}_X \oplus \mathfrak{D}_Z, \mathfrak{K} \oplus \mathfrak{G} \oplus \mathfrak{D}_Y \oplus \tilde{\mathfrak{D}}_Z)$. Thus defect spaces for A and A^* are given by

$$\tilde{\mathfrak{D}}_A = \mathfrak{D}_Y \oplus \tilde{\mathfrak{D}}_Z \qquad \text{and} \qquad \mathfrak{D}_A = \mathfrak{D}_X \oplus \mathfrak{D}_Z.$$

Now the index identities read:

$$\left.\begin{aligned} \operatorname{ind}_{\pm}(1-A^*A) &= \operatorname{ind}_{\pm}(1-YY^*)+\operatorname{ind}_{\pm}(1-Z^*Z),\\ \operatorname{ind}_{\pm}(1-AA^*) &= \operatorname{ind}_{\pm}(1-XX^*)+\operatorname{ind}_{\pm}(1-ZZ^*),\\ \operatorname{ind}_{\pm}\mathfrak{H}+\operatorname{ind}_{\pm}(1-TT^*) &= \operatorname{ind}_{\pm}\mathfrak{K}+\operatorname{ind}_{\pm}(1-T^*T),\\ \operatorname{ind}_{\pm}\mathfrak{F}+\operatorname{ind}_{\pm}(1-XX^*) &= \operatorname{ind}_{\pm}(1-TT^*)+\operatorname{ind}_{\pm}(1-X^*X),\\ \operatorname{ind}_{\pm}\mathfrak{G}+\operatorname{ind}_{\pm}(1-YY^*) &= \operatorname{ind}_{\pm}(1-T^*T)+\operatorname{ind}_{\pm}(1-Y^*Y),\\ \operatorname{ind}_{\pm}(1-X^*X)+\operatorname{ind}_{\pm}(1-ZZ^*) &= \operatorname{ind}_{\pm}(1-Y^*Y)+\operatorname{ind}_{\pm}(1-Z^*Z). \end{aligned}\right\} \tag{2.19}$$

Proofs of Theorems 2.11, 2.12, and 2.13. The matrices (2.14), (2.16), and (2.18) are unitary because each is the product of unitary operators. By inspection the 12-entries and adjoints of the 21-entries have zero kernels, and thus the conditions for a Julia operator are met. □

Addendum: Adjoints of contractions

Adjoints of contractions also have special mapping properties with respect to negative subspaces (Dritschel and Rovnyak [1990]), and these too can be studied by the method of isometric extensions. We illustrate the idea in an elaboration of Theorem 2.8 that includes a condition on adjoints. We first prove a part of this result (Theorem 2.8′ below) for the special case of isometries.

Lemma 2.14 *Let $V \in \mathfrak{L}(\mathfrak{H},\mathfrak{K})$ be an isometry. If V^* maps some maximal negative subspace $\mathfrak{N}$ in a one-to-one way into a negative subspace, then V is a bicontraction.*

Proof. By Theorem 1.7, V^* is isometric on the range of V. To prove that V is a bicontraction, it is sufficient to show that $(\operatorname{ran} V)^{\perp} = \ker\ V^*$ is a Hilbert space.

Since $\ker\ V^*$ is a Kreĭn subspace of $\mathfrak{K}$, it has a fundamental decomposition $\ker\ V^* = \mathfrak{E}_+ \oplus \mathfrak{E}_-$. Put

$$\begin{aligned} \mathfrak{K}' &= \operatorname{ran} V \oplus \mathfrak{E}_-,\\ \mathfrak{N}' &= \operatorname{Pr}_{\mathfrak{K}'}\mathfrak{N},\\ V' &= V|_{\mathfrak{K}'} \in \mathfrak{L}(\mathfrak{H},\mathfrak{K}'). \end{aligned}$$

It is easy to see that the hypotheses of the lemma are satisfied with $V, N, \mathfrak{N}$ replaced by $V', N', \mathfrak{N}'$ (the space $\mathfrak{N}'$ is maximal negative by the comments preceding Theorem 1.6). Since $\ker\ V'^* = \mathfrak{E}_-$, by a change of notation we may assume in the original situation that $\ker\ V^*$ is uniformly negative.

We show that under the assumption that $\ker\ V^*$ is uniformly negative, it contains no nonzero element. Let $u \in \ker\ V^*$. For any $g \in \mathfrak{N}$ and any scalar c,

$$0 \geq \langle V^*g, V^*g\rangle_{\mathfrak{H}} = \langle VV^*(cu+g), VV^*(cu+g)\rangle_{\mathfrak{K}} \geq \langle cu+g, cu+g\rangle_{\mathfrak{K}},$$

where the second inequality results from the assumption that $\ker\ V^*$ is uniformly negative. Since $\mathfrak{N}$ is maximal negative, $u \in \mathfrak{N}$. Since V^* is one-to-one on $\mathfrak{N}$ and $V^*u = 0$, we obtain $u = 0$. Thus $\ker\ V^* = \{0\}$, and the result follows. □

Theorem 2.8′ *If $T \in \mathfrak{L}(\mathfrak{H}, \mathfrak{K})$ is a contraction, the following are equivalent:*

(1) *T is a bicontraction;*
(2) *αT^* is a contraction for some scalar $\alpha \neq 0$;*
(3) *T maps some maximal negative subspace of $\mathfrak{H}$ onto a maximal negative subspace of $\mathfrak{K}$;*
(4) *T maps every maximal negative subspace of $\mathfrak{H}$ onto a maximal negative subspace of $\mathfrak{K}$;*
(5) *T^* maps some maximal negative subspace of $\mathfrak{K}$ in a one-to-one way into a negative subspace of $\mathfrak{H}$;*
(6) *if $\tilde{D} \in \mathfrak{L}(\tilde{\mathfrak{D}}, \mathfrak{H})$ is a defect operator for T, the isometry*

$$V = \begin{pmatrix} T \\ \tilde{D}^* \end{pmatrix}$$

is a bicontraction.

In this case, T maps every maximal uniformly negative subspace of $\mathfrak{H}$ onto a maximal uniformly negative subspace of $\mathfrak{K}$.

Proof. The equivalence of (1), (3), (4) and last assertion are shown in Theorem 2.8. The proof is completed by first establishing (1) $\Leftrightarrow$ (6). Then (2) is added to the list of equivalences by showing (1) $\Rightarrow$ (2) and (2) $\Rightarrow$ (4). Finally, (5) is added by showing (2) $\Rightarrow$ (5) and (5) $\Rightarrow$ (6).

(1) $\Leftrightarrow$ (6) By Theorem 2.3, we may choose a Julia operator

$$\begin{pmatrix} T & D \\ \tilde{D}^* & -L^* \end{pmatrix} \in \mathfrak{L}(\mathfrak{H} \oplus \mathfrak{D}, \mathfrak{K} \oplus \tilde{\mathfrak{D}})$$

for T such that $\tilde{D} \in \mathfrak{L}(\tilde{\mathfrak{D}}, \mathfrak{H})$ is the defect operator which is given in (6). Writing

$$V = \begin{pmatrix} T \\ \tilde{D}^* \end{pmatrix} \qquad \text{and} \qquad D_V = \begin{pmatrix} D \\ -L^* \end{pmatrix},$$

we obtain a Julia operator $\begin{pmatrix} V & D_V \end{pmatrix} \in \mathfrak{L}(\mathfrak{H} \oplus \mathfrak{D}, (\mathfrak{K} \oplus \tilde{\mathfrak{D}}) \oplus \{0\})$ for V with defect spaces $\tilde{\mathfrak{D}}_V = \{0\}$ and $\mathfrak{D}_V = \mathfrak{D}$. Thus V is a bicontraction if and only if $\mathfrak{D}$ is a Hilbert space, that is, T^* is a contraction. Since we assume that T is a contraction, this proves the equivalence of (1) and (6).

(1) $\Rightarrow$ (2) Take $\alpha = 1$.

(2) $\Rightarrow$ (4) Assume that αT^* is a contraction for some $\alpha \neq 0$. Let $D_\alpha \in \mathfrak{L}(\mathfrak{D}_\alpha, \mathfrak{K})$ be a defect operator for αT^*, so

$$V_\alpha = \begin{pmatrix} \alpha T^* \\ D_\alpha^* \end{pmatrix} \in \mathfrak{L}(\mathfrak{K}, \mathfrak{H} \oplus \mathfrak{D}_\alpha)$$

is an isometry. Since αT^* is a contraction, $\mathfrak{D}_\alpha$ is a Hilbert space. Therefore any maximal negative subspace $\mathfrak{M}$ of $\mathfrak{H}$ is maximal negative in $\mathfrak{H} \oplus \mathfrak{D}_\alpha$. Now

$$V_\alpha^*|_{\mathfrak{M}} = \alpha T|_{\mathfrak{M}},$$

and since T is a contraction, αT maps $\mathfrak{M}$ in a one-to-one way into a negative subspace by Theorem 2.7. By Lemma 2.14, V_α is a bicontraction. It follows that $T\mathfrak{M} = V_\alpha^* \mathfrak{M}$ is a maximal negative subspace of $\mathfrak{K}$, proving (4).

(2) $\Rightarrow$ (5) By Theorem 2.7, if αT^* is a contraction, then αT^* maps any maximal negative subspace of $\mathfrak{K}$ in a one-to-one way onto a negative subspace of $\mathfrak{H}$. Since $\alpha \neq 0$, T^* has the same property.

(5) $\Rightarrow$ (6) By assumption, there is a maximal negative subspace $\mathfrak{N}$ of $\mathfrak{K}$ which T^* maps in a one-to-one way into a negative subspace of $\mathfrak{H}$. Let

$$V = \begin{pmatrix} T \\ \tilde{D}^* \end{pmatrix} \in \mathfrak{L}(\mathfrak{H}, \mathfrak{K} \oplus \tilde{\mathfrak{D}})$$

as in (6) for some defect operator $\tilde{D} \in \mathfrak{L}(\tilde{\mathfrak{D}}, \mathfrak{H})$ for T. Since T is a contraction, $\tilde{\mathfrak{D}}$ is a Hilbert space, and therefore $\mathfrak{N}$ is a maximal negative subspace of $\mathfrak{K} \oplus \tilde{\mathfrak{D}}$. Therefore $V^*|_{\mathfrak{N}} = T^*|_{\mathfrak{N}}$ maps $\mathfrak{N}$ in a one-to-one way into a negative subspace. By Lemma 2.14, V is a bicontraction. $\square$

Lecture 3: Extension and Completion Problems

Key ideas:

- Extension and completion problems for operators on Kreĭn spaces often have the same form as in the Hilbert space case, even in very general situations. The theory is particularly simple for contractions.
- The problems are reduced to the isometric case via defect and Julia operators, and this reduction allows us to apply Hilbert space methods.

Kreĭn space operator theory comes very close to matching the completeness of the Hilbert space case in some areas, notably the extension properties of operators. This is the first of several lectures which will explore such problems.

Extension and completion problems come in numerous variations, leading to commutant lifting theorems, the Schur algorithm, and interpolation theory, with their infinite possibilities (see Foiaş and Frazho [1990]). Here we focus on a single but basic type of problem, which in many ways serves as a prototype.

Given a fixed contraction $T \in \mathfrak{L}(\mathfrak{H}, \mathfrak{K})$, we wish to determine all contractive operators of the form

$$\begin{pmatrix} T & F \end{pmatrix} \in \mathfrak{L}(\mathfrak{H} \oplus \mathfrak{F}, \mathfrak{K}), \tag{3.1}$$

$$\begin{pmatrix} T \\ G \end{pmatrix} \in \mathfrak{L}(\mathfrak{H}, \mathfrak{K} \oplus \mathfrak{G}), \tag{3.2}$$

and

$$\begin{pmatrix} T & F \\ G & H \end{pmatrix} \in \mathfrak{L}(\mathfrak{H} \oplus \mathfrak{F}, \mathfrak{K} \oplus \mathfrak{G}). \tag{3.3}$$

Variations of the problem replace contractions T by operators such that

$$\operatorname{ind}_{-}(1 - T^*T) < \infty \qquad \text{or} \qquad \operatorname{ind}_{-}(1 - TT^*) < \infty. \tag{3.4}$$

Then we seek a description of row, column, and two-by-two extensions which have similar properties.

Particular extensions have already been encountered, namely,

$$\begin{pmatrix} T & D_T \end{pmatrix} \in \mathfrak{L}(\mathfrak{H} \oplus \mathfrak{D}_T, \mathfrak{K}),$$

$$\begin{pmatrix} T \\ \tilde{D}_T^* \end{pmatrix} \in \mathfrak{L}(\mathfrak{H}, \mathfrak{K} \oplus \tilde{\mathfrak{D}}_T),$$

and

$$\begin{pmatrix} T & D_T \\ \tilde{D}_T^* & -L_T^* \end{pmatrix} \in \mathfrak{L}(\mathfrak{H} \oplus \mathfrak{D}_T, \mathfrak{K} \oplus \tilde{\mathfrak{D}}_T),$$

where the last operator is a Julia operator for T.

In the Hilbert space case, contractive row, column, and two-by-two completions take the form

$$\begin{pmatrix} T & D_T X \end{pmatrix} \in \mathfrak{L}(\mathfrak{H} \oplus \mathfrak{F}, \mathfrak{K}), \tag{3.5}$$

$$\begin{pmatrix} T \\ Y^* \tilde{D}_T^* \end{pmatrix} \in \mathfrak{L}(\mathfrak{H}, \mathfrak{K} \oplus \mathfrak{G}), \tag{3.6}$$

$$\begin{pmatrix} T & D_T X \\ Y^* \tilde{D}_T^* & -Y^* L_T^* X + \tilde{D}_Y Z \tilde{D}_X^* \end{pmatrix} \in \mathfrak{L}(\mathfrak{H} \oplus \mathfrak{F}, \mathfrak{K} \oplus \mathfrak{G}), \tag{3.7}$$

where X, Y, Z are contractions on appropriate spaces. The Hilbert space theory originates in work of Davis [1976] and Parrott [1978] and has been much studied (see Dritschel and Rovnyak [1990] for references).

An example shows some of the problems that can arise with indefinite inner product spaces.

Example:

Let $T \in \mathfrak{L}(\mathfrak{H}, \mathfrak{K})$ be an isometry on a Hilbert space $\mathfrak{H}$ into a Hilbert space $\mathfrak{K}$. Let

$$C = \begin{pmatrix} T \\ N \end{pmatrix} \in \mathfrak{L}(\mathfrak{H}, \mathfrak{K} \oplus \mathfrak{G})$$

be a column extension of T such that $N \in \mathfrak{L}(\mathfrak{H}, \mathfrak{G})$ is a **neutral operator**, which we define as an operator such that $N^*N = 0$. Such an operator is constructed simply by arranging that ran N is a neutral subspace of $\mathfrak{G}$. For example, we can find a neutral operator $N \neq 0$ if $\mathfrak{G} = \mathbf{M}^2$. Then

$$C^*C = T^*T + N^*N = T^*T = 1.$$

Thus C is a contractive (even isometric) column extension of T. It does not have the form (3.6) because $1 - T^*T = 0$ and hence $\tilde{\mathfrak{D}}_T = \{0\}$ and $\tilde{D}_T = 0$.

Nevertheless, it is desirable to try to derive positive results in which the forms (3.5)–(3.7) are obtained. For example, when these representations hold, the Julia operators for the extensions are given by the matrix formulas in Theorems 2.11–2.13. All of the index formulas stated there are valid. We can easily read off conditions on the parameters X, Y, Z that the operators are isometries or contractions, and so forth. This information is useful for more structured problems, such as those involving commutant lifting or the Schur algorithm (Lecture 4).

We follow a simplified method given in Dritschel [1993c]. The row extension problem will first be solved for isometries.

Theorem 3.1 *Let*

$$V = \begin{pmatrix} V_1 & V_2 \end{pmatrix} \in \mathfrak{L}(\mathfrak{H} \oplus \mathfrak{F}, \mathfrak{K}) \tag{3.8}$$

be an isometry, and let $D_{V_1} \in \mathfrak{L}(\mathfrak{D}_{V_1}, \mathfrak{K})$ be a defect operator for V_1^*. Then

$$V_2 = D_{V_1} W, \tag{3.9}$$

where $W = D_{V_1}^* V_2 \in \mathfrak{L}(\mathfrak{F}, \mathfrak{D}_{V_1})$ is an isometry. If V is unitary and V_2 is written in any way in the form (3.9) with $W \in \mathfrak{L}(\mathfrak{F}, \mathfrak{D}_{V_1})$ an isometry, then W is unitary.

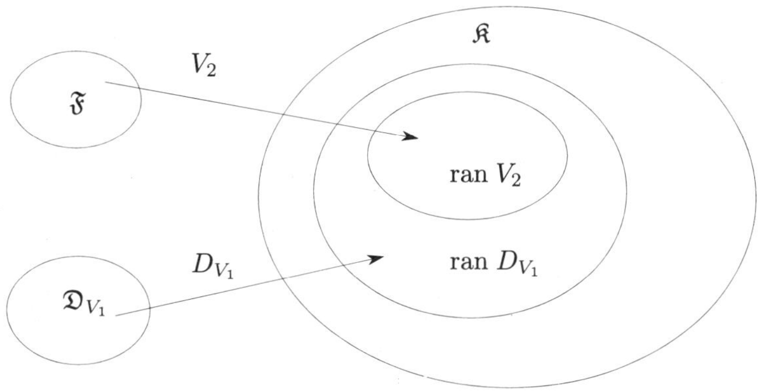

Figure 5

Proof. Since V is isometric, so are its restrictions $V_1 = V|_{\mathfrak{H}}$ and $V_2 = V|_{\mathfrak{F}}$. The identity

$$0 = 1 - V^*V = \begin{pmatrix} * & -V_1^*V_2 \\ * & * \end{pmatrix}$$

implies that $V_1^*V_2 = 0$, and so ran $V_2 \subseteq$ ker V_1^*. Since V_1 is an isometry,

$$D_{V_1} D_{V_1}^* = 1 - V_1 V_1^*$$

is a projection with range $(\text{ran } V_1)^{\perp} = \text{ker } V_1^*$. Thus

$$D_{V_1} D_{V_1}^* D_{V_1} D_{V_1}^* = D_{V_1} D_{V_1}^*,$$

and hence $D_{V_1}^* D_{V_1} = 1$, because D_{V_1} has zero kernel and $D_{V_1}^*$ has dense range. Therefore D_{V_1} is an isometry with ran $D_{V_1} =$ ker $V_1^* \supseteq$ ran V_2 (see figure 5). Hence $W = D_{V_1}^* V_2$ defines an isometry from $\mathfrak{F}$ into $\mathfrak{D}_{V_1}$. Since $D_{V_1} D_{V_1}^*$ acts like the identity on ran D_{V_1},

$$D_{V_1} W = D_{V_1} D_{V_1}^* V_2 = V_2,$$

as was to be shown.

Assume V is unitary and V_2 is written in the form (3.9) with $W \in \mathfrak{L}(\mathfrak{F}, \mathfrak{D}_{V_1})$ isometric. Then

$$0 = 1 - VV^* = 1 - V_1 V_1^* - D_{V_1} W W^* D_{V_1}^* = D_{V_1}(1 - WW^*) D_{V_1}^*,$$

and $1 - WW^* = 0$ since D_{V_1} has zero kernel and $D_{V_1}^*$ has dense range. Since $W^*W = 1$ by assumption, W is unitary. □

We shall first state the main results in their simplest form — for contractions. In Theorems 3.2–3.4, we make these assumptions on underlying spaces:

(1) The main spaces $\mathfrak{H}$ and $\mathfrak{K}$ are Kreĭn spaces.

(2) The extension space $\mathfrak{F}$ is a Kreĭn space, but $\mathfrak{G}$ is a Hilbert space.

Notice that the example above is excluded then from Theorem 3.3 because the space $\mathfrak{G}$ in the example is not a Hilbert space.

Theorem 3.2 *Let $T \in \mathfrak{L}(\mathfrak{H}, \mathfrak{K})$ be a contraction, and let $D_T \in \mathfrak{L}(\mathfrak{D}_T, \mathfrak{K})$ be a defect operator for T^*. Every contractive row extension*

$$R = \begin{pmatrix} T & F \end{pmatrix} \in \mathfrak{L}(\mathfrak{H} \oplus \mathfrak{F}, \mathfrak{K})$$

of T has the form

$$R = \begin{pmatrix} T & D_T X \end{pmatrix}, \tag{3.10}$$

where $X \in \mathfrak{L}(\mathfrak{F}, \mathfrak{D}_T)$ is a contraction. Conversely, if $X \in \mathfrak{L}(\mathfrak{F}, \mathfrak{D}_T)$ is a contraction, then the operator R defined by (3.10) *is a contraction.*

Theorem 3.3 *Let $T \in \mathfrak{L}(\mathfrak{H}, \mathfrak{K})$ be a contraction, and let $\tilde{D}_T \in \mathfrak{L}(\tilde{\mathfrak{D}}_T, \mathfrak{H})$ be a defect operator for T. Every contractive column extension*

$$C = \begin{pmatrix} T \\ G \end{pmatrix} \in \mathfrak{L}(\mathfrak{H}, \mathfrak{K} \oplus \mathfrak{G})$$

of T with $\mathfrak{G}$ a Hilbert space has the form

$$C = \begin{pmatrix} T \\ Y^* \tilde{D}_T^* \end{pmatrix}, \tag{3.11}$$

where $Y \in \mathfrak{L}(\mathfrak{G}, \tilde{\mathfrak{D}}_T)$ is a contraction. Conversely, if $Y \in \mathfrak{L}(\mathfrak{G}, \tilde{\mathfrak{D}}_T)$ is a contraction, then the operator C defined by (3.11) *is a contraction.*

Theorem 3.4 *Let $T \in \mathfrak{L}(\mathfrak{H}, \mathfrak{K})$ be a contraction with Julia operator*

$$\begin{pmatrix} T & D_T \\ \tilde{D}_T^* & -L_T^* \end{pmatrix} \in \mathfrak{L}(\mathfrak{H} \oplus \mathfrak{D}_T, \mathfrak{K} \oplus \tilde{\mathfrak{D}}_T).$$

Every contractive extension

$$\begin{pmatrix} T & F \\ G & H \end{pmatrix} \in \mathfrak{L}(\mathfrak{H} \oplus \mathfrak{F}, \mathfrak{K} \oplus \mathfrak{G})$$

of T with $\mathfrak{G}$ a Hilbert space has the form

$$\begin{pmatrix} T & F \\ G & H \end{pmatrix} = \begin{pmatrix} T & D_T X \\ Y^* \tilde{D}_T^* & -Y^* L_T^* X + \tilde{D}_Y Z \tilde{D}_X^* \end{pmatrix}, \tag{3.12}$$

where

(i) *$X \in \mathfrak{L}(\mathfrak{F}, \mathfrak{D}_T)$ is a contraction with defect operator $\tilde{D}_X \in \mathfrak{L}(\tilde{\mathfrak{D}}_X, \mathfrak{F})$,*
(ii) *$Y \in \mathfrak{L}(\mathfrak{G}, \tilde{\mathfrak{D}}_T)$ is a contraction with defect operator $\tilde{D}_Y \in \mathfrak{L}(\tilde{\mathfrak{D}}_Y, \mathfrak{G})$,*
(iii) *$Z \in \mathfrak{L}(\tilde{\mathfrak{D}}_X, \tilde{\mathfrak{D}}_Y)$ is a contraction.*

Conversely, given X, Y, Z satisfying (i), (ii), (iii), *then the operator defined by* (3.12) *is a contraction.*

The method of proof of Theorems 3.2–3.4 applies with no extra input to operators satisfying (3.4). We now generalize each of these results, with no restriction on underlying spaces:

(1) The main spaces $\mathfrak{H}$ and $\mathfrak{K}$ are Kreĭn spaces.
(2) The extension spaces $\mathfrak{F}$ and $\mathfrak{G}$ are Kreĭn spaces.

Theorem 3.2 (Generalized Form) *Let* $T \in \mathfrak{L}(\mathfrak{H}, \mathfrak{K})$*, and let* $D_T \in \mathfrak{L}(\mathfrak{D}_T, \mathfrak{K})$ *be a defect operator for* T^**. Let*

$$R = \begin{pmatrix} T & F \end{pmatrix} \in \mathfrak{L}(\mathfrak{H} \oplus \mathfrak{F}, \mathfrak{K})$$

be a row extension of T *satisfying at least one of the conditions*

$$\begin{aligned} \operatorname{ind}_-(1 - RR^*) + \operatorname{ind}_- \mathfrak{F} &= \operatorname{ind}_-(1 - TT^*) < \infty, \\ \operatorname{ind}_-(1 - R^*R) &= \operatorname{ind}_-(1 - T^*T) < \infty. \end{aligned}$$

Then

$$R = \begin{pmatrix} T & D_T X \end{pmatrix}, \tag{3.13}$$

where $X \in \mathfrak{L}(\mathfrak{F}, \mathfrak{D}_T)$ *is a contraction. Conversely, if* R *has the form* (3.13) *where* $X \in \mathfrak{L}(\mathfrak{F}, \mathfrak{D}_T)$ *is a contraction, then the operator* R *defined by* (3.13) *satisfies both of the identities*

$$\begin{aligned} \operatorname{ind}_-(1 - RR^*) + \operatorname{ind}_- \mathfrak{F} &= \operatorname{ind}_-(1 - TT^*), \\ \operatorname{ind}_-(1 - R^*R) &= \operatorname{ind}_-(1 - T^*T). \end{aligned}$$

Theorem 3.3 (Generalized Form) *Let* $T \in \mathfrak{L}(\mathfrak{H}, \mathfrak{K})$*, and let* $\tilde{D}_T \in \mathfrak{L}(\tilde{\mathfrak{D}}_T, \mathfrak{H})$ *be a defect operator for* T*. Let*

$$C = \begin{pmatrix} T \\ G \end{pmatrix} \in \mathfrak{L}(\mathfrak{H}, \mathfrak{K} \oplus \mathfrak{G})$$

be a column extension of T *satisfying at least one of the conditions*

$$\begin{aligned} \operatorname{ind}_-(1 - C^*C) + \operatorname{ind}_- \mathfrak{G} &= \operatorname{ind}_-(1 - T^*T) < \infty, \\ \operatorname{ind}_-(1 - CC^*) &= \operatorname{ind}_-(1 - TT^*) < \infty. \end{aligned}$$

Then

$$C = \begin{pmatrix} T \\ Y^* \tilde{D}_T^* \end{pmatrix}, \tag{3.14}$$

where $Y \in \mathfrak{L}(\mathfrak{G}, \tilde{\mathfrak{D}}_T)$ *is a contraction. Conversely, if* C *has the form* (3.14) *where* $Y \in \mathfrak{L}(\mathfrak{G}, \tilde{\mathfrak{D}}_T)$ *is a contraction, then the operator* C *defined by* (3.14) *satisfies both of the identities*

$$\begin{aligned} \operatorname{ind}_-(1 - C^*C) + \operatorname{ind}_- \mathfrak{G} &= \operatorname{ind}_-(1 - T^*T), \\ \operatorname{ind}_-(1 - CC^*) &= \operatorname{ind}_-(1 - TT^*). \end{aligned}$$

Theorem 3.4 (Generalized Form) *Let* $T \in \mathfrak{L}(\mathfrak{H}, \mathfrak{K})$ *have Julia operator*

$$\begin{pmatrix} T & D_T \\ \tilde{D}_T^* & -L_T^* \end{pmatrix} \in \mathfrak{L}(\mathfrak{H} \oplus \mathfrak{D}_T, \mathfrak{K} \oplus \tilde{\mathfrak{D}}_T),$$

and let

$$A = \begin{pmatrix} T & F \\ G & H \end{pmatrix} \in \mathfrak{L}(\mathfrak{H} \oplus \mathfrak{F}, \mathfrak{K} \oplus \mathfrak{G})$$

be a 2×2 *matrix extension of* T *satisfying at least one of the conditions*

$$\begin{aligned} \operatorname{ind}_-(1 - A^*A) + \operatorname{ind}_- \mathfrak{G} &= \operatorname{ind}_-(1 - T^*T) < \infty, \\ \operatorname{ind}_-(1 - AA^*) + \operatorname{ind}_- \mathfrak{F} &= \operatorname{ind}_-(1 - TT^*) < \infty. \end{aligned}$$

Then

$$A = \begin{pmatrix} T & D_T X \\ Y^* \tilde{D}_T^* & -Y^* L_T^* X + \tilde{D}_Y Z \tilde{D}_X^* \end{pmatrix}, \tag{3.15}$$

where

(i) $X \in \mathfrak{L}(\mathfrak{F}, \mathfrak{D}_T)$ *is a contraction with defect operator* $\tilde{D}_X \in \mathfrak{L}(\tilde{\mathfrak{D}}_X, \mathfrak{F})$,
(ii) $Y \in \mathfrak{L}(\mathfrak{G}, \tilde{\mathfrak{D}}_T)$ *is a contraction with defect operator* $\tilde{D}_Y \in \mathfrak{L}(\tilde{\mathfrak{D}}_Y, \mathfrak{G})$,
(iii) $Z \in \mathfrak{L}(\tilde{\mathfrak{D}}_X, \tilde{\mathfrak{D}}_Y)$ *is a contraction.*

Conversely, given X, Y, Z *satisfying* (i), (ii), (iii), *then the operator defined by* (3.15) *satisfies both of the identities*

$$\begin{aligned} \operatorname{ind}_- (1 - A^*A) + \operatorname{ind}_- \mathfrak{G} &= \operatorname{ind}_- (1 - T^*T), \\ \operatorname{ind}_- (1 - AA^*) + \operatorname{ind}_- \mathfrak{F} &= \operatorname{ind}_- (1 - TT^*). \end{aligned}$$

It is possible to give the complete proofs in just a few pages, and they are included here. It is sufficient to prove the generalized forms of the theorems.

Proof of Theorem 3.2 (Generalized Form). Assume that

$$R = \begin{pmatrix} T & F \end{pmatrix} \in \mathfrak{L}(\mathfrak{H} \oplus \mathfrak{F}, \mathfrak{K})$$

and that one of the two conditions holds, say the first:

$$\operatorname{ind}_- (1 - RR^*) + \operatorname{ind}_- \mathfrak{F} = \operatorname{ind}_- (1 - TT^*) < \infty.$$

Choose a defect operator $D_R \in \mathfrak{L}(\mathfrak{D}_R, \mathfrak{K})$ for R^*. Then

$$\begin{pmatrix} R^* \\ D_R^* \end{pmatrix} = \begin{pmatrix} T^* \\ F^* \\ D_R^* \end{pmatrix} \in \mathfrak{L}(\mathfrak{K}, \mathfrak{H} \oplus \mathfrak{F} \oplus \mathfrak{D}_R)$$

is an isometry. The operator

$$\begin{pmatrix} T^* \\ D_T^* \end{pmatrix} \in \mathfrak{L}(\mathfrak{K}, \mathfrak{H} \oplus \mathfrak{D}_T)$$

is also an isometry. Thus

$$\left\langle \begin{pmatrix} T^* \\ F^* \\ D_R^* \end{pmatrix} f, \begin{pmatrix} T^* \\ F^* \\ D_R^* \end{pmatrix} g \right\rangle_{\mathfrak{H} \oplus \mathfrak{F} \oplus \mathfrak{D}_R} = \left\langle \begin{pmatrix} T^* \\ D_T^* \end{pmatrix} f, \begin{pmatrix} T^* \\ D_T^* \end{pmatrix} g \right\rangle_{\mathfrak{H} \oplus \mathfrak{D}_T}$$

for all $f, g \in \mathfrak{K}$. Therefore

$$\left\langle \begin{pmatrix} F^* \\ D_R^* \end{pmatrix} f, \begin{pmatrix} F^* \\ D_R^* \end{pmatrix} g \right\rangle_{\mathfrak{F} \oplus \mathfrak{D}_R} = \langle D_T^* f, D_T^* g \rangle_{\mathfrak{D}_T}$$

for all $f, g \in \mathfrak{K}$. Since D_T has zero kernel, ran D_T^* is dense in $\mathfrak{D}_T$. Our hypotheses imply that

$$\operatorname{ind}_- \mathfrak{D}_T = \operatorname{ind}_- (\mathfrak{F} \oplus \mathfrak{D}_R) < \infty,$$

and hence $\mathfrak{F} \oplus \mathfrak{D}_R$ and $\mathfrak{D}_T$ are Pontryagin spaces having the same negative index. Hence by Shmul'yan's theorem (see Theorem 2.10 (Alternative Form)), there is an isometry $V \in \mathfrak{L}(\mathfrak{D}_T, \mathfrak{F} \oplus \mathfrak{D}_R)$ such that

$$\begin{pmatrix} F^* \\ D_R^* \end{pmatrix} = V D_T^*.$$

Define $X \in \mathfrak{L}(\mathfrak{F}, \mathfrak{D}_T)$ by

$$X^* = \Pr_{\mathfrak{F}} V \in \mathfrak{L}(\mathfrak{D}_T, \mathfrak{F}).$$

By Corollary 2.5, V is a bicontraction. The operator $X = V^*|_{\mathfrak{F}}$ is thus a contraction. By construction, $F = D_T X$, and we have obtained the conclusion in case the first of the two conditions holds.

Suppose that the second condition holds, that is,

$$\operatorname{ind}_-(1 - R^*R) = \operatorname{ind}_-(1 - T^*T) < \infty,$$

or equivalently $\operatorname{ind}_- \tilde{\mathfrak{D}}_R = \operatorname{ind}_- \tilde{\mathfrak{D}}_T < \infty$. Let $\tilde{D}_R \in \mathfrak{L}(\tilde{\mathfrak{D}}_R, \mathfrak{H} \oplus \mathfrak{F})$ be a defect operator for R, and choose a Julia operator

$$\begin{pmatrix} T & D_T \\ \tilde{D}_T^* & -L_T^* \end{pmatrix} \in \mathfrak{L}(\mathfrak{H} \oplus \mathfrak{D}_T, \mathfrak{K} \oplus \tilde{\mathfrak{D}}_T).$$

The operator

$$V = \begin{pmatrix} R \\ \tilde{D}_R^* \end{pmatrix} \in \mathfrak{L}(\mathfrak{H} \oplus \mathfrak{F}, \mathfrak{K} \oplus \tilde{\mathfrak{D}}_R)$$

is an isometry and may be written in the form

$$V = \begin{pmatrix} T & F \\ \tilde{D}_1^* & \tilde{D}_2^* \end{pmatrix} \in \mathfrak{L}(\mathfrak{H} \oplus \mathfrak{F}, \mathfrak{K} \oplus \tilde{\mathfrak{D}}_R) \tag{3.16}$$

where $\tilde{D}_R^* = \begin{pmatrix} \tilde{D}_1^* & \tilde{D}_2^* \end{pmatrix}$. The first column in (3.16),

$$V_1 = \begin{pmatrix} T \\ \tilde{D}_1^* \end{pmatrix} \in \mathfrak{L}(\mathfrak{H}, \mathfrak{K} \oplus \tilde{\mathfrak{D}}_R),$$

is an isometric extension of T, implying that $\operatorname{ind}_-(1 - V_1^* V_1) = 0$. It is easy to see that the row extension

$$V_1^* = \begin{pmatrix} T^* & \tilde{D}_1 \end{pmatrix} \in \mathfrak{L}(\mathfrak{K} \oplus \tilde{\mathfrak{D}}_R, \mathfrak{H})$$

of T^* satisfies the first of the two conditions of the theorem (the index formula that is needed reduces to $\operatorname{ind}_- \tilde{\mathfrak{D}}_R = \operatorname{ind}_- \tilde{\mathfrak{D}}_T < \infty$), and the result has already been established in this case. Therefore

$$\tilde{D}_1^* = Y^* \tilde{D}_T^*,$$

where $Y \in \mathfrak{L}(\tilde{\mathfrak{D}}_R, \tilde{\mathfrak{D}}_T)$ is a contraction. Now apply Theorem 3.1 to the isometry

$$V = \begin{pmatrix} V_1 & V_2 \end{pmatrix} = \left(\begin{pmatrix} T \\ \tilde{D}_1^* \end{pmatrix} \begin{pmatrix} F \\ \tilde{D}_2^* \end{pmatrix} \right) = \left(\begin{pmatrix} T \\ Y^* \tilde{D}_T^* \end{pmatrix} \begin{pmatrix} F \\ \tilde{D}_2^* \end{pmatrix} \right).$$

By Theorem 2.12, a defect operator for V_1^* is given by

$$D_{V_1} = \begin{pmatrix} D_T & 0 \\ -Y^* L_T^* & \tilde{D}_Y \end{pmatrix} \in \mathfrak{L}(\mathfrak{D}_T \oplus \tilde{\mathfrak{D}}_Y, \mathfrak{K} \oplus \tilde{\mathfrak{D}}_R).$$

Therefore by Theorem 3.1,

$$\begin{pmatrix} F \\ \tilde{D}_2^* \end{pmatrix} = \begin{pmatrix} D_T & 0 \\ -Y^* L_T^* & \tilde{D}_Y \end{pmatrix} \begin{pmatrix} X \\ X' \end{pmatrix},$$

where

$$W = \begin{pmatrix} X \\ X' \end{pmatrix} \in \mathfrak{L}(\mathfrak{F}, \mathfrak{D}_T \oplus \tilde{\mathfrak{D}}_Y)$$

is an isometry. This yields

$$F = D_T X,$$

where $X \in \mathfrak{L}(\mathfrak{F}, \mathfrak{D}_T)$. Since $W^*W = 1$,

$$1 - X^*X = X'^*X' \geq 0,$$

because $X' \in \mathfrak{L}(\mathfrak{F}, \tilde{\mathfrak{D}}_Y)$ and $\tilde{\mathfrak{D}}_Y$ is a Hilbert space (since Y is a contraction). Thus X is a contraction, and we have obtained the conclusion in case the second of the two conditions holds.

Conversely, assume that $X \in \mathfrak{L}(\mathfrak{F}, \mathfrak{D}_T)$ is a contraction, and define R by (3.13). Choose Julia operators for T and X as in Theorem 2.11. By that result, R has defect spaces

$$\begin{aligned} \tilde{\mathfrak{D}}_R &= \tilde{\mathfrak{D}}_T \oplus \tilde{\mathfrak{D}}_X, \\ \mathfrak{D}_R &= \mathfrak{D}_X. \end{aligned}$$

Since $\tilde{\mathfrak{D}}_X$ is a Hilbert space, $\operatorname{ind}_-(1 - X^*X) = 0$, and by the index formulas (2.15),

$$\begin{aligned} \operatorname{ind}_-(1 - R^*R) &= \operatorname{ind}_-(1 - T^*T), \\ \operatorname{ind}_-(1 - RR^*) &= \operatorname{ind}_-(1 - XX^*), \\ \operatorname{ind}_- \mathfrak{F} + \operatorname{ind}_-(1 - XX^*) &= \operatorname{ind}_-(1 - TT^*). \end{aligned}$$

The last two identities in the statement of the theorem follow. □

Proof of Theorem 3.3 (Generalized Form). On considering adjoints, this follows from Theorem 3.2. □

Proof of Theorem 3.4 (Generalized Form). It is enough to treat the case

$$\operatorname{ind}_-(1 - A^*A) + \operatorname{ind}_- \mathfrak{G} = \operatorname{ind}_-(1 - T^*T) < \infty, \tag{3.17}$$

since if the other condition holds we can apply this result to the adjoint of A. Form an isometric column extension

$$\begin{pmatrix} A \\ \tilde{D}_A \end{pmatrix} \in \mathfrak{L}(\mathfrak{H} \oplus \mathfrak{F}, \mathfrak{K} \oplus \mathfrak{G} \oplus \tilde{\mathfrak{D}}_A)$$

using some choice of a defect operator $\tilde{D}_A \in \mathfrak{L}(\tilde{\mathfrak{D}}_A, \mathfrak{H} \oplus \mathfrak{F})$ for A. Write

$$\begin{pmatrix} A \\ \tilde{D}_A \end{pmatrix} = \begin{pmatrix} C & C' \\ \tilde{D}_1^* & \tilde{D}_2^* \end{pmatrix}, \tag{3.18}$$

where

$$C = \begin{pmatrix} T \\ G \end{pmatrix} \in \mathfrak{L}(\mathfrak{H}, \mathfrak{K} \oplus \mathfrak{G})$$

is the first column of A and C' is the second column of A. The first column in (3.18), that is,

$$K = \begin{pmatrix} C \\ \tilde{D}_1^* \end{pmatrix} = \begin{pmatrix} \begin{pmatrix} T \\ G \end{pmatrix} \\ \tilde{D}_1^* \end{pmatrix} \in \mathfrak{L}(\mathfrak{H}, \mathfrak{K} \oplus \mathfrak{G} \oplus \tilde{\mathfrak{D}}_A),$$

is an isometry because it is the restriction of an isometry. Viewing K as a column extension of T, we may apply Theorem 3.3 (Generalized Form) because (3.17) insures that the first of the two conditions required in that result is met. In fact, since K is an isometry, $\operatorname{ind}_-(1-K^*K)=0$, and so

$$\begin{aligned}\operatorname{ind}_-(1-K^*K)\; &+ \operatorname{ind}_-(\mathfrak{G}\oplus\tilde{\mathfrak{D}}_A)\\ &= 0+\big(\operatorname{ind}_-\mathfrak{G}+\operatorname{ind}_-(1-A^*A)\big)=\operatorname{ind}_-(1-T^*T)<\infty.\end{aligned}$$

Hence

$$\begin{pmatrix} G \\ \tilde{D}_1^* \end{pmatrix} = V^*\tilde{D}_T^*,$$

where $V\in\mathfrak{L}(\mathfrak{G}\oplus\tilde{\mathfrak{D}}_A,\tilde{\mathfrak{D}}_T)$ is a contraction. This implies that

$$C=\begin{pmatrix} T \\ G \end{pmatrix}=\begin{pmatrix} T \\ Y^*\tilde{D}_T^* \end{pmatrix},$$

where $Y^*=\operatorname{Pr}_{\mathfrak{G}}V^*$. Here $Y=V|_{\mathfrak{G}}\in\mathfrak{L}(\mathfrak{G},\tilde{\mathfrak{D}}_T)$ is the restriction of a contraction, and so Y is a contraction. Choose a defect operator $\tilde{D}_Y\in\mathfrak{L}(\tilde{\mathfrak{D}}_Y,\mathfrak{G})$ for Y.

Additional information is obtained by viewing A as a row extension of C:

$$A=\begin{pmatrix} T & F \\ G & H \end{pmatrix}=\left(\begin{pmatrix} T \\ G \end{pmatrix}\ \begin{pmatrix} F \\ H \end{pmatrix}\right)=\left(C\ \begin{pmatrix} F \\ H \end{pmatrix}\right).$$

By (3.17) and Theorem 3.3 (Generalized Form),

$$\operatorname{ind}_-(1-A^*A)=\operatorname{ind}_-(1-T^*T)-\operatorname{ind}_-\mathfrak{G}=\operatorname{ind}_-(1-C^*C)<\infty.$$

The hypotheses of Theorem 3.2 (Generalized Form) are met, and therefore

$$A=\begin{pmatrix} C & D_CX_C \end{pmatrix},$$

where $X_C\in\mathfrak{L}(\mathfrak{F},\mathfrak{D}_C)$ is a contraction. Here we choose $D_C\in\mathfrak{L}(\mathfrak{D}_C,\mathfrak{K}\oplus\mathfrak{G})$ as in Theorem 2.12, that is, $\mathfrak{D}_C=\mathfrak{D}_T\oplus\tilde{\mathfrak{D}}_Y$ and

$$D_C=\begin{pmatrix} D_T & 0 \\ -Y^*L_T^* & \tilde{D}_Y \end{pmatrix}\in\mathfrak{L}(\mathfrak{D}_T\oplus\tilde{\mathfrak{D}}_Y,\mathfrak{K}\oplus\mathfrak{G}).$$

Then X_C has the form

$$X_C=\begin{pmatrix} X \\ X' \end{pmatrix},$$

where $X=\operatorname{Pr}_{\mathfrak{D}_T}X_C\in\mathfrak{L}(\mathfrak{F},\mathfrak{D}_T)$ is a contraction because $\tilde{\mathfrak{D}}_Y$ is a Hilbert space. Let $\tilde{D}_X\in\mathfrak{L}(\tilde{\mathfrak{D}}_X,\mathfrak{F})$ be a defect operator for X. Applying Theorem 3.3 (in either its simple form or generalized form) to X_C, we can write, finally,

$$X'=Z\tilde{D}_X^*,$$

where $Z\in\mathfrak{L}(\tilde{\mathfrak{D}}_X,\tilde{\mathfrak{D}}_Y)$ is a Hilbert space contraction. With these choices of X,Y,Z,

$$D_CX_C=\begin{pmatrix} D_T & 0 \\ -Y^*L_T^* & \tilde{D}_Y \end{pmatrix}\begin{pmatrix} X \\ Z\tilde{D}_X^* \end{pmatrix}=\begin{pmatrix} D_TX \\ -Y^*L_T^*X+\tilde{D}_YZD_X^* \end{pmatrix},$$

and therefore A has the required form.

The other direction follows from Theorem 2.13 and the index formulas (2.15), (2.17), and (2.19). □

Example:
Let us return to the example at the beginning of the lecture:

$$C = \begin{pmatrix} T \\ N \end{pmatrix} \in \mathfrak{L}(\mathfrak{H}, \mathfrak{K} \oplus \mathfrak{G}),$$

where $T \in \mathfrak{L}(\mathfrak{H}, \mathfrak{K})$ is an isometry with $\mathfrak{H}$ and $\mathfrak{K}$ Hilbert spaces and N is a nonzero operator satisfying $N^*N = 0$. We showed that C is a contractive column extension of T but not of the form given in Theorem 3.3 (Generalized Form). It is not hard to see why this result is not applicable. The hypotheses state that at least one of the conditions

$$\begin{aligned} \operatorname{ind}_-(1 - C^*C) + \operatorname{ind}_- \mathfrak{G} &= \operatorname{ind}_-(1 - T^*T) < \infty, \\ \operatorname{ind}_-(1 - CC^*) &= \operatorname{ind}_-(1 - TT^*) < \infty, \end{aligned}$$

holds. However, in this example,

$$\begin{aligned} \operatorname{ind}_-(1 - C^*C) + \operatorname{ind}_- \mathfrak{G} &> \operatorname{ind}_-(1 - T^*T), \\ \operatorname{ind}_-(1 - CC^*) &> \operatorname{ind}_-(1 - TT^*). \end{aligned}$$

The first relation holds because C and T are contractions and $\mathfrak{G}$ is not a Hilbert space (a nonzero operator N with the required properties cannot exist if $\mathfrak{G}$ is a Hilbert space). To verify the second relation, notice that $\operatorname{ind}_-(1 - TT^*) = 0$ because T^* is a contraction. It suffices to show that C^* is not a contraction. One way to do this to argue that ker C^* is not a Hilbert space (see Theorem 2.7). In fact, ker $C^* =$ ker $\begin{pmatrix} T^* & N^* \end{pmatrix}$ contains $\{0\} \oplus \operatorname{ran} N$, which is a neutral subspace, and so ker C^* is not a Hilbert space.

There is a more general theory which covers the preceding example and many more situations (see Dritschel [1993a]). Neutral operators, that is, operators N such that $N^*N = 0$, appear systematically in these more general results. It can be shown that Theorems 3.2–3.4 (Generalized Forms) give exactly the cases in which the neutral operators are not present.

As indicated at the outset, we have treated just one kind of extension problem. For related material on isometric dilations, commutant lifting, as well as other methods of approach, see, for example, Arov and Grossman [1983], [1992], Constantinescu and Gheondea [1989], [1992], [1993], Dijksma, Marcantognini, and de Snoo [1994], Dijksma, Dritschel, Marcantognini, and de Snoo [1993], Dritschel and Rovnyak [1990], and Marcantognini [1992]. In the Hilbert space case, such extension and completion problems are closely associated with interpolation theory. These problems are more complex in the indefinite case. They have been successfully treated by invariant subspace methods in a series of papers by J. A. Ball and J. W. Helton, including Ball and Helton [1983].

Lecture 4:
The Schur Algorithm

Key ideas:

- Given Kreĭn spaces $\mathfrak{A}$ and $\mathfrak{B}$, a Schur class $\mathbf{S}(\mathfrak{A}, \mathfrak{B})$ is defined whose elements are holomorphic functions with values in $\mathfrak{L}(\mathfrak{A}, \mathfrak{B})$ such that the Schur algorithm and Carathéodory-Fejér problem have generalizations.
- The approach uses a formal calculus of Julia operators. The calculus is equivalent to the choice sequence method of describing extensions of operators.

In complex function theory, the **Schur class S** is defined as the set of complex-valued analytic functions $s(z) = \sum_{n=0}^{\infty} s_n z^n$ which are defined and bounded by one in the unit disk $D = \{z : |z| < 1\}$ in the complex plane.

Classical Schur algorithm: Given $s(z)$ in $\mathbf{S}$, define $s_0(z), s_1(z), s_2(z), \ldots$ in $\mathbf{S}$ by writing $s_0(z) = s(z)$ and

$$\begin{aligned} s_1(z) &= \frac{1}{z} \frac{s_0(z) - s_0(0)}{1 - \bar{s}_0(0) s_0(z)}, \\ s_2(z) &= \frac{1}{z} \frac{s_1(z) - s_1(0)}{1 - \bar{s}_1(0) s_1(z)}, \\ &\ldots \end{aligned}$$

The sequence is well defined. For by assumption, $s_0(z)$ belongs to $\mathbf{S}$. If $s_n(z)$ belongs to $\mathbf{S}$ for some $n \geq 0$, then $s_{n+1}(z)$ is in $\mathbf{S}$ by Schwarz's lemma. Here $s_n(0)$ has modulus at most one because $s_n(z)$ is in $\mathbf{S}$; if $|s_n(0)| = 1$, we take $s_k(z) \equiv 0$ for all $k > n$. The numbers $\alpha_n = s_n(0)$, $n \geq 0$, are called the **Schur parameters** of $s(z)$. They form a sequence $\{\alpha_n\}_{n=0}^{\infty}$ in $\bar{D}$ such that if $|\alpha_r| = 1$ for some r, then $\alpha_k = 0$ for all $k > r$. Conversely, every such sequence occurs as the Schur parameters of a unique $s(z)$ in $\mathbf{S}$. The Schur algorithm thus labels $\mathbf{S}$ by the set of sequences $\{\alpha_n\}_{n=0}^{\infty}$ of the above type.

The Schur algorithm was introduced by Schur [1917] to study the Carathéodory-Fejér problem, and this is still one of its most striking applications. Assume that numbers $s_0, \ldots, s_n$ are given. The Carathéodory-Fejér problem asks for conditions that there exists a function $s(z)$ in $\mathbf{S}$ such that the first $n+1$ coefficients in the Taylor series for $s(z)$ are the given numbers. Schur's theorem is that such a function exists if and only if the lower triangular matrix

$$\begin{pmatrix} s_0 & & & \\ s_1 & s_0 & 0 & \\ & \ddots & \ddots & \\ s_n & & s_1 & s_0 \end{pmatrix}$$

acts as a contraction on $\mathbf{C}^{n+1}$. The Schur algorithm is also closely related to Szegő's theory of orthogonal polynomials on the unit circle and interpolation theory.

Hilbert space generalizations of these classical theories have been known for a long time, and they play an important role in operator theory and its applications. If $\mathfrak{A}$ and $\mathfrak{B}$ are Hilbert spaces, the Schur class $\mathbf{S}(\mathfrak{A}, \mathfrak{B})$ is defined to be the set of analytic functions $S(z)$ on D whose values are contractions in $\mathfrak{L}(\mathfrak{A}, \mathfrak{B})$. The scalar theory can be written in such a way that large portions carry over essentially intact to the Hilbert space case. This definition of the Schur class is not suitable for Kreĭn spaces, however. For example, it is not natural to require that functions be analytic in the full unit disk in the indefinite case.

Problem. *How should the theory of the Schur class be formulated for functions whose values are Kreĭn space operators?*

The generalized Schur class is defined in Definition 4.6, after suitable preparation. Our approach is essentially the choice sequence method of Bakonyi and Constantinescu [1992], with the main constructions encoded in a formal calculus of Julia operators. In one key place, we use the extension theory from Lecture 3. Full details may be found in Christner and Rovnyak [to appear].

The Schur algorithm was studied in an indefinite setting by Ivanchenko [1979]. An approach by reproducing kernel spaces is given in Alpay and Dym [1986], [1992]. Our interest in indefinite generalizations of the Schur algorithm was sparked, in part, from an appearance of a variant of the algorithm in a coefficient problem for univalent functions (Christner, Li, and Rovnyak [1994]); indefinite inner products are central here also, but there are differences with the present situation, such as the use of composition operators. For the Hilbert space generalization of the Schur algorithm, see the books of Bakonyi and Constantinescu [1992] and Dubovoj, Fritzsche, and Kirstein [1992]. Additional references may be found in the historical accounts in Herglotz, Schur, et al. [1991] and the book review Rovnyak [1994].

To begin, assume that we are given Kreĭn spaces $\mathfrak{A}_0, \mathfrak{A}_1, \ldots$ and $\mathfrak{B}_0, \mathfrak{B}_1, \ldots$ and a sequence of Julia operators

$$\left.\begin{aligned}
&\begin{pmatrix} \alpha_0 & \beta_0 \\ \gamma_0 & \delta_0 \end{pmatrix} \in \mathfrak{L}(\mathfrak{A}_0 \oplus \mathfrak{B}_1, \mathfrak{B}_0 \oplus \mathfrak{A}_1), \\
&\begin{pmatrix} \alpha_1 & \beta_1 \\ \gamma_1 & \delta_1 \end{pmatrix} \in \mathfrak{L}(\mathfrak{A}_1 \oplus \mathfrak{B}_2, \mathfrak{B}_1 \oplus \mathfrak{A}_2), \\
&\qquad\quad \cdots \\
&\begin{pmatrix} \alpha_j & \beta_j \\ \gamma_j & \delta_j \end{pmatrix} \in \mathfrak{L}(\mathfrak{A}_j \oplus \mathfrak{B}_{j+1}, \mathfrak{B}_j \oplus \mathfrak{A}_{j+1}), \\
&\qquad\quad \cdots
\end{aligned}\right\} \tag{4.1}$$

Put

$$\mathfrak{A} = \mathfrak{A}_0 \quad \text{and} \quad \mathfrak{B} = \mathfrak{B}_0.$$

No other assumptions are made at this point: the key hypothesis is the particular way in which the domains of the Julia operators are linked.

The sequence (4.1) is used in Theorems 4.1–4.4. For simplicity, we assume that the sequence is infinite. However, the main constructions are recursive and depend only on segments, so essentially everything that we say applies to sequences of the form (4.1) which terminate. Define a Julia operator

$$\begin{pmatrix} R_k & D_{R_k} \\ \tilde{D}^*_{R_k} & -L^*_{R_k} \end{pmatrix}$$

$$= \begin{pmatrix} \alpha_1 & \beta_1 & & & 0 \\ \gamma_1 & \delta_1 & & & \\ & & 1 & & \\ & & & 1 & \\ 0 & & & & \ddots \\ & & & & & 1 \end{pmatrix} \begin{pmatrix} 1 & & & & \\ & \alpha_2 & \beta_2 & & 0 \\ & \gamma_2 & \delta_2 & & \\ & & & 1 & \\ & 0 & & & \ddots \\ & & & & & 1 \end{pmatrix} \cdots \begin{pmatrix} 1 & & & & 0 \\ & 1 & & & \\ & & 1 & & \\ & & & \ddots & \\ 0 & & & & \alpha_k & \beta_k \\ & & & & \gamma_k & \delta_k \end{pmatrix}$$

$$= \begin{pmatrix} \begin{pmatrix} \alpha_1 & \beta_1\alpha_2 & \beta_1\beta_2\alpha_3 & \ldots & \beta_1\beta_2\cdots\beta_{k-1}\alpha_k \end{pmatrix} & \begin{pmatrix} \beta_1\beta_2\cdots\beta_k \end{pmatrix} \\ \begin{pmatrix} \gamma_1 & \delta_1\alpha_2 & \delta_1\beta_2\alpha_3 & \ldots & \delta_1\beta_2\cdots\beta_{k-1}\alpha_k \\ 0 & \gamma_2 & \delta_2\alpha_3 & \ldots & \delta_2\beta_3\cdots\beta_{k-1}\alpha_k \\ \vdots & \vdots & \vdots & & \vdots \\ 0 & 0 & 0 & \ldots & \gamma_k \end{pmatrix} & \begin{pmatrix} \delta_1\beta_2\cdots\beta_k \\ \delta_2\beta_3\cdots\beta_k \\ \vdots \\ \delta_k \end{pmatrix} \end{pmatrix}$$

with domain $(\mathfrak{A}_1 \oplus \cdots \oplus \mathfrak{A}_k) \oplus \mathfrak{D}_{R_k} = (\mathfrak{A}_1 \oplus \cdots \oplus \mathfrak{A}_k) \oplus \mathfrak{B}_{k+1}$ and range

$$\mathfrak{B}_1 \oplus \tilde{\mathfrak{D}}_{R_k} = \mathfrak{B}_1 \oplus (\mathfrak{A}_2 \oplus \cdots \oplus \mathfrak{A}_{k+1}) .$$

Similarly,

$$\begin{pmatrix} C_k & D_{C_k} \\ \tilde{D}^*_{C_k} & -L^*_{C_k} \end{pmatrix}$$

$$= \begin{pmatrix} \begin{pmatrix} \alpha_1 \\ \alpha_2\gamma_1 \\ \alpha_3\gamma_2\gamma_1 \\ \vdots \\ \alpha_k\gamma_{k-1}\cdots\gamma_1 \end{pmatrix} & \begin{pmatrix} \beta_1 & 0 & \ldots & 0 \\ \alpha_2\delta_1 & \beta_2 & \ldots & 0 \\ \alpha_3\gamma_2\delta_1 & \alpha_3\delta_2 & \ldots & 0 \\ \vdots & \vdots & & \vdots \\ \alpha_k\gamma_{k-1}\cdots\gamma_2\delta_1 & \alpha_k\gamma_{k-1}\cdots\gamma_3\delta_2 & \ldots & \beta_k \end{pmatrix} \\ \begin{pmatrix} \gamma_k\gamma_{k-1}\cdots\gamma_1 \end{pmatrix} & \begin{pmatrix} \gamma_k\gamma_{k-1}\cdots\gamma_2\delta_1 & \gamma_k\gamma_{k-1}\cdots\gamma_3\delta_2 & \ldots & \delta_k \end{pmatrix} \end{pmatrix}$$

$$= \begin{pmatrix} 1 & & & & 0 \\ & 1 & & & \\ & & 1 & & \\ & & & \ddots & \\ 0 & & & & \alpha_k & \beta_k \\ & & & & \gamma_k & \delta_k \end{pmatrix} \cdots \begin{pmatrix} 1 & & & & \\ & \alpha_2 & \beta_2 & & 0 \\ & \gamma_2 & \delta_2 & & \\ & & & 1 & \\ & 0 & & & \ddots \\ & & & & & 1 \end{pmatrix} \begin{pmatrix} \alpha_1 & \beta_1 & & & 0 \\ \gamma_1 & \delta_1 & & & \\ & & 1 & & \\ & & & 1 & \\ 0 & & & & \ddots \\ & & & & & 1 \end{pmatrix}$$

is a Julia operator with domain $\mathfrak{A}_1 \oplus \mathfrak{D}_{C_k} = \mathfrak{A}_1 \oplus (\mathfrak{B}_2 \oplus \cdots \oplus \mathfrak{B}_{k+1})$ and range

$$(\mathfrak{B}_1 \oplus \cdots \oplus \mathfrak{B}_k) \oplus \tilde{\mathfrak{D}}_{C_k} = (\mathfrak{B}_1 \oplus \cdots \oplus \mathfrak{B}_k) \oplus \mathfrak{A}_{k+1}.$$

Theorem 4.1 *In the situation just described, construct Julia operators*

$$\begin{pmatrix} T_k & D_{T_k} \\ \tilde{D}^*_{T_k} & -L^*_{T_k} \end{pmatrix} \in \mathfrak{L}(\mathfrak{A}^{k+1} \oplus \mathfrak{D}_{T_k}, \mathfrak{B}^{k+1} \oplus \tilde{\mathfrak{D}}_{T_k}),$$
$$\tilde{\mathfrak{D}}_{T_k} = \mathfrak{A}_1 \oplus \cdots \oplus \mathfrak{A}_{k+1},$$
$$\mathfrak{D}_{T_k} = \mathfrak{B}_1 \oplus \cdots \oplus \mathfrak{B}_{k+1},$$

for $0 \le k \le r$ *by setting*

$$\begin{pmatrix} T_0 & D_{T_0} \\ \tilde{D}^*_{T_0} & -L^*_{T_0} \end{pmatrix} = \begin{pmatrix} \alpha_0 & \beta_0 \\ \gamma_0 & \delta_0 \end{pmatrix}$$

and

$$\begin{pmatrix} T_k & D_{T_k} \\ \tilde{D}^*_{T_k} & -L^*_{T_k} \end{pmatrix} = \begin{pmatrix} 1 & 0 & 0 & 0 \\ 0 & \alpha_0 & \beta_0 & 0 \\ 0 & \gamma_0 & \delta_0 & 0 \\ 0 & 0 & 0 & 1 \end{pmatrix} \begin{pmatrix} 1 & 0 & 0 & 0 \\ 0 & 1 & 0 & 0 \\ 0 & 0 & R_k & D_{R_k} \\ 0 & 0 & \tilde{D}^*_{R_k} & -L^*_{R_k} \end{pmatrix} \begin{pmatrix} T_{k-1} & 0 & D_{T_{k-1}} & 0 \\ 0 & 1 & 0 & 0 \\ \tilde{D}^*_{T_{k-1}} & 0 & -L^*_{T_{k-1}} & 0 \\ 0 & 0 & 0 & 1 \end{pmatrix}$$

for $1 \le k \le r$. *Then for* $1 \le k \le r$,

$$\begin{pmatrix} T_k & D_{T_k} \\ \tilde{D}^*_{T_k} & -L^*_{T_k} \end{pmatrix} = \begin{pmatrix} 1 & 0 & 0 & 0 \\ 0 & T_{k-1} & D_{T_{k-1}} & 0 \\ 0 & \tilde{D}^*_{T_{k-1}} & -L^*_{T_{k-1}} & 0 \\ 0 & 0 & 0 & 1 \end{pmatrix} \begin{pmatrix} 1 & 0 & 0 & 0 \\ 0 & 1 & 0 & 0 \\ 0 & 0 & C_k & D_{C_k} \\ 0 & 0 & \tilde{D}^*_{C_k} & -L^*_{C_k} \end{pmatrix} \begin{pmatrix} \alpha_0 & 0 & \beta_0 & 0 \\ 0 & 1 & 0 & 0 \\ \gamma_0 & 0 & \delta_0 & 0 \\ 0 & 0 & 0 & 1 \end{pmatrix}$$

The theorem is proved by an inductive argument (Christner and Rovnyak [to appear]). It would be interesting to have a more conceptual proof. For example, is there a systems explanation?

On expanding the matrix products, we obtain

$$\begin{pmatrix} T_k & D_{T_k} \\ \tilde{D}^*_{T_k} & -L^*_{T_k} \end{pmatrix} = \begin{pmatrix} \begin{pmatrix} T_{k-1} & 0 \\ \beta_0 R_k \tilde{D}^*_{T_{k-1}} & \alpha_0 \end{pmatrix} & \begin{pmatrix} D_{T_{k-1}} & 0 \\ -\beta_0 R_k L^*_{T_{k-1}} & \beta_0 D_{R_k} \end{pmatrix} \\ \begin{pmatrix} \delta_0 R_k \tilde{D}^*_{T_{k-1}} & \gamma_0 \\ \tilde{D}^*_{R_k} \tilde{D}^*_{T_{k-1}} & 0 \end{pmatrix} & \begin{pmatrix} -\delta_0 R_k L^*_{T_{k-1}} & \delta_0 D_{R_k} \\ -\tilde{D}^*_{R_k} L^*_{T_{k-1}} & L_{R_k} \end{pmatrix} \end{pmatrix}$$

and

$$\begin{pmatrix} T_k & D_{T_k} \\ \tilde{D}^*_{T_k} & -L^*_{T_k} \end{pmatrix} = \begin{pmatrix} \begin{pmatrix} \alpha_0 & 0 \\ D_{T_{k-1}} C_k \gamma_0 & T_{k-1} \end{pmatrix} & \begin{pmatrix} \beta_0 & 0 \\ D_{T_{k-1}} C_k \delta_0 & D_{T_{k-1}} D_{C_k} \end{pmatrix} \\ \begin{pmatrix} -L^*_{T_{k-1}} C_k \gamma_0 & \tilde{D}^*_{T_{k-1}} \\ \tilde{D}^*_{C_k} \gamma_0 & 0 \end{pmatrix} & \begin{pmatrix} -L^*_{T_{k-1}} C_k \delta_0 & -L^*_{T_{k-1}} D_{C_k} \\ \tilde{D}^*_{C_k} \delta_0 & -L^*_{C_k} \end{pmatrix} \end{pmatrix}.$$

These relations concisely summarize many useful identities. For example, they give recursive formulas for the operators T_k, D_{T_k}, and $\tilde{D}_{T_k}$. They give two different sets of formulas for these operators, and the fact that the sets are identical gives other relations.

It is not apparent in Theorem 4.1 why the operators $T_0, T_1, T_2, \dots$ should be of interest. The reason is that they have a Toeplitz structure.

Theorem 4.2 *In Theorem* 4.1, *there exist operators* $S_0, \dots, S_r$ *in* $\mathfrak{L}(\mathfrak{A}, \mathfrak{B})$ *such that*

$$T_k = \begin{pmatrix} S_0 & & & \\ S_1 & S_0 & 0 & \\ & \ddots & \ddots & \\ S_k & & S_1 & S_0 \end{pmatrix}$$

for each $k = 0, \dots, r$.

This is also proved by induction. The recursive formulas for T_k, D_{T_k}, and $\tilde{D}_{T_k}$ described above lead to similar formulas for the operators $S_0, S_1, \dots$. Closed form expressions for $S_0, S_1, \dots$ can be given. Define formal matrices

$$\begin{pmatrix} \widehat{\alpha}_0 & \widehat{\beta}_0 \\ \widehat{\gamma}_0 & \widehat{\delta}_0 \end{pmatrix} = \begin{pmatrix} (\alpha_0) & (\beta_0 \ \ 0 \ \ 0 \ \ \dots) \\ \begin{pmatrix} \gamma_0 \\ 0 \\ 0 \\ \vdots \end{pmatrix} & \begin{pmatrix} \delta_0 & 0 & 0 & \dots \\ 0 & 1 & 0 & \dots \\ 0 & 0 & 1 & \dots \\ & \dots & & \ddots \end{pmatrix} \end{pmatrix}$$

and

$$X = \begin{pmatrix} \alpha_1 & \beta_1 & & & \\ \gamma_1 & \delta_1 & & 0 & \\ & & 1 & & \\ & & & 1 & \\ & 0 & & & 1 \\ & & & & & \ddots \end{pmatrix} \begin{pmatrix} 1 & & & & \\ & \alpha_2 & \beta_2 & & \\ & \gamma_2 & \delta_2 & 0 & \\ & & & 1 & \\ & & 0 & & 1 \\ & & & & & \ddots \end{pmatrix} \cdots.$$

The definition of X is meaningful because each term in the expansion of the product involves only a finite number of algebraic expressions.

Theorem 4.3 *Let* $S(z) = S_0 + S_1 z + S_2 z^2 + \cdots$, *where* $S_0, S_1, S_2, \dots$ *are the operators in Theorem* 4.2. *Then*

$$S(z) = \widehat{\alpha}_0 + z\widehat{\beta}_0 X \left(1 - z\widehat{\delta}_0 X\right)^{-1} \widehat{\gamma}_0$$

in the sense of formal power series.

The theorem asserts that

$$S_0 = \widehat{\alpha}_0$$

and

$$S_n = \widehat{\beta}_0 X \left(\widehat{\delta}_0 X\right)^{n-1} \widehat{\gamma}_0, \qquad n \geq 1.$$

The proof is mainly a matter of algebra proceeding from the recursive construction of the coefficients.

The structure of the coefficients $S_0, S_1, \ldots$ in $S(z)$ described in Theorem 4.3 leads to a linear fractional representation.

Theorem 4.4 *Define $S(z)$ from the Julia operators*

$$\begin{pmatrix} \alpha_0 & \beta_0 \\ \gamma_0 & \delta_0 \end{pmatrix}, \begin{pmatrix} \alpha_1 & \beta_1 \\ \gamma_1 & \delta_1 \end{pmatrix}, \begin{pmatrix} \alpha_2 & \beta_2 \\ \gamma_2 & \delta_2 \end{pmatrix}, \ldots$$

in the manner described in Theorem 4.3. *In a similar way, define $S'(z)$ from the shifted sequence*

$$\begin{pmatrix} \alpha_1 & \beta_1 \\ \gamma_1 & \delta_1 \end{pmatrix}, \begin{pmatrix} \alpha_2 & \beta_2 \\ \gamma_2 & \delta_2 \end{pmatrix}, \ldots .$$

Then $S(z)$ and $S'(z)$ are related by

$$S(z) = \alpha_0 + z\beta_0 S'(z)\left(1 - z\delta_0 S'(z)\right)^{-1}\gamma_0.$$

Theorems 4.1–4.4 are much more general than we need for the application to the Schur algorithm. In (4.1) we assume that $\mathfrak{A}_0, \mathfrak{A}_1, \mathfrak{A}_2, \ldots$ and $\mathfrak{B}_0, \mathfrak{B}_1, \mathfrak{B}_2, \ldots$ are Kreĭn spaces. An interesting special case arises when $\mathfrak{A}_1, \mathfrak{A}_2, \ldots$ and $\mathfrak{B}_1, \mathfrak{B}_2, \ldots$ are Hilbert spaces. In this situation, $\mathfrak{A} = \mathfrak{A}_0$ and $\mathfrak{B} = \mathfrak{B}_0$ are Kreĭn spaces having the same negative index. In fact, by the unitarity of the first Julia operator in (4.1),

$$\operatorname{ind}_- \mathfrak{A}_0 = \operatorname{ind}_- (\mathfrak{A}_0 \oplus \mathfrak{B}_1) = \operatorname{ind}_- (\mathfrak{A}_0 \oplus \mathfrak{B}_1) = \operatorname{ind}_- \mathfrak{B}_0$$

because $\mathfrak{A}_1$ and $\mathfrak{B}_1$ are Hilbert spaces

Theorem 4.5 *Let $\mathfrak{A} = \mathfrak{A}_0$ and $\mathfrak{B} = \mathfrak{B}_0$ be Kreĭn spaces having the same negative index, and let $\mathfrak{A}_1, \mathfrak{A}_2, \ldots$ and $\mathfrak{B}_1, \mathfrak{B}_2, \ldots$ be Hilbert spaces. Let* (4.1) *be a sequence of Julia operators using these spaces, and define*

$$S(z) = S_0 + S_1 z + S_2 z^2 + \cdots$$

as in Theorems 4.1–4.4. *Then for each $k = 0, 1, 2, \ldots$,*

$$T_k = \begin{pmatrix} S_0 & & & 0 \\ S_1 & S_0 & & \\ & \ddots & \ddots & \\ S_k & & S_1 & S_0 \end{pmatrix}$$

is a bicontraction on $\mathfrak{A}^{k+1}$ to $\mathfrak{B}^{k+1}$.

In the statement of the theorem, $\mathfrak{A}^{k+1}$ and $\mathfrak{B}^{k+1}$ denote $(k+1)$-fold Cartesian products of $\mathfrak{A}$ and $\mathfrak{B}$. The proof of the theorem is very easy.

Proof. Let k be any nonnegative integer. By Theorem 4.1, we may choose a Julia operator for T_k with defect spaces given by

$$\begin{aligned} \tilde{\mathfrak{D}}_{T_k} &= \mathfrak{A}_1 \oplus \cdots \oplus \mathfrak{A}_{k+1}, \\ \mathfrak{D}_{T_k} &= \mathfrak{B}_1 \oplus \cdots \oplus \mathfrak{B}_{k+1}. \end{aligned}$$

Our hypotheses imply that these are Hilbert spaces, and so T_k is a bicontraction by Theorem 2.3. □

In view of these results, we are now able to propose a definition of the Schur class $\mathbf{S}(\mathfrak{A},\mathfrak{B})$ when $\mathfrak{A}$ and $\mathfrak{B}$ are Kreĭn spaces:

Definition 4.6 *Let $\mathfrak{A}$ and $\mathfrak{B}$ be Kreĭn spaces having the same negative index. By the* **Schur class** $\mathbf{S}(\mathfrak{A},\mathfrak{B})$ *we mean the set of all formal power series*

$$S(z) = \sum_{n=0}^{\infty} S_n z^n$$

with coefficients in $\mathfrak{L}(\mathfrak{A},\mathfrak{B})$ such that for each $k = 0, 1, 2, \ldots$, the matrix

$$T_k = \begin{pmatrix} S_0 & & & \\ S_1 & S_0 & 0 & \\ & \ddots & \ddots & \\ S_k & & S_1 & S_0 \end{pmatrix}$$

is a bicontraction on $\mathfrak{A}^{k+1}$ to $\mathfrak{B}^{k+1}$.

The condition that $\mathfrak{A}$ and $\mathfrak{B}$ have the same negative index is necessary to insure that the Schur class $\mathbf{S}(\mathfrak{A},\mathfrak{B})$ is nontrivial. For if $\mathbf{S}(\mathfrak{A},\mathfrak{B})$ contains a single element $S(z) = \sum_{n=0}^{\infty} S_n z^n$, then

$$\operatorname{ind}_- \mathfrak{A} = \operatorname{ind}_- \mathfrak{B}$$

by Theorem 2.8, because $T_0 = S_0$ is a bicontraction on $\mathfrak{A}$ to $\mathfrak{B}$. When the condition is satisfied, Theorem 4.5 provides a rich class of examples of elements $S(z)$ of $\mathbf{S}(\mathfrak{A},\mathfrak{B})$.

A number of questions arise.

Question (1): Does every function $S(z)$ in $\mathbf{S}(\mathfrak{A},\mathfrak{B})$ have the form described in Theorem 4.5? The answer is affirmative:

Every $S(z)$ in $\mathbf{S}(\mathfrak{A},\mathfrak{B})$ is obtained as in Theorem 4.5 for some sequence of Julia operators

$$\begin{pmatrix} \alpha_n & \beta_n \\ \gamma_n & \delta_n \end{pmatrix} \in \mathfrak{L}(\mathfrak{A}_n \oplus \mathfrak{B}_{n+1}, \mathfrak{B}_n \oplus \mathfrak{A}_{n+1}), \qquad n \geq 0, \tag{4.2}$$

such that $\mathfrak{A}_1, \mathfrak{A}_2, \ldots$ and $\mathfrak{B}_1, \mathfrak{B}_2, \ldots$ are Hilbert spaces.

The proof uses an extension theorem of the type discussed in Lecture 3: Let $A_1 \in \mathfrak{L}(\mathfrak{H}_1)$ and $A_2 \in \mathfrak{L}(\mathfrak{H}_2)$ be bicontractions, and choose defect operators $\tilde{D}_{A_1} \in \mathfrak{L}(\tilde{\mathfrak{D}}_{A_1}, \mathfrak{H}_1)$ and $D_{A_2} \in \mathfrak{L}(\mathfrak{D}_{A_2}, \mathfrak{H}_2)$ for A_1^* and A_2. An operator

$$\begin{pmatrix} A_1 & 0 \\ Q & A_2 \end{pmatrix} \in \mathfrak{L}(\mathfrak{H}_1 \oplus \mathfrak{H}_2, \mathfrak{K}_1 \oplus \mathfrak{K}_2)$$

is a bicontraction if and only if

$$Q = D_{A_2} Y \tilde{D}_{A_1}^*,$$

where $Y \in \mathfrak{L}(\tilde{\mathfrak{D}}_{A_1}, \mathfrak{D}_{A_2})$ is a bicontraction. This result is an easy consequence of theorems in Lecture 3.

Now suppose that $S(z) = S_0 + S_1 z + S_2 z^2 + \cdots$. We choose $\alpha_0 = S_0$ and take the first term of the sequence (4.2) to be any Julia operator

$$\begin{pmatrix} \alpha_0 & \beta_0 \\ \gamma_0 & \delta_0 \end{pmatrix} \in \mathfrak{L}(\mathfrak{A}_0 \oplus \mathfrak{B}_1, \mathfrak{B}_0 \oplus \mathfrak{A}_1)$$

for α_0. Since α_0 is bicontraction by assumption, $\mathfrak{A}_1$ and $\mathfrak{B}_1$ are Hilbert spaces. The next Julia operator is obtained by applying the extension theorem to the triangular matrix

$$\begin{pmatrix} A_1 & 0 \\ Q & A_2 \end{pmatrix} = \begin{pmatrix} S_0 & 0 \\ S_1 & S_0 \end{pmatrix}.$$

The extension theorem gives a factorization

$$S_1 = \beta_0 \alpha_1 \gamma_0,$$

where $\alpha_1 \in \mathfrak{L}(\mathfrak{A}_1, \mathfrak{B}_1)$ is a bicontraction (in fact, a Hilbert space contraction). For the second term of the sequence (4.2), we choose any Julia operator

$$\begin{pmatrix} \alpha_1 & \beta_1 \\ \gamma_1 & \delta_1 \end{pmatrix} \in \mathfrak{L}(\mathfrak{A}_1 \oplus \mathfrak{B}_2, \mathfrak{B}_1 \oplus \mathfrak{A}_2)$$

for α_1. The new spaces $\mathfrak{A}_2$ and $\mathfrak{B}_2$ are Hilbert spaces because α_1 is a Hilbert space contraction.

The construction proceeds recursively. In the general step, we apply the extension theorem to the triangular matrix

$$\begin{pmatrix} A_1 & 0 \\ Q & A_2 \end{pmatrix} = \left(\begin{matrix} S_0 & 0 \\ \begin{pmatrix} S_1 \\ S_2 \\ \vdots \\ S_{k+1} \end{pmatrix} & \begin{pmatrix} S_0 & & & \\ S_1 & S_0 & & 0 \\ & \ddots & \ddots & \\ S_k & & S_1 & S_0 \end{pmatrix} \end{matrix} \right).$$

A certain amount of algebra is needed to see that the pattern persists (see Christner and Rovnyak [to appear], Th. 4.3).

Question (2): Can two essentially different sequences of Julia operators lead to the same element $S(z)$ of $\mathbf{S}(\mathfrak{A}, \mathfrak{B})$ in Theorem 4.5? The answer here is negative. An element $S(z)$ of $\mathbf{S}(\mathfrak{A}, \mathfrak{B})$ determines the sequence of Julia operators up to isomorphisms between the defect spaces. See Christner and Rovnyak [to appear], Th. 4.4.

Question (3): Does the definition of $\mathbf{S}(\mathfrak{A}, \mathfrak{B})$ agree with the previous one when $\mathfrak{A}$ and $\mathfrak{B}$ are Hilbert spaces? Here the answer is affirmative and a consequence of known Hilbert space theorems as well as results given below.

Generalized Schur algorithm:

One can ask more broadly if the Schur class $\mathbf{S}(\mathfrak{A}, \mathfrak{B})$ has properties analogous to the scalar and Hilbert space cases. The Schur algorithm is one test. We shall confirm that this test is passed.

First note that Theorems 4.1–4.4 apply in particular to the Schur class $\mathbf{S}(\mathfrak{A}, \mathfrak{B})$. The information can be put into this form.

Theorem 4.7 (Christner [1993]) *Let $\mathfrak{A}$ and $\mathfrak{B}$ be Kreĭn spaces having the same negative index. Let $S_0 \in \mathfrak{L}(\mathfrak{A}, \mathfrak{B})$ be a bicontraction with Julia operator*

$$\begin{pmatrix} S_0 & D_0 \\ \tilde{D}_0^* & -L_0^* \end{pmatrix} \in \mathfrak{L}(\mathfrak{A} \oplus \mathfrak{D}_0, \mathfrak{B} \oplus \tilde{\mathfrak{D}}_0).$$

(1) *If $S'(z)$ belongs to $\mathbf{S}(\tilde{\mathfrak{D}}_0, \mathfrak{D}_0)$, then*

$$S(z) = S_0 + zD_0 S'(z) \left(1 + zL_0^* S'(z)\right)^{-1} \tilde{D}_0^*$$

defines an element of $\mathbf{S}(\mathfrak{A}, \mathfrak{B})$ such that $S(0) = S_0$. (2) *Every $S(z)$ in $\mathbf{S}(\mathfrak{A}, \mathfrak{B})$ with $S(0) = S_0$ has this form, and the choice of $S'(z)$ in $\mathbf{S}(\tilde{\mathfrak{D}}_0, \mathfrak{D}_0)$ is unique when the Julia operator for S_0 is fixed.*

One consequence is that the elements $S(z)$ of the Schur class $\mathbf{S}(\mathfrak{A}, \mathfrak{B})$ represent analytic functions in a neighborhood of the origin. For since $\tilde{\mathfrak{D}}_0$ and $\mathfrak{D}_0$ are Hilbert spaces, $\mathbf{S}(\tilde{\mathfrak{D}}_0, \mathfrak{D}_0)$ is the set of analytic functions on the unit disk whose values are contraction operators in $\mathfrak{L}(\tilde{\mathfrak{D}}_0, \mathfrak{D}_0)$. Analyticity in a neighborhood of the origin therefore follows from the linear fractional representation in part (1) of the theorem.

The Schur algorithm in the Kreĭn space setting exactly follows the Hilbert space case, which is given in Bakonyi and Constantinescu [1992] and Dubovoj, Fritzsche, and Kirstein [1992]. Let $\mathfrak{A}$ and $\mathfrak{B}$ be Kreĭn spaces having the same negative index, and assume that $S(z)$ belongs to $\mathbf{S}(\mathfrak{A}, \mathfrak{B})$. Set $\mathfrak{A}_0 = \mathfrak{A}$ and $\mathfrak{B}_0 = \mathfrak{B}$. Define $S_0(z) \in \mathbf{S}(\mathfrak{A}_0, \mathfrak{B}_0)$ and $\alpha_0 \in \mathfrak{L}(\mathfrak{A}_0, \mathfrak{B}_0)$ by

$$S_0(z) = S(z) \qquad \text{and} \qquad \alpha_0 = S_0(0),$$

and choose a Julia operator

$$\begin{pmatrix} \alpha_0 & \beta_0 \\ \gamma_0 & \delta_0 \end{pmatrix} \in \mathfrak{L}(\mathfrak{A}_0 \oplus \mathfrak{B}_1, \mathfrak{B}_0 \oplus \mathfrak{A}_1).$$

The defect spaces $\mathfrak{A}_1$ and $\mathfrak{B}_1$ are Hilbert spaces because α_0 is a bicontraction. By Theorem 4.7, there is a unique solution $S_1(z) \in \mathbf{S}(\mathfrak{A}_1, \mathfrak{B}_1)$ of the functional equation

$$S_0(z) = \alpha_0 + z\beta_0 S_1(z) \left(1 - z\delta_0 S_1(z)\right)^{-1} \gamma_0.$$

Set $\alpha_1 = S_1(0)$, and then choose a Julia operator for α_1 with defect Hilbert spaces $\mathfrak{A}_2$ and $\mathfrak{B}_2$ and repeat the construction. For the inductive step, suppose that for each $k = 0, \dots, n$, we have chosen $S_k(z) \in \mathbf{S}(\mathfrak{A}_k, \mathfrak{B}_k)$ and a Julia operator

$$\begin{pmatrix} \alpha_k & \beta_k \\ \gamma_k & \delta_k \end{pmatrix} \in \mathfrak{L}(\mathfrak{A}_k \oplus \mathfrak{B}_{k+1}, \mathfrak{B}_k \oplus \mathfrak{A}_{k+1})$$

such that $\alpha_k = S_k(0)$ and $\mathfrak{A}_{k+1}$ and $\mathfrak{B}_{k+1}$ are Hilbert spaces. Use Theorem 4.7 to choose $S_{k+1}(z) \in \mathbf{S}(\mathfrak{A}_{k+1}, \mathfrak{B}_{k+1})$ such that

$$S_k(z) = \alpha_k + z\beta_k S_{k+1}(z) \left(1 - z\delta_k S_{k+1}(z)\right)^{-1} \gamma_k.$$

Set $\alpha_{k+1} = S_{k+1}(0)$ and choose a Julia operator for α_{k+1} with new defect Hilbert spaces $\mathfrak{A}_{k+2}$ and $\mathfrak{B}_{k+2}$ to complete the inductive construction. The construction produces a sequence

$$S_n(z) \in \mathbf{S}(\mathfrak{A}_n, \mathfrak{B}_n), \qquad n \geq 0.$$

The operators

$$\begin{pmatrix} \alpha_n & \beta_n \\ \gamma_n & \delta_n \end{pmatrix} \in \mathfrak{L}(\mathfrak{A}_n \oplus \mathfrak{B}_{n+1}, \mathfrak{B}_n \oplus \mathfrak{A}_{n+1}), \qquad n = 0, 1, 2, \dots ,$$

which arise in the construction are called the **Schur parameters** of $S(z)$. For a parallel with the classical case, we also call

$$\alpha_0, \alpha_1, \alpha_2, \dots$$

the Schur parameters of $S(z)$ with a choice of Julia operators understood.

The Schur parameters of an element $S(z)$ of $\mathbf{S}(\mathfrak{A}, \mathfrak{B})$ are defined up to replacement of domain spaces by isomorphic copies. If we identify such sequences, then our results show that there is a one-to-one correspondence between the Schur class $\mathbf{S}(\mathfrak{A}, \mathfrak{B})$ and the set of sequences of Julia operators

$$\begin{pmatrix} \alpha_n & \beta_n \\ \gamma_n & \delta_n \end{pmatrix} \in \mathfrak{L}(\mathfrak{A}_n \oplus \mathfrak{B}_{n+1}, \mathfrak{B}_n \oplus \mathfrak{A}_{n+1}), \qquad n = 0, 1, 2, \dots ,$$

such that $\mathfrak{A}_0 = \mathfrak{A}$, $\mathfrak{B}_0 = \mathfrak{B}$, and $\mathfrak{A}_1, \mathfrak{A}_2, \dots$ and $\mathfrak{B}_1, \mathfrak{B}_2, \dots$ are Hilbert spaces.

This version of the Schur algorithm can be used as in the scalar and Hilbert space cases. A great number of familiar constructions can be extended to the indefinite setting. For example, we obtain a Kreĭn space generalization of Schur's theorem.

Theorem 4.8 *Let $S_0, \dots, S_r$ be given operators in $\mathfrak{L}(\mathfrak{A}, \mathfrak{B})$, where $\mathfrak{A}$ and $\mathfrak{B}$ are Kreĭn spaces having the same negative index. The operators occur as the initial segment of coefficients of an element $S(z)$ in $\mathbf{S}(\mathfrak{A}, \mathfrak{B})$ if and only if*

$$T_r = \begin{pmatrix} S_0 & & & \\ S_1 & S_0 & & 0 \\ & \ddots & \ddots & \\ S_r & & S_1 & S_0 \end{pmatrix}$$

is a bicontraction on $\mathfrak{A}^{r+1}$ to $\mathfrak{B}^{r+1}$.

It is instructive to see how the classical Schur algorithm is recovered from the generalized Schur algorithm. As before, write $\mathbf{S}$ for $\mathbf{S}(\mathfrak{A}, \mathfrak{B})$ when $\mathfrak{A}$ and $\mathfrak{B}$ coincide with the complex numbers in the Euclidean metric. Start with an element $s_0(z) = s(z)$ of $\mathbf{S}$. Then $\alpha_0 = s_0(0)$ is a number of modulus at most one. Choose the Julia operator

$$\begin{pmatrix} \alpha_0 & \beta_0 \\ \gamma_0 & \delta_0 \end{pmatrix} = \begin{pmatrix} \alpha_0 & \beta_0 \\ \bar{\beta}_0 & -\bar{\alpha}_0 \end{pmatrix},$$

where β_0 is any complex number such that $|\beta_0|^2 = 1 - |\alpha_0|^2$. Then $s_1(z)$ is the solution of the equation

$$\begin{aligned} s_0(z) &= \alpha_0 + z\beta_0 s_1(z) \left(1 - z\delta_0 s_1(z)\right)^{-1} \gamma_0 \\ &= \alpha_0 + z|\beta_0|^2 s_1(z) \left(1 + z\bar{\alpha}_0 s_1(z)\right)^{-1}, \end{aligned}$$

and therefore

$$\frac{s_0(z) - \alpha_0}{z} = \frac{|\beta_0|^2 s_1(z)}{1 + z\bar{\alpha}_0 s_1(z)}.$$

Solving, we get

$$s_1(z) = \frac{1}{z} \frac{s_0(z) - s_0(0)}{1 - \bar{s}_0(0)s_0(z)},$$

in agreement with the construction at the beginning of this lecture. The Kreĭn space Schur algorithm thus reduces to the classical Schur algorithm in the scalar case.

In the Hilbert space case, the main ideas in this lecture also appear in the context of commutant lifting. For example, see Foiaş and Frazho [1990]. A Kreĭn space generalization of the commutant lifting theorem is known (see, for example, Dritschel and Rovnyak [1990]). We believe that commutant lifting has the potential to provide a viable approach to the Schur theory in the indefinite setting as well.

In Lecture 5 we present another approach to the Schur class, based on reproducing kernel Pontryagin spaces and colligations, which gives new information and leads to further generalizations.

Lecture 5:
Reproducing Kernel Pontryagin Spaces and Colligations

Key ideas:

- For each nonnegative integer κ, a generalized Schur class $\mathbf{S}_\kappa(\mathfrak{A}, \mathfrak{B})$ is defined in such a way that $\mathbf{S}_0(\mathfrak{A}, \mathfrak{B}) = \mathbf{S}(\mathfrak{A}, \mathfrak{B})$ is the Schur class defined in Lecture 4. The approach uses kernels such as $K_S(w, z) = [1 - S(z)S(w)^*]/(1 - z\bar{w})$.
- Such kernels and associated Hilbert and Pontryagin reproducing kernel spaces are studied by means of representations $S(z) = \Theta_V(z)$ as characteristic functions of colligations V.

The purpose of this lecture is to arrive at a definition of the generalized Schur class $\mathbf{S}_\kappa(\mathfrak{A}, \mathfrak{B})$, where κ is any nonnegative integer. In the course of doing this, we shall simultaneously show a means to derive the main properties of such classes.

Vector-valued functions will take their values in Pontryagin spaces which are denoted $\mathfrak{A}$ and $\mathfrak{B}$ and called **coefficient spaces**. For simplicity, we assume that

$$\mathrm{ind}_- \mathfrak{A} = \mathrm{ind}_- \mathfrak{B}.$$

A coefficient space $\mathfrak{C}$ is also used and can be thought of as either $\mathfrak{A}$ or $\mathfrak{B}$, or their direct sum. For any coefficient space $\mathfrak{C}$, if $b \in \mathfrak{C}$, define b^* to be the linear functional on $\mathfrak{C}$ such that

$$b^* a = \langle a, b \rangle_{\mathfrak{C}}, \qquad a \in \mathfrak{C}.$$

If $\mathfrak{C}$ is a finite-dimensional space of column vectors, we can interpret b^* as the conjugate transpose of b and b^*a as a matrix product. The theory is mainly written for a more general situation in which coefficient spaces are Kreĭn spaces, and it is only in a few places where there is any difference. However, it will simplify the exposition to restrict the discussion throughout to Pontryagin spaces $\mathfrak{A}$ and $\mathfrak{B}$ having the same negative index.

Let $K(w, z)$ be a function with values in $\mathfrak{L}(\mathfrak{C})$ defined on $\Omega \times \Omega$ for some region Ω in the complex plane (a region is an open connected set). Assume that

$$K(w, z)^* = K(z, w), \qquad w, z \in \Omega,$$

and that $K(w, z)$ is analytic in z for each fixed w and analytic in $\bar{w}$ for each fixed z. We shall call such a function a **hermitian analytic kernel**, or for the purpose of this lecture just a **kernel**. We say that $K(w, z)$ has κ **negative squares on** Ω if for any finite set $w_1, \dots, w_n \in \Omega$ and vectors $c_1, \dots, c_n \in \mathfrak{C}$, the matrix

$$\left(c_i^* K(w_j, w_i) c_j\right)_{i,j=1}^n$$

has at most κ negative eigenvalues, and at least one such matrix has exactly κ negative eigenvalues. Eigenvalues are counted according to multiplicity. Write

$$\mathrm{sq}_- K = \kappa$$

if the condition holds and $\mathrm{sq}_- K = \infty$ if no such κ exists. A kernel $K(w,z)$ is said to be **positive** if $\mathrm{sq}_- K = 0$, that is,

$$\sum_{i,j=1}^{n} c_i^* K(w_j, w_i) c_j \geq 0$$

for any finite set $w_1, \ldots, w_n \in \Omega$ and vectors $c_1, \ldots, c_n \in \mathfrak{C}$.

Let $\mathfrak{H}$ be a Pontryagin space of $\mathfrak{C}$-valued holomorphic functions on a region Ω in the complex plane. Assume that for any point $w \in \Omega$,

$$(af + bg)(w) = af(w) + bg(w)$$

for all complex numbers a, b and all $f, g \in \mathfrak{H}$. A kernel $K(w,z)$ on $\Omega \times \Omega$ with values in $\mathfrak{L}(\mathfrak{C})$ is said to be a **reproducing kernel** for $\mathfrak{H}$ if for each $w \in \Omega$ and $c \in \mathfrak{C}$, $K(w,z)c$ belongs to $\mathfrak{H}$ as a function of z, and

$$\langle h(\cdot), K(w,\cdot)c \rangle_{\mathfrak{H}} = c^* h(w)$$

for every $h(\cdot)$ in $\mathfrak{H}$.

Theorem 5.1 (1) *A Pontryagin space $\mathfrak{H}$ of holomorphic $\mathfrak{C}$-valued functions on a region Ω has a reproducing kernel $K(w,z)$ if and only if for each $w \in \Omega$, the evaluation mapping $h \to h(w)$ acts continuously from $\mathfrak{H}$ into $\mathfrak{C}$. A reproducing kernel is unique when it exists.*

(2) *Let $K(w,z)$ be a kernel on $\Omega \times \Omega$ with values in $\mathfrak{L}(\mathfrak{C})$ having a finite number of negative squares. Then there is a unique Pontryagin space $\mathfrak{H}$ of holomorphic $\mathfrak{C}$-valued functions on Ω with reproducing kernel $K(w,z)$, and* $\mathrm{ind}_- \mathfrak{H} = \mathrm{sq}_- K$.

Most of the details are straightforward and follow the Hilbert space case. We prove one part where the indefiniteness of the inner product comes into play. See Sorjonen [1975] for a full treatment.

Proof, existence part of (2). We are given a kernel $K(w,z)$ having a finite number of negative squares, say κ, and we wish to construct a Pontryagin space $\mathfrak{H}$ for which $K(w,z)$ is a reproducing kernel and $\mathrm{sq}_- K = \kappa$. We take this as known in the case $\kappa = 0$ and reduce the general case to this.

Let $\mathfrak{H}_0$ be the linear span of functions $K(w,\cdot)c$ with w in Ω and $c \in \mathfrak{C}$. Define an inner product on $\mathfrak{H}_0$ by requiring that

$$\left\langle \sum_{j=1}^{n} K(w_j,\cdot)c_j, \sum_{i=1}^{n} K(w_i,\cdot)c_i \right\rangle_{\mathfrak{H}_0} = \sum_{i,j=1}^{n} c_i^* K(w_j, w_i) c_j$$

whenever $w_1, \ldots, w_n \in \Omega$ and $c_1, \ldots, c_n \in \mathfrak{C}$. As in the positive case, it is not hard to see that the inner product is well defined.

Since $K(w,z)$ has κ negative squares, we may choose $\alpha_1, \ldots, \alpha_p \in \Omega$ and $\varphi_1, \ldots, \varphi_p \in \mathfrak{C}$ such that the matrix $\left(\varphi_i^* K(\alpha_j, \alpha_i)\varphi_j\right)_{i,j=1}^{p}$ has κ negative eigenvalues $\lambda_1, \ldots, \lambda_\kappa$. Let

$$\begin{pmatrix} \gamma_{11} \\ \vdots \\ \gamma_{1p} \end{pmatrix}, \ldots, \begin{pmatrix} \gamma_{\kappa 1} \\ \vdots \\ \gamma_{\kappa p} \end{pmatrix}$$

be orthonormal eigenvectors for these eigenvalues. Then

$$\begin{aligned} u_1(\cdot) &= \gamma_{11}K(\alpha_1,\cdot)\varphi_1 + \cdots + \gamma_{1p}K(\alpha_p,\cdot)\varphi_p, \\ &\cdots \\ u_\kappa(\cdot) &= \gamma_{\kappa 1}K(\alpha_1,\cdot)\varphi_1 + \cdots + \gamma_{\kappa p}K(\alpha_p,\cdot)\varphi_p, \end{aligned}$$

belong to $\mathfrak{H}_0$ and satisfy

$$\langle u_j(\cdot), u_i(\cdot)\rangle_{\mathfrak{H}_0} = \lambda_j\, \delta_{ij}, \qquad i,j = 1,\dots,\kappa.$$

Hence these functions span a κ-dimensional subspace $\mathfrak{H}_-$ of $\mathfrak{H}_0$. Taken in the inner product of $\mathfrak{H}_0$, $\mathfrak{H}_-$ is the antispace of a Hilbert space with reproducing kernel

$$K_-(w,z) = \sum_{j=1}^{\kappa} \frac{1}{\lambda_j}\, u_j(z)\, u_j(w)^*, \qquad w,z \in \Omega.$$

In a similar way, we can show that because $\mathrm{sq}_- K = \kappa$, no subspace of $\mathfrak{H}_0$ of dimension greater than κ can be the antispace of a Hilbert space in the inner product of $\mathfrak{H}_0$.

Now define

$$K_+(w,z) = K(w,z) - K_-(w,z),$$

and let $\mathfrak{H}_{0+}$ be the span of functions $K_+(w,\cdot)c$ with w in Ω and c in $\mathfrak{C}$. This subspace is orthogonal to $\mathfrak{H}_-$ in $\mathfrak{H}_0$. If

$$f(\cdot) = \sum_{j=1}^{n} K_+(w_j,\cdot)a_j, \qquad g(\cdot) = \sum_{i=1}^{n} K_+(w_i,\cdot)b_i$$

where $w_1,\dots,w_n$ belong to Ω and $a_1,\dots,a_n,b_1,\dots,b_n$ are in $\mathfrak{C}$, then

$$\begin{aligned} \langle f(\cdot), g(\cdot)\rangle_{\mathfrak{H}_0} &= \Big\langle \sum_{j=1}^{n} K(w_j,\cdot)a_j - \sum_{j=1}^{n} K_-(w_j,\cdot)a_j, \\ &\qquad \sum_{i=1}^{n} K(w_i,\cdot)b_i - \sum_{i=1}^{n} K_-(w_i,\cdot)b_i \Big\rangle_{\mathfrak{H}_0} \\ &= \sum_{i,j=1}^{n} b_i^* K(w_j,w_i)a_j - \sum_{i,j=1}^{n} b_i{}^* K_-(w_j,w_i)a_j \\ &\qquad - \sum_{i,j=1}^{n} b_i{}^* K_-(w_j,w_i)a_j + \sum_{i,j=1}^{n} b_i{}^* K_-(w_j,w_i)a_j \\ &= \sum_{i,j=1}^{n} b_i^* K_+(w_j,w_i)a_j. \end{aligned}$$

If $K_+(w,z)$ is not positive, we can find $\beta_1,\dots,\beta_q$ in Ω and $\psi_1,\dots,\psi_q$ in $\mathfrak{C}$ such that the element $u_{\kappa+1}(\cdot) = \sum_{i=1}^{q} K_+(\beta_i,\cdot)\psi_i$ of $\mathfrak{H}_{0+}$ satisfies

$$\langle u_{\kappa+1}(\cdot), u_{\kappa+1}(\cdot)\rangle_{\mathfrak{H}_0} = \sum_{i,j=1}^{q} \psi_i^* K_+(\beta_i,\beta_j)\psi_j < 0.$$

Then the span of $\mathfrak{H}_-$ and $u_{\kappa+1}(\cdot)$ is a $(\kappa+1)$-dimensional subspace of $\mathfrak{H}_0$ which is the antispace of a Hilbert space. As noted above, this contradicts our hypothesis that $\mathrm{sq}_- K = \kappa$. Therefore $K_+(w,z)$ is positive.

Since we assume that the assertion is true in the case $\kappa = 0$, $\mathfrak{H}_{0+}$ is contained as a dense set in a Hilbert space $\mathfrak{H}_+$ of holomorphic functions on Ω with reproducing kernel $K_+(w,z)$. Define a Pontryagin space $\mathfrak{H}$ such that $\mathfrak{H}_\pm$ are isometrically contained in $\mathfrak{H}$ and $\mathfrak{H} = \mathfrak{H}_+ \oplus \mathfrak{H}_-$. Since $\mathfrak{H}_\pm$ have reproducing kernels $K_\pm(w,z)$, $\mathfrak{H}$ has reproducing kernel

$$K(w,z) = K_+(w,z) + K_-(w,z).$$

This completes the existence part of the proof. □

We are concerned with kernels determined by a holomorphic function $S(z)$ which is defined on a region $\Omega(S)$ in the complex plane whose values are operators in $\mathfrak{L}(\mathfrak{A},\mathfrak{B})$. Recall our standing assumption that $\mathfrak{A}$ and $\mathfrak{B}$ are Pontryagin spaces having the same negative index. We assume also that $\Omega(S)$ contains the origin and is a subset of the unit disk. Define

$$\tilde{S}(z) = S(\bar{z})^*, \qquad z \in \Omega(\tilde{S}),$$

where $\Omega(\tilde{S})$ is the set of conjugates of numbers in $\Omega(S)$. Then in a neighborhood of the origin,

$$\tilde{S}(z) = S_0^* + S_1^* z + S_2^* z^2 + \cdots$$

if $S(z) = S_0 + S_1 z + S_2 z^2 + \cdots$. Set

$$\begin{aligned} K_S(w,z) &= \frac{1 - S(z)S(w)^*}{1 - z\bar{w}}, \\ K_{\tilde{S}}(w,z) &= \frac{1 - \tilde{S}(z)\tilde{S}(w)^*}{1 - z\bar{w}}, \end{aligned}$$

and

$$D_S(w,z) = \begin{pmatrix} K_S(w,z) & \dfrac{S(z) - S(\bar{w})}{z - \bar{w}} \\ \dfrac{\tilde{S}(z) - \tilde{S}(\bar{w})}{z - \bar{w}} & K_{\tilde{S}}(w,z) \end{pmatrix},$$

$$D_{\tilde{S}}(w,z) = \begin{pmatrix} K_{\tilde{S}}(w,z) & \dfrac{\tilde{S}(z) - \tilde{S}(\bar{w})}{z - \bar{w}} \\ \dfrac{S(z) - S(\bar{w})}{z - \bar{w}} & K_S(w,z) \end{pmatrix}.$$

For example, the domain of $K_S(w,z)$ is the set of all pairs (w,z) with $w,z \in \Omega(S)$, and the values of $K_S(w,z)$ are operators in $\mathfrak{L}(\mathfrak{B})$. The domain of $D_S(w,z)$ is the set of all pairs (w,z) with $w,z \in \Omega(S) \cap \Omega(\tilde{S})$, and the values of $D_S(w,z)$ are operators in $\mathfrak{L}(\mathfrak{B} \oplus \mathfrak{A})$. Observe that

$$\mathrm{sq}_- D_{\tilde{S}} = \mathrm{sq}_- D_S \leq \infty$$

and

$$\begin{aligned} \mathrm{sq}_- K_S &\leq \mathrm{sq}_- D_S, \\ \mathrm{sq}_- K_{\tilde{S}} &\leq \mathrm{sq}_- D_{\tilde{S}}. \end{aligned}$$

Definition 5.2 *In the preceding notation, define Pontryagin spaces*
(1) $\mathfrak{H}(S)$ *with reproducing kernel* $K_S(w,z)$ *if* $\mathrm{sq}_- K_S < \infty$, *and*
(2) $\mathfrak{D}(S)$ *with reproducing kernel* $D_S(w,z)$ *if* $\mathrm{sq}_- D_S < \infty$.
Spaces $\mathfrak{H}(\tilde{S})$ *and* $\mathfrak{D}(\tilde{S})$ *with reproducing kernels* $K_{\tilde{S}}(w,z)$ *and* $D_{\tilde{S}}(w,z)$ *are defined analogously when* $\mathrm{sq}_- K_{\tilde{S}} < \infty$ *and* $\mathrm{sq}_- D_{\tilde{S}} < \infty$.

Our account of these spaces follows a manuscript by Alpay, Dijksma, and de Snoo [1994]. A new version will appear in Alpay, Dijksma, Rovnyak, and de Snoo [in preparation]. The approach is based on the notion of a colligation. In operator theory, the term "colligation" has come to mean a particular group of operators which arise both in the study of nonselfadjoint operators and in the theory of linear systems. A standard source is the survey article by M. S. Brodskiĭ [1978]. Ćurgus, Dijksma, Langer, and de Snoo [1988] and Dijksma, Langer, and de Snoo [1986] adapt the theory to the indefinite setting.

An **operator colligation** is a quadruple $(\mathfrak{H}, \mathfrak{F}, \mathfrak{G}, V)$ consisting of three Kreĭn spaces, the **state space** $\mathfrak{H}$, the **inner** or **incoming space** $\mathfrak{F}$, and the **outer** or **outgoing space** $\mathfrak{G}$, and a **connecting operator**

$$V = \begin{pmatrix} T & F \\ G & H \end{pmatrix} \in \mathfrak{L}(\mathfrak{H} \oplus \mathfrak{F}, \mathfrak{H} \oplus \mathfrak{G}).$$

We call T the **main operator** for the colligation. We say that the colligation is **isometric, coisometric, or unitary** according as the connecting operator V is isometric, coisometric, or unitary. When there is no possibility of ambiguity, we sometimes identify the colligation $(\mathfrak{H}, \mathfrak{F}, \mathfrak{G}, V)$ by its connecting operator V. The characteristic function of the colligation is

$$\Theta_V(z) = H + zG(1 - zT)^{-1}F$$

which is a holomorphic function with values in $\mathfrak{L}(\mathfrak{F}, \mathfrak{G})$ defined in a neighborhood of the origin. The precise domain of the characteristic function is taken to be the component of the origin in the set of complex numbers z such that

$$(1 - zT)^{-1}$$

exists as a continuous everywhere defined operator. A colligation $(\mathfrak{H}, \mathfrak{F}, \mathfrak{G}, V)$ is said to be **closely inner connected**, **closely outer connected**, and **closely connected** according as

$$\mathfrak{H} = \overline{\mathrm{span}}\,\{\mathrm{ran}\, T^m F : m \geq 0\},$$
$$\mathfrak{H} = \overline{\mathrm{span}}\,\{\mathrm{ran}\, T^{*n} G^* : n \geq 0\},$$
$$\mathfrak{H} = \overline{\mathrm{span}}\,\{\mathrm{ran}\, T^m F,\ \mathrm{ran}\, T^{*n} G^* : m, n \geq 0\}.$$

The terms "inner" and "outer" refer to the inner and outer spaces $\mathfrak{F}$ and $\mathfrak{G}$ of the colligation. In systems theory, the terms "controllable" and "observable" are also used for "closely inner connected" and "closely outer connected." The characteristic function of a colligation is also called a "transfer function."

We use colligations to construct examples of kernels having a finite number of negative squares.

Theorem 5.3 *Let*

$$V = \begin{pmatrix} T & F \\ G & H \end{pmatrix} \in \mathfrak{L}(\mathfrak{H} \oplus \mathfrak{A}, \mathfrak{H} \oplus \mathfrak{B}).$$

be a colligation such that $\mathfrak{H}$ *is a Pontryagin space. Assume that*

$$S(z) = \Theta_V(z), \qquad z \in \Omega(S),$$

is the restriction of the characteristic function of V *to a subregion* $\Omega(S)$ *of the unit disk which contains the origin.*

(1) *If* V^* *is isometric, then*

$$K_S(w,z) = G(1-zT)^{-1}(1-\bar{w}T^*)^{-1}G^*, \qquad w,z \in \Omega(S),$$

and $\operatorname{sq}_- K_S \le \operatorname{ind}_- \mathfrak{H}$. *Equality holds if the colligation is closely outer connected, and in this case there is an isomorphism* W *mapping* $\mathfrak{H}$ *onto* $\mathfrak{H}(S)$ *such that*

$$W : (1-\bar{w}T^*)^{-1}G^*b \to K_S(w,\cdot)b, \qquad w \in \Omega(S),\ b \in \mathfrak{B}.$$

(2) *If* V *is isometric, then*

$$K_{\tilde{S}}(w,z) = F^*(1-zT^*)^{-1}(1-\bar{w}T)^{-1}F, \qquad w,z \in \Omega(\tilde{S}),$$

and $\operatorname{sq}_- K_{\tilde{S}} \le \operatorname{ind}_- \mathfrak{H}$. *Equality holds if the colligation is closely inner connected, and then there is an isomorphism* W *mapping* $\mathfrak{H}$ *onto* $\mathfrak{H}(\tilde{S})$ *such that*

$$W : (1-\bar{w}T)^{-1}Fa \to K_{\tilde{S}}(w,\cdot)a, \qquad w \in \Omega(\tilde{S}),\ a \in \mathfrak{A}.$$

(3) *If* V *is unitary, then for all* $w,z \in \Omega(S) \cap \Omega(\tilde{S})$,

$$D_S(w,z) = \begin{pmatrix} G(1-zT)^{-1} \\ F^*(1-zT^*)^{-1} \end{pmatrix} \begin{pmatrix} (1-\bar{w}T^*)^{-1}G^* & (1-\bar{w}T)^{-1}F \end{pmatrix},$$
$$D_{\tilde{S}}(w,z) = \begin{pmatrix} F^*(1-zT^*)^{-1} \\ G(1-zT)^{-1} \end{pmatrix} \begin{pmatrix} (1-\bar{w}T)^{-1}F & (1-\bar{w}T^*)^{-1}G^* \end{pmatrix},$$

and $\operatorname{sq}_- D_S = \operatorname{sq}_- D_{\tilde{S}} \le \operatorname{ind}_- \mathfrak{H}$. *Equality holds if the colligation is closely connected, and then there are isomorphisms* W *and* $\tilde{W}$ *mapping* $\mathfrak{H}$ *onto* $\mathfrak{D}(S)$ *and* $\mathfrak{D}(\tilde{S})$ *such that*

$$W : \begin{pmatrix} (1-\bar{w}T^*)^{-1}G^* & (1-\bar{w}T)^{-1}F \end{pmatrix} \begin{pmatrix} b \\ a \end{pmatrix} \to D_S(w,\cdot) \begin{pmatrix} b \\ a \end{pmatrix},$$
$$\tilde{W} : \begin{pmatrix} (1-\bar{w}T)^{-1}F & (1-\bar{w}T^*)^{-1}G^* \end{pmatrix} \begin{pmatrix} a \\ b \end{pmatrix} \to D_{\tilde{S}}(w,\cdot) \begin{pmatrix} a \\ b \end{pmatrix},$$

for all $w,z \in \Omega(S) \cap \Omega(\tilde{S})$, $a \in \mathfrak{A}$, *and* $b \in \mathfrak{B}$.

Proof of identities. Since V^* is isometric,

$$\begin{aligned} TT^* + FF^* &= 1, \\ GG^* + HH^* &= 1, \\ GT^* + HF^* &= 0. \end{aligned}$$

Therefore

$$\begin{aligned} &(1 - z\bar{w})K_S(w, z) \\ &\quad = 1 - \Big[H + zG(1 - zT)^{-1}F\Big]\Big[H^* + \bar{w}F^*(1 - \bar{w}T^*)^{-1}G^*\Big] \\ &\quad = 1 - HH^* - \bar{w}HF^*(1 - \bar{w}T^*)^{-1}G^* - zG(1 - zT)^{-1}FH^* \\ &\qquad -z\bar{w}G(1 - zT)^{-1}FF^*(1 - \bar{w}T^*)^{-1}G^* \\ &\quad = GG^* + \bar{w}GT^*(1 - \bar{w}T^*)^{-1}G^* + zG(1 - zT)^{-1}TG^* \\ &\qquad -z\bar{w}G(1 - zT)^{-1}(1 - TT^*)(1 - \bar{w}T^*)^{-1}G^* \\ &\quad = G(1 - zT)^{-1}\Big\{(1 - zT)(1 - \bar{w}T^*) + (1 - zT)\bar{w}T^* \\ &\qquad zT(1 - \bar{w}T^*) - z\bar{w}(1 - TT^*)\Big\}(1 - \bar{w}T^*)^{-1}G^* \\ &\quad = (1 - z\bar{w})G(1 - zT)^{-1}(1 - \bar{w}T^*)^{-1}G^* \end{aligned}$$

for all $w, z \in \Omega(S)$. This proves the identity in (1).

The identity in (2) follows on applying (1) with V replaced by V^*.

The formulas for $D_S(w, z)$ and $D_{\tilde{S}}(w, z)$ in (3) follow from (1) and (2) together with the identities

$$\begin{aligned} \frac{S(z) - S(\bar{w})}{z - \bar{w}} &= G(1 - zT)^{-1}(1 - \bar{w}T)^{-1}F, \\ \frac{\tilde{S}(z) - \tilde{S}(\bar{w})}{z - \bar{w}} &= F^*(1 - zT^*)^{-1}(1 - \bar{w}T^*)^{-1}G^*, \end{aligned}$$

which we derive in a similar way. □

Given $S(z)$, the question arises if there exists a colligation V such that

$$S(z) = \Theta_V(z).$$

This is the realization problem. The existence question asks if such a colligation can be found which is isometric, coisometric, or unitary. The corresponding uniqueness question is to determine if two such colligations are essentially the same.

Theorem 5.4 *Let $S(z)$ be a holomorphic function defined on a subregion $\Omega(S)$ of the unit disk containing the origin and taking values in $\mathfrak{L}(\mathfrak{A}, \mathfrak{B})$. If $K_S(w, z)$ has a finite number of negative squares, there exists a coisometric and closely outer connected colligation*

$$V = \begin{pmatrix} T & F \\ G & H \end{pmatrix} \in \mathfrak{L}(\mathfrak{H}(S) \oplus \mathfrak{A}, \mathfrak{H}(S) \oplus \mathfrak{B})$$

such that

$$\begin{aligned}(Th)(z) &= \frac{h(z)-h(0)}{z},\\ (Fa)(z) &= \frac{S(z)-S(0)}{z}a,\\ Gh &= h(0),\\ Ha &= S(0)a,\end{aligned}$$

for all h in $\mathfrak{H}(S)$ and a in $\mathfrak{A}$. The characteristic function of the colligation coincides with $S(z)$ on the set where both are defined:

$$S(z) = \Theta_V(z), \qquad z \in \Omega(S) \cap \Omega(\Theta_V). \tag{5.1}$$

For all h in $\mathfrak{H}(S)$, the relation

$$((1-wT)^{-1}h)(z) = \frac{zh(z)-wh(w)}{z-w}$$

holds for all $z \in \Omega(S)$ and $w \in \Omega(S) \cap \Omega(\Theta_V)$. Moreover

$$(1-\bar{w}T^*)^{-1}G^*b = K_S(w,\cdot)b$$

for all w in $\Omega(S) \cap \Omega(\Theta_V)$ and $b \in \mathfrak{B}$.

The realization (5.1) is of the type which is described in Theorem 5.3(1). In a certain natural sense which we shall not make precise, there is an essentially unique coisometric closely outer connected colligation V whose characteristic function has the form (5.1). The colligation constructed in Theorem 5.4 is called the **canonical coisometric colligation** associated with $S(z)$.

Proof. Let $\mathbf{R}$ be the linear relation in $\big(\mathfrak{H}(S) \oplus \mathfrak{B}\big) \times \big(\mathfrak{H}(S) \oplus \mathfrak{A}\big)$ spanned by the pairs

$$\left(\begin{pmatrix} K_S(\alpha,\cdot)u_1 \\ u_2 \end{pmatrix}, \begin{pmatrix} \dfrac{K_S(\alpha,\cdot)-K_S(0,\cdot)}{\bar{\alpha}}u_1 + K_S(0,\cdot)u_2 \\ \dfrac{S(\alpha)^*-S(0)^*}{\bar{\alpha}}u_1 + S(0)^*u_2 \end{pmatrix}\right)$$

with $0 \neq \alpha \in \Omega(S)$ and $u_1, u_2 \in \mathfrak{B}$. Let

$$\left(\begin{pmatrix} K_S(\beta,\cdot)v_1 \\ v_2 \end{pmatrix}, \begin{pmatrix} \dfrac{K_S(\beta,\cdot)-K_S(0,\cdot)}{\bar{\beta}}v_1 + K_S(0,\cdot)v_2 \\ \dfrac{S(\beta)^*-S(0)^*}{\bar{\beta}}v_1 + S(0)^*v_2 \end{pmatrix}\right)$$

be a second pair with $0 \neq \beta \in \Omega(S)$ and $v_1, v_2 \in \mathfrak{B}$. Straightforward algebra yields the identity

$$\left\langle \begin{pmatrix} \dfrac{K_S(\alpha,\cdot) - K_S(0,\cdot)}{\bar{\alpha}} u_1 + K_S(0,\cdot)u_2 \\ \dfrac{S(\alpha)^* - S(0)^*}{\bar{\alpha}} u_1 + S(0)^* u_2 \end{pmatrix}, \right.$$

$$\left. \begin{pmatrix} \dfrac{K_S(\beta,\cdot) - K_S(0,\cdot)}{\bar{\beta}} v_1 + K_S(0,\cdot)v_2 \\ \dfrac{S(\beta)^* - S(0)^*}{\bar{\beta}} v_1 + S(0)^* v_2 \end{pmatrix} \right\rangle_{\mathfrak{H}(S)\oplus\mathfrak{A}}$$

$$= \frac{1}{\bar{\alpha}\,\beta}\Big[v_1^* K_S(\alpha,\beta)u_1 - v_1^* K_S(0,\beta)u_1 - v_1^* K_S(\alpha,0)u_1 + v_1^* K_S(0,0)u_1\Big]$$

$$+\frac{1}{\beta}\Big[v_1^* K_S(0,\beta)u_2 - v_1^* K_S(0,0)u_2\Big] + \frac{1}{\bar{\alpha}}\Big[v_2^* K_S(\alpha,0)u_1 - v_2^* K_S(0,0)u_1\Big]$$

$$+v_2^* K_S(0,0)u_2 + v_1^* \frac{S(\beta) - S(0)}{\beta}\,\frac{S(\alpha)^* - S(0)^*}{\bar{\alpha}} u_1$$

$$+v_1^* \frac{S(\beta) - S(0)}{\beta} S(0)^* u_2 + v_2^* S(0)\,\frac{S(\alpha)^* - S(0)^*}{\bar{\alpha}} u_1 + v_2^* S(0)S(0)^* u_2$$

$$= v_1^* K_S(\alpha,\beta)u_1 + v_2^* u_2$$

$$= \left\langle \begin{pmatrix} K_S(\alpha,\cdot)u_1 \\ u_2 \end{pmatrix}, \begin{pmatrix} K_S(\beta,\cdot)v_1 \\ v_2 \end{pmatrix} \right\rangle_{\mathfrak{H}(S)\oplus\mathfrak{B}}.$$

It follows that $\mathbf{R}$ is an isometric relation in

$$\big(\mathfrak{H}(S) \oplus \mathfrak{B}\big) \times \big(\mathfrak{H}(S) \oplus \mathfrak{A}\big).$$

The domain of $\mathbf{R}$ is easily seen to be dense in $\mathfrak{H}(S) \oplus \mathfrak{B}$. Since we assume that $\mathfrak{A}$ and $\mathfrak{B}$ are Pontryagin spaces having the same negative index, by the version of Shmul′yan's theorem given in Theorem 2.10 (alternative form), the closure of $\mathbf{R}$ is the graph of an isometry

$$V^* = \begin{pmatrix} T^* & G^* \\ F^* & H^* \end{pmatrix} \in \mathfrak{L}(\mathfrak{H}(S) \oplus \mathfrak{B}, \mathfrak{H}(S) \oplus \mathfrak{A}).$$

This completes the definition of the colligation.

Suppose that

$$\begin{pmatrix} T & F \\ G & H \end{pmatrix} : \begin{pmatrix} h \\ a \end{pmatrix} \to \begin{pmatrix} k \\ b \end{pmatrix},$$

where $h, k \in \mathfrak{H}(S)$, $a \in \mathfrak{A}$, and $b \in \mathfrak{B}$. Then for any $0 \neq \alpha \in \Omega(\Theta)$ and $u_1, u_2 \in \mathfrak{B}$,

$$u_1^* k(\alpha) + u_2^* b = \left\langle \begin{pmatrix} k(\cdot) \\ b \end{pmatrix}, \begin{pmatrix} K_S(\alpha,\cdot)u_1 \\ u_2 \end{pmatrix} \right\rangle_{\mathfrak{H}(S)\oplus\mathfrak{B}}$$

$$= \left\langle \begin{pmatrix} h(\cdot) \\ a \end{pmatrix}, \begin{pmatrix} \dfrac{K_S(\alpha,\cdot) - K_S(0,\cdot)}{\bar\alpha} u_1 + K_S(0,\cdot)u_2 \\ \dfrac{S(\alpha)^* - S(0)^*}{\bar\alpha} u_1 + S(0)^* u_2 \end{pmatrix} \right\rangle_{\mathfrak{H}(S)\oplus\mathfrak{A}}$$

$$= u_1^* \frac{h(\alpha) - h(0)}{\alpha} + u_2^* h(0) + u_1^* \frac{S(\alpha) - S(0)}{\alpha} a + u_2^* S(0) a.$$

By the arbitrariness of u_1, u_2, the formulas for T, F, G, H in the statement of the theorem follow.

To see that V is closely outer connected, we must show that the only element h in $\mathfrak{H}(S)$ which is orthogonal to the range of $T^{*n}G^*$ for all $n \geq 0$ is the zero element. The conditions imply that

$$T^n h \perp K_S(0,\cdot)b, \qquad n \geq 0,\ b \in \mathfrak{B}.$$

Since T acts as the difference quotient operator at the origin, this implies that h and all of its derivatives are zero at the origin. Therefore $h \equiv 0$ in a neighborhood of the origin, and since $\Omega(S)$ is a region and hence connected, h is the zero element of $\mathfrak{H}(S)$. Thus V is closely outer connected.

If $0 \neq w \in \Omega(S) \cap \Omega(\Theta_V)$ and h in $\mathfrak{H}(S)$ and $(1 - wT)^{-1}h = g$, then

$$h(z) = g(z) - w\frac{g(z) - g(0)}{z}$$

on $\Omega(S)$. Setting $z = w$, we find that $g(0) = h(w)$ and hence

$$g(z) = \frac{zh(z) - wh(w)}{z - w}$$

on $\Omega(S)$. The same conclusion holds trivially for $w = 0$. Thus for any $h \in \mathfrak{H}(S)$, $b \in \mathfrak{B}$, and $w \in \Omega(S) \cap \Omega(\Theta_V)$,

$$\begin{aligned} \langle h(\cdot), (1 - \bar{w}T^*)^{-1}G^*b \rangle_{\mathfrak{H}(S)} &= \langle G(1 - wT)^{-1}h, b \rangle_{\mathfrak{B}} \\ &= b^* \left\{ \frac{zh(z) - wh(w)}{z - w} \right\}_{z=0} \\ &= b^* h(w). \end{aligned}$$

This proves the two formulas at the end of the theorem.

It remains to prove (5.1). For any $w \in \Omega(\Theta) \cap \Omega(\Theta_V)$ and $a \in \mathfrak{A}$,

$$\begin{aligned}\Theta_V(w)a &= \left(H + wG(1-wT)^{-1}F\right)a \\ &= S(0)a + w\left\{\frac{z\,\dfrac{S(z)-S(0)}{z}\,a - w\,\dfrac{S(w)-S(0)}{w}\,a}{z-w}\right\}_{z=0} \\ &= S(0)a + \frac{S(w)-S(0)}{w}\,a \\ &= S(w)a,\end{aligned}$$

as asserted. □

The **canonical isometric colligation** associated with $S(z)$ is obtained as an immediate consequence and does not need a separate construction. It is described in the next result.

Theorem 5.5 *Let $S(z)$ be a holomorphic function defined on a subregion $\Omega(S)$ of the unit disk containing the origin and taking values in $\mathfrak{L}(\mathfrak{A}, \mathfrak{B})$. If $K_{\tilde{S}}(w,z)$ has a finite number of negative squares, there exists an isometric and closely inner connected colligation*

$$V = \begin{pmatrix} T & F \\ G & H \end{pmatrix} \in \mathfrak{L}(\mathfrak{H}(\tilde{S}) \oplus \mathfrak{A}, \mathfrak{H}(\tilde{S}) \oplus \mathfrak{B})$$

such that

$$\begin{aligned}(T^*h)(z) &= \frac{h(z)-h(0)}{z}, \\ (G^*b)(z) &= \frac{\tilde{S}(z)-\tilde{S}(0)}{z}b, \\ F^*h &= h(0), \\ H^*b &= \tilde{S}(0)b,\end{aligned}$$

for all h in $\mathfrak{H}(\tilde{S})$ and b in $\mathfrak{B}$. The characteristic function of the colligation coincides with $S(z)$ on the set where both are defined:

$$S(z) = \Theta_V(z), \qquad z \in \Omega(S) \cap \Omega(\Theta_V). \tag{5.2}$$

For all h in $\mathfrak{H}(\tilde{S})$, the relation

$$((1-wT^*)^{-1}h)(z) = \frac{zh(z)-wh(w)}{z-w}$$

holds for all $z \in \Omega(\tilde{S})$ and $w \in \Omega(\tilde{S}) \cap \overline{\Omega(\Theta_V)}$. Moreover

$$(1-\bar{w}T)^{-1}Fa = K_{\tilde{S}}(w,\cdot)a$$

for all w in $\Omega(\tilde{S}) \cap \overline{\Omega(\Theta_V)}$ and $a \in \mathfrak{A}$.

Proof. Let

$$U = \begin{pmatrix} T^* & G^* \\ F^* & H^* \end{pmatrix} \in \mathfrak{L}(\mathfrak{H}(\tilde{S}) \oplus \mathfrak{B}, \mathfrak{H}(\tilde{S}) \oplus \mathfrak{A})$$

be the canonical coisometric colligation associated with $\tilde{S}(z)$, as constructed in Theorem 5.4. Define

$$V = U^* = \begin{pmatrix} T & F \\ G & H \end{pmatrix} \in \mathfrak{L}(\mathfrak{H}(\tilde{S}) \oplus \mathfrak{A}, \mathfrak{H}(\tilde{S}) \oplus \mathfrak{B}).$$

Then V is an isometric and closely inner connected colligation. By Theorem 5.4 and the definition of a characteristic function of a colligation,

$$\Theta_U(z) = H^* + zF^*(1 - zT^*)^{-1}G^*,$$

and so

$$S(z) = \tilde{\Theta}_U(z) = H + zG(1 - zT)^{-1}F = \Theta_V(z)$$

whenever the functions are defined. The remaining assertions of the theorem follow from Theorem 5.4. □

Theorem 5.6 *Under the assumptions of Theorem* 5.5, *there is a Hilbert space* $\mathfrak{E}$ *and a closely connected unitary colligation of the form*

$$U = \begin{pmatrix} \begin{pmatrix} A & 0 \\ B & T \end{pmatrix} & \begin{pmatrix} 0 \\ F \end{pmatrix} \\ \begin{pmatrix} C & G \end{pmatrix} & \begin{pmatrix} H \end{pmatrix} \end{pmatrix} \in \mathfrak{L}((\mathfrak{E} \oplus \mathfrak{H}(\tilde{S})) \oplus \mathfrak{A}, (\mathfrak{E} \oplus \mathfrak{H}(\tilde{S})) \oplus \mathfrak{B})$$

such that

$$S(z) = \Theta_U(z), \qquad z \in \Omega(S) \cap \Omega(\Theta_U).$$

We omit the proof of Theorem 5.6. A similar result is proved for Hilbert spaces by Dijksma, Langer, and de Snoo [1986], Lemma 6.2.

We are now able to derive a special case of a result of Alpay, Dijksma, van der Ploeg, and de Snoo [1992].

Theorem 5.7 *Let* $S(z)$ *be a holomorphic function defined on a subregion* $\Omega(S)$ *of the unit disk containing the origin and taking values in* $\mathfrak{L}(\mathfrak{A}, \mathfrak{B})$, *where* $\mathfrak{A}$ *and* $\mathfrak{B}$ *are Pontryagin spaces having the same negative index. If either* $K_S(w,z)$ *or* $K_{\tilde{S}}(w,z)$ *has a finite number of negative squares, then all four of the kernels*

$$K_S(w,z),\ K_{\tilde{S}}(w,z),\ D_S(w,z),\ D_{\tilde{S}}(w,z)$$

have a finite number of squares, and

$$\mathrm{sq}_-\, K_S = \mathrm{sq}_-\, K_{\tilde{S}} = \mathrm{sq}_-\, D_S = \mathrm{sq}_-\, D_{\tilde{S}}. \tag{5.3}$$

In this case, the Pontryagin spaces $\mathfrak{H}(S), \mathfrak{H}(\tilde{S}), \mathfrak{D}(S), \mathfrak{D}(\tilde{S})$ have the same negative index. In the statement of Theorem 5.7, for emphasis we have repeated our standing hypothesis that $\mathfrak{A}$ and $\mathfrak{B}$ are Pontryagin spaces having the same negative index. The reason is that there is a version of Theorem 5.7 for Kreĭn spaces, and the statement is not exactly the same.

Proof. Suppose that $K_{\tilde{S}}(w,z)$ has a finite number of negative squares. Construct V as in Theorem 5.5 and U as in Theorem 5.6. By Theorem 5.3(3),

$$\text{sq}_-\, D_S = \text{sq}_-\, D_{\tilde{S}} = \text{ind}_-\, (\mathfrak{C} \oplus \mathfrak{H}(\tilde{S})).$$

Since $\mathfrak{C}$ is a Hilbert space,

$$\text{ind}_-\, (\mathfrak{C} \oplus \mathfrak{H}(\tilde{S})) = \text{ind}_-\, \mathfrak{H}(\tilde{S}) = \text{sq}_-\, K_{\tilde{S}}.$$

Thus

$$\text{sq}_-\, D_S = \text{sq}_-\, D_{\tilde{S}} = \text{sq}_-\, K_{\tilde{S}}.$$

As noted before Definition 5.2, $\text{sq}_-\, K_S \le \text{sq}_-\, D_S$, and applying what we have shown with the roles of $S(z)$ and $\tilde{S}(z)$ interchanged, we obtain the equality of all four indices. □

Finally we arrive at:

Definition 5.8 *For any nonnegative integer κ, we define the* **generalized Schur class** $\mathbf{S}_\kappa(\mathfrak{A},\mathfrak{B})$ *to be the set of all holomorphic functions $S(z)$ defined on a subregion $\Omega(S)$ of the unit disk containing the origin and taking values in $\mathfrak{L}(\mathfrak{A},\mathfrak{B})$ such that the indices in* (5.3) *have the common value κ.*

The idea for such a definition is due to Kreĭn and Langer [1973], [1977], [1981]. The next result relates the generalized Schur class to the Schur class introduced in Lecture 4.

Theorem 5.9 *The generalized Schur class* $\mathbf{S}_0(\mathfrak{A},\mathfrak{B})$ *coincides with the Schur class* $\mathbf{S}(\mathfrak{A},\mathfrak{B})$ *defined in Lecture* 4.

Sketch of proof. Suppose that $S(z)$ is in $\mathbf{S}_0(\mathfrak{A},\mathfrak{B})$, so $K_S(w,z)$ is positive. Write

$$K_S(w,z) = \frac{1 - S(z)S(w)^*}{1 - \bar{w}z} = \sum_{m,n=0}^{\infty} C_{mn}\, z^m \bar{w}^n.$$

An algebraic calculation shows that for any nonnegative integer k,

$$[C_{mn}]_{m,n=0}^{k} = 1 - T_k T_k^*,$$

where

$$T_k = \begin{pmatrix} S_0 & & & \\ S_1 & S_0 & 0 & \\ & \ddots & \ddots & \\ S_k & & S_1 & S_0 \end{pmatrix}.$$

The positivity of the kernel shows that the Toeplitz matrices have contractive adjoints. Since we assume that $\mathfrak{A}$ and $\mathfrak{B}$ are Pontryagin spaces having the same negative index, the Toeplitz matrices are bicontractions, so $S(z)$ belongs to $\mathbf{S}(\mathfrak{A},\mathfrak{B})$.

The other direction follows on reversing the argument to show that if $S(z)$ belongs to $\mathbf{S}(\mathfrak{A},\mathfrak{B})$, then $K_S(w,z)$ is positive. Hence by Theorem 5.7, $S(z)$ belongs to $\mathbf{S}_0(\mathfrak{A},\mathfrak{B})$. □

The Schur class $\mathbf{S}(\mathfrak{A}, \mathfrak{B})$ is defined for any Kreĭn spaces $\mathfrak{A}$ and $\mathfrak{B}$, and so it is natural to extend Definition 5.8 to this generality. We define $\mathbf{S}_\kappa(\mathfrak{A}, \mathfrak{B})$ for any Kreĭn spaces $\mathfrak{A}$ and $\mathfrak{B}$ to be the set of all holomorphic functions $S(z)$ defined on a subregion $\Omega(S)$ of the unit disk containing the origin and taking values in $\mathfrak{L}(\mathfrak{A}, \mathfrak{B})$ such that the indices of the four kernels in (5.3) are equal and have the common value κ.

The proof of Theorem 5.9 is not quite adequate to obtain the conclusion when $\mathfrak{A}$ and $\mathfrak{B}$ are Kreĭn spaces. What is missing is a proof that the kernels $D_S(w, z)$ and $D_{\tilde{S}}(w, z)$ are positive if $S(z)$ belongs to $\mathbf{S}(\mathfrak{A}, \mathfrak{B})$. This may be seen in various ways. One is to use the Kreĭn space generalization of Theorem 5.7. The results at hand are sufficient, however, if we write the linear fractional representation of $S(z)$ given in Theorem 4.7 in the form

$$S(z) = \alpha + z\beta S'(z)\big(1 - z\delta S'(z)\big)^{-1}\gamma,$$

where $\alpha = S_0$,

$$\begin{pmatrix} \alpha & \beta \\ \gamma & \delta \end{pmatrix} \in \mathfrak{L}(\mathfrak{A} \oplus \mathfrak{B}', \mathfrak{B} \oplus \mathfrak{A}')$$

is a Julia operator, and $S'(z)$ is in the Schur class $\mathbf{S}(\mathfrak{A}', \mathfrak{B}')$. Here $\mathfrak{A}'$ and $\mathfrak{B}'$ are Hilbert spaces because α is a bicontraction. A short calculation shows that

$$\begin{aligned}
D_S(w, z) \;&=\; \begin{pmatrix} \dfrac{1 - S(z)S(w)^*}{1 - z\bar{w}} & \dfrac{S(z) - S(\bar{w})}{z - \bar{w}} \\ \dfrac{\tilde{S}(z) - \tilde{S}(\bar{w})}{z - \bar{w}} & \dfrac{1 - \tilde{S}(z)\tilde{S}(w)^*}{1 - z\bar{w}} \end{pmatrix} \\
&=\; \begin{pmatrix} \beta\big(1 - zS'(z)\delta\big)^{-1} & 0 \\ 0 & \gamma^*\big(1 - z\tilde{S}'(z)\delta^*\big)^{-1} \end{pmatrix} \\
&\qquad \times \begin{pmatrix} \dfrac{1 - zS'(z)\bar{w}S'(w)^*}{1 - z\bar{w}} & \dfrac{zS'(z) - \bar{w}S'(\bar{w})}{z - \bar{w}} \\ \dfrac{z\tilde{S}'(z) - \bar{w}\tilde{S}'(\bar{w})}{z - \bar{w}} & \dfrac{1 - z\tilde{S}'(z)\bar{w}\tilde{S}'(w)^*}{1 - z\bar{w}} \end{pmatrix} \\
&\qquad \times \begin{pmatrix} \big(1 - \bar{w}\delta^* S'(w)^*\big)^{-1}\beta^* & 0 \\ 0 & \big(1 - \bar{w}\delta S'(\bar{w})\big)^{-1}\gamma \end{pmatrix}.
\end{aligned}$$

Now we can appeal to the Hilbert space case of Theorem 5.7 to conclude that the kernels $D_S(w, z)$ and $D_{\tilde{S}}(w, z)$ are positive. Therefore Theorem 5.9 holds when $\mathfrak{A}$ and $\mathfrak{B}$ are Kreĭn spaces.

Our purpose in Lectures 4 and 5 has been to motivate generalizations of the Schur class in the indefinite setting. Many issues arise which we have not discussed. Some of these will be taken up in Alpay, Dijksma, Rovnyak, and de Snoo [in preparation]. In other areas, we do not have the answers at this time, and it seems likely that generalizations of the Schur class will remain an interesting subject for future work.

Spaces $\mathfrak{H}(S)$ and $\mathfrak{D}(S)$ are studied in many places for the indefinite case as well as the scalar and Hilbert space cases. A geometric viewpoint based on a concept of complementation is used in de Branges and Rovnyak [1966a], [1966b], de Branges [1988b], [1991], [1994], and Yang [1994]. The theory of complementation in Kreĭn spaces is due to de Branges [1988a] and is further developed by Dritschel and Rovnyak [1991]. The approach of Ball and Cohen [1991] is operator theoretic and features colligations. Operator methods are also used in Andô [1990] and Marcantognini [1990]. Sarason [1994] treats the scalar case and shows applications to function theory on the unit disk. The interaction with canonical models is explored in Ball and Kriete [1987] and Nikol′skiĭ and Vasyunin [1986], [1989]. Similar notions arise in the model theory of noncontractions on Hilbert spaces, as, for example, in McEnnis [1981]. Dym [1989] approaches the subject from the point of view of reproducing kernel spaces and shows many applications. For connections with engineering and linear systems, see also Bart, Gohberg, and Kaashoek [1979] and Helton [1987]. Applications in interpolation theory are given in Ball, Gohberg, and Rodman [1990].

Lecture 6: Invariant Subspaces

Key ideas:

- There is a natural homomorphism between commutants of selfadjoint operators which is useful in the study of hyperinvariant subspaces and maximal dual pairs of invariant subspaces for positive and definitizable operators.
- Certain compact perturbations of definitizable operators are shown to have nontrivial hyperinvariant subspaces. This implies the existence of nontrivial hyperinvariant subspaces for Hilbert space operators of the form $A(B+K)$ with A selfadjoint, B positive, and K selfadjoint and compact and in a Schatten class.

There is a long tradition of invariant subspace theorems in operator theory on spaces with indefinite inner product (see, for example, Azizov and Iokhvidov [1989], Bognár [1974], and Iokhvidov, Kreĭn, and Langer [1982]). We shall not attempt a survey of this large field, but instead highlight techniques found in Dritschel [1993b] and Dritschel [to appear]. The lecture is divided into three parts:

Section A. The commutant of a selfadjoint operator.

Section B. Dual pairs of invariant subspaces for definitizable operators.

Section C. Compact perturbations of definitizable operators.

Proofs of results in Section A are mostly complete and self-contained. Some results in Section B and Section C have complete proofs, but others draw on outside material and are only sketched.

A. The commutant of a selfadjoint operator

A generalization of the notion of a Julia operator is used to construct special homomorphisms of the commutant of a selfadjoint operator (Theorem 6.4). Some examples of the use of such homomorphisms in the study of invariant subspaces will appear in Theorems 6.5 and 6.6.

Our starting point is again the Bognár-Krámli factorization, but an additional property is needed. We do not know if the property holds for every factorization, but it holds for those produced by the method of proof of Theorem 1.1.

Theorem 6.1 *Let $A \in \mathfrak{L}(\mathfrak{H})$ be a selfadjoint operator on a Kreĭn space $\mathfrak{H}$, and construct a factorization*

$$A = DD^*, \qquad D \in \mathfrak{L}(\mathfrak{D}, \mathfrak{H}),\ ker\ D = \{0\}, \tag{6.1}$$

by the method of proof of Theorem 1.1. If $X \in \mathfrak{L}(\mathfrak{H})$ and XA is selfadjoint, there is a unique selfadjoint $Y \in \mathfrak{L}(\mathfrak{D})$ such that $XD = DY$.

A factorization (6.1) of a selfadjoint operator $A \in \mathfrak{L}(\mathfrak{H})$ which is constructed by the method of proof of Theorem 1.1 will be called a **polar factorization**.

Proof. Let $X \in \mathfrak{L}(\mathfrak{H})$ and assume that XA is selfadjoint. Assume for the moment that we can factor

$$XA = EE^*, \qquad E \in \mathfrak{L}(\mathfrak{E}, \mathfrak{H}),\ \ker\ E = \{0\}, \tag{6.2}$$

such that $E = DZ$ for some $Z \in \mathfrak{L}(\mathfrak{E}, \mathfrak{D})$. This gives

$$XDD^* = XA = EE^* = DZZ^*D^*.$$

Setting $Y = ZZ^*$, we get $XD = DY$ since D^* has dense range. Clearly Y is selfadjoint. It is unique because D has zero kernel.

It remains to show that (6.2) can be so chosen. In the notation of the proof of Theorem 1.1, ran $D =$ ran $R^{1/2}$, where $AJ_{\mathfrak{H}} = UR$ is the polar decomposition of the Hilbert space selfadjoint operator $AJ_{\mathfrak{H}}$. Construct $XA = EE^*$ in the same way with the same fundamental symmetry. Then ran $E =$ ran $S^{1/2}$, where $XAJ_{\mathfrak{H}} = VS$ is the polar decomposition of the Hilbert space selfadjoint operator $XAJ_{\mathfrak{H}}$. Since

$$VS = XAJ_{\mathfrak{H}} = XUR,$$

we obtain

$$S^2 = SV^{\times}VS = RU^{\times}X^{\times}XUR \leq \|X\|^2 R^2,$$

and therefore $S \leq \|X\| R$ by Loewner's theorem (Rosenblum and Rovnyak [1985], p. 41). Factor $S^{1/2} = R^{1/2}C$ as in Douglas' lemma. The required operator Z is the restriction of C to the closure of the range of R. □

Theorem 6.2 *Let $A \in \mathfrak{L}(\mathfrak{H})$ and $B \in \mathfrak{L}(\mathfrak{K})$ be selfadjoint operators on Kreĭn spaces $\mathfrak{H}$ and $\mathfrak{K}$ having polar factorizations*

$$A = \tilde{D}\tilde{D}^* \qquad \text{and} \qquad B = DD^*$$

with $\tilde{D} \in \mathfrak{L}(\tilde{\mathfrak{D}}, \mathfrak{H})$ and $D \in \mathfrak{L}(\mathfrak{D}, \mathfrak{K})$. For any $S, T \in \mathfrak{L}(\mathfrak{H}, \mathfrak{K})$ such that

$$TA = BS,$$

there is a unique $L \in \mathfrak{L}(\tilde{\mathfrak{D}}, \mathfrak{D})$ satisfying

$$T\tilde{D} = DL \qquad \text{and} \qquad S^*D = \tilde{D}L^*.$$

If $S = T$, then also

$$L\hat{A} = \hat{B}L,$$

where $\hat{A} = \tilde{D}^*\tilde{D}$ and $\hat{B} = D^*D$.

Proof. Set

$$X = \begin{pmatrix} 0 & S^* \\ T & 0 \end{pmatrix} \qquad \text{and} \qquad C = \begin{pmatrix} A & 0 \\ 0 & B \end{pmatrix}.$$

Our hypotheses imply that C and XC are selfadjoint operators on $\mathfrak{H} \oplus \mathfrak{K}$. We apply Theorem 6.1 to the polar factorization

$$C = \begin{pmatrix} \tilde{D} & 0 \\ 0 & D \end{pmatrix} \begin{pmatrix} \tilde{D}^* & 0 \\ 0 & D^* \end{pmatrix}.$$

Matrix multiplication shows that the operator Y in Theorem 6.1 has the form

$$Y = \begin{pmatrix} 0 & L^* \\ L & 0 \end{pmatrix},$$

where $L \in \mathfrak{L}(\tilde{\mathfrak{D}}, \mathfrak{D})$. The equations $T\tilde{D} = DL$ and $S^*D = \tilde{D}L^*$ now follow. The last statement is a consequence of the first. □

Theorem 6.2 generalizes the notion of a Julia operator. This is most easily seen when $S = T$. In this case, put

$$U = U_{A,B}(T) = \begin{pmatrix} T & D \\ \tilde{D}^* & -L^* \end{pmatrix}.$$

The relations $T\tilde{D} = DL$ and $T^*D = \tilde{D}L^*$ then merely express the fact that the off diagonal terms of U^*U and UU^* are zero. We recover the usual Julia operator for T with $A = 1 - T^*T$ and $B = 1 - TT^*$.

For any operator $A \in \mathfrak{L}(\mathfrak{H})$, we write $\{A\}'$ for the **commutant** of A, that is, the set of operators $T \in \mathfrak{L}(\mathfrak{H})$ such that $AT = TA$. The **double commutant** of A is the set $\{A\}''$ of operators $S \in \mathfrak{L}(\mathfrak{H})$ such that $ST = TS$ for all $T \in \{A\}'$.

Corollary 6.3 *Let $A \in \mathfrak{L}(\mathfrak{H})$ be a selfadjoint operator on a Kreĭn space $\mathfrak{H}$, and let $A = DD^*$ and $\hat{A} = D^*D$ for some polar factorization* (6.1). *Then there is a mapping*

$$\pi : \{A\}' \to \{\hat{A}\}' \tag{6.3}$$

such that for each $T \in \{A\}'$, $\pi(T) = \hat{T}$ is the unique operator in $\mathfrak{L}(\mathfrak{D})$ such that

$$TD = D\hat{T} \qquad \text{and} \qquad T^*D = D\hat{T}^*.$$

Proof. Apply Theorem 6.2 with $A = B = DD^*$ and $S = T$. Continuity follows from the closed graph theorem. □

Theorem 6.4 *Let $A \in \mathfrak{L}(\mathfrak{H})$ be a selfadjoint operator on a Kreĭn space $\mathfrak{H}$, and let $A = DD^*$ and $\hat{A} = D^*D$ for some polar factorization* (6.1). *Then the mapping* (6.3) *is a $*$-homomorphism: for all $X, Y \in \{A\}'$ and all scalars a and b,*

(1) $\pi(aX + bY) = a\pi(X) + b\pi(Y)$,
(2) $\pi(XY) = \pi(X)\pi(Y)$,
(3) $\pi(X^*) = \pi(X)^*$.

Moreover, $\pi(1_{\mathfrak{H}}) = 1_{\mathfrak{D}}$, and $\pi(A) = \hat{A}$. The mapping takes isometric, partially isometric, unitary, selfadjoint, and projection operators to operators of the same type. For $T \in \{A\}'$, $\sigma(\pi(T)) \subseteq \sigma(T)$ and $\sigma_p(\pi(T)) \subseteq \sigma_p(T)$. If further $\varphi(T)$ is defined by the Riesz functional calculus for some holomorphic function φ, then $\varphi(\pi(T))$ is defined and $\pi(\varphi(T)) = \varphi(\pi(T))$.

We write $\sigma(T)$ and $\rho(T)$ for the spectrum and resolvent set for an operator T. The point spectrum is denoted $\sigma_p(T)$.

Proof. Most of the statements are immediate. We only prove the last two assertions.

Fix $T \in \{A\}'$. If $\lambda \in \rho(T)$, then $(T-\lambda)^{-1} \in \{A\}'$. Hence there is an $X \in \{\hat{A}\}'$ such that $(T-\lambda)^{-1}D = DX$ and $(T^* - \bar{\lambda})^{-1}D = DX^*$. If $\hat{T} = \pi(T)$, then

$$D = (T-\lambda)(T-\lambda)^{-1}D = D(\hat{T}-\lambda)X,$$

and since D has zero kernel, X is a right inverse for $\hat{T}-\lambda$. It is similarly a left inverse. Thus $\lambda \in \rho(\pi(T))$ and $\pi((T-\lambda)^{-1}) = X = (\pi(T)-\lambda)^{-1}$. In particular, $\sigma(\pi(T)) \subseteq \sigma(T)$. On the other hand, if there exists $f \neq 0$ such that $\hat{T}f = \lambda f$, then $Df \neq 0$ and $TDf = D\hat{T}f = \lambda Df$. Therefore $\sigma_p(\pi(T)) \subseteq \sigma_p(T)$.

The last assertion is verified above when $\varphi(z) = 1/(z-\lambda)$ for some $\lambda \in \rho(T)$. Suppose that φ is holomorphic on an open set containing $\sigma(T)$, so that

$$\varphi(T) = \frac{1}{2\pi i}\int_\Gamma \varphi(z)(z-T)^{-1}\,dz$$

is defined. Here Γ is a suitable system of curves as in Conway [1990], p. 201. Since $\sigma(\pi(T)) \subseteq \sigma(T)$, $\varphi(\pi(T))$ is also defined and given by

$$\varphi(\pi(T)) = \frac{1}{2\pi i}\int_\Gamma \varphi(z)(z-\pi(T))^{-1}\,dz.$$

We then get $\pi(\varphi(T)) = \varphi(\pi(T))$ by approximating the integrals by Riemann sums and using the result for the special case of resolvents. □

It can be shown that in cases of interest, the mapping π in Theorem 6.4 is essentially unique. What essential uniqueness means in this context is that if π' is a second such mapping defined from a polar factorization

$$A = D'D'^*, \qquad D' \in \mathfrak{L}(\mathfrak{D}', \mathfrak{H}),\ \ker\ D' = \{0\}, \tag{6.4}$$

then

$$\pi'(T) = W^*\pi(T)W, \qquad T \in \{A\}', \tag{6.5}$$

for some Kreĭn space isomorphism W from $\mathfrak{D}'$ onto $\mathfrak{D}$. A sufficient condition for the essential uniqueness of π is that A is a positive operator, that is, $A \geq 0$. In this case, the space $\mathfrak{D}$ in any polar factorization (6.1) is a Hilbert space, and standard operator methods can be used to show that (6.1) and (6.4) are related by

$$D' = DW$$

for some isomorphism W from $\mathfrak{D}'$ onto $\mathfrak{D}$. Now suppose that $T \in \{A\}'$, and put

$$\hat{T} = \pi(T) \qquad \text{and} \qquad \hat{T}' = \pi'(T).$$

From the relations $TD = D\hat{T}$ and $TD' = D'\hat{T}'$, we then get $TDW = DW\hat{T}'$ and so $W\hat{T}'W^* = \hat{T}$, yielding (6.5). More generally, there will be essential uniqueness if $\mathfrak{D}$ is a Pontryagin space for some polar factorization (6.1), or, more generally still, if ran D^* contains a maximal uniformly definite subspace (see Dritschel [1993a]). We make no use of these results and omit their proofs.

Let $\mathfrak{M}$ be a closed subspace of a Kreĭn space $\mathfrak{H}$. The subspace $\mathfrak{M}$ is said to be **proper** if it is not the zero subspace or the whole space. We say that $\mathfrak{M}$ is **invariant** for an operator $A \in \mathfrak{L}(\mathfrak{H})$ if $A\mathfrak{M} \subseteq \mathfrak{M}$ and **hyperinvariant** if $\mathfrak{M}$ is invariant for each operator in the commutant of A. We call $\mathfrak{M}$ **reducing** for A if it is invariant for A and A^*, or, equivalently, the subspace and its orthogonal companion are invariant for A. A hyperinvariant subspace for a selfadjoint operator

$A \in \mathfrak{L}(\mathfrak{H})$ is a reducing subspace for every $T \in \{A\}'$, because the commutant of a selfadjoint operator A contains T^* whenever it contains T.

The following result illustrates the use of Theorem 6.4 in the study of invariant subspaces.

Theorem 6.5 *Let $A \in \mathfrak{L}(\mathfrak{H})$ be a selfadjoint operator on a Kreĭn space $\mathfrak{H}$, not a constant multiple of the identity. Let $A = DD^*$ and $\hat{A} = D^*D$ for some polar factorization* (6.1).

(1) *The operator A has a proper closed hyperinvariant subspace if $p(\hat{A}) = 0$ for some nonzero polynomial p.*

(2) *The operator A has a proper closed hyperinvariant subspace if $\varphi(\hat{A})$ has such a subspace for some holomorphic function φ which is defined on an open set containing the spectrum of A.*

(3) *The operator A has a pair of orthogonal proper closed hyperinvariant subspaces if $\varphi(\hat{A})$ has such a pair for some holomorphic function φ which is defined on an open set containing the spectrum of A.*

Proof. (1) For any nonnegative integer n, $D\hat{A}^n D^* = D(D^*D)^n D^* = A^{n+1}$, and so $Ap(A) = Dp(\hat{A})D^* = 0$. Since $q(A) = 0$ for a nonconstant polynomial q, A has a proper closed hyperinvariant subspace by an elementary argument.

(2) We adopt the notation of Corollary 6.3 and Theorem 6.4. Let $\mathfrak{M}$ be a proper closed hyperinvariant subspace for $\varphi(\hat{A})$, where φ is a holomorphic function which is defined on an open set containing the spectrum of A. Set $\mathfrak{N} = \overline{D\mathfrak{M}}$. We show that $\mathfrak{N}$ is hyperinvariant for A. Suppose $T \in \{A\}'$ and $\pi(T) = \hat{T}$. Then T commutes with $\varphi(A)$, and therefore $\hat{T} = \pi(T)$ commutes with $\varphi(\hat{A}) = \pi(\varphi(A))$ by Theorem 6.4. Since $\mathfrak{M}$ is hyperinvariant for $\varphi(\hat{A})$, $\hat{T}\mathfrak{M} \subseteq \mathfrak{M}$. Thus for $f \in \mathfrak{M}$,

$$TDf = D\hat{T}f \in D\mathfrak{M} \subseteq \mathfrak{N}.$$

By the continuity of T, $T\mathfrak{N} \subseteq \mathfrak{N}$. This shows that $\mathfrak{N}$ is hyperinvariant for A.

Observe that $\mathfrak{N}$ is nonzero since $\mathfrak{M}$ is nonzero and $\ker D = \{0\}$. If $\mathfrak{N} = \mathfrak{H}$, then $D\mathfrak{M}$ is dense in $\mathfrak{H}$ and hence $D^*D\mathfrak{M}$ is dense in $\mathfrak{D}$. But $D^*D\mathfrak{M} = \hat{A}\mathfrak{M} \subseteq \mathfrak{M}$ since $\mathfrak{M}$ is hyperinvariant for $\varphi(\hat{A})$ and $\hat{A}$ commutes with $\varphi(\hat{A})$. This contradicts the assumption that $\mathfrak{M}$ is proper. Therefore $\mathfrak{N}$ is a proper subspace of $\mathfrak{H}$.

(3) Now suppose that $\mathfrak{M}_1$ and $\mathfrak{M}_2$ are orthogonal proper closed hyperinvariant subspaces for $\varphi(\hat{A})$ for φ as above. By the first part of the proof, $\mathfrak{N}_1 = \overline{D\mathfrak{M}_1}$ and $\mathfrak{N}_2 = \overline{D\mathfrak{M}_2}$ are proper closed hyperinvariant subspaces for A. If $f \in \mathfrak{M}_1$ and $g \in \mathfrak{M}_2$, then

$$\langle Df, Dg \rangle_{\mathfrak{H}} = \langle D^*Df, g \rangle_{\mathfrak{D}} = 0,$$

since $\mathfrak{M}$ is invariant for $\hat{A} = D^*D$. It follows that $\mathfrak{N}_1 \perp \mathfrak{N}_2$. □

Theorem 6.5 can be used to give a quick proof that a positive operator has nontrivial hyperinvariant subspaces. More is true in this case.

Theorem 6.6 *Let $A \in \mathfrak{L}(\mathfrak{H})$ be a positive operator, not a constant multiple of the identity. Then A has a proper closed hyperinvariant subspace. Moreover, there exist closed hyperinvariant subspaces $\mathfrak{N}_+$ and $\mathfrak{N}_-$ which are positive and negative, respectively, such that*

(1) $\mathfrak{N}_- \perp \mathfrak{N}_+$,

(2) $\mathfrak{N}_+ + \mathfrak{N}_- \supseteq \operatorname{ran} A$, *and*

(3) $\mathfrak{N}_\pm^\perp \cap \operatorname{ran} A \subseteq \mathfrak{N}_\mp$.

Proof. Let $A = DD^*$, $D \in \mathfrak{L}(\mathfrak{D}, \mathfrak{H})$, be a polar factorization, and let $\hat{A} = D^*D$. Since A is positive, $\mathfrak{D}$ is a Hilbert space. The projections in the range of the spectral measure $E(\cdot)$ for $\hat{A}$ commute with every operator T that commutes with $\hat{A}$ (Halmos [1951], p. 68). If $\hat{A}$ is not a constant multiple of the identity, we obtain proper closed hyperinvariant subspaces for A by part (2) of Theorem 6.5. If $\hat{A}$ is a constant multiple of the identity, the same conclusion follows from part (1) of Theorem 6.5.

For any real Borel set Δ, the closure of the range of $DE(\Delta)$ is a hyperinvariant subspace for A by the proof of Theorem 6.5. Let $\mathfrak{N}_+$ and $\mathfrak{N}_-$ be these spaces for $\Delta_+ = [0, \infty)$ and $\Delta_- = (-\infty, 0]$. If $f \in \mathfrak{D}$, then

$$\langle DE(\Delta_+)f, DE(\Delta_+)f \rangle_{\mathfrak{H}} = \left\langle \hat{A}E(\Delta_+)f, E(\Delta_+)f \right\rangle_{\mathfrak{D}} \geq 0.$$

Thus $\mathfrak{N}_+$ is a positive subspace. In a similar way, we see that $\mathfrak{N}_-$ is a negative subspace. If $f, g \in \mathfrak{D}$, then

$$\langle DE(\Delta_+)f, DE(\Delta_-)g \rangle_{\mathfrak{H}} = \left\langle \hat{A}E(\Delta_+)f, E(\Delta_-)g \right\rangle_{\mathfrak{D}} = 0,$$

so (1) holds. Since the span of the ranges of $E(\Delta_+)$ and $E(\Delta_-)$ is all of $\mathfrak{D}$,

$$\mathfrak{N}_+ + \mathfrak{N}_- \supseteq \operatorname{ran} D \supseteq \operatorname{ran} A,$$

which yields (2). Suppose $f \in \mathfrak{N}_+^\perp \cap \operatorname{ran} A$. Then $f = Ag = DD^*g$ and $f \perp DE(\Delta_+)u$ for all $u \in \mathfrak{D}$. Thus for all $u \in \mathfrak{D}$,

$$0 = \langle f, DE(\Delta_+)u \rangle_{\mathfrak{H}} = \langle D^*g, D^*DE(\Delta_+)u \rangle_{\mathfrak{D}} = \left\langle D^*g, \hat{A}E(\Delta_+)u \right\rangle_{\mathfrak{D}}.$$

It follows that $D^*g \in \operatorname{ran} E(\Delta_-)$, and therefore $f = DD^*g \in \mathfrak{N}_-$. We have shown that $\mathfrak{N}_+^\perp \cap \operatorname{ran} A \subseteq \mathfrak{N}_-$. In a similar way, $\mathfrak{N}_-^\perp \cap \operatorname{ran} A \subseteq \mathfrak{N}_+$, and (3) follows. □

Theorem 6.7 *Let $\mathfrak{H}$ be a Kreĭn space, and let $A \in \mathfrak{L}(\mathfrak{H})$ be a positive operator. Assume that ker $A = \{0\}$ and $\sigma(A) \subseteq [0, \infty)$. Then $\mathfrak{H}$ is a Hilbert space.*

Proof. Let $A = DD^*$, $D \in \mathfrak{L}(\mathfrak{D}, \mathfrak{H})$, be a polar factorization. Then $\mathfrak{D}$ is a Hilbert space, and $\hat{A} = D^*D \in \mathfrak{L}(\mathfrak{D})$ is a positive operator because $\hat{A}$ is selfadjoint and $\sigma(\hat{A}) \subseteq \sigma(A) \subseteq [0, \infty)$ by Theorem 6.4. For any $f \in \mathfrak{D}$,

$$\langle Df, Df \rangle_{\mathfrak{H}} = \langle \hat{A}f, f \rangle_{\mathfrak{D}} \geq 0.$$

Since A has zero kernel, it has dense range. Therefore D has dense range, and so $\mathfrak{H}$ is a Hilbert space. □

Norm estimates are implicit in Theorems 6.1, 6.2, and Corollary 6.3. Recall that any two operator norms on $\mathfrak{L}(\mathfrak{H}, \mathfrak{K})$ are equivalent. Operator norms are determined by fundamental decompositions, and sometimes these can be chosen to produce special effects. With slightly greater care, the proof of Theorem 6.1 produces the inequality

$$\|Y\| \leq \|X\|$$

for suitable operator norms on $\mathfrak{L}(\mathfrak{H})$ and $\mathfrak{L}(\mathfrak{D})$. In the proof of Theorem 6.2 it is possible to show that there exist operator norms on $\mathfrak{L}(\mathfrak{H}, \mathfrak{K})$ and $\mathfrak{L}(\tilde{\mathfrak{D}}, \mathfrak{D})$ such that

$$\|L\| \leq \max \;(\|S\|, \|T\|).$$

In Corollary 6.3, the polar factorization of A similarly determines operator norms on $\mathfrak{L}(\mathfrak{H})$ and $\mathfrak{L}(\mathfrak{D})$ such that

$$\|\pi(T)\| \le \|T\|$$

for every $T \in \{A\}'$.

A particular instance of Theorem 6.2, as noted in the remarks following the proof, yields another proof of the existence of a Julia operator (2.2) for any given operator $T \in \mathfrak{L}(\mathfrak{H}, \mathfrak{K})$. By this method, operator norms may be chosen such that in addition

$$\|L\| \le \|T\|,$$
$$\max \left(\|D\|, \|\tilde{D}\|\right) \le \left(1 + \|T\|^2\right)^{1/2}.$$

Such estimates were obtained by Arsene, Constantinescu, and Gheondea [1987] in their original construction of a Julia operator (see Dritschel and Rovnyak [1990], p. 296).

B. Dual pairs of invariant subspaces for definitizable operators

For applications it is often useful to know if an operator on a Kreĭn space has an invariant subspace which is maximal negative or maximal positive. This is a large field, going back to L. S. Pontryagin [1944]. An extensive account with bibliographic notes may be found in Azizov and Iokhvidov [1989]. In this section, we examine some of the ideas in Langer [1965], [1971], [1982] from the perspective of the constructions in §A. The main results of the section are due to Langer and involve, in particular, the spectral theory of definitizable operators. The methods of §A give an alternative approach to these ideas.

Define a **dual pair** of subspaces of a Kreĭn space $\mathfrak{H}$ as a pair consisting of a closed positive subspace $\mathfrak{M}_+$ and a closed negative subspace $\mathfrak{M}_-$ of $\mathfrak{H}$ such that $\mathfrak{M}_+ \perp \mathfrak{M}_-$. A **maximal dual pair** is a dual pair consisting of a maximal positive subspace and a maximal negative subspace. A theorem of R. S. Phillips [1961] assures the existence of a maximal dual pair extending any given dual pair. A particular case was previously verified in Theorem 1.6(5).

Phillips' Theorem. *For any dual pair $\mathfrak{N}_+$, $\mathfrak{N}_-$ of subspaces of a Kreĭn space $\mathfrak{H}$, there is a maximal dual pair $\mathfrak{M}_+$, $\mathfrak{M}_-$ such that $\mathfrak{N}_\pm \subseteq \mathfrak{M}_\pm$.*

Proof. (Arsene and Gheondea [1982]). Let $\mathfrak{H} = \mathfrak{H}_+ \oplus \mathfrak{H}_-$ be a fundamental decomposition. By the graph representation of definite subspaces, we may write

$$\mathfrak{H}_+ = \mathfrak{S}_+ \oplus \mathfrak{R}_+, \qquad |\mathfrak{H}_-| = \mathfrak{S}_- \oplus \mathfrak{R}_- \tag{6.6}$$

and choose Hilbert space contractions

$$C : \mathfrak{S}_+ \to |\mathfrak{H}_-| = \mathfrak{S}_- \oplus \mathfrak{R}_-,$$
$$R^\times : \mathfrak{S}_- \to \mathfrak{H}_+ = \mathfrak{S}_+ \oplus \mathfrak{R}_+,$$

such that

$$\mathfrak{N}_+ = \left\{ \begin{pmatrix} f \\ Cf \end{pmatrix} : f \in \mathfrak{S}_+ \right\} \qquad \text{and} \qquad \mathfrak{N}_- = \left\{ \begin{pmatrix} R^\times g \\ g \end{pmatrix} : g \in \mathfrak{S}_- \right\}.$$

Write $R = \begin{pmatrix} T & F \end{pmatrix}$ and $C^\times = \begin{pmatrix} S^\times & G^\times \end{pmatrix}$ relative to the decompositions (6.6). The orthogonality of $\mathfrak{N}_+$ and $\mathfrak{N}_-$ means that for all $f \in \mathfrak{S}_+$ and $g \in \mathfrak{S}_-$,

$$0 = \left\langle \begin{pmatrix} f \\ Cf \end{pmatrix}, \begin{pmatrix} R^\times g \\ g \end{pmatrix} \right\rangle_{\mathfrak{H}} = \langle f, R^\times g \rangle_{\mathfrak{H}_+} - \langle Cf, g \rangle_{|\mathfrak{H}_-|} = \langle f, T^\times g \rangle_{\mathfrak{S}_+} - \langle Sf, g \rangle_{\mathfrak{S}_-},$$

and so $S = T$. By the Hilbert space cases of Theorems 3.2–3.4, we may find an operator $H \in \mathfrak{L}(\mathfrak{R}_+, \mathfrak{R}_-)$ such that

$$A = \begin{pmatrix} T & F \\ G & H \end{pmatrix} \in \mathfrak{L}(\mathfrak{S}_+ \oplus \mathfrak{R}_+, \mathfrak{S}_- \oplus \mathfrak{R}_-)$$

is a contraction. Without difficulty we check that the subspaces

$$\mathfrak{M}_+ = \left\{ \begin{pmatrix} f \\ Af \end{pmatrix} : f \in \mathfrak{H}_+ \right\} \qquad \text{and} \qquad \mathfrak{M}_- = \left\{ \begin{pmatrix} A^\times g \\ g \end{pmatrix} : g \in |\mathfrak{H}_-| \right\}$$

have the required properties. $\square$

We are interested in criteria for the existence of maximal positive or maximal negative subspaces which are invariant for a selfadjoint operator on a Kreĭn space. Such subspaces automatically occur in dual pairs. For example, if $\mathfrak{M}_-$ is a maximal negative subspace of a Kreĭn space $\mathfrak{H}$ which is invariant for a selfadjoint operator $A \in \mathfrak{L}(\mathfrak{H})$, then $\mathfrak{M}_+ = \mathfrak{M}_-^\perp$ is a maximal positive subspace which is invariant for A. The roles of positive and negative subspaces can obviously be interchanged here. It is thus natural to speak of maximal dual pairs of invariant subspaces of a selfadjoint operator, and we state existence criteria in these terms (Theorems 6.8 and 6.11).

In one situation, Theorem 6.6 assures the existence of nontrivial dual pairs of invariant subspaces, which, however, are not necessarily maximal. Using Phillips' theorem, we then obtain maximal dual pairs of invariant subspaces. This method follows Langer [1982], p. 44, except that in place of spectral theory we employ Theorem 6.6.

Theorem 6.8 *Every positive operator A on a Kreĭn space $\mathfrak{H}$ has a maximal dual pair of invariant subspaces.*

Proof. The conclusion is trivial if A is a constant multiple of the identity, so assume that this is not the case. Then by Theorem 6.6, there exists a dual pair of invariant subspaces $\mathfrak{N}_+$, $\mathfrak{N}_-$ such that $\mathfrak{N}_\pm^\perp \cap \operatorname{ran} A \subseteq \mathfrak{N}_\mp$. Use Phillips' theorem to extend to a maximal dual pair $\mathfrak{M}_\pm \supseteq \mathfrak{N}_\pm$. Then

$$A\mathfrak{M}_+ \subseteq A(\mathfrak{N}_-^\perp) \subseteq \mathfrak{N}_-^\perp \cap \operatorname{ran} A \subseteq \mathfrak{N}_+ \subseteq \mathfrak{M}_+.$$

In a similar way, $A\mathfrak{M}_- \subseteq \mathfrak{M}_-$. $\square$

The full class of Kreĭn space selfadjoint operators has no structure theory which is comparable to the Hilbert space case. It is not difficult to understand why this should be so. Let $\mathfrak{H} = \mathfrak{K} \oplus \mathfrak{K}$, where $\mathfrak{K}$ is a Hilbert space. We view $\mathfrak{H}$ as a Kreĭn space with fundamental symmetry

$$J = \begin{pmatrix} 0 & 1 \\ 1 & 0 \end{pmatrix}.$$

Any selfadjoint operator on $\mathfrak{H}$ is easily seen to have the form

$$A = \begin{pmatrix} T & B \\ C & T^{\times} \end{pmatrix} \tag{6.7}$$

where $B, C \in \mathfrak{L}(\mathfrak{K})$ are Hilbert space selfadjoint operators, $T \in \mathfrak{L}(\mathfrak{K})$, and $T^{\times}$ is the Hilbert space adjoint of T. Considering such operators with $B = C = 0$, we observe that knowing the structure of Kreĭn space selfadjoint operators is equivalent to knowing the structure of an arbitrary bounded operator on a Hilbert space.

Special classes of selfadjoint operators do have a detailed structure theory, however. The positive operators provide an example. In fact, the structure of positive operators can be explicitly described in a manner similar to the Hilbert space case. The analysis lends itself to a broader class of operators. A selfadjoint operator A on a Kreĭn space $\mathfrak{H}$ is said to be **definitizable** if there is a nonconstant polynomial p having real coefficients such that $p(A) \geq 0$. We call any such polynomial p a **definitizing polynomial** for A. A definitizable operator thus commutes with a positive operator.

Figure 6 shows a typical picture for the spectrum of a definitizable operator. The nonreal spectrum is a subset of the zeros of the definitizing polynomial. To describe this picture, we need some standard notions from spectral theory. Assume $A \in \mathfrak{L}(\mathfrak{H})$ for some Kreĭn space $\mathfrak{H}$, and let λ_0 be an isolated point in $\sigma(A)$. We call

$$P_{\lambda_0} = \frac{1}{2\pi i} \int_{\gamma} (z - A)^{-1}\, dz$$

the **Riesz idempotent** for the point λ_0 (the term "projection" is reserved for a selfadjoint idempotent as in Lecture 1). Here γ is any sufficiently small circle around λ_0 with counterclockwise orientation. The Riesz idempotent P_{λ_0} satisfies $P_{\lambda_0}^2 = P_{\lambda_0}$ but is not necessarily selfadjoint. The range $\mathfrak{M}_{\lambda_0}$ of P_{λ_0} is a closed subspace of $\mathfrak{H}$ which is invariant under A. We say that λ_0 has finite **Riesz index** equal to the positive integer ν if $(\lambda_0 - A)^{\nu}\mathfrak{M}_{\lambda_0} = \{0\}$ and no smaller positive integer has this property. Such a point is necessarily an eigenvalue, but it could have infinite multiplicity. If no such integer ν exists, the Riesz index is considered to be infinite. For example, see Conway [1990], pp. 210–211.

Theorem 6.9 *Let $A \in \mathfrak{L}(\mathfrak{H})$ be a selfadjoint definitizable operator with definitizing polynomial p. The nonreal spectrum of A consists of a finite number of conjugate pairs of points $\lambda_j, \bar{\lambda}_j$, $j = 1, \ldots, r$, which are zeros of p. Each such point is an eigenvalue of finite Riesz index not greater than the order of the point as a zero of p. If λ is one of the nonreal eigenvalues, let P_λ be the associated Riesz projection and $\mathfrak{M}_\lambda$ its range.*

(1) *For each nonreal eigenvalue λ, $P_\lambda^* = P_{\bar{\lambda}}$ and $P_\lambda P_{\bar{\lambda}} = P_{\bar{\lambda}} P_\lambda = 0$. The subspaces $\mathfrak{M}_\lambda, \mathfrak{M}_{\bar{\lambda}}$ are neutral, and their sum $\mathfrak{M}_\lambda + \mathfrak{M}_{\bar{\lambda}}$ is a Kreĭn subspace which reduces A and coincides with the range of the projection $P_\lambda + P_{\bar{\lambda}}$.*

(2) *The subspace $\mathfrak{M}_{\lambda_j} + \mathfrak{M}_{\bar{\lambda}_j}$ is orthogonal to $\mathfrak{M}_{\lambda_k} + \mathfrak{M}_{\bar{\lambda}_k}$ whenever $j \neq k$, and*

$$\mathfrak{H} = \mathfrak{H}_0 \oplus \sum_{j=1}^{r} \left(\mathfrak{M}_{\lambda_j} + \mathfrak{M}_{\bar{\lambda}_j} \right),$$

where $\mathfrak{H}_0$ reduces A and $A_0 = A|_{\mathfrak{H}_0}$ is definitizable with a definitizing polynomial p_0 which has only real zeros.

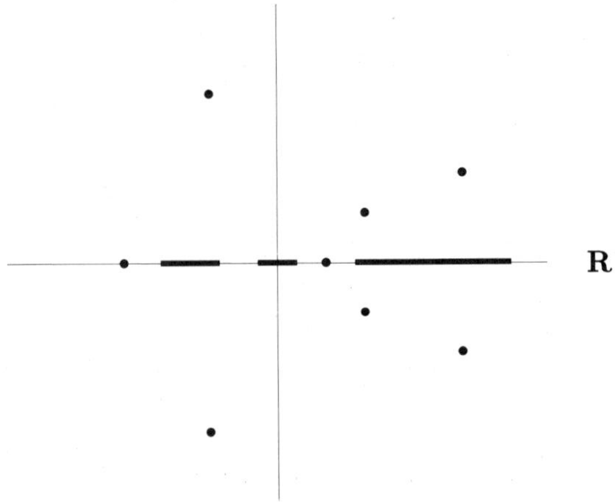

Figure 6

Proof. If $\sigma(A)$ contains a nonreal point, so does $\sigma_{ap}(A)$, the approximate point spectrum, since the boundary of $\sigma(A)$ is contained in $\sigma_{ap}(A)$ (Conway [1990], p. 210; $\sigma_{ap}(A)$ is defined as in the Hilbert space case for any choice of norm for the Kreĭn space). The first assertion will be proved by showing that a nonreal point λ in $\sigma_{ap}(A)$ is a zero of p.

Let $p(A) = DD^*$, $D \in \mathfrak{L}(\mathfrak{D}, \mathfrak{H})$, be a polar factorization. Since $p(A)$ is positive by hypothesis, $\mathfrak{D}$ is a Hilbert space. Define π as in Corollary 6.3, and set $\hat{A} = \pi(A)$. Then $D^*(A - \lambda) = (\hat{A} - \lambda)D^*$. Since $\hat{A}$ is selfadjoint, the spectrum of $\hat{A}$ is real, and so $\hat{A} - \lambda$ is invertible. Choose a norm for $\mathfrak{H}$ and a sequence $\{f_n\}_{n=1}^{\infty}$ of unit vectors such that $\|(A - \lambda)f_n\| \to 0$. Since $p(A) = D(\hat{A} - \lambda)^{-1}D^*(A - \lambda)$, we have $\|p(A)f_n\| \to 0$. Write $p(z) = c(z - \mu_1)\cdots(z - \mu_r)$, where $\mu_1, \ldots, \mu_r$ are roots of p and c is a constant. If $p_1(z)$ is the polynomial $p(z)/(z - \mu_1)$, then

$$\|p(A)f_n\| = \|p_1(A)[(A - \lambda) + (\lambda - \mu_1)]f_n\| \to 0.$$

Since $\|p_1(A)(A - \lambda)f_n\| \to 0$, either $\lambda = \mu_1$ or $\|p_1(A)f_n\| \to 0$. In the latter case, we repeat the argument. Continuing in this way, we see that λ must be one of the roots $\mu_1, \ldots, \mu_r$, since otherwise we obtain the contradiction that $\|f_n\| \to 0$. The first assertion has thus been proved.

Let λ be any nonreal point of $\sigma(A)$. Its Riesz idempotent

$$P_\lambda = \frac{1}{2\pi i} \int_{\gamma(\lambda)} (z - A)^{-1}\, dz$$

can be formed by integrating over any circle centered at λ and containing no other point of $\sigma(A)$ in its interior. Choosing this circle so that it does not meet the real axis, we obtain by Theorem 6.4,

$$\pi(P_\lambda) = \frac{1}{2\pi i} \int_{\gamma(\lambda)} (z - \hat{A})^{-1}\, dz = 0,$$

since $(z - \hat{A})^{-1}$ is holomorphic off the real axis. Therefore

$$p(A)P_\lambda = P_\lambda DD^* = D\pi(P_\lambda)D^* = 0.$$

According to the Riesz functional calculus, if λ is a root of p of order k, we can factor $p(A) = (A - \lambda)^k q(A)$, where q is a polynomial and the restriction of $q(A)$ to $\mathfrak{M}_\lambda$ is invertible. Thus $(A - \lambda)^k \mathfrak{M}_\lambda = \{0\}$, and so λ has Riesz index not greater than k. As a consequence, λ is an eigenvalue for A.

In the same situation, by the selfadjointness of A, $\bar{\lambda}$ also belongs to $\sigma(A)$. The relations $P_\lambda^* = P_{\bar{\lambda}}$ and $P_\lambda P_{\bar{\lambda}} = P_{\bar{\lambda}} P_\lambda = 0$ follow from the Riesz functional calculus. In particular, $P_{\bar{\lambda}}^* P_\lambda = P_{\bar{\lambda}} P_\lambda = 0$, and hence $\mathfrak{M}_\lambda$ is a neutral subspace of $\mathfrak{H}$. The same assertion holds for $\mathfrak{M}_{\bar{\lambda}}$. The sum $P_\lambda + P_{\bar{\lambda}}$ is a selfadjoint idempotent and hence a projection. Its range is $\mathfrak{M}_\lambda + \mathfrak{M}_{\bar{\lambda}}$, which is therefore a Kreĭn subspace of $\mathfrak{H}$. The subspace reduces A because each of the summands is invariant under A and A is selfadjoint. This proves (1).

The assertions in (2) follow from straightforward arguments using the Riesz functional calculus. □

It is therefore sufficient to restrict attention to definitizable operators whose definitizing polynomials have only real zeros. Such operators have real spectrum as in the Hilbert space case.

Theorem 6.10 *Let $A \in \mathfrak{L}(\mathfrak{H})$ be a definitizable selfadjoint operator with definitizing polynomial p having only real zeros. Let $\mathfrak{Z}$ be the set of zeros of p, and let Ω be the Boolean algebra of all Borel subsets Δ of $\mathbf{R}$ such that $\partial\Delta \cap \mathfrak{Z} = \emptyset$. Then there exist a function $E : \Omega \to \mathfrak{L}(\mathfrak{H})$ and positive operators $N_\nu \in \mathfrak{L}(\mathfrak{H})$, $\nu \in \mathfrak{Z}$, such that $N_\nu N_\mu = 0$ for all $\mu, \nu \in \mathfrak{Z}$ (including $\mu = \nu$), and $E(\Delta)N_\nu = N_\nu E(\Delta) = 0$ whenever $\Delta \in \Omega$, $\nu \in \mathfrak{Z}$, and $\nu \notin \Delta$, and such that the following properties hold:*

1. *For $\Delta \in \Omega$, $E(\Delta)$ is a projection in the the double commutant $\{A\}''$ of A, $E(\emptyset) = 0$, and $E(\mathbf{R}) = 1$.*
2. *For all $\Delta, \Delta' \in \Omega$, $E(\Delta \cap \Delta') = E(\Delta)E(\Delta')$.*
3. *If $\{\Delta_n\}_{n=1}^\infty \subset \Omega$ are pairwise disjoint and $\Delta = \cup_{n=1}^\infty \Delta_n \in \Omega$, then $E(\Delta) = \sum_{n=1}^\infty E(\Delta_n)$ with convergence in the strong operator topology.*
4. *If $\Delta \in \Omega$ and $p > 0$ ($p < 0$) on Δ, then the range of $E(\Delta)$ is uniformly positive (uniformly negative).*
5. *The spectrum of the restriction of A to the range of $E(\Delta)$ is contained in $\bar{\Delta}$ for each $\Delta \in \Omega$.*
6. *If φ is a Borel measurable function which is bounded on $\sigma(A)$, then*

$$\int_{\sigma(A)\setminus\mathfrak{Z}} \varphi(\lambda)p(\lambda)\, E(d\lambda) = \lim_{\epsilon \downarrow 0} \int_{S_\epsilon} \varphi(\lambda)p(\lambda)\, E(d\lambda) \tag{6.8}$$

exists in the strong operator topology, where S_ϵ is the set of points in $\sigma(A)$ at a distance greater than ϵ from $\mathfrak{Z}$. An operator $\varphi(A)p(A)$ may be defined for all such φ and expressed as

$$\varphi(A)p(A) = \int_{\sigma(A)\setminus\mathfrak{Z}} \varphi(\lambda)p(\lambda)\, E(d\lambda) + \sum_{\nu\in\mathfrak{Z}} \varphi(\nu)N_\nu. \tag{6.9}$$

The definition of $\varphi(A)p(A)$ is made in such a way that if φ is holomorphic on an open set containing $\sigma(A)$ and $\varphi(A)$ is defined by the Riesz functional calculus, $\varphi(A)p(A)$ is the product of $\varphi(A)$ and $p(A)$.

The function $E(\cdot)$ is unique and called the **spectral function** for A. The functional calculus is adequate for our purposes, but it has an obvious limitation in that it does not define $\varphi(A)$ directly but only speaks of $\varphi(A)p(A)$. See Jonas [1981] for a more refined analysis which overcomes this difficulty for functions which are sufficiently smooth in a neighborhood of what are called "critical points" for A (the critical points are a subset of the zeros of the definitizing polynomial). Also, there may be choices of definitizing polynomial with fewer zeros. The article of Langer [1982] contains further discussions of this point.

Sketch of proof. By assumption, $p(A)$ is positive. Therefore it has a polar factorization $p(A) = DD^*$, $D \in \mathfrak{L}(\mathfrak{D}, \mathfrak{H})$, where $\mathfrak{D}$ is a Hilbert space. Let π be the associated mapping constructed in Corollary 6.3. The operator $\hat{A} = \pi(A)$ is Hilbert space selfadjoint and therefore has a spectral representation

$$\hat{A} = \int t\, F(dt)$$

for some spectral measure $F(\cdot)$ on the Borel subsets of the real line.

Every $\Delta \in \Omega$ can be written as a disjoint union $\Delta = I_{\nu_1} \cup \cdots \cup I_{\nu_r} \cup \Delta_0$, where each I_{ν_k} is a closed interval containing a single point of $\mathfrak{Z}$ in its interior and Δ_0 is a Borel set at a positive distance from $\mathfrak{Z}$. Possibly there are no intervals I_{ν_k}, in which case that part of the union is the empty set. We show how to define $E(\Delta')$ if Δ' is one of the sets in the union and then take

$$E(\Delta) = E(I_{\nu_1}) + \cdots + E(I_{\nu_k}) + E(\Delta_0)$$

in the general case. The representation Δ as a union of special sets is not unique, and a full account should include a verification that the definition is independent of the representation.

For sets of the form Δ_0, we may write $\Delta_0 = \Delta_0^+ \cup \Delta_0^-$, where $p(t) > \delta$ for all $t \in \Delta_0^+$ and $p(t) < -\delta$ for all $t \in \Delta_0^-$ for some positive number δ. Here essential use is made of the fact that Δ_0 lies at a positive distance from the zeros of p. Put $\mathfrak{F}(\Delta_0^\pm) = \text{ran } F(\Delta_0^\pm)$. For $f \in \mathfrak{F}(\Delta_0^+)$,

$$\langle Df, Df \rangle_{\mathfrak{H}} = \langle D^*Df, f \rangle_{\mathfrak{D}} = \langle p(\hat{A})f, f \rangle_{\mathfrak{D}} \geq \delta \langle f, f \rangle_{\mathfrak{D}} \geq \delta \|D\|^{-2} \|Df\|^2.$$

It follows that $\mathfrak{E}(\Delta_0^+) = D\mathfrak{F}(\Delta_0^+)$ is a closed uniformly positive subspace of $\mathfrak{H}$. In a similar way, $\mathfrak{E}(\Delta_0^-) = D\mathfrak{F}(\Delta_0^-)$ is closed and uniformly negative. We easily check that these subspaces are orthogonal and hyperinvariant for A (similar arguments appear in the proofs of Theorems 6.5 and 6.6). Therefore

$$\mathfrak{E}(\Delta_0) = \mathfrak{E}(\Delta_0^+) \oplus \mathfrak{E}(\Delta_0^-)$$

is a Kreĭn subspace of $\mathfrak{H}$ which is hyperinvariant for A. Define $E(\Delta_0)$ to be the projection with range $\mathfrak{E}(\Delta_0)$. Since A is selfadjoint and $\mathfrak{E}(\Delta_0)$ is hyperinvariant for A, $E(\Delta_0)$ is in the double commutant of A. By construction,

$$E(\Delta_0)Df = DF(\Delta_0)f, \qquad f \in \mathfrak{F}(\Delta_0^\pm).$$

It is not hard to see that $E(\Delta_0)Df = 0 = DF(\Delta_0)f$ for f orthogonal to $\mathfrak{F}(\Delta_0^\pm)$, and therefore

$$E(\Delta_0)D = DF(\Delta_0).$$

In other words, $\pi(E(\Delta_0)) = F(\Delta_0)$. The subspaces in the decomposition

$$\mathfrak{H} = \mathfrak{E}(\Delta_0) \oplus \mathfrak{E}(\Delta_0)^\perp$$

reduce A. The restrictions of A to the summands can be shown to have spectra contained in $\overline{\Delta_0}$ and $\overline{\mathbf{R} \setminus \Delta_0}$, respectively.

These constructions apply in particular when Δ_0 is the complement of $\cup_{\nu \in \mathfrak{Z}} I_\nu$, where I_ν, $\nu \in \mathfrak{Z}$, are disjoint closed intervals each of which contains exactly one zero of p in its interior (it is important here to exhaust the zeros of p so that Δ_0 lies at a positive distance from $\mathfrak{Z}$). Then the spectrum of $A|_{\mathfrak{E}(\Delta_0)^\perp}$ is contained in $\cup_{\nu \in \mathfrak{Z}} I_\nu$. The appropriate definition of $E(I_{\nu_k})$ turns out to be the projection obtained from the Riesz functional calculus using a symmetric contour that surrounds I_{ν_k} and no other interval.

Let φ be a bounded Borel measurable function on $\sigma(A)$. By Theorem 6.4, $\sigma(\hat{A}) \subseteq \sigma(A)$. Therefore $\varphi(\hat{A})$ is defined by the functional calculus for selfadjoint operators on a Hilbert space. In this situation, we define

$$\varphi(A)p(A) = D\varphi(\hat{A})D^*.$$

For any $\nu \in \mathfrak{Z}$, define

$$N_\nu = D\varphi_\nu(\hat{A})D^*,$$

where φ_ν is the characteristic function of the singleton $\{\nu\}$. For any two zeros μ and ν of p (possibly $\mu = \nu$),

$$N_\mu N_\nu = D\varphi_\mu(\hat{A})D^* D\varphi_\nu(\hat{A})D^* = D\varphi_\mu(\hat{A})p(\hat{A})\varphi_\nu(\hat{A})D^* = 0.$$

With these definitions, the proof is completed by checking the various assertions. For example, (6.9) holds when φ is the characteristic function of a point in $\mathfrak{Z}$ by the definition of the operators N_ν, $\nu \in \mathfrak{Z}$. Using $\varphi = \chi_{\Delta_0}$ for a Borel set Δ_0 at a positive distance from $\mathfrak{Z}$, we obtain

$$\chi_{\Delta_0}(A)p(A) = D\chi_{\Delta_0}(\hat{A})D^* = DF(\Delta_0)D^* = E(\Delta_0)DD^* = E(\Delta_0)p(A).$$

This verifies (6.9) when $\varphi = \chi_{\Delta_0}$.

For additional details, see Dritschel [1993b]. Also of interest is Bognár [1987], which foreshadows this work. □

Theorem 6.11 *Every definitizable selfadjoint operator $A \in \mathfrak{L}(\mathfrak{H})$ has a maximal dual pair of invariant subspaces.*

Sketch of proof. We first make some reductions. Let p be a definitizing polynomial, and let $\mathfrak{Z}$ be the set of zeros of p. Let $\Delta = \cup_{\nu \in \mathfrak{Z}} I_\nu$, where I_ν, $\nu \in \mathfrak{Z}$, are disjoint closed intervals each of which contains exactly one zero of p in its interior. Writing $\mathfrak{E}(\Delta) = \text{ran } E(\Delta)$ with $E(\cdot)$ as in Theorem 6.10, we obtain a decomposition

$$\mathfrak{H} = \mathfrak{E}(\Delta) \oplus \mathfrak{E}(\Delta)^\perp$$

into Kreĭn subspaces which reduce A. The conclusion is easily obtained for the restriction of A to $\mathfrak{E}(\Delta)^\perp$, and it is thus sufficient to consider the case in which the spectrum of A is contained in $\Delta = \cup_{\nu \in \mathfrak{Z}} I_\nu$.

By considering a decomposition corresponding to the separate intervals that make up Δ, we may assume further that the spectrum of A is contained in a single interval, say I_μ. There is only one zero of p in I_μ, but p may have other zeros not in I_μ. We show that p can be replaced by a monomial. Let λ be a zero of p not in I_μ, and suppose that its order is m. For definiteness, suppose that λ lies to the left of I_μ. Then the spectrum of $(A - \lambda)^m$ is contained in $(0, \infty)$. Write

$p(z) = q(z)(z-\lambda)^m$. Let $q(A) = \tilde{D}\tilde{D}^*$, $\tilde{D} \in \mathfrak{L}(\tilde{\mathfrak{D}}, \mathfrak{H})$, be a polar factorization. Define π as in Corollary 6.3. Since

$$p(A) = (A-\lambda)^m q(A) = \tilde{D}\pi((A-\lambda)^m)\tilde{D}^*$$

is a positive operator on $\mathfrak{H}$ and $\tilde{D}^*$ has dense range, $\pi((A-\lambda)^m)$ is a positive operator on $\tilde{\mathfrak{D}}$. By Theorem 6.4, the spectrum of $\pi((A-\lambda)^m)$ is contained in the spectrum of $(A-\lambda)^m$ and hence in $(0,\infty)$. By Theorem 6.7, $\tilde{\mathfrak{D}}$ is a Hilbert space, and thus $q(A)$ is positive. In the case that λ lies to the right of I_μ, then a similar argument can be applied to $-q(A)$. In this way, we reduce the theorem to the case in which the definitizing polynomial is a monomial. Without loss of generality, we may suppose that A^n is positive for some $n \geq 2$ (the case $n = 1$ is handled by Theorem 6.8).

For any selfadjoint operator $C \in \mathfrak{L}(\mathfrak{H})$ there is a neutral invariant subspace $\mathfrak{N}$ which is maximal in the sense that there is no other neutral invariant subspace strictly containing it. If J is a fundamental symmetry on $\mathfrak{H}$, then $\mathfrak{K} = \mathfrak{N} + J\mathfrak{N}$ is a Kreĭn subspace of $\mathfrak{H}$. The compression $\tilde{C}$ of C to $\tilde{\mathfrak{H}} = \mathfrak{K}^\perp$ is selfadjoint, and if $\mathfrak{M}$ is an invariant subspace for $\tilde{C}$, then $\mathfrak{M}+\mathfrak{N}$ is invariant for C. In particular, $\tilde{C}$ has no nonzero neutral invariant subspace. Also, $\mathfrak{M}+\mathfrak{N}$ is maximal positive or maximal negative in $\mathfrak{H}$ if $\mathfrak{M}$ is maximal positive or maximal negative in $\tilde{\mathfrak{H}}$. Furthermore, if $C^n \geq 0$, then $\tilde{C}^n \geq 0$. Hence the problem is reduced to the case where A has no nonzero neutral invariant subspace. Then both ker A and (ker $A)^\perp$ are invariant under A, and so is their intersection, which is a neutral subspace. Thus

$$\ker\ A \cap \overline{\operatorname{ran} A} = \{0\}.$$

This can be used to show that ker A+ran A is dense in $\mathfrak{H}$, and that ker $A^k =$ ker A and $\overline{\operatorname{ran} A^k} = \overline{\operatorname{ran} A}$, $k \geq 1$.

If n is even, then

$$\left\langle A^{n/2}f, A^{n/2}f\right\rangle \geq 0$$

for all $f \in \mathfrak{H}$, and so $\overline{\operatorname{ran} A}$ is a positive subspace. Also ker A is either strictly positive or strictly negative, since A has no nonzero invariant neutral subspace. If ker A is positive, then since ker A + ran A is dense in $\mathfrak{H}$, $\mathfrak{H}$ is a Hilbert space. The result is trivially true in this case, since then the whole space $\mathfrak{H}$ is maximal positive and invariant for A. If ker A is negative, then it is maximal negative, since (ker $A)^\perp = \overline{\operatorname{ran} A}$ is positive in $\mathfrak{H}$ (this follows from the graph representation). It then follows that $\overline{\operatorname{ran} A}$ is maximal positive. In either case, we have a maximal dual pair of invariant subspaces for A.

If n is odd, then for all $f \in \mathfrak{H}$, $g \in$ ker A,

$$\left\langle A(A^{(n-1)/2}f + g), A^{(n-1)/2}f + g\right\rangle \geq 0,$$

and since the vectors $A^{(n-1)/2}f + g$ span a dense subspace of $\mathfrak{H}$, it follows that $A \geq 0$. The result is then true by Theorem 6.8. □

C. Compact perturbations of definitizable operators

Some of the results for definitizable operators discussed in §B are true for other classes including certain compact perturbations of definitizable operators. This area has also been intensively investigated. For example, see Jonas [1982], [1988], [1993], Jonas and Langer [1979], [1983], Langer and Najman [1983], and Najman [1980]. We sketch the following result from Dritschel [to appear].

Compact operators and subclasses such as the Schatten class $\mathfrak{C}_p$ on a Kreĭn space are defined as in the Hilbert space case. The definitions can be made relative to any Hilbert space norm for the Kreĭn space, and since any two such norms are equivalent the definitions are independent of choice of norm.

Theorem 6.12 *Let $\mathfrak{H}$ be a Kreĭn space, A a definitizable selfadjoint operator in $\mathfrak{L}(\mathfrak{H})$, K a compact operator in $\mathfrak{C}_p$, $1 \leq p < \infty$. Then if $T = A + K$ is not a constant multiple of the identity, there is a proper closed subspace of $\mathfrak{H}$ which is hyperinvariant for T.*

To prove this we use an idea in Radjavi and Rosenthal [1973], Chapter 6, which is based in part on work of J. T. Schwartz. The following theorems will be needed. They appear in Radjavi and Rosenthal [1973], pp. 7, 97, 158, for Hilbert spaces, and they apply without change when the underlying spaces are Kreĭn spaces.

Weyl's Theorem. *If $A \in \mathfrak{L}(\mathfrak{H})$ and K is compact, then*

$$\sigma(A + K) \subseteq \sigma(A) \cup \sigma_p(A + K).$$

Lomonosov's Theorem. *Every operator commuting with a nonzero compact operator and which is not a constant multiple of the identity has a proper closed hyperinvariant subspace.*

Theorem on Exposed Arcs. *Let $T \in \mathfrak{L}(\mathfrak{H})$ be such that $\sigma(T)$ contains an exposed arc Γ and let k be a positive integer. Suppose that for each point $z_0 \in \Gamma$ and each closed line segment Λ which meets $\sigma(T)$ only at $\{z_0\}$ and which is not tangent to Γ, there exists a constant $\tilde{C}$ such that*

$$\left\|(T - z)^{-1}\right\| \leq \exp\left(\tilde{C}\,|z - z_0|^{-k}\right)$$

for all $z \in \Lambda$ not equal to z_0. Then T has a proper closed hyperinvariant subspace.

A few explanations are needed to make the last theorem clear. Some choice of operator norm is understood in the resolvent estimate. A **smooth Jordan arc** is (the range of) a one-to-one function $z(t) = x(t) + iy(t)$ mapping the interval $(0, 1)$ to the complex numbers which is twice differentiable on $(0, 1)$. A bounded operator T is said to have an **exposed arc** Γ in its spectrum if there is an open disk Δ such that $\Delta \cap \sigma(T) = \Gamma$ is a smooth Jordan arc.

Sketch of proof of Theorem 6.12. It may be assumed that $\sigma_p(T) = \emptyset$, since if T has an eigenvalue the conclusion is elementary.

As a first case, suppose that $\sigma(T)$ contains more than one point. We may take $\sigma(T)$ to be connected, because otherwise a nontrivial Riesz projection exists, and the range of such a projection is a proper closed hyperinvariant subspace for T. By Weyl's theorem $\sigma(T) \subseteq \sigma(A)$, and so, by Theorem 6.9, $\sigma(T)$ must be a closed and bounded interval with nonempty interior. Thus the spectrum of T has an exposed arc. The resolvent estimate required to apply the Theorem on Exposed Arcs holds. In fact, in the situation required by that theorem,

$$\left\|(T-z)^{-1}\right\| \le \exp\left(\tilde{C}\,|z-z_0|^{-p(k_A+1)-1}\right),$$

where k_A is the degree of a definitizing polynomial for A. The derivation of this estimate follows a method in Radjavi and Rosenthal [1973] (proof of Lemma 1.7) and is justified by estimates in Dunford and Schwartz which are cited in the same place. The only difference is that we use an estimate of the form

$$\left\|(A-z)^{-1}\right\| \le M|z-z_0|^{-k_A-1},$$

which follows from the spectral theory for definitizable selfadjoint operators (see Langer [1982]). Thus the Theorem on Exposed Arcs applies, and the result follows in the case that $\sigma(T)$ contains more than one point.

Before taking up the case in which the spectrum of T is a single point, we make a preliminary observation that if $C \in \mathfrak{L}(\mathfrak{K})$ is a positive operator on a Kreĭn space $\mathfrak{K}$ and $\sigma(C) = \{\lambda\}$, then either $\lambda \neq 0$ and $C = \lambda$ or $\lambda = 0$ and $C^2 = 0$. For C has a polar factorization $C = DD^*$, $D \in \mathfrak{L}(\mathfrak{D}, \mathfrak{K})$. If $\hat{C} = D^*D$, then by Theorem 6.4, $\sigma(\hat{C}) \subseteq \sigma(C) = \{\lambda\}$. Thus $\hat{C}$ is a Hilbert space selfadjoint operator with spectrum $\{\lambda\}$, and so $\hat{C} = \lambda$. Hence

$$CD = \lambda D.$$

If $\lambda \neq 0$, then C is invertible, the range of D is all of $\mathfrak{K}$, and $C = \lambda$. If $\lambda = 0$, then $C^2 = D\hat{C}D^* = 0$, which proves the claim.

We also use some properties of the Calkin algebra in the Kreĭn space setting. The Calkin algebra is defined as $\mathfrak{L}(\mathfrak{H})/\mathfrak{C}(\mathfrak{H})$, where $\mathfrak{C}(\mathfrak{H})$ is the set of compact operators on $\mathfrak{H}$. The quotient mapping $\varphi : \mathfrak{L}(\mathfrak{H}) \to \mathfrak{L}(\mathfrak{H})/\mathfrak{C}(\mathfrak{H})$ takes any fundamental symmetry on $\mathfrak{H}$ into a selfadjoint and unitary element of the algebra. If we represent the algebra as a space of operators on a Hilbert space $\mathfrak{K}$, we induce in this way a fundmental symmetry and hence a Kreĭn space structure for $\mathfrak{K}$. In this representation, the image of a Kreĭn space selfadjoint or positive operator on $\mathfrak{H}$ is again Kreĭn space selfadjoint or positive.

Now suppose that the spectrum of T reduces to a point, say $\sigma(T) = \{\lambda\}$. It can be assumed that $\mathfrak{H}$ is infinite dimensional, since otherwise the result is clear. For any polynomial p, $p(T) = p(A) + S$ with S compact, and hence $\varphi(p(A)) = \varphi(p(T))$. The spectrum of $p(T)$ is $\{p(\lambda)\}$, and since the Calkin algebra is nontrivial under our assumptions, the spectrum of $\varphi(p(T))$ is nonempty. Since $\sigma(\varphi(p(T))) \subseteq \sigma(p(T))$, it follows that $\sigma(\varphi(p(T))) = \{p(\lambda)\}$. Thus

$$\sigma(\varphi(p(A))) = \sigma(\varphi(p(T))) = \{p(\lambda)\}.$$

The choice $p(z) = z$ shows that $\sigma(\varphi(A)) = \{\lambda\}$. Since $\varphi(A)$ is selfadjoint, λ is real. Next take p to be a definitizing polynomial for A. Then $\varphi(p(A))$ is positive and its spectrum is $\{p(\lambda)\}$. By the preliminary observation about positive Kreĭn space operators with such spectra, either $p(A) - p(\lambda)$ or $p(A)^2$ is compact, since in both cases the image under φ is zero. But then T is polynomially compact, and the result follows from Lomonosov's theorem. □

Theorem 6.12 provides a natural framework for studying invariant subspace problems for Hilbert space operators as well. The next theorem generalizes a result of Radjavi and Rosenthal [1974].

Theorem 6.13 *Suppose that A and B are bounded operators on a Hilbert space $\mathfrak{K}$ with A selfadjoint and $B = C + K$, where $C \geq 0$ and K is in $\mathfrak{C}_p$, $1 \leq p < \infty$. Assume that AB is not a constant multiple of the identity, and that either B is selfadjoint or both A and B have dense range and zero kernel. Then AB and BA have proper closed hyperinvariant subspaces.*

Proof. For the case in which B is selfadjoint, if AB fails to have dense range or zero kernel, it is not hard to produce proper closed hyperinvariant subspaces for AB and $BA = (AB)^*$. Hence we may assume that AB has dense range and zero kernel. Then by the selfadjointness of A and B, both A and B have dense range and zero kernel. So we assume that this is the case.

Use the Bognár-Krámli factorization (Theorem 1.1) to write $A = DD^*$, where $D \in \mathfrak{L}(\mathfrak{D}, \mathfrak{K})$ for some Kreĭn space $\mathfrak{D}$ and $\ker\ D = \{0\}$. Then

$$(AB)D = D(D^*BD) \qquad\qquad D^*B(AB) = (D^*BD)D^*B,$$

and

$$(BA)BD = BD(D^*BD) \qquad\qquad D^*(BA) = (D^*BD)D^*.$$

Our hypotheses imply that the operators D, D^*, BD, D^*B have dense range and zero kernel. Therefore the relations show that AB and BA are both quasisimilar to D^*BD. Since $DD^*BD = ABD$ and AB is not a constant multiple of the identity, D^*BD is not a constant multiple of the identity. Now

$$D^*BD = D^*CD + D^*KD,$$

where D^*CD is a positive selfadjoint operator on the Kreĭn space $\mathfrak{D}$ and D^*KD in the same Schatten p-class as K. Hence by Theorem 6.12, D^*BD has a proper closed hyperinvariant subspace. By a theorem of Sz.-Nagy and Foiaş (see Radjavi Rosenthal [1973], p. 108), this implies that AB and BA also have proper closed hyperinvariant subspaces. □

Theorem 6.14 *Suppose that A and B are bounded operators on a Hilbert space $\mathfrak{K}$ with A selfadjoint and B the sum of a positive operator and a compact operator in $\mathfrak{C}_p$, $1 \leq p < \infty$. Assume that A and B are not both zero. Set*

$$T = \begin{pmatrix} 0 & A \\ B & 0 \end{pmatrix}.$$

Then T has a proper closed hyperinvariant subspace.

Proof. The nontrivial case is when T has zero kernel and dense range. Then A and B both have dense range and zero kernel. We may assume that AB is not a constant multiple of the identity. For if $AB = c$, then $ABA = cA$ and so $BA = c$; thus $T^2 - c = 0$, and the result is elementary in this case.

By the identities

$$\begin{pmatrix} 0 & A \\ B & 0 \end{pmatrix}\begin{pmatrix} 0 & 1 \\ B & 0 \end{pmatrix} = \begin{pmatrix} 0 & 1 \\ B & 0 \end{pmatrix}\begin{pmatrix} 0 & 1 \\ AB & 0 \end{pmatrix}$$

and

$$\begin{pmatrix} 0 & A \\ AB & 0 \end{pmatrix} \begin{pmatrix} 0 & A \\ B & 0 \end{pmatrix} = \begin{pmatrix} 0 & 1 \\ AB & 0 \end{pmatrix} \begin{pmatrix} 0 & A \\ AB & 0 \end{pmatrix},$$

T is quasisimilar to

$$R = \begin{pmatrix} 0 & 1 \\ AB & 0 \end{pmatrix}.$$

We show that R has a proper closed hyperinvariant subspace. By Theorem 6.13, AB has a proper closed hyperinvariant subspace $\mathfrak{N}$. We show that $\mathfrak{N} \oplus \mathfrak{N}$ is hyperinvariant for R. In fact, suppose that

$$X = \begin{pmatrix} X_1 & X_2 \\ X_3 & X_4 \end{pmatrix}$$

commutes with R. Then

$$X_1 = X_4, \qquad X_4 AB = ABX_1, \qquad X_3 = ABX_2, \qquad X_2 AB = X_3.$$

So X_1, X_2, X_3, and X_4 all commute with AB and consequently leave $\mathfrak{N}$ invariant. Thus X leaves $\mathfrak{N} \oplus \mathfrak{N}$ invariant, and the result is proved. □

The operator matrix which appears in Theorem 6.14 may also be viewed as a compact perturbation of a Kreĭn space selfadjoint operator (see (6.7)). Such operator matrices are considered by Jonas [1993] in the study of the Klein-Gordon equation.

In Theorem 6.13, if $A = J$ is selfadjoint and unitary and B is selfadjoint, then AB is Kreĭn space selfadjoint relative to the Kreĭn space structure induced by J. In view of the very general nature of Kreĭn space selfadjoint operators (again see (6.7)), this points out an apparent need for some strong hypothesis in order to say anything meaningful about structural properties of products of Hilbert space selfadjoint operators.

Addendum: Nondefinitizable operators

There exist selfadjoint operators which are not definitizable and, in fact, which are not in the commutant of any nonzero positive operator. Examples can be constructed which further have a rich invariant subspace structure. Such examples underscore the difference between Hilbert space and Kreĭn space selfadjoint operators.

Let T be a normal operator on a Hilbert space $\mathfrak{K}$. Let $\mathfrak{H} = \mathfrak{K} \oplus \mathfrak{K}$ viewed as a Kreĭn space with fundamental symmetry

$$J = \begin{pmatrix} 0 & 1 \\ 1 & 0 \end{pmatrix}.$$

The operator defined on $\mathfrak{H}$ by

$$A = \begin{pmatrix} T & 0 \\ 0 & T^\times \end{pmatrix}$$

is a special case of (6.7) and is therefore selfadjoint. Observe that the spectrum of A is the union of the spectrum of T and the spectrum of $T^\times$, which consists of conjugates of all points in the spectrum of T. Assume that the spectrum of T does not meet the real axis. Then the spectrum of A does not meet the real axis either. Now suppose that $B \in \mathfrak{L}(\mathfrak{H})$ is a positive operator and $A \in \{B\}'$. Choose a polar factorization $B = DD^*$, $D \in \mathfrak{L}(\mathfrak{D}, \mathfrak{H})$. Let $\hat{A} = \pi(A)$, where π is the mapping defined in Corollary 6.3. Then $\mathfrak{D}$ is a Hilbert space since B is positive, and therefore $\hat{A} \in \mathfrak{L}(\mathfrak{D})$ is Hilbert space selfadjoint by Theorem 6.4. By the same result, $\sigma(\hat{A}) \subseteq \sigma(A) \cap \mathbf{R}$. This means that $\hat{A}$ has empty spectrum. It follows that $\mathfrak{D} = \{0\}$, and this implies that $B = 0$. In other words, A does not commute with any nonzero positive operator.

More generally, the same conclusion follows if $\sigma(T) \cap \mathbf{R}$ is a finite set and these points are not eigenvalues for T. For in the preceding notation, $\sigma(A) \cap \mathbf{R} = \sigma(T) \cap \mathbf{R}$, and no point in this set is an eigenvalue for A. Since $\sigma(\hat{A}) \subseteq \sigma(A) \cap \mathbf{R}$, the spectrum of $\hat{A}$ consists of eigenvalues, and by Theorem 6.4, $\sigma_p(\hat{A}) \subseteq \sigma_p(A) \cap \mathbf{R}$. Therefore the spectrum of $\hat{A}$ is empty, and the conclusion follows as before.

Bibliography

Alpay, D., Dijksma, A., van der Ploeg, J. and de Snoo, H. S. V. [1992] *Holomorphic operators between Krein spaces and the number of squares of associated kernels*, Oper. Theory: Adv. Appl., vol. 59, Birkhäuser, Basel-Boston, pp. 11–29.

Alpay, D., Dijksma, A., Rovnyak, J. and de Snoo, H. S. V. [in preparation] *Schur functions, operator colligations, and reproducing kernel Pontryagin spaces.*

Alpay, D., Dijksma, A. and de Snoo, H. S. V. [1994] *Reproducing kernel Pontryagin spaces, operator colligations and operator models for quasicontractions*, preprint.

Alpay, D. and Dym, H. [1986] *On applications of reproducing kernel spaces to the Schur algorithm and rational J unitary factorization*, Oper. Theory: Adv. Appl., vol. 18, Birkhäuser, Basel-Boston, pp. 89–159; MR 89g:46051.

——— [1992] *On reproducing kernel spaces, the Schur algorithm, and interpolation in a general class of domains*, Oper. Theory: Adv. Appl., vol. 59, Birkhäuser, Basel-Boston, pp. 30–37; MR 94j:46034.

Andô, T. [1979] *Linear Operators on Krein Spaces*, Hokkaido University, Research Institute of Applied Electricity, Division of Applied Mathematics, Sapporo.

——— [1990] *de Branges Spaces and Analytic Operator Functions*, Hokkaido University, Research Institute of Applied Electricity, Division of Applied Mathematics, Sapporo.

Arov, D. Z. and Grossman, I. V. [1983] *Scattering matrices in the theory of extensions of isometric operators*, Dokl. Akad. Nauk. SSSR **270**, 17–20; English transl. in Soviet Math. Dokl. **27** (1983), 518–522; MR 85c:47008.

——— [1992] *Scattering matrices in the theory of unitary extensions of isometric operators*, Math. Nachr. **157**, 105–123; MR 94i:47019.

Arsene, Gr., Constantinescu, T. and Gheondea, A. [1987] *Lifting of operators and prescribed numbers of negative squares*, Michigan Math. J. **34**, 201–216; MR 88j:47008.

Arsene, Gr. and Gheondea, A. [1982] *Completing matrix contractions*, J. Operator Theory **7**, 179–189; MR 83i:47010.

Azizov, T. Ya. and Iokhvidov, I. S. [1989] *Linear Operators in Spaces with an Indefinite Metric*, "Nauka", Moscow, 1986; MR 88g:47070; Engl. transl., Wiley, New York; MR 90j:47042.

Bakonyi, M. and Constantinescu, T. [1992] *Schur's Algorithm and Several Applications*, Pitman Research Notes, vol. 261, Longman Scientific and Technical, Harlow (co-publisher, Wiley, New York); MR 94b:47022.

Ball, J. A. and Cohen, N. [1991] *de Branges-Rovnyak operator models and systems theory: a survey*, Oper. Theory: Adv. Appl., vol. 50, Birkhäuser, Basel-Boston, pp. 93–136; MR 93a:47011.

Ball, J. A., Gohberg, I. and Rodman, L. [1990] *Interpolation of Rational Matrix Functions*, Oper. Theory: Adv. Appl., vol. 45, Birkhäuser, Basel-Boston; MR 92m:47027.

Ball, J. A. and Helton, J.W. [1983] *A Beurling-Lax theorem for the Lie group $U(m, n)$ which contains most interpolation theory*, J. Operator Theory **9**, 107–142; MR 84m:47046.

Ball, J. A. and Kriete, T. L. III [1987] *Operator-valued Nevanlinna-Pick kernels and the functional models for contraction operators*, Integral Equations and Operator Theory **10**, 17–61; MR 88a:47013.

Bart, H., Gohberg, I. and Kaashoek, M. A. [1979] *Minimal factorization of matrix and operator functions*, Oper. Theory: Adv. Appl., vol. 1, Birkhäuser, Basel-Boston; MR 81a:47001.

Bognár, J. [1974] *Indefinite Inner Product Spaces*, Ergeb. Math. Grenzgeb., Bd. 78, Springer-Verlag, Berlin-New York; MR 57#7125.

_____ [1987] *An approach to the spectral decomposition of J-positizable operators*, J. Operator Theory **17**, 309–326 MR 88g:47071.

de Branges, L. [1988a] *Complementation in Krein spaces*, Trans. Amer. Math. Soc. **305**, 277–291; MR 89c:46034.

_____ [1988b] *Krein spaces of analytic functions*, J. Funct. Anal. **81**, 219–259; MR 90c:47060.

_____ [1991] *A construction of Krein spaces of analytic functions*, J. Funct. Anal. **98**, 1–41; MR 92k:46031.

_____ [1994] *Factorization in Krein spaces*, J. Funct. Anal. **124**, 228–262.

de Branges, L. and Rovnyak, J. [1966a] *Square Summable Power Series*, Holt, Rinehart and Winston, New York; MR 35#5909.

_____ [1966b] *Canonical models in quantum scattering theory*, Perturbation Theory and its Applications in Quantum Mechanics (Proc. Adv. Sem. Math. Res. Center, U.S. Army, Theoret. Chem. Inst., Univ. of Wisconsin, Madison, Wis., 1965), pp. 295–392; MR 39#6109.

Brodskiĭ, M. S. [1978] *Unitary operator colligations and their characteristic functions*, Uspekhi Mat. Nauk **33**, no. 4, 141–168; English transl. Russian Math. Surveys **33** (1978), no. 4, 159–191; MR 80e:47010.

Christner, G. [1993] *Applications of the Extension Properties of Operators on Krein Spaces*, Dissertation, University of Virginia.

Christner, G., Li, Kin Y. and Rovnyak, J. [1994] *Julia Operators and coefficient problems*, Oper. Theory Adv. Appl., vol. 73, Birkhäuser, Basel-Boston, pp. 138–181.

Christner, G. and Rovnyak, J. [to appear] *Julia operators and the Schur algorithm*, Proceedings of International Conference on Operator Theory and Harmonic Analysis (Caracas, 1994).

Constantinescu, T. and Gheondea, A. [1989] *Minimal signature in lifting of operators.* I, J. Operator Theory **22**, 345–367; MR 91b:47013.

——— [1992] *Minimal signature in lifting of operators.* II, J. Funct. Anal. **103**, 317–351; MR 93c:47041.

——— [1993] *Elementary rotations of linear operators in Krein spaces*, J. Operator Theory **29**, 167–203.

Conway, J. B. [1990] *A Course in Functional Analysis*, 2nd ed., Springer-Verlag, Berlin-New York; MR 91e:46001.

Ćurgus, B., Dijksma, A., Langer, H. and de Snoo, H. S. V. [1988] *Characteristic functions of unitary colligations and of bounded operators in Krein spaces*, Oper. Theory: Adv. Appl., vol. 41, Birkhäuser, Basel-Boston, pp. 125–152; MR 91c:47020.

Davis, Ch. [1970] *J-unitary dilation of a general operator*, Acta Sci. Math. (Szeged) **31**, 75–86; MR 41#9032.

——— [1976] *An extremal problem for extensions of a sesquilinear form*, Linear Algebra and Appl. **13**, 91–102; MR 52#15068.

Dijksma, A., Dritschel, M. A., Marcantognini, S. and de Snoo, H. S. V. [1993] *The commutant lifting theorem for contractions on Krein spaces*, Oper. Theory: Adv. Appl., vol. 61, Birkhäuser, Basel-Boston, pp. 65–83.

Dijksma, A., Langer, H. and de Snoo, H. S. V. [1986] *Characteristic functions of unitary operator colligations in* Π_κ*-spaces*, Oper. Theory: Adv. Appl., vol. 19, Birkhäuser, Basel-Boston, pp. 125–194; MR 89a:47053.

Dijksma, A., Marcantognini, S. A. M. and de Snoo, H. S. V. [1994] *A Schur type analysis of the minimal unitary Hilbert space extensions of a Kreĭn space isometry whose defect subspaces are Hilbert spaces*, Z. Anal. Anwendungen **13**, 233–260.

Dritschel, M. A. [1993a] *The essential uniqueness property for linear operators in Kreĭn spaces*, J. Funct. Anal. **118**, 198–248; MR 94j:47051.

——— [1993b] *A method for constructing invariant subspaces for some operators on Krein spaces*, Oper. Theory: Adv. Appl., vol. 61, Birkhäuser, Basel-Boston, pp. 85–113.

——— [1993c] *Commutant lifting when the intertwining operator is not necessarily a contraction*, unpublished manuscript.

——— [to appear] *Compact perturbations of operators on Kreĭn spaces*, Proceedings of International Conference on Operator Theory and Harmonic Analysis (Caracas, 1994).

Dritschel, M. A. and Rovnyak, J. [1990] *Extension theorems for contraction operators on Kreĭn spaces*, Oper. Theory: Adv. Appl., vol. 47, Birkhäuser, Basel-Boston, pp. 221–305; MR 92m:47068.

_____ [1991] *Julia operators and complementation in Kreĭn spaces*, Indiana Univ. Math. J. **40**, 885–901; MR 93b:46042.

Dubovoj, V. K., Fritzsche, B. and Kirstein, B. [1992] *Matricial Version of the Classical Schur Problem*, Teubner-Texte zur Mathematik, vol. 129, B. G. Teubner Verlagsgesellschaft, Stuttgart; MR 93e:47021.

Dym, H. [1989] *J Contractive Matrix Functions, Reproducing Kernel Hilbert Spaces, and Interpolation*, CBMS Regional Conference Series in Mathematics, vol. 71, Amer. Math. Soc., Providence, RI; MR 90g:47003.

Foiaş, C. and Frazho, A. [1990] *The Commutant Lifting Approach to Interpolation Problems*, Oper. Theory: Adv. Appl., vol. 44, Birkhäuser, Basel-Boston; MR 92k:47033.

Gohberg, I., Lancaster, P. and Rodman, L. [1983] *Matrices and Indefinite Scalar Products*, Oper. Theory: Adv. Appl., vol. 8, Birkhäuser, Basel-Boston; MR 87j:15001.

Halmos, P. R. [1951] *Introduction to Hilbert Spaces*, Chelsea, New York; MR 13,563a.

Hassibi, B., Sayed, A. H. and Kailath, T. [1994] *Kalman filtering in Krein spaces-Parts* I*: Theory;* II*: Applications*, preprint.

Helton, J. W. with the assistance of J. A. Ball, C. R. Johnson and J. N. Palmer [1987] *Operator Theory, Analytic Functions, Matrices, and Electrical Engineering*, CBMS Regional Conference Series in Mathematics, vol. 68, Amer. Math. Soc., Providence, RI; MR 89f:47001.

Herglotz, G., Schur, I. et al. [1991] *Ausgewählte Arbeiten zu den Ursprüngen der Schur-Analysis* (B. Fritzsche and B. Kirstein, eds.), B. G. Teubner, Stuttgart; MR 93d:01095.

Iokhvidov, I. S., Kreĭn, M. G. and Langer, H. [1982] *Introduction to the Spectral Theory of Operators in Spaces with an Indefinite Metric*, Akademie-Verlag, Berlin; MR 85g:47050.

Ivanchenko, T. [1979] *The Schur problem for J-contractive matrix functions*, Technical Report, 31/7/79, VINITI #2868–79.

Jonas, P. [1981] *On the functional calculus and the spectral function for definitizable operators in Kreĭn space*, Beiträge Anal. **16**, 121–135; MR 83i:47047.

_____ [1982] *Compact perturbations of definitizable operators.* II, J. Operator Theory **8**, 3–18; MR 84d:47046.

_____ [1988] *On a class of selfadjoint operators in Krein space and their compact perturbations*, Integral Equations and Operator Theory **11**, 351–384; MR 89f:47052.

_____ [1993] *On the spectral theory of operators associated with perturbed Klein-Gordon and wave type equations*, J. Operator Theory **29**, 207–224.

Jonas, P. and Langer, H. [1979] *Compact perturbations of definitizable operators*, J. Operator Theory **2**, 63–77; MR 81c:47039.

_____ [1983] *Some questions in the perturbation theory of J-nonnegative operators in Kreĭn spaces*, Math. Nachr. **114**, 205–226; MR 85j:47045.

Kreĭn , M. G. and Langer, H. [1973] *Über die Q-Funktion eines π-hermiteschen Operators im Raume* Π_κ, Acta Sci. Math. (Szeged) **34**, 191–230; MR 47#7504.

——— [1977] *Über einige Fortsetzungsprobleme, die eng mit der Theorie hermitescher Operatoren im Raume* Π_κ *zusammenhangen.* I. *Einige Funktionenklassen und ihre Darstellungen*, Math. Nachr. **77**, 187–236; MR 57#1173; II. *Verallgemeinerte Resolventen, u-Resolventen und ganze Operatoren*, J. Funct. Anal. **30**, 390–447; MR 80h:47045; III. *Indefinite analogues of the Hamburger and Stieltjes moment problems. Part* I, Beiträge Anal. **14**, 25–40; MR 83b:47047a; III. *Indefinite analogues of the Hamburger and Stieltjes moment problems. Part* II, Beiträge Anal. **15**, 27–45; MR 83b: 47047b; IV. *Continuous analogues of orthogonal polynomials on the unit circle with respect to an indefinite weight and related continuation problems for some classes of functions*, J. Operator Theory **13**, 299–417; MR 87h:47084.

——— [1981] *Some propositions on analytic matrix functions related to the theory of operators in the space* Π_κ, Acta Sci. Math. (Szeged) **43**, 181–205; MR 82i:47053.

Langer, H. [1965] *Spektraltheorie Linearer Operatoren in J-Räumen und Einige Anwendungen auf die Schar* $L(\lambda) = \lambda^2 I + \lambda B + C$, Habilitationsschrift, Fakultät für Mathematik und Naturwissenschaften der Technischen Universität Dresden, Dresden.

——— [1971] *Invariante Teilräume definisierbarer J-selbstadjungierter Operatoren*, Ann. Acad. Sci. Fenn. Ser. A I Math., no. 475, 23 pages; MR 44#4553.

——— [1982] *Spectral functions of definitizable operators in Krein spaces*, Functional Analysis (Dubrovnik, 1981) Lecture Notes in Math., vol. 948, Springer-Verlag, Berlin-New York, pp. 1–46; MR 84g:47034.

Langer, H. and Najman, B. [1983] *Perturbation theory for definitizable operators in Krein spaces*, J. Operator Theory **9**, 297–317; MR 85e:47053.

McEnnis, B. [1981] *Models for operators with bounded characteristic function*, Acta Sci. Math. (Szeged) **43**, 71–90; MR 83b:47012.

——— [1990] *Shifts on Krein spaces*, Operator Theory: Operator Algebras and Applications, Part 2 (Durham, NH, 1988), Amer. Math. Soc., Providence, RI, pp. 201–211; MR 91j:47038.

Marcantognini, S. A. M. [1990] *Unitary colligations of operators in Krein spaces*, Integral Equations and Operator Theory **13**, 701–727; MR 92f:47010.

——— [1992] *The commutant lifting theorem in the Krein space setting: a proof based on the coupling method*, Indiana Univ. Math. J. **41**, 1303–1314; MR 93k:47009.

Najman, B. [1980] *Perturbation theory for selfadjoint operators in Pontrjagin spaces.* I, Glas. Mat. Ser. III **15(35)**, 351–371; MR 82e:47048.

Nikol′skiĭ , N. K. and Vasyunin, V. I. [1986] *Notes on two function models*, The Bieberbach Conjecture (West Lafayette, Ind., 1985), Amer. Math. Soc., Providence, RI, pp. 113–141; MR 91c:47017.

——— [1989] *A unified approach to function models, and the transcription problem*, Oper. Theory: Adv. Appl., vol. 41, Birkhäuser, Basel-Boston, pp. 405–434; MR 91c:47017.

Parrott, S. [1978] *On a quotient norm and the Sz.-Nagy-Foiaş lifting theorem*, J. Funct. Anal. **30**, 311–328; MR 81h:47006.

Phillips, R. S. [1961] *The extension of dual subspaces invariant under an algebra*, Proc. Internat. Sympos. Linear Spaces (Jerusalem, 1960), Jerusalem Academic Press, Jerusalem (Pergamon, Oxford), pp. 366–398; MR 24#A3512.

Pontryagin, L. S. [1944] *Hermitian operators in spaces with indefinite metric*, Izv. Akad. Nauk SSSR Ser. Mat. **8**, 243–280; MR 6,273.

Radjavi, H. and Rosenthal, P. [1973] *Invariant Subspaces*, Ergeb. Math. Grenzgeb., Bd. 77, Springer-Verlag, New York-Heidelberg; MR 51#3924.

——— [1974] *Invariant subspaces for products of Hermitian operators*, Proc. Amer. Math. Soc. **43**, 483–484; MR 58#23663.

Rosenblum, M. and Rovnyak, J. [1985] *Hardy Classes and Operator Theory*, Oxford Mathematical Monographs, Oxford University Press, New York; MR 87e:47001.

Rovnyak, J. [1994] Review of *Schur's Algorithm and Several Applications* by M. Bakonyi and T. Constantinescu, Bull. Amer. Math. Soc. (N.S.) **30**, 270–276.

Rudin, W. [1974] *Real and Complex Analysis*, 2nd ed., McGraw-Hill, New York; MR 49#8783.

Sarason, D. [1994] *Sub-Hardy Hilbert Spaces in the Unit Disk*, University of Arkansas Lecture Notes in the Mathematical Sciences, Wiley, New York.

Schur, I. [1917] *Über Potenzreihen, die im Innern des Einheitskreises beschränkt sind* I, J. Reine Angew. Math. **147**, 205–232; II, ibid. **148** (1918), Gesammelte Abhandlungen, vol. II, nos. 29, 30.

Shmul′yan, Yu. L. [1967] *Division in the class of J-expansive operators*, Mat. Sb. (N.S.) **74 (116)**, 516–525; Math. USSR-Sbornik **3**, 471–479; MR 37#784.

Sorjonen, P. [1975] *Pontrjaginräume mit einem reproduzierenden kern*, Ann. Acad. Sci. Fenn. Ser. A I Math., no. 594, 30 pages; MR 53#8875.

Sz.-Nagy, B. and Foiaş, C. [1970] *Harmonic Analysis of Operators on Hilbert Space*, North Holland, New York; MR 43#947.

Yang, A. [1994] *A construction of unitary linear systems*, Integral Equations and Operator Theory **19**, 477–499.

LECTURE SERIES 4

STATE SPACE THEORY OF RATIONAL MATRIX FUNCTIONS AND APPLICATIONS

M. A. Kaashoek
Faculteit Wiskunde en Informatica
Vrije Universiteit
De Boelelaan 1081a
1081 HV Amsterdam, The Netherlands

Introduction

During the last ten to fifteen years a new method has been developed to deal with analytic problems involving rational matrix functions. This method, which nowadays is called the *state space method*, has its roots in the mathematical system theory of the sixties and is based on the fact that rational matrix functions may be realized as transfer functions of time-invariant input-output systems with a finite dimensional state space. In particular, if $W(\lambda)$ is a rational $m \times r$ matrix function which is analytic at infinity, then $W(\lambda)$ can be written in the form

$$W(\lambda) = D + C(\lambda - A)^{-1}B, \tag{$*$}$$

where A is a square (complex) matrix of order n, say, and B, C and D are matrices of sizes $n \times r$, $m \times n$ and $m \times r$, respectively.

Representations of the type $(*)$ have proved to be very useful. They can be used to reduce problems involving rational matrix functions to linear algebra problems involving only the four matrices appearing in $(*)$, and on the basis of this connection a number of new constructive operator theory methods have been developed. The connection has also been and still is a source of new questions about finite and infinite dimensional operators.

In the present lecture notes some of the highlights of this state space theory are presented. The emphasis is on regular rational matrix functions, i.e., square matrix functions of which the determinant does not vanish identically. Two kinds of applications are considered in detail, namely the ones involving problems of canonical factorization and/or problems of interpolation.

The text consists of six chapters. In the first chapter the state space method is introduced and used to construct canonical factorizations. The results are illustrated on the problem of inverting a system of Wiener-Hopf integral equations with a rational matrix symbol. The second chapter deals with rational matrix functions that are J-unitary with respect to the imaginary axis. Such functions are analyzed in terms of minimal realizations. The state space theorem for canonical factorization of the first chapter is used to produce explicit J-inner-outer factorizations for rational matrix functions that have their poles in the open left half plane. The third chapter analyzes the zeros of a regular rational matrix function, and the corresponding eigenvectors and generalized eigenvectors in terms of null functions. Canonical right and left null pairs are introduced, and their connection with minimal realizations of the Laurent principal parts is clarified. The fourth chapter solves three inverse problems involving null pairs. The first is a Lagrange type interpolation problem, and asks for the construction of a regular rational matrix function W, analytic on the open right half plane, with a prescribed null pair in the open right half plane. The second deals with the same question but now under the additional condition that the interpolant is required to be J-inner. The third problem is the left tangential Nevanlinna-Pick interpolation problem for rational matrix functions. For each of the three problems the solutions are given explicitly using representations of the type $(*)$. In the fifth chapter the theory of null pairs is developed further. Now, the poles are also taken into account. The coupling transformation, which intertwines the zero and pole data, is introduced, and null-pole triples are defined. Null-pole triples are described in terms of realizations, and, for the case of a single singularity, also in terms a local Smith-McMillan form. The final chapter treats three inverse problems involving null-pole triples. The problems are of the same type as those considered in the fourth chapter, but now the solutions may have zeros as well as poles. The first problem seeks regular rational matrix functions that have a prescribed null-pole triple on a given set. The second problem deals with existence and construction of a J-inner rational matrix function with a prescribed null-pole triple on the open right half plane. The third problem is the bitangential Nudelman interpolation problem for rational matrix functions.

The text is largely selfcontained. The basic theorems from mathematical sytems theory in the first section and the solution of the bitangential Nudelman interpolation problem in the last section are given without proof; in all other sections the details of the proofs are provided.

The material presented here involves the work of many mathematicians; references are given in the notes at the end of each chapter. The books Bart-Gohberg-Kaashoek [1979], Ball-Gohberg-Rodman [1990], and various chapters from Gohberg-Goldberg-Kaashoek [1990], [1993] served as the main sources for the present notes. The informal lecture notes Kaashoek-Ran [1991] also helped to organize the material.

A few words about notation. The symbol $M \oplus N$ denotes the orthogonal direct sum of the spaces M and N, and $M \dot{+} N$ stands for an algebraic direct sum. Throughout these notes we identify a $p \times q$ matrix A with the linear operator from $\mathbb{C}^q$ into $\mathbb{C}^p$ defined by multiplying the column vectors in $\mathbb{C}^q$ from the left by the matrix A. The set of eigenvalues of a square matrix A and the spectrum of an operator A are denoted by $\sigma(A)$. The symbol I stands for an identity matrix or an indentity operator. Instead of $\lambda I - A$ we often write $\lambda - A$. For an invertible matrix A we use A^{-*} in place of $(A^{-1})^*$. A rational $m \times m$ matrix function W is called *biregular* on a set Ω (or at a point λ) if W is analytic on Ω (or at λ) and $\det W(\zeta) \neq 0$ for each $\zeta \in \Omega$ (or $\det W(\lambda) \neq 0$).

I thank Joe Ball, Israel Gohberg, Peter Lancaster, André Ran, and Freek van Schagen for their comments and suggestions for improvements. Derk Pik and Sjoerd Verduyn Lunel read the entire manuscript; I am grateful for their remarks and corrections. Finally, I like to thank the Fields Institute for the opportunity to lecture on the state space theory of rational matrix functions and for publishing these notes.

Amsterdam, January 1995 M.A. Kaashoek

CHAPTER 1

Canonical factorization and the state space method

In this chapter the state space method is introduced and applied to the problem of canonical factorization. The first section reviews the elements of mathematical system theory that will be used throughout these notes. The second section concerns the problem of inverting systems of Wiener-Hopf integral equations and motivates the notion of canonical factorization. In the third section the state space method is used to obtain necessary and sufficient conditions in order that a rational matrix function admits a canonical factorization, and is used for the construction of such factorizations if they exist. The last section specifies the factorization result for Wiener-Hopf equations with a rational matrix symbol.

1.1 Elements of mathematical system theory

In this section we review the elements of mathematical system theory that will be used throughout these notes.

A mathematical system may be viewed as a set of quantities, which depend on and evolve in time. Between these quantities, which are usually called variables, various relations exist. For input-output systems, there are two types of variables; namely input variables (inputs), which often one may choose freely, and output variables (outputs), which are determined by the given inputs. Here we shall only consider systems that have a finite number of input variables $u_1(t), \ldots, u_r(t)$, say, and a finite number of output variables $y_1(t), \ldots, y_m(t)$, say. Here t denotes the time. As usual the input variables and output variables are stacked as column vectors, as follows

$$u(t) = \begin{pmatrix} u_1(t) \\ \vdots \\ u_r(t) \end{pmatrix} \qquad y(t) = \begin{pmatrix} y_1(t) \\ \vdots \\ y_m(t) \end{pmatrix}.$$

A linear, time-invariant, finite dimensional state space model for an input-output system is (in the continuous time case) a set of equations of the form

$$\Sigma \left\{ \begin{array}{ll} x'(t) & = Ax(t) + Bu(t), \quad t \geq 0 \\ y(t) & = Cx(t) + Du(t), \\ x(0) & = x_0. \end{array} \right. \tag{1.1.1}$$

Here

$$A : \mathcal{X} \to \mathcal{X}, \quad B : \mathbb{C}^r \to \mathcal{X}, \quad C : \mathcal{X} \to \mathbb{C}^m, \quad D : \mathbb{C}^r \to \mathbb{C}^m \tag{1.1.2}$$

are linear operators, and the space $\mathcal{X}$, which is called the *state space*, is assumed to be a finite dimensional inner product space. The vector $x(t)$ is the *state vector* which relates the *input vector* $u(t)$ to the *output vector* $y(t)$. One refers to A as the *state* or *main operator*; B and C are called *the input and output coefficients*, respectively, and D is the *feedthrough coefficient* (or *external coefficient*). In the sequel we identify the system and the state space model Σ, and we shall use the word "system" also for the set of equations Σ.

Let us assume that the system Σ is at rest at time $t = 0$ (i.e., the initial state x_0 is zero), and let us apply the Laplace transform,

$$\widetilde{v}(\lambda) = \int_0^\infty e^{-\lambda t} v(t)\, dt,$$

where λ is a complex variable. Then the system equations take the following form:

$$\begin{cases} \lambda\widetilde{x}(\lambda) &= A\widetilde{x}(\lambda) + B\widetilde{u}(\lambda), \\ \widetilde{y}(\lambda) &= C\widetilde{x}(\lambda) + D\widetilde{u}(\lambda). \end{cases} \tag{1.1.3}$$

The equations (1.1.3) describe the connection between input and output in the so-called frequency domain. Notice that it is simple to solve the Laplace transform of the state vector from the first equation in (1.1.3). In fact, $\widetilde{x}(\lambda) = (\lambda - A)^{-1}B\widetilde{u}(\lambda)$. By inserting this expression for $\widetilde{x}(\lambda)$ into the second equation in (1.1.3), we see that in the frequency domain the connection between input and output is given by multiplication by a matrix function, namely $\widetilde{y}(\lambda) = W(\lambda)\widetilde{u}(\lambda)$, where

$$W(\lambda) = D + C(\lambda - A)^{-1}B. \tag{1.1.4}$$

The function in (1.1.4) is called the *transfer function* of the system Σ. Since all spaces involved are finite dimensional, the transfer function $W(\lambda)$ is an $m \times r$ matrix function (here we use the convention that a linear operator from $\mathbb{C}^r$ into $\mathbb{C}^m$ is identified with its matrix relative to the standard bases in $\mathbb{C}^r$ and $\mathbb{C}^m$), which is rational (i.e., each of the entries is a quotient of scalar polynomials). Furthermore, since

$$\lim_{\lambda\to\infty} (\lambda - A)^{-1} = 0, \tag{1.1.5}$$

we see that the transfer function $W(\lambda)$ is analytic at ∞. Rational matrix functions with the latter property are called *proper*. Thus the transfer function of the system Σ is a proper rational matrix function.

It turns out (see the next proposition) that any proper rational matrix function appears as the transfer function of a system of the type Σ.

Proposition 1.1.1 *Let $W(\lambda)$ be a proper $m \times r$ rational matrix function. Put $D = W(\infty)$, and write*

$$W(\lambda) = D + \frac{1}{q(\lambda)}P(\lambda), \tag{1.1.6}$$

where $q(\lambda)$ is a monic scalar polynomial of degree ℓ, say, and $P(\lambda)$ is a $m \times r$ matrix polynomial of degree at most $\ell - 1$,

$$q(\lambda) = q_0 + \lambda q_1 + \cdots + \lambda^{\ell-1}q_{\ell-1} + \lambda^\ell, \quad P(\lambda) = P_0 + \lambda P_1 + \cdots + \lambda^{\ell-1}P_{\ell-1}.$$

Set $Q_j = q_j I_r$, *where* I_r *is the identity operator on* $\mathbb{C}^r$, *and put*

$$A = \begin{pmatrix} 0 & I_r & & \\ & & \ddots & \\ & & & I_r \\ -Q_0 & -Q_1 & \dots & -Q_{\ell-1} \end{pmatrix}, \qquad B = \begin{pmatrix} 0 \\ \vdots \\ 0 \\ I_r \end{pmatrix},$$

$$C = \begin{pmatrix} P_0 & P_1 & \dots & P_{\ell-1} \end{pmatrix}.$$

Then $W(\lambda) = D + C(\lambda - A)^{-1}B$, *and the eigenvalues of* A *coincide with the zeros of the polynomial* q.

Proof First, note that $W(\lambda)$ can be represented in the form (1.1.6). Indeed, write each entry of $W(\lambda) - D$ as a quotient of two coprime scalar polynomials, and take q to be the least common multiple of the denominators. We may normalize q to be monic. Notice that $q(\lambda)\big(W(\lambda) - D\big)$ is a matrix polynomial $P(\lambda)$, say. Since $W(\lambda) - D$ vanishes at infinity, the degree of $P(\lambda)$ is strictly less than the degree of $q(\lambda)$.

Put $L(\lambda) = Q_0 + \lambda Q_1 + \cdots + \lambda^{\ell-1}Q_{\ell-1} + \lambda^\ell I_r = q(\lambda)I_r$. It is well-known that $\det(\lambda - A) = \det L(\lambda)$ (see, e.g., Section 1.1 in Gohberg-Lancaster-Rodman [1982], where it is shown that $\lambda - A$ is a linearization of $L(\lambda)$). It follows that

$$\det(\lambda - A) \neq 0 \Longleftrightarrow \det L(\lambda) \neq 0 \Longleftrightarrow q(\lambda) \neq 0,$$

and hence the eigenvectors of A coincide with the zeros of q. Furthermore (see Proposition 1.2 in Gohberg-Lancaster-Rodman [1982]), for $\lambda \in \mathbb{C}$ which is not an eigenvalue of A the following equality holds

$$L(\lambda)^{-1} = R(\lambda - A)^{-1}B, \tag{1.1.7}$$

where $R = \begin{pmatrix} I & 0 & \dots & 0 \end{pmatrix}$. Now, partition $(\lambda - A)^{-1}B$ as

$$(\lambda - A)^{-1}B = \begin{pmatrix} C_1(\lambda) \\ \vdots \\ C_{\ell-1}(\lambda) \\ C_\ell(\lambda) \end{pmatrix},$$

where each $C_j(\lambda)$ is a $r \times r$ matrix function. From (1.1.7) we see that $C_1(\lambda) = L(\lambda)^{-1}$. Since

$$(\lambda - A)\begin{pmatrix} C_1(\lambda) \\ \vdots \\ C_{\ell-1}(\lambda) \\ C_\ell(\lambda) \end{pmatrix} = \begin{pmatrix} 0 \\ \vdots \\ 0 \\ I \end{pmatrix},$$

the special block form of A yields $C_j(\lambda) = \lambda^{j-1}C_1(\lambda)$ for $j = 1, \dots, \ell$. Therefore,

$$C(\lambda - A)^{-1}B = \begin{pmatrix} P_0 & P_1 \dots & P_{\ell-1} \end{pmatrix} \begin{pmatrix} L(\lambda)^{-1} \\ \lambda L(\lambda)^{-1} \\ \vdots \\ \lambda^{\ell-1}L(\lambda)^{-1} \end{pmatrix} = \frac{1}{q(\lambda)}P(\lambda),$$

as desired. □

Let W be a proper $m \times r$ rational matrix function. A quadruple of operators $[A, B, C, D]$, where A, B, C, D are as in (1.1.2), is called a *realization* of W if

$$W(\lambda) = D + C(\lambda - A)^{-1}B. \tag{1.1.8}$$

The term *realization* will also be used for the right hand side of (1.1.8). The space $\mathcal{X}$ on which A acts is called the *state* or *main space* of the realization. From Proposition 1.1.1 we know that a realization always exists, and from the remarks made in the first paragraph of the proof of Proposition 1.1.1 we see that we may choose the realization so that the eigenvalues of A coincide with the poles of W. Recall that $\lambda \in \mathbb{C}$ is said to be a *pole* of a rational matrix function if λ is a pole of at least one of its entries.

Realizations are not unique. If $[A, B, C, D]$ is a realization of W, then we can make new realizations by using the following two basic operations. The first involves a similarity transformation. Let $S : \mathcal{X} \to \widetilde{\mathcal{X}}$ be an invertible linear operator, where $\mathcal{X}$ is the state space of the given realization and $\widetilde{\mathcal{X}}$ is a second (possibly different) finite dimensional inner product space. Consider the operators

$$\widetilde{A} = SAS^{-1}, \qquad \widetilde{B} = SB, \qquad \widetilde{C} = CS^{-1}. \tag{1.1.9}$$

Then

$$\widetilde{C}(\lambda - \widetilde{A})^{-1}\widetilde{B} = C(\lambda - A)^{-1}B, \tag{1.1.10}$$

and hence $[\widetilde{A}, \widetilde{B}, \widetilde{C}, D]$ is again a realization of W. In this case the two realizations $[A, B, C, D]$ and $[\widetilde{A}, \widetilde{B}, \widetilde{C}, D]$ are said to be *similar*.

The second operation enlarges the state space dimension. Let $[A, B, C, D]$ be a realization of W. We say that $[\widetilde{A}, \widetilde{B}, \widetilde{C}, \widetilde{D}]$ is a *dilation* of $[A, B, C, D]$ if $\widetilde{D} = D$ and the state space $\widetilde{\mathcal{X}}$ admits a direct sum decomposition, $\widetilde{\mathcal{X}} = \mathcal{X}_1 \dot{+} \mathcal{X} \dot{+} \mathcal{X}_2$ such that, relative to this decomposition, the operators $\widetilde{A}$, $\widetilde{B}$, $\widetilde{C}$ admit the following partitioning:

$$\widetilde{A} = \begin{pmatrix} * & * & * \\ 0 & A & * \\ 0 & 0 & * \end{pmatrix} : \mathcal{X}_1 \dot{+} \mathcal{X} \dot{+} \mathcal{X}_2 \to \mathcal{X}_1 \dot{+} \mathcal{X} \dot{+} \mathcal{X}_2, \tag{1.1.11}$$

$$\widetilde{B} = \begin{pmatrix} * \\ B \\ 0 \end{pmatrix} : \mathbb{C}^r \to \mathcal{X}_1 \dot{+} \mathcal{X} \dot{+} \mathcal{X}_2, \tag{1.1.12}$$

$$\widetilde{C} = \begin{pmatrix} 0 & C & * \end{pmatrix} : \mathcal{X}_1 \dot{+} \mathcal{X} \dot{+} \mathcal{X}_2 \to \mathbb{C}^m. \tag{1.1.13}$$

The $*$'s in the above operator matrices stand for entries that we do not specify any further. Notice that the state space $\mathcal{X}$ of $[A, B, C, D]$ is a subspace of the state space $\widetilde{\mathcal{X}}$ of the dilation $[\widetilde{A}, \widetilde{B}, \widetilde{C}, \widetilde{D}]$, and that this subspace $\mathcal{X}$ is semi-invariant under $\widetilde{A}$. Formulas (1.1.11)–(1.1.13) again yield (1.1.10), and the dilation $[\widetilde{A}, \widetilde{B}, \widetilde{C}, \widetilde{D}]$ is also a realization of W.

To give a full description of the freedom one has in the choice of a realization, we need the notion of minimality. A realization $[A, B, C, D]$ of the rational matrix function W is said to be *minimal* if its state space dimension is as small as possible among all realizations of W.

Theorem 1.1.2 *Each realization of W is a dilation of a minimal realization of W.*

Theorem 1.1.3 *Two minimal realizations $[\widetilde{A}, \widetilde{B}, \widetilde{C}, \widetilde{D}]$ and $[A, B, C, D]$ of W are similar. Moreover, the invertible operator S that provides the similarity, as in (1.1.9), is uniquely determined by the two realizations.*

Theorem 1.1.3 is known as the *state space similarity theorem*. The following corollary about the eigenvalues of the state operator in a minimal realization will be useful later.

Corollary 1.1.4 *If $[A, B, C, D]$ is a minimal realization of W, then the eigenvalues of A coincide with the poles of W.*

Proof Proposition 1.1.1 shows that we may choose a realization $[\widetilde{A}, \widetilde{B}, \widetilde{C}, D]$ so that the eigenvalues of $\widetilde{A}$ coincide with the poles of W. According to Theorem 1.1.2 the realization $[\widetilde{A}, \widetilde{B}, \widetilde{C}, D]$ appears as a dilation of a minimal realization $[A_0, B_0, C_0, D]$, say. From the definition of a dilation it is clear that each eigenvalue of A_0 is an eigenvalue of $\widetilde{A}$, and thus the eigenvalues of A_0 are poles of W. By Theorem 1.1.2 the two realizations $[A, B, C, D]$ and $[A_0, B_0, C_0, D]$ are similar. In particular, A and A_0 are similar linear operators, and hence A and A_0 have the same eigenvalues. So we have now proved that each eigenvalue of A is a pole of W_0. Since $W(\lambda) = D + (\lambda - A)^{-1}B$, the reverse implication is trivially true. □

We need one more theorem about minimal realizations. For this purpose we return to the system Σ in (1.1.1).

The system Σ in (1.1.1) is said to be *controllable* if from each initial state x_0 any other state x_1 can be reached over a finite time interval with an appropriate choice of the input u. The property of controllability does not involve the output coefficient and the feedthrough coefficient. In fact, the system Σ is controllable if and only if

$$\operatorname{Im}\begin{pmatrix} B & AB & \dots & A^{n-1}B \end{pmatrix} = \mathcal{X}. \tag{1.1.14}$$

Here $n = \dim \mathcal{X}$. In the sequel, a pair of linear operators $A : \mathcal{X} \to \mathcal{X}$ and $B : \mathbb{C}^r \to \mathcal{X}$ will be called a *full-range pair* if for some n the identity (1.1.14) holds.

The system Σ in (1.1.1) is said to be *observable* if given a zero input the initial state x_0 is uniquely determined by the output y over a finite time interval. Observability does not involve the input coefficient and the feedthrough coefficient. In fact, the system Σ is observable if and only if

$$\operatorname{Ker}\begin{pmatrix} C \\ CA \\ \vdots \\ CA^{n-1} \end{pmatrix} = \{0\}, \tag{1.1.15}$$

where $n = \dim \mathcal{X}$. Any pair of linear operators $A : \mathcal{X} \to \mathcal{X}$ and $C : \mathcal{X} \to \mathbb{C}^m$ satisfying (1.1.15) for some n will be called a *null-kernel pair*.

Theorem 1.1.5 *A realization $[A, B, C, D]$ of W is minimal if and only if the pair (A, B) is a full-range pair and the pair (C, A) is a null-kernel pair.*

Proofs of Theorems 1.1.2, 1.1.3, 1.1.5 may be found in Chapters 3 and 4 of Bart-Gohberg-Kaashoek [1979]; for the characterizations of controllability and observability by (1.1.14) and (1.1.15), respectively, we refer to Wonham [1979].

1.2 Inverting systems of Wiener-Hopf integral equations

This section concerns the problem of inverting a system of Wiener-Hopf integral equations, and it introduces the notion of a canonical factorization.

Consider the following Wiener-Hopf integral equation

$$f(t) - \int_0^\infty k(t-s)f(s)\,ds = g(t), \qquad 0 \le t < \infty. \tag{1.2.1}$$

We think about (1.2.1) as representing a system of m equations. Hence g and f are $\mathbb{C}^m$-valued vector functions and the kernel function k is an $m \times m$ matrix function. We shall assume that the entries of k are in $L_1(\mathbb{R})$. The given function g belongs to $L_2^m([0,\infty))$, that is, g is a $\mathbb{C}^m$-valued function of which the components are square (Lebesgue) integrable on $[0,\infty)$. The problem is to find $f \in L_2^m([0,\infty))$ such that (1.2.1) is fulfilled.

In the study of equation (1.2.1) an important role is played by the *symbol* of (1.2.1), which, by definition, is the $m \times m$ matrix function W given by

$$W(\lambda) = I_m - \int_{-\infty}^\infty e^{i\lambda t}k(t)\,dt, \qquad \lambda \in \mathbb{R}. \tag{1.2.2}$$

Since the entries of the kernel function k are in $L_1(\mathbb{R})$, the symbol W is continuous on $\mathbb{R}$, and, by the Riemann-Lebesgue lemma, we have

$$\lim_{\lambda\in\mathbb{R},|\lambda|\to\infty} W(\lambda) = I_m. \tag{1.2.3}$$

The following inversion theorem is due to I. Gohberg and M.G. Krein [1960].

Theorem 1.2.1 *In order that for each $g \in L_2^m([0,\infty))$ equation* (1.2.1) *has a unique solution $f \in L_2^m([0,\infty))$ it is necessary and sufficient that the symbol W of equation* (1.2.1) *admits a factorization,*

$$W(\lambda) = W_-(\lambda)W_+(\lambda), \qquad \lambda \in \mathbb{R}, \tag{1.2.4}$$

such that

$$W_+(\lambda) = I_m + \int_0^\infty e^{i\lambda t}h_1(t)\,dt, \quad \det W_+(\lambda) \ne 0 \quad (Im\ \lambda \ge 0), \tag{1.2.5}$$

$$W_-(\lambda) = I_m + \int_{-\infty}^0 e^{i\lambda t}h_2(t)\,dt, \quad \det W_-(\lambda) \ne 0 \quad (Im\ \lambda \le 0), \tag{1.2.6}$$

where h_1, h_2 are $m \times m$ matrix functions with entries in $L_1([0,\infty))$ and $L_1((-\infty,0])$, respectively.

Furthermore, in this case the unique solution f of equation (1.2.1) may be obtained in the following way. Choose $m \times m$ matrix functions γ_+, γ_- with entries in $L_1([0,\infty))$ and $L_1((-\infty,0])$, respectively, such that

$$W_+(\lambda)^{-1} = I_m + \int_0^\infty e^{i\lambda t}\gamma_+(t)\,dt, \qquad Im\ \lambda \geq 0, \tag{1.2.7}$$

$$W_-(\lambda)^{-1} = I_m + \int_{-\infty}^0 e^{i\lambda t}\gamma_-(t)\,dt, \qquad Im\ \lambda \leq 0, \tag{1.2.8}$$

Then the solution f is given by

$$f(t) = g(t) + \int_0^\infty \gamma(t,s)g(s)\,ds, \qquad 0 \leq t \leq \infty,$$

where

$$\gamma(t,s) = \begin{cases} \gamma_+(t-s) + \int_0^s \gamma_+(t-r)\gamma_-(r-s)\,dr, & t > s \geq 0, \\ \gamma_-(t-s) + \int_0^t \gamma_+(t-r)\gamma_-(r-s)\,dr, & s > t \geq 0. \end{cases}$$

A factorization (1.2.4) with the factors satisfying (1.2.5) and (1.2.6) is called a *(right) canonical factorization* of W with respect to the real line. If such a factorization exists, then the continuous analogue of the matrix version of Wiener's theorem yields (cf., Goldberg-Gohberg-Kaashoek [1993], pages 836, 837) the existence of the matrix functions γ_+ and γ_- in (1.2.7) and (1.2.8). To find these functions explicitly is another matter. In fact, for the above inversion theorem to be effective we need

1. conditions that guarantee the existence of canonical factorizations,

and methods which allow us

2. to construct canonical factorizations explicitly,
3. to invert the factors,
4. to compute the inverse Fourier transforms of the inverted factors.

In general, for the matrix case these are non-trivial problems.

In the scalar case ($m = 1$) the symbol $W(\lambda)$ admits a canonical factorization if and only if the curve parametrized by $\lambda \mapsto W(\lambda)$, $-\infty < \lambda < \infty$, does not go through zero and has winding number zero around zero. Furthermore, if this winding number condition is fulfilled, then the plus-factor in the canonical factorization is given by

$$W_+(\lambda) = \exp\big(\mathcal{P}\log W(\lambda)\big), \qquad \lambda \in \mathbb{R},$$

where $\mathcal{P}$ is the projection operator defined by

$$\mathcal{P}\Big(\int_{-\infty}^\infty e^{i\lambda t}h(t)\,dt\Big) = \int_0^\infty e^{i\lambda t}h(t)\,dt.$$

In particular, if the scalar symbol W is rational (quotient of two polynomials), then the (unique) canonical factorization may be constructed in terms of the zeros and poles of W. Indeed, in this case we may write $W(\lambda) = p(\lambda)/q(\lambda)$, where p and q are coprime polynomials. The fact that W is a symbol implies that W is continuous on $\mathbb{R}$, and hence q has no zero on $\mathbb{R}$. From (1.2.3) we may conclude that p and q have the same degree and the same leading coefficient. The winding number condition mentioned above reduces to the condition that p has no zero on $\mathbb{R}$ and that the numbers of zeros of p and q in the upper half plane are equal. If these conditions are fulfilled, then the canonical factorization $W = W_-W_+$ is obtained by taking

$$W_-(\lambda) = \frac{\Pi_{j=1}^{n_+}(\lambda - a_j^+)}{\Pi_{j=1}^{n_+}(\lambda - b_j^+)}, \qquad W_+(\lambda) = \frac{\Pi_{j=1}^{n_-}(\lambda - a_j^-)}{\Pi_{j=1}^{n_-}(\lambda - b_j^-)},$$

where a_j^+ and b_j^+ $(j = 1, \ldots, n_+)$ are, respectively, the zeros of p and q in the open upper half plane, and a_j^- and b_j^- $(j = 1, \ldots, n_-)$ are the corresponding zeros in the open lower half plane. Notice that in the scalar rational case it is easy to invert the factors and to compute the inverse Fourier transforms of the inverted factors. In fact, we only need to know that

$$\left(\frac{\lambda - a}{\lambda - b}\right)^{-1} = 1 + \begin{cases} i(a-b)\displaystyle\int_{-\infty}^{0} e^{i\lambda t}e^{-iat}\,dt, & \text{Im } a > 0, \\ -i(a-b)\displaystyle\int_{0}^{\infty} e^{i\lambda t}e^{-iat}\,dt, & \text{Im } a < 0. \end{cases}$$

For matrix symbols the situation is much more complex. There is no simple winding number condition. For rational matrix symbols one has a factorization algorithm, using elementary row and column operations, which leads in a finite number of steps to a canonical factorization provided such a factorization exists. In the late seventies and the beginning of the eighties a method has been developed which allows one to deal with factorization problems for rational matrix functions in terms of the idea of realization, which we met in the previous section. This method is called the *state space method.* We shall see (in the next two sections) that with the state space method the canonical factorization problem for rational matrix symbols can be handled with the same level of explicitness as was done above for the rational scalar case.

1.3 Canonical factorization of rational matrix functions

Throughout this section W is a proper rational $m \times m$ matrix function without poles on the real line. We are interested to factor W as

$$W(\lambda) = W_-(\lambda)W_+(\lambda), \qquad \lambda \in \mathbb{R}, \tag{1.3.1}$$

where W_- and W_+ are proper rational $m \times m$ matrix functions satisfying

1. W_+ has no poles on Im $\lambda \geq 0$, and $\det W_+(\lambda) \neq 0$ for Im $\lambda \geq 0$,
2. W_- has no poles on Im $\lambda \leq 0$, and $\det W_-(\lambda) \neq 0$ for Im $\lambda \leq 0$,
3. $\det W_\pm(\infty) \neq 0$.

Such a factorization of W is called a *(right) canonical factorization* of W with respect to the real line in the class of proper rational matrix functions. A canonical factorization of W (assuming it exists) is basically unique. The freedom one has is in the value of the factors at infinity. More precisely, for $W_+(\infty)$ one may choose any invertible $m \times m$ matrix, and as soon as the value of W_+ at infinity is fixed, there is no more freedom in the choice of the factors.

From Section 1.1 we know that a proper rational matrix function W appears as the transfer function of an input-output system, and hence W can be represented in the form

$$W(\lambda) = D + C(\lambda - A)^{-1}B. \tag{1.3.2}$$

In our case W has no poles on the real line, and therefore we know from Proposition 1.1.1 that we may choose the state operator A so that A has no eigenvalue on $\mathbb{R}$. The next theorem, which is the main result of this section, tells us in terms of the realization (1.3.2) when and how one may construct a canonical factorization for W.

Theorem 1.3.1 *Let W be a proper rational $m \times m$ matrix function without poles on the real line, and let W be given by the realization* (1.3.2)*, where A has no eigenvalue on the real line. Then W admits a right canonical factorization with respect to the real line in the class of proper rational matrix functions if and only if the following two conditions are fulfilled:*

(i) *D is non-singular, and $A^\times = A - BD^{-1}C$ has no real eigenvalue,*

(ii) $\mathcal{X} = \operatorname{Im} P \dot{+} \operatorname{Ker} P^\times$.

Here $\mathcal{X}$ is the state space of the realization (1.3.2)*, and the operators P and $P^\times$ are the Riesz projections of A and $A^\times$, respectively, corresponding to the eigenvalues in the open upper half plane. If conditions* (*i*) *and* (*ii*) *are fulfilled, then a canonical factorization $W = W_- W_+$ is obtained by taking*

$$\begin{aligned} W_-(\lambda) &= D + C(\lambda - A)^{-1}(I - \Pi)B, \\ W_+(\lambda) &= I + D^{-1}C\Pi(\lambda - A)^{-1}B, \end{aligned}$$

where Π is the projection of $\mathcal{X}$ along $\operatorname{Im} P$ *onto* $\operatorname{Ker} P^\times$. *Moreover, in this case*

$$\begin{aligned} W_-(\lambda)^{-1} &= D^{-1} - D^{-1}C(I - \Pi)(\lambda - A^\times)^{-1}BD^{-1}, \\ W_+(\lambda)^{-1} &= I - D^{-1}C(\lambda - A^\times)^{-1}\Pi B. \end{aligned}$$

The Riesz projections P and $P^\times$ in Theorem 1.3.1 are usually defined in terms of contour integrals (see, e.g., Chapter I of Gohberg-Goldberg-Kaashoek [1990]). In this case the underlying space $\mathcal{X}$ is finite dimensional, and hence P and $P^\times$ may also be defined as the sum of the residues in the open upper half plane of $(\lambda - A)^{-1}$ and $(\lambda - A^\times)^{-1}$, respectively. From the properties of Riesz projections, it follows that $\operatorname{Im} P$ is the maximal A-invariant subspace M of $\mathcal{X}$ such that the eigenvalues of $A|M$ are in the open upper half plane. Similarly, $\operatorname{Ker} P^\times$ is the maximal $A^\times$-invariant subspace N of $\mathcal{X}$ such that the eigenvalues of $A^\times|N$ are in open lower half plane. Here we assume that $A^\times$ has no eigenvalues on $\mathbb{R}$.

A full proof of Theorem 1.3.1 may be found in Section XIII.6 of Gohberg-Goldberg-Kaashoek [1990]. Here we only make a few remarks to clarify the conditions of the theorem, and we sketch the proof of the sufficiency of the conditions (i) and (ii) including the derivation of the formulas.

Recall that W in Theorem 1.3.1 is proper and has no poles on the real line. In other words, W is analytic on $\mathbb{R}$ and at infinity. We claim that condition (i) in Theorem 1.3.1 is equivalent to the requirement that $\det W(\lambda) \neq 0$ for each $\lambda \in \mathbb{R} \cup \{\infty\}$. To see this, notice that $D = W(\infty)$, and if D is invertible, then for

$\lambda \in \mathbb{R}$ we have

$$\begin{aligned}\det W(\lambda) &= \det D \det\{I + D^{-1}C(\lambda - A)^{-1}B\} \\ &= \det D \det\{I + (\lambda - A)^{-1}BD^{-1}C\} \\ &= \det D \frac{\det(\lambda - A^{\times})}{\det(\lambda - A)}.\end{aligned}$$

Thus, if D is invertible, then $\det W(\lambda) \neq 0$ for $\lambda \in \mathbb{R}$ if and only if $A^{\times}$ has no eigenvalue on $\mathbb{R}$, which proves our claim.

Next, we remark that

$$W(\lambda)^{-1} = D^{-1} - D^{-1}C(\lambda - A^{\times})^{-1}BD^{-1}, \tag{1.3.3}$$

whenever the feedthrough coefficient D is invertible. To see this, remember that W is the transfer function of the system

$$\Sigma \left\{ \begin{array}{rcl} x'(t) &=& Ax(t) + Bu(t), \quad t \geq 0, \\ y(t) &=& Cx(t) + Du(t), \\ x(0) &=& 0. \end{array} \right.$$

Since D is invertible, we may reverse the roles of input and output. Indeed, apply BD^{-1} to the second equation and substract the resulting equation from the first. This yields the inverse system

$$\Sigma^{\times} \left\{ \begin{array}{rcl} x'(t) &=& A^{\times}x(t) + BD^{-1}y(t), \quad t \geq 0, \\ u(t) &=& -D^{-1}Cx(t) + D^{-1}y(t), \\ x(0) &=& 0. \end{array} \right.$$

From the construction of $\Sigma^{\times}$ it follows that its transfer function is equal to $W(\cdot)^{-1}$. On the other hand, by definition, the transfer function of $\Sigma^{\times}$ is the function

$$W^{\times}(\lambda) = D^{-1} - D^{-1}C(\lambda - A^{\times})^{-1}BD^{-1}.$$

Hence (1.3.3) holds.

Proof of the sufficiency part of Theorem 1.3.1 Assume conditions (i) and (ii) in Theorem 1.3.1 are fulfilled. We have to produce a canonical factorization of W. To do this write A, B and C as operator matrices relative to the decomposition $\mathcal{X} = \operatorname{Im} P \dot{+} \operatorname{Ker} P^{\times}$, as follows

$$\begin{aligned} A &= \begin{pmatrix} A_{11} & A_{12} \\ 0 & A_{22} \end{pmatrix} : \operatorname{Im} P \dot{+} \operatorname{Ker} P^{\times} \to \operatorname{Im} P \dot{+} \operatorname{Ker} P^{\times}, \\ C &= \begin{pmatrix} C_1 & C_2 \end{pmatrix} : \operatorname{Im} P \dot{+} \operatorname{Ker} P^{\times} \to \mathbb{C}^m, \\ B &= \begin{pmatrix} B_1 \\ B_2 \end{pmatrix} : \mathbb{C}^m \to \operatorname{Im} P \dot{+} \operatorname{Ker} P^{\times}. \end{aligned}$$

We also need the analogous partitionings of $A^{\times}$ and the projection Π:

$$\begin{aligned} A^{\times} &= \begin{pmatrix} A_{11}^{\times} & 0 \\ A_{21}^{\times} & A_{22}^{\times} \end{pmatrix} : \operatorname{Im} P \dot{+} \operatorname{Ker} P^{\times} \to \operatorname{Im} P \dot{+} \operatorname{Ker} P^{\times}, \\ \Pi &= \begin{pmatrix} 0 & 0 \\ 0 & I \end{pmatrix} : \operatorname{Im} P \dot{+} \operatorname{Ker} P^{\times} \to \operatorname{Im} P \dot{+} \operatorname{Ker} P^{\times}. \end{aligned}$$

The zero in the left lower corner of the operator matrix for A and the zero in the right upper corner of the operator matrix for $A^\times$ are justified by the fact that $\operatorname{Im} P$ is invariant under A and $\operatorname{Ker} P^\times$ is invariant under $A^\times$. Since $A^\times = A - BD^{-1}C$, we have

$$A_{11}^\times = A_{11} - B_1D^{-1}C_1, \quad A_{22}^\times = A_{22} - B_2D^{-1}C_2, \quad A_{12} = B_1D^{-1}C_2. \tag{1.3.4}$$

Consider the matrix functions,

$$W_1(\lambda) = D + C(\lambda - A)^{-1}(I - \Pi)B, \qquad W_2(\lambda) = I + D^{-1}C\Pi(\lambda - A)^{-1}B.$$

From the above operator matrix representations it follows that

$$W_1(\lambda) = D + C_1(\lambda - A_{11})^{-1}B_1, \qquad W_2(\lambda) = I + D^{-1}C_2(\lambda - A_{22})^{-1}B_2,$$

and hence (using the third identity in (1.3.4))

$$\begin{aligned} W_1(\lambda)W_2(\lambda) &= D + C_1(\lambda - A_{11})^{-1}B_1 + C_2(\lambda - A_{22})^{-1}B_2 \\ &\qquad + C_1(\lambda - A_{11})^{-1}B_1D^{-1}C_2(\lambda - A_{22})^{-1}B_2 \\ &= D + \begin{pmatrix} C_1 & C_2 \end{pmatrix} \begin{pmatrix} \lambda - A_{11} & -B_1D^{-1}C_2 \\ 0 & \lambda - A_{22} \end{pmatrix}^{-1} \begin{pmatrix} B_1 \\ B_2 \end{pmatrix} \\ &= D + C(\lambda - A)^{-1}B = W(\lambda). \end{aligned}$$

Since $\operatorname{Im} P$ is the maximal A-invariant subspace M of $\mathcal{X}$ such that the eigenvalues of $A|M$ are in the open upper half plane, we have

$$\sigma(A_{11}) \subset \mathbb{C}^+ = \{\lambda \in \mathbb{C} \mid \operatorname{Im}\ \lambda > 0\}, \quad \sigma(A_{22}) \subset \mathbb{C}^- = \{\lambda \in \mathbb{C} \mid \operatorname{Im}\ \lambda < 0\}.$$

Here $\sigma(A_{11})$ and $\sigma(A_{22})$ denote the sets of eigenvalues of A_{11} and A_{22}, respectively. It follows that W_1 has no poles in $\operatorname{Im}\ \lambda \le 0$ and W_2 has no poles in $\operatorname{Im}\ \lambda \ge 0$.

Next, we use (1.3.3) and the first two identities in (1.3.4) to conclude that

$$\begin{aligned} W_1(\lambda)^{-1} &= D^{-1} - D^{-1}C_1(\lambda - A_{11}^\times)^{-1}B_1D^{-1}, \\ W_2(\lambda)^{-1} &= I - D^{-1}C_2(\lambda - A_{22}^\times)^{-1}B_2. \end{aligned}$$

Recall that $\operatorname{Ker} P^\times$ is the maximal $A^\times$-invariant subspace N of $\mathcal{X}$ such that $A^\times|N$ has its eigenvalues in $\mathbb{C}^-$. It follows that $\sigma(A_{11}^\times) \subset \mathbb{C}^+$ and $\sigma(A_{22}^\times) \subset \mathbb{C}^-$, and therefore $W_1(\cdot)^{-1}$ has no poles in $\operatorname{Im}\ \lambda \le 0$, and $W_2(\cdot)^{-1}$ has no poles in $\operatorname{Im}\ \lambda \ge 0$. From the partitionings of $A^\times$, B, C and Π we see that $W_1(\lambda)^{-1}$ and $W_2(\lambda)^{-1}$ may be rewritten as

$$\begin{aligned} W_1(\lambda)^{-1} &= D^{-1} - D^{-1}C(I - \Pi)(\lambda - A^\times)^{-1}BD^{-1}, \\ W_2(\lambda)^{-1} &= I - D^{-1}C(\lambda - A^\times)^{-1}\Pi B. \end{aligned}$$

Therefore, with $W_- = W_1$ and $W_+ = W_2$ we have $W = W_-W_+$ and this factorization is a canonical factorization. □

1.4 Wiener-Hopf integral equations revisited

In this section we return to the Wiener-Hopf integral equation

$$f(t) - \int_0^\infty k(t-s)f(s)\,dt = g(t), \quad 0 \le t < \infty. \tag{1.4.1}$$

We assume now that the symbol W is a rational matrix function. Since the symbol W is continuous on the real line and satisfies (1.2.3), we can use Proposition 1.1.1 to show that W admits the realization

$$W(\lambda) = I + C(\lambda - A)^{-1}B, \qquad \lambda \in \mathbb{R}, \tag{1.4.2}$$

where A has no eigenvalue on the real line. The following theorem is the main result of this section.

Theorem 1.4.1 *Assume that the symbol of the Wiener-Hopf integral equation* (1.4.1) *is the* $m \times m$ *rational matrix function* W *given by the realization* (1.4.2), *where* A *has no eigenvalue on the real line. Then for each* $g \in L_2^m([0,\infty))$ *the equation* (1.4.1) *has a unique solution* $f \in L_2^m([0,\infty))$ *if and only if the following two conditions are fulfilled:*

(i) $A^\times := A - BC$ *has no real eigenvalue,*

(ii) $\mathcal{X} = \operatorname{Im} P \dot{+} \operatorname{Ker} P^\times$.

Here $\mathcal{X}$ *is the state space of the realization* (1.4.2), *and the operators* P *and* $P^\times$ *are the Riesz projections of* A *and* $A^\times$ *respectively, corresponding to the eigenvalues in the open upper half plane. Furthermore, if conditions* (i) *and* (ii) *are fulfilled, then the unique solution* f *of* (1.4.1) *is given by*

$$f(t) = g(t) + \int_0^\infty \gamma(t,s)g(s)\,ds, \quad 0 \le t < \infty,$$

where

$$\gamma(t,s) = \begin{cases} iCe^{-itA^\times}\Pi e^{isA^\times}B, & 0 \le s < t < \infty, \\ -iCe^{-itA^\times}(I-\Pi)e^{isA^\times}B, & 0 \le t < s < \infty, \end{cases} \tag{1.4.3}$$

with Π *being the projection of* $\mathcal{X}$ *along* $\operatorname{Im} P$ *onto* $\operatorname{Ker} P^\times$.

Proof Assume that equation (1.4.1) is uniquely solvable in $L_2^m([0,\infty))$. Then, by Theorem 1.2.1, the symbol W admits a factorization

$$W(\lambda) = W_-(\lambda)W_+(\lambda), \quad \lambda \in \mathbb{R}, \tag{1.4.4}$$

so that the factors W_+ and W_- satisfy (1.2.5) and (1.2.6), respectively. First, we show that the factors W_- and W_+ are rational matrix functions. To do this, notice that (1.2.5) implies that W_+ is analytic on the open upper half plane and continuous on its closure $\overline{\mathbb{C}^+}$, where the closure is taken in $\mathbb{C} \cup \{\infty\}$. Similarly, by (1.2.6), the function W_- is analytic on the open lower half plane $\mathbb{C}^-$ and continuous on $\overline{\mathbb{C}^-}$. Since W is analytic on $\mathbb{R}$ and at infinity, the factorization (1.4.4) and the above mentioned properties of W_+ and W_- imply that W_+ and W_- are also analytic on $\mathbb{R}$ and at infinity. On the other hand, we see from $W_-(\lambda) = W_+(\lambda)^{-1}W(\lambda)$ that W_- can have only poles in $\mathbb{C}^+$, and the number of poles has to be finite. So W_- is rational, and hence W_+ is also rational.

Thus the factorization (1.4.4) is a right canonical factorization in the class of rational matrix functions. But then we can apply Theorem 1.3.1 to show that (i) and (ii) are fulfilled.

Conversely, assume that (i) and (ii) are fulfilled. Put

$$W_-(\lambda) = I + C(\lambda - A)^{-1}(I - \Pi)B, \qquad W_+(\lambda) = I + C\Pi(\lambda - A)^{-1}B,$$

where Π is the projection of $\mathcal{X}$ along $\operatorname{Im} P$ onto $\operatorname{Ker} P^\times$. Then, by Theorem 1.3.1, we have $W = W_- W_+$, and this factorization is a canonical factorization of W in the class of rational matrix functions. Notice that

$$\begin{aligned} W_-(\lambda) &= I + i\int_{-\infty}^{0} e^{i\lambda t} C e^{-itA}(I - \Pi)B\,dt, \\ W_+(\lambda) &= I - i\int_{0}^{\infty} e^{i\lambda t} C\Pi e^{-itA}B\,dt, \end{aligned}$$

which implies that the factorization $W = W_- W_+$ is a canonical factorization in the sense of Section 1.2. Thus equation (1.4.1) is uniquely solvable in $L_2^m([0,\infty))$, because of Theorem 1.2.1.

To compute the solution f of (1.4.1) we apply the method described in Theorem 1.2.1. From Theorem 1.3.1 we know that

$$W_-(\lambda)^{-1} = I - C(I - \Pi)(\lambda - A^\times)^{-1}B, \qquad W_+(\lambda)^{-1} = I - C(\lambda - A^\times)^{-1}\Pi B.$$

It follows that

$$\begin{aligned} W_-(\lambda)^{-1} &= I - i\int_{-\infty}^{0} e^{i\lambda t} C(I - \Pi)e^{-itA^\times}B\,dt, \\ W_+(\lambda)^{-1} &= I + i\int_{0}^{\infty} e^{i\lambda t} C e^{-itA^\times}\Pi B\,dt. \end{aligned}$$

Thus, in this case the matrix functions γ_+ and γ_- appearing in (1.2.7) and (1.2.8), respectively, are given by

$$\begin{aligned} \gamma_+(t) &= iCe^{-itA^\times}\Pi B, \quad t \geq 0, \\ \gamma_-(t) &= -iC(I - \Pi)e^{-itA^\times}B, \quad t \leq 0. \end{aligned}$$

To compute the resolvent kernel $\gamma(t,s)$ we use the formula for $\gamma(t,s)$ in Theorem 1.2.1. A straightforward computation, using

$$\Pi BC(I - \Pi) = A^\times \Pi - \Pi A^\times,$$

yields the desired formula (1.4.3). □

The proof of Theorem 1.4.1 shows that for a rational matrix symbol W there is no distinction between a canonical factorization in the class of rational matrix functions and a canonical factorization of the type appearing in Theorem 1.2.1. For that reason in the sequel the phrase "in the class of rational matrix functions" is omitted.

We conclude this section with an example. Consider the (scalar) equation

$$f(t) - \tfrac{c}{2}\int_0^\infty e^{-|t-s|}f(s)\,ds = g(t), \quad 0 \leq t < \infty. \tag{1.4.5}$$

Here c is a real parameter, which we shall specify later. The symbol of equation (1.4.5) is the function $(\lambda^2+1-c)(\lambda^2+1)^{-1}$. This function can be written in the form (1.4.2) by taking

$$A=\begin{pmatrix}0 & 1\\ -1 & 0\end{pmatrix}, \qquad B=\begin{pmatrix}0\\ 1\end{pmatrix}, \qquad C=\begin{pmatrix}-c & 0\end{pmatrix}.$$

Notice that A has no eigenvalue on the real line, and

$$A^\times = A-BC=\begin{pmatrix}0 & 1\\ c-1 & 0\end{pmatrix}.$$

It follows that $A^\times$ has no real eigenvalue if and only if $c<1$. Let us assume that the latter condition is satisfied. In this case $A^\times$ has two different eigenvalues, namely $\lambda=\pm i\sqrt{1-c}$ with corresponding eigenvector

$$\begin{pmatrix}1\\ \pm i\sqrt{1-c}\end{pmatrix}.$$

It follows that

$$\operatorname{Im} P=\operatorname{span}\Big\{\begin{pmatrix}1\\ i\end{pmatrix}\Big\}, \qquad \operatorname{Ker} P^\times=\operatorname{span}\Big\{\begin{pmatrix}1\\ -i\sqrt{1-c}\end{pmatrix}\Big\},$$

where P and $P^\times$ are the Riesz projections of A and $A^\times$ corresponding to the eigenvalues in the open upper half plane. We see that $\mathbb{C}^2=\operatorname{Im} P\dot{+}\operatorname{Ker} P^\times$. Hence equation (1.4.5) is uniquely solvable in $L_2([0,\infty))$ if and only if $c<1$. To find the solution for $c<1$, we apply the second part of Theorem 1.4.1. Put $\alpha=\sqrt{1-c}$. The projection Π of $\mathbb{C}^2$ along $\operatorname{Im} P$ onto $\operatorname{Ker} P^\times$ is given by

$$\Pi=\tfrac{1}{\alpha+1}\begin{pmatrix}1 & i\\ -i\alpha & \alpha\end{pmatrix}.$$

A straight forward calculation now shows that for $c<1$ the unique solution of (1.4.5) is given by

$$f(t)=g(t)+\int_0^\infty \gamma(t,s)g(s)\,ds, \quad 0\le t<\infty,$$

where

$$\gamma(t,s)=\tfrac{1}{2\alpha}\{(1-\alpha^2)e^{-\alpha|t-s|}-(1-\alpha)^2e^{-\alpha(t+s)}\}.$$

1.5 Notes

Theorems 1.1.2, 1.1.3, 1.1.5, and the characterizations of controllability and observability by (1.1.14) and (1.1.15), respectively, are the basic theorems in the Kalman state space theory for input-output systems which was developed at the end of the fifties and in the beginning of the sixties (see Kalman [1960], [1961]).The presentation in Section 1.1 follows Bart-Gohberg-Kaashoek [1979]. Proofs may also be found in text books on system theory (see, e.g., Kalman-Falb-Arbib [1969], Kailath [1980], Wonham [1979], and the more recent Callier-Desoer [1991] and Sontag [1990]). The terms "full-range pair" and "null-kernel pair" are mathematical terms which were introduced in Gohberg-Lancaster-Rodman [1986]. The material in Section 1.1 may also be seen as a finite dimensional affine version of the theory of unitary colligations which was initiated by M.S. Livsic in the early fifties. For this and other connections with the theory of characteristic operator functions we refer to Chapter 1 of Bart-Gohberg-Kaashoek [1979] and to Ball-Cohen [1991].

The literature on Wiener-Hopf equations is extensive, see, e.g., the books Gohberg-Fel'dman [1974], Bottcher-Silbermann [1990], Mikhlin-Prossdorf [1986], Gohberg-Krupnik [1992a], [1992b], and the references therein. Theorem 1.2.1 is taken from Gohberg-Krein [1960]; the theorem is also proved in Gohberg-Fel'dman [1974] (see also Gohberg-Goldberg-Kaashoek [1993], Section XXX.10). There is an analogous theorem for block Toeplitz operators (also due to Gohberg-Krein [1960]).

For a minimal realization Theorem 1.3.1 is proved in Bart-Gohberg-Kaashoek [1979]; the non-minimal case is treated in Bart-Gohberg-Kaashoek [1986a]. The latter paper also contains the construction of a non-canonical Wiener-Hopf factorization by using the state space method. A full proof of Theorem 1.3.1 is also presented in Section XIII.6 of Gohberg-Goldberg-Kaashoek [1990]. Theorem 1.3.1 remains true if the real line is replaced by other contours, and the theorem has an infinite dimensional version involving realizations with infinite dimensional operator coefficients (see Theorem I.2.1 in Bart-Gohberg-Kaashoek [1986a]). The analogue of Theorem 1.3.1 for non-proper rational matrix functions and with the real line replaced by the circle may be found in Gohberg-Kaashoek [1988b]; see also Section XXIV.7 in Gohberg-Goldberg-Kaashoek [1993]. The state space theory has also succesfully been used to analyze and construct explicitly other type of factorizations. See, e.g., Bart-Gohberg-Kaashoek [1979] and Bart-Gohberg-Kaashoek-Van Dooren [1980] for minimal factorizations, Francis [1987], Green[1987], and Green-Glover-Limebeer-Doyle [1990] for spectral and J-spectral factorizations, Roozemond [1986],[1987] for speudo-spectral and pseudo-canonical factorizations, and the next chapter for J-inner-outer factorizations.

A first version of Theorem 1.4.1 (without the formula for the resolvent kernel) appears in Section 4.5 of Bart-Gohberg-Kaashoek [1979]. The full theorem appears in Bart-Gohberg-Kaashoek [1982a] (see also Section XIII.7 in Gohberg-Goldberg-Kaashoek [1990]). For a class of Wiener-Hopf equations with a non-rational matrix symbol the analogue of Theorem 1.4.1 may be found in Bart-Gohberg-Kaashoek [1986c]. For block Toeplitz operators the analogue of Theorem 1.4.1 is in Gohberg-Kaashoek [1986b] (see also Section XXIV.8 in Gohberg-Goldberg-Kaashoek [1993]), for singular integral equations in Gohberg-Kaashoek [1991b], and for a class of integral-differential equations of convolution type in Kuijper [1992]. There are many other interesting applications of the state space method to Wiener-Hopf and Toeplitz operators. Here we only mention Gohberg-Kaashoek-Van Schagen [1987] which deals with the computation of Szegö-Widom and Kac-Achiezer limits in terms of realizations, and Vreugdenhil [1990], [1991], [1994] which give explicit formulas for the spectral measures of selfadjoint Wiener-Hopf and Toeplitz operators with rational matrix symbols. Also, it is quite useful to think about operators as input-output operators of sytems. See Kaashoek [1984] for a survey on results in this direction and the more recent papers Ben-Artzi-Gohberg-Kaashoek [1993], [1994].

CHAPTER 2

J-unitary rational matrix functions

This chapter deals with rational matrix functions that are J-unitary on the imaginary axis. In the first section such functions are analyzed in terms of minimal realizations. In the second section the results are specified further for rational matrix functions that are J-inner. In the last section the state space method for canonical factorization is used to produce certain J-inner-outer factorizations.

2.1 J-unitary rational matrix functions

Throughout this chapter J is an $m \times m$ *signature matrix*, that is, J is both unitary and Hermitian. So $J = J^* = J^{-1}$. For example, J may be the matrix

$$\begin{pmatrix} I_p & 0 \\ 0 & -I_q \end{pmatrix}. \tag{2.1.1}$$

In fact, up to a unitary basis transformation any signature matrix J is of the form (2.1.1).

An $m \times m$ matrix A is said to be *J-unitary* if $A^*JA = J$. Since J is non-singular, the identity $A^*JA = J$ implies that A is non-singular, and hence it can be rewritten in the equivalent form $JA^*J = A^{-1}$. In other words, A is J-unitary if and only if A is unitary with respect to the indefinite inner product $[x, y] = \langle Jx, y\rangle$.

Let W be a rational $m \times m$ matrix function. By definition the *dual* of W with respect to the imaginary axis is the function $W^{\#}(\lambda) = W(-\bar{\lambda})^*$. In particular, $W^{\#}(s) = W(s)^*$ for $s \in i\mathbb{R}$ and s not a pole of W. We say that W is *J-unitary* with respect to the imaginary axis if $W^{\#}(\lambda)JW(\lambda) = J$ whenever λ is not a pole of W. Equivalently, W is J-unitary if the values of W on the imaginary axis are J-unitary matrices.

The next theorem characterizes J-unitary rational matrix functions in terms of minimal realizations.

Theorem 2.1.1 *Let $W(\lambda) = D + C(\lambda - A)^{-1}B$ be a minimal realization with state space $\mathcal{X}$. Then W is J-unitary if and only if the feedthrough coefficient D is a J-unitary matrix and there exists a (unique) invertible linear operator $H : \mathcal{X} \to \mathcal{X}$ such that $H^* = H$ and*

$$HA + A^*H = -C^*JC, \qquad HB = -C^*JD.$$

Proof We first prove the necessity of the conditions. Assume $W^*(\lambda)JW(\lambda) = J$. Notice that

$$W^{\#}(\lambda) = D^* - B^*(\lambda + A^*)^{-1}C^*.$$

Taking the limit for $\lambda \to \infty$ in $W^{\#}(\lambda)JW(\lambda) = J$ yields $D^*JD = J$, and thus D is a J-unitary matrix. In particular, D is invertible. So (cf., formula (1.3.3)), we may write

$$W(\lambda)^{-1} = D^{-1} - D^{-1}C(\lambda - A^{\times})^{-1}BD^{-1}, \tag{2.1.2}$$

where $A^\times = A - BD^{-1}C$. Also, $W(\lambda)^{-1} = JW^\#(\lambda)J$. Thus

$$W(\lambda)^{-1} = JD^*J + JB^*(\lambda + A^*)^{-1}(-C^*J). \tag{2.1.3}$$

The right hand sides of (2.1.2) and (2.1.3) are minimal realizations of $W(\cdot)^{-1}$. So the state space similarity theorem (Theorem 1.1.3) tells us that the two realizations are similar, and the corresponding similarity is unique. Hence there exists a unique invertible linear operator $H : \mathcal{X} \to \mathcal{X}$ such that

$$H(A - BD^{-1}C)H^{-1} = -A^*, \quad HBD^{-1} = -C^*J, \quad -D^{-1}CH^{-1} = JB^*. \tag{2.1.4}$$

The first two equalities in (2.1.4) are equivalent to

$$HA + A^*H = -C^*JC, \qquad HB = -C^*JD. \tag{2.1.5}$$

The third equality in (2.1.4) and $D^{-1} = JD^*J$ yield $H^*B = -C^*JD$. So (2.1.5) remains true if H in (2.1.5) is replaced by H^*. Furthermore, the second equation in (2.1.4) yields the third with H^* in place of H. So (2.1.4) remains true if H is replaced by H^*. But the state space similarity is unique. Therefore, $H = H^*$.

Next, we prove the sufficiency of the conditions and the uniqueness of H. Assume D is J-unitary and (2.1.2) holds, where $H : \mathcal{X} \to \mathcal{X}$ is an invertible selfadjoint operator. It follows (by reversing the arguments in the previous paragraph) that (2.1.4) holds. In other words, the realizations $[A^\times, BD^{-1}, -D^{-1}C, D^{-1}]$ and $[-A^*, -C^*J, JB^*, JD^*J]$ are similar, with H being the similarity. Hence the corresponding transfer functions $W(\lambda)^{-1}$ and $JW^\#(\lambda)J$ are equal, and therefore W is J-unitary. The uniqueness of H follows from the uniqueness of the similarity in the state space similarity theorem. □

2.2 J-inner rational matrix functions

We begin this section with two definitions. An $m \times m$ matrix A is said to be *J-contractive* if $A^*JA \leq J$ or, equivalently, if A is contractive with respect to the indefinite inner product $[x, y] = \langle Jx, y\rangle$ on $\mathbb{C}^m$. A rational $m \times m$ matrix function W is called *J-inner* with respect to the imaginary axis if W is proper and has no poles on the imaginary axis, W is J-unitary with respect to the imaginary axis, and $W(\lambda)$ is J-contractive for each λ in the open right half plane which is not a pole of W.

The next theorem further specifies Theorem 2.1.1 for J-inner rational matrix functions. Notice that the equations in (2.1.2) and (2.2.1) are the same; the difference is that in the J-inner case the solution H is required to be positive definite.

Theorem 2.2.1 *Let $W(\lambda) = D + C(\lambda - A)^{-1}B$ be a minimal realization with state space $\mathcal{X}$. Then W is J-inner if and only if A has no eigenvalue on the imaginary axis, the feedthrough coefficient D is a J-unitary matrix, and there exists a (unique) linear operator $H : \mathcal{X} \to \mathcal{X}$ such that H is positive definite and*

$$HA + A^*H = -C^*JC, \qquad HB = -C^*JD. \tag{2.2.1}$$

To prove Theorem 2.2.1 we need an auxilary result about J-contractive matrices for the case when the signature matrix J has the form

$$J = \begin{pmatrix} I_p & 0 \\ 0 & -I_q \end{pmatrix}. \tag{2.2.2}$$

Lemma 2.2.2 *Let A be a J-contractive matrix, where J is given by (2.2.2). Partition A according to the partitioning of J as follows*

$$A = \begin{pmatrix} A_{11} & A_{12} \\ A_{21} & A_{22} \end{pmatrix} : \mathbb{C}^p \oplus \mathbb{C}^q \to \mathbb{C}^p \oplus \mathbb{C}^q. \tag{2.2.3}$$

Then A_{22} is non-singular and the matrix

$$T(A) = \begin{pmatrix} A_{11} - A_{12}A_{22}^{-1}A_{21} & A_{12}A_{22}^{-1} \\ -A_{22}^{-1}A_{21} & A_{22}^{-1} \end{pmatrix}. \tag{2.2.4}$$

is contractive. Moreover, the operator $A_{22}^{-1}A_{21}$ is a strict contraction.

Proof By specifying the assumption $A^*JA \leq J$ for the $(2,2)$-block entries we see that

$$A_{12}^*A_{12} - A_{22}^*A_{22} \leq -I_q.$$

Thus $A_{22}^*A_{22} \geq I + A_{12}^*A_{12}$, and hence $A_{22}^*A_{22}$ is positive definite. This together with the fact that A_{22} is a square matrix implies that A_{22} is invertible. Next note that

$$A\begin{pmatrix} x_1 \\ x_2 \end{pmatrix} = \begin{pmatrix} y_1 \\ y_2 \end{pmatrix} \iff T(A)\begin{pmatrix} x_1 \\ y_2 \end{pmatrix} = \begin{pmatrix} y_1 \\ x_2 \end{pmatrix} \tag{2.2.5}$$

Since A is J-contractive, the left hand side of (2.2.5) yields

$$\begin{aligned} \|y_1\|^2 - \|y_2\|^2 &= \begin{pmatrix} y_1^* & y_2^* \end{pmatrix} J \begin{pmatrix} y_1 \\ y_2 \end{pmatrix} &&= \begin{pmatrix} x_1^* & x_2^* \end{pmatrix} A^*JA \begin{pmatrix} x_1 \\ x_2 \end{pmatrix} \\ &\leq \begin{pmatrix} x_1^* & x_2^* \end{pmatrix} J \begin{pmatrix} x_1 \\ x_2 \end{pmatrix} &&= \|x_1\|^2 - \|x_2\|^2, \end{aligned}$$

and thus

$$\|T(A)\begin{pmatrix} x_1 \\ y_2 \end{pmatrix}\|^2 = \|y_1\|^2 + \|x_2\|^2 \leq \|x_1\|^2 + \|y_2\|^2 = \|\begin{pmatrix} x_1 \\ y_2 \end{pmatrix}\|^2,$$

which shows that $T(A)$ is contractive.

Since $T(A)$ is a contraction, $T(A)^*$ is also a contraction, and thus $T(A)T(A)^* \leq I$. From the $(2,2)$-block entries in the latter inequality we see that

$$A_{22}^{-1}A_{21}A_{21}^*A_{22}^{-*} + A_{22}^{-1}A_{22}^{-*} \leq I_q.$$

Now use that $A_{22}^{-1}A_{22}^{-*}$ is positive definite to conclude that $A_{22}^{-1}A_{21}A_{21}^*A_{22}^{-*} \leq (1-\varepsilon)I_q$ for some $0 < \varepsilon < 1$. Hence $\|A_{21}^*A_{22}^{-*}\| < 1$, and $A_{21}^*A_{22}^{-*}$ is a strict contraction. By duality, the same is true for $A_{22}^{-1}A_{21}$. □

The map $A \mapsto T(A)$ defined by (2.2.4) is known as the *Redheffer transform* or as the *Ginzburg-Potapov transform* when the entries of A are matrix functions. Notice that $T(A)$ is well-defined whenever the $(2,2)$-entry A_{22} in (2.2.3) is non-singular. Moreover, in this case, $T\big(T(A)\big) = A$.

Proof of Theorem 2.2.1 Let $W(\lambda) = D + C(\lambda - A)^{-1}B$ be a minimal realization with state space $\mathcal{X}$. Clearly, W is proper, and the minimality of the realization implies (use Corallary 1.1.4) that W has no pole on the imaginary axis if and only if A has no eigenvalue on the imaginary axis. Furthermore, by Theorem 2.1.1, the function W is J-unitary if and only if D is a J-unitary matrix and there exists a (unique) invertible linear operator $H : \mathcal{X} \to \mathcal{X}$ such that $H^* = H$ and (2.2.1) holds. It remains to show that $W(\lambda)$ is J-contractive for each λ in the open right half plane which is not a pole of W if and only if the operator H in (2.2.1) is positive definite. To do this we may assume that D is J-unitary and that (2.2.1) hold with H invertible and H selfadjoint. It will be convenient to split the remaining part of the proof into two parts.

Part (a). In this part we assume that $W(\lambda)$ is J-contractive for each λ in the open right half plane which is not a pole of W. Let us denote the set of these λ's by Ω^+. We have to show that H is positive definite. Since H is invertible, it suffices to show that $H \geq 0$.

Recall that D is a J-unitary matrix. By replacing $W(\lambda)$ by $W(\lambda)D^{-1}$, we see that without loss of generality we may assume that $D = I$. Furthermore, there exists a unitary matrix U such that $\widetilde{J} := UJU^{-1}$ is equal to the right hand side of (2.2.2) for a suitable choice of p and q. Notice that $\widetilde{W}(\lambda) := UW(\lambda)U^{-1}$ is a $\widetilde{J}$-unitary rational matrix function, and for each $\lambda \in \Omega^+$ the matrix $\widetilde{W}(\lambda)$ is $\widetilde{J}$-contractive. Moreover, (2.2.1) with $D = I$ remains valid if J is replaced by $\widetilde{J}$, the matrix B by BU^{-1}, and C by UC. It follows that without loss of generality we may also assume that J is given by (2.2.2).

So, take $D = I$, and let J be as in (2.2.2). Partition $W(\lambda)$ according to the partitioning of J as follows

$$W(\lambda) = \begin{pmatrix} W_{11}(\lambda) & W_{12}(\lambda) \\ W_{21}(\lambda) & W_{22}(\lambda) \end{pmatrix} : \mathbb{C}^p \oplus \mathbb{C}^q \to \mathbb{C}^p \oplus \mathbb{C}^q.$$

Furthermore, put

$$B = \begin{pmatrix} B_1 & B_2 \end{pmatrix} : \mathbb{C}^p \oplus \mathbb{C}^q \to \mathcal{X}, \quad C = \begin{pmatrix} C_1 \\ C_2 \end{pmatrix} : \mathcal{X} \to \mathbb{C}^p \oplus \mathbb{C}^q.$$

It follows that $W_{22}(\lambda) = I + C_2(\lambda - A)^{-1}B_2$. In particular, $\det W_{22}(\lambda)$ does not vanish identically, and we may consider $V(\lambda) = T(W(\lambda))$, where T is the Redheffer transform defined by (2.2.4). Put

$$\widetilde{A} = A - B_2C_2 = A + \tfrac{1}{2}B(J - I)C. \tag{2.2.6}$$

Straightforward computations, using that $B_2C_2 = (\lambda - \widetilde{A}) - (\lambda - A)$, yield

$$\begin{aligned}
&W_{22}(\lambda)^{-1} = I_q - C_2(\lambda - \widetilde{A})^{-1}B_2, \\
&W_{12}(\lambda)W_{22}(\lambda)^{-1} = C_1(\lambda - \widetilde{A})^{-1}B_2, \\
&W_{22}(\lambda)^{-1}W_{21}(\lambda) = C_2(\lambda - \widetilde{A})^{-1}B_1, \\
&W_{11}(\lambda) - W_{12}(\lambda)W_{22}(\lambda)^{-1}W_{21}(\lambda) = I_p + C_1(\lambda - \widetilde{A})^{-1}B_1,
\end{aligned}$$

and hence

$$V(\lambda) = I + JC(\lambda - \widetilde{A})^{-1}B. \tag{2.2.7}$$

Since $I + C(\lambda - A)^{-1}B$ is a minimal realization of W, the right hand side of (2.2.7) is a minimal realization of V. Indeed, if this is not the case, then V has a realization with a smaller state space dimension. Since $W(\lambda) = T(V(\lambda))$, we can repeat the computations in the previous paragraph (with the roles of V and W reversed) using this new realization of V, and this would give a realization of W of which the state space dimension is smaller than dim $\mathcal{X}$. The latter is impossible, and thus the right hand side of (2.2.7) is a minimal realization.

By Lemma 2.2.2, for $\lambda \in \Omega^+$ we have $\|V(\lambda)\| \leq 1$. Notice that up to a finite number of points Ω^+ is equal to the open right half plane Π^+. Since V is analytic and uniformly bounded on Ω^+, if follows that V does not have poles on its closure $\overline{\Omega^+}$. Thus V is analytic at all points of $\overline{\Pi^+}$, and therefore the poles of V are in the open left half plane Π^-. But then, because of the minimality of the realization (2.2.7), we may use Corollary 1.1.4 to show that the eigenvalues of $\widetilde{A}$ are in Π^-.

From (2.2.1) with $D = I$ and (2.2.6) it follows that $H\widetilde{A} + \widetilde{A}^*H = -C^*C$. Since $\widetilde{A}$ has its eigenvalues in the open left half plane, this yields

$$H = \int_0^\infty e^{t\widetilde{A}^*} C^*Ce^{t\widetilde{A}}\, dt,$$

and hence $H \geq 0$.

Part (b). In this part we assume that H in (2.2.1) is positive definite, and we show that for each $\lambda \in \Omega^+$ the matrix $W(\lambda)$ is J-contractive. Fix $\lambda \in \Omega^+$. Since the realization $W(\lambda) = I + C(\lambda - A)^{-1}B$ is minimal, therefore $\lambda - A$ is invertible (by Corollary 1.1.4). Notice that

$$\begin{aligned} J - W(\lambda)^*JW(\lambda) \quad &= -JC(\lambda - A)^{-1}B - B^*(\bar{\lambda} - A^*)^{-1}C^*J - \\ &\quad -B^*(\bar{\lambda} - A^*)^{-1}C^*JC(\lambda - A)^{-1}B. \end{aligned}$$

We may use the first identity in (2.2.1) to rewrite C^*JC in the following form

$$C^*JC = H(\lambda - A) + (\bar{\lambda} - A^*)H - (\lambda + \bar{\lambda})H.$$

Hence

$$\begin{aligned} (\bar{\lambda} - A^*)^{-1}C^*JC(\lambda - A)^{-1} \quad &= (\bar{\lambda} - A^*)^{-1}H + H(\lambda - A)^{-1} - \\ &\quad -(\lambda + \bar{\lambda})(\bar{\lambda} - A^*)^{-1}H(\lambda - A)^{-1}. \end{aligned}$$

Also, $HB = -C^*J$. Thus

$$J - W(\lambda)^*JW(\lambda) = (\lambda + \bar{\lambda})B^*(\bar{\lambda} - A^*)^{-1}H(\lambda - A)^{-1}B. \qquad (2.2.8)$$

Recall that $\lambda \in \Omega^+ \subset \Pi^+$. Thus $\lambda + \bar{\lambda} > 0$. Also, H is positive definite. We see that the right hand side of (2.2.8) is positive semi-definite, and therefore $W(\lambda)$ is J-contractive. □

The following corollary (which does not require the realization to be minimal) will be used in the next section.

Corollary 2.2.3 *Let $W(\lambda) = I + C(\lambda - A)^{-1}B$, where A has its eigenvalues in the open left half plane, and let H be a selfadjoint operator such that*

$$HA + A^*H = -C^*JC, \qquad HB = -C^*J. \qquad (2.2.9)$$

Then W is J-unitary, and W is J-inner if and only if H is positive semi-definite.

Proof First we show that without loss of generality we may assume that the operator H is invertible.

Let $\mathcal{X}$ be the state space of the realization. Put $\mathcal{X}_0 = \operatorname{Ker} H$, and $\mathcal{X}_1 = \operatorname{Im} H$. Then

$$H = \begin{pmatrix} 0 & 0 \\ 0 & H_1 \end{pmatrix} : \mathcal{X}_0 \oplus \mathcal{X}_1 \to \mathcal{X}_0 \oplus \mathcal{X}_1, \tag{2.2.10}$$

where H_1 is an invertible selfadjoint operator. Partition A, B and C according to the partitioning of H in (2.2.10) as follows

$$\begin{aligned} A &= \begin{pmatrix} A_{00} & A_{01} \\ A_{10} & A_{11} \end{pmatrix} : \mathcal{X}_0 \oplus \mathcal{X}_1 \to \mathcal{X}_0 \oplus \mathcal{X}_1, \\ B &= \begin{pmatrix} B_0 \\ B_1 \end{pmatrix} : \mathbb{C}^m \to \mathcal{X}_0 \oplus \mathcal{X}_1, \\ C &= \begin{pmatrix} C_0 & C_1 \end{pmatrix} : \mathcal{X}_0 \oplus \mathcal{X}_1 \to \mathbb{C}^m. \end{aligned}$$

Take $y \in \mathcal{X}_0 = \operatorname{Ker} H$. Then (2.2.9) yields

$$Cy = -JB^*Hy = 0, \qquad HAy = -A^*Hy - C^*JCy = 0.$$

Thus $C_0 = 0$ and $A_{10} = 0$. Therefore, the realization $[A, B, C, I]$ is a dilation of $[A_{11}, B_1, C_1, I]$. It follows that $W(\lambda) = I + C_1(\lambda - A_1)^{-1}B_1$ and the eigenvalues of A_{11} are in the open left half plane Π^-. Furthermore, one computes that

$$H_1A_{11} + A_{11}^*H_1 = -C_1^*JC_1, \qquad H_1B_1 = -C_1^*J.$$

So, in order to prove the corollary we may as well assume that H is invertible.

Assume H is invertible. We shall show that in this case our conditions imply that the given realization of W is minimal. Put $A^\times = A - BC$. Then

$$HA^\times + A^*H = HA + A^*H - HBC = -C^*JC - (-C^*J)C = 0,$$

and therefore $HA^\times H^{-1} = -A^*$. The latter implies that the eigenvalues of $A^\times$ are in the open right half plane. Hence $A^\times$ and A have no common eigenvalues. It follows that the realization $[A, B, C, I]$ is minimal. Indeed, assume not. Then, by Theorem 1.1.2, the realization $[A, B, C, I]$ is a proper dilation of a minimal realization $[A_0, B_0, C_0, I]$, say. The operators A, B, C and A_0, B_0, C_0 are related in the following way:

$$A = \begin{pmatrix} A_1 & * & * \\ 0 & A_0 & * \\ 0 & 0 & A_2 \end{pmatrix}, \quad B = \begin{pmatrix} * \\ B_0 \\ 0 \end{pmatrix}, \quad C = \begin{pmatrix} 0 & C_0 & * \end{pmatrix}, \tag{2.2.11}$$

where at least one of the operators A_1, A_2 act on a non-zero space. (The $*$'s in (2.2.11) denote unspecified operator entries.) From (2.2.11) we see that $A^\times$ has the same structure as A; more precisely,

$$A^\times = \begin{pmatrix} A_1 & * & * \\ 0 & A_0^\times & * \\ 0 & 0 & A_2 \end{pmatrix},$$

where $A_0^\times = A_0 - B_0C_0$. In particular, the eigenvalues of A_1 and A_2 are eigenvalues of both A and $A^\times$, which contradicts the fact that A and $A^\times$ have no common eigenvalues. Thus the realizaton $[A, B, C, I]$ is minimal.

So, we may assume that H is an invertible selfadjoint operator, and that $W(\lambda) = I + C(\lambda - A)^{-1}B$ is a minimal realization. Then (2.2.9) implies that

W is J-unitary (by Theorem 2.1.1), and we can use Theorem 2.2.1 to show that W is J-inner if and only if $H \geq 0$. □

2.3 J-inner-outer factorization in $\mathcal{R}at(H_\infty)$

Throughout this section Π^- stands for the open left half plane, and Π^+ for the open right half plane.

We say that a matrix function G belongs to the class $\mathcal{R}at(H_\infty)$ if G is a rational matrix function with the additional property that the poles of G are in Π^-. To explain the notation, notice that in this case the entries of G are bounded analytic functions on Π^+, and hence these entries belong to the Hardy space H_∞. Conversely, if G is a rational matrix function of which the entries are in H_∞, then G cannot have poles in $\overline{\Pi^+}$ and at infinity, and hence $G \in \mathcal{R}at(H_\infty)$. Often one refers to the elements of $\mathcal{R}at(H_\infty)$ as *stable* rational matrix functions. We write $G \in \mathcal{R}at(H_\infty^{p\times q})$ if $G \in \mathcal{R}at(H_\infty)$ and has size $p \times q$.

A matrix function $R \in \mathcal{R}at(H_\infty^{m\times m})$ is called *outer* if $\det R(\lambda) \neq 0$ for each $\lambda \in \overline{\Pi^+}$ and for $\lambda = \infty$. In this case, we also say that R is *biregular* on $\overline{\Pi^+}$ and at infinity. Notice that a rational $m \times m$ matrix function R is outer if and only if $R(\cdot)$ and $R(\cdot)^{-1}$ are both in $\mathcal{R}at(H_\infty)$.

A matrix function $G \in \mathcal{R}at(H_\infty^{m\times m})$ is said to admit a *J-inner-outer factorization* if $G(\lambda) = \Theta(\lambda)R(\lambda)$, where Θ is J-inner and R is outer. In this case, $\Theta(\lambda) = G(\lambda)R(\lambda)^{-1}$, and hence Θ has to be in $\mathcal{R}at(H_\infty)$. In this section we are interested in constructing such factorizations.

Let $G = \Theta R$ be a J-inner-outer factorization with $G \in \mathcal{R}at(H_\infty)$. Since Θ is J-inner, Θ has no poles on the imaginary axis and at infinity. Moreover, the values of Θ on $i\mathbb{R}$ and at infinity are J-unitary matrices, and hence $\det \Theta(s) \neq 0$ for $s \in i\mathbb{R} \cup \{\infty\}$. Also, $\det R(s) \neq 0$ for $s \in i\mathbb{R} \cup \{\infty\}$, because R is biregular on $\overline{\Pi^+}$ and at infinity. We see that necessarily

$$\det G(s) \neq 0, \qquad s \in i\mathbb{R} \cup \{\infty\}. \tag{2.3.1}$$

Furthermore,

$$G^\#(\lambda)JG(\lambda) = R^\#(\lambda)\Theta^\#(\lambda)J\Theta(\lambda)R(\lambda) = R^\#(\lambda)JR(\lambda).$$

Now, put $W_+(\lambda) = JR(\lambda)$ and $W_-(\lambda) = R^\#(\lambda)$. Then

$$G^\#(\lambda)JG(\lambda) = W_-(\lambda)W_+(\lambda), \tag{2.3.2}$$

and the factors W_+ and W_- are proper rational $m \times m$ matrix functions satisfying

1. W_+ has no poles on Re $\lambda \geq 0$ and $\det W_+(\lambda) \neq 0$ for Re $\lambda \geq 0$,
2. W_- has no poles on Re $\lambda \leq 0$ and $\det W_-(\lambda) \neq 0$ for Re $\lambda \leq 0$,
3. $\det W_\pm(\infty) \neq 0$.

Here Re λ stands for the real part of λ. It follows that (2.3.2) is a right canonical factorization with respect to the imaginary axis.

The remarks made in the previous paragraph suggest the following method to construct a J-inner-outer factorization for a matrix function $G \in \mathcal{R}at(H_\infty^{m\times m})$. First one constructs (if possible) a right canonical factorization of the left hand side of (2.3.2) by using the state space procedure summarized by Theorem 1.3.1 with the real line replaced by the imaginary axis. This determines the outer factor R up to multiplication on the left by a non-singular matrix. Next, one computes Θ

from the identity $\Theta = GR^{-1}$, and one checks that Θ has the desired properties. It turns out that this method works. The following theorem is the main result of this section.

Theorem 2.3.1 *Let $G(\lambda) = D + C(\lambda - A)^{-1}B$ be a realization with D non-singular, and assume that A has its eigenvalues in Π^-. Put $A^\times = A - BD^{-1}C$, and assume that $A^\times$ has no eigenvalue on the imaginary axis. Then G admits a J-inner-outer factorization if and only if the algebraic Riccati equation*

$$XA^\times + (A^\times)^*X - XBD^{-1}JD^{-*}B^*X = 0 \tag{2.3.3}$$

has a solution X satisfying the following two conditions:

(i) *X is positive semi-definite,*

(ii) *$K := A^\times - BD^{-1}JD^{-*}B^*X$ has its eigenvalues in Π^-.*

Furthermore, if these two conditions are satisfied, then a J-inner-outer factorization is obtained by taking

$$R(\lambda) = D + (C + JD^{-*}B^*X)(\lambda - A)^{-1}B, \tag{2.3.4}$$

$$\Theta(\lambda) = I - JD^{-*}B^*X(\lambda - K)^{-1}BD^{-1}. \tag{2.3.5}$$

Proof We split the proof into 5 parts. In the first four parts we assume that G admits a J-inner-outer factorization.

Part (a). In this part we construct a particular solution of equation (2.3.3). Put $W(\lambda) = G^\#(\lambda)JG(\lambda)$. From the given realization for G we derive a realization for W. Put

$$\alpha = \begin{pmatrix} -A^* & C^*JC \\ 0 & A \end{pmatrix}, \quad \beta = \begin{pmatrix} C^*JD \\ B \end{pmatrix}, \quad \gamma = \begin{pmatrix} -B^* & D^*JC \end{pmatrix}, \quad \delta = D^*JD.$$

Then

$$W(\lambda) = \delta + \gamma(\lambda - \alpha)^{-1}\beta. \tag{2.3.6}$$

Note that the realization (2.3.6) has state space $\mathcal{Y} = \mathcal{X} \oplus \mathcal{X}$, where $\mathcal{X}$ is the state space of the given realization for G. Since D is non-singular, the same is true for δ, and hence we may consider

$$\alpha^\times = \alpha - \beta\delta^{-1}\gamma = \begin{pmatrix} -(A^\times)^* & 0 \\ BD^{-1}JD^{-*}B^* & A^\times \end{pmatrix}.$$

The fact that $A^\times$ has no eigenvalue on $i\mathbb{R}$, implies that $\alpha^\times$ has the same property.

Since G admits a J-inner-outer factorization, we know (cf., formula (2.3.2)) that W has a right canonical factorization with respect to the imaginary axis. This allows us to apply Theorem 1.3.1 (with the imaginary axis in place of the real line). It follows that

$$\mathcal{Y} = \operatorname{Im} P \dot{+} \operatorname{Ker} P^\times, \tag{2.3.7}$$

where P and $P^\times$ are the Riesz projections of α and $\alpha^\times$, respectively, corresponding to the eigenvalues in Π^+. From the definition of α it follows that

$$(\lambda - \alpha)^{-1} = \begin{pmatrix} (\lambda + A^*)^{-1} & (\lambda + A^*)^{-1}C^*JC(\lambda - A)^{-1} \\ 0 & (\lambda - A)^{-1} \end{pmatrix},$$

and hence, because of the fact that A has its eigenvalues in Π^-, we have

$$P = \begin{pmatrix} I & * \\ 0 & 0 \end{pmatrix} : \mathcal{X} \oplus \mathcal{X} \to \mathcal{X} \oplus \mathcal{X}.$$

In particular,

$$\operatorname{Im} P = \operatorname{Im} \begin{pmatrix} I \\ 0 \end{pmatrix} = \{ \begin{pmatrix} x \\ 0 \end{pmatrix} \mid x \in \mathcal{X} \}. \tag{2.3.8}$$

But then the matching in (2.3.7) tells us that there exists a (unique) linear operator $X : \mathcal{X} \to \mathcal{X}$ such that

$$\operatorname{Ker} P^\times = \operatorname{Im} \begin{pmatrix} -X \\ I \end{pmatrix} = \{ \begin{pmatrix} -Xx \\ x \end{pmatrix} \mid x \in \mathcal{X} \}. \tag{2.3.9}$$

Now, let us show that this operator X satisfies (2.3.3). Since $\operatorname{Ker} P^\times$ is invariant under $\alpha^\times$, we must have

$$\begin{pmatrix} -(A^\times)^* & 0 \\ BD^{-1}JD^{-*}B^* & A^\times \end{pmatrix} \begin{pmatrix} -X \\ I \end{pmatrix} = \begin{pmatrix} -X \\ I \end{pmatrix} K \tag{2.3.10}$$

for some linear operator $K : \mathcal{X} \to \mathcal{X}$. By comparing the second rows in the left and right hand sides of (2.3.10), we see that

$$K = A^\times - BD^{-1}JD^{-*}B^*X. \tag{2.3.11}$$

The first rows yield $(A^\times)^*X + XK = 0$, and hence X must satisfy (2.3.3).

Part (b). In this part we show that the operator K in (2.3.11) has its eigenvalues in Π^-. Define

$$S : \mathcal{X} \to \operatorname{Ker} P^\times, \qquad Sx = \begin{pmatrix} -X \\ I \end{pmatrix} x.$$

According to (2.3.9) the operator S is bijective. Formula (2.3.10) shows that

$$SK = (\alpha^\times | \operatorname{Ker} P^\times)S,$$

and hence K and $\alpha^\times | \operatorname{Ker} P^\times$ have the same eigenvalues. Since $P^\times$ is the Riesz projection of $\alpha^\times$ corresponding to the eigenvalues in Π^+, the eigenvalues of $\alpha^\times | \operatorname{Ker} P^\times$ are in Π^+, and thus the same is true for the eigenvalues of K.

Part (c). In this part we show that operator X defined by (2.3.9) is selfadjoint. Put

$$M = \begin{pmatrix} -X \\ I \end{pmatrix}, \qquad Q = \begin{pmatrix} 0 & I \\ -I & 0 \end{pmatrix}.$$

Obviously, $Q^* = -Q$. Furthermore,

$$Q\alpha^\times = \begin{pmatrix} BD^{-1}JD^{-*}B^* & A^\times \\ (A^\times)^* & 0 \end{pmatrix},$$

and hence $Q\alpha^\times$ is selfadjoint. By (2.3.10), we have $\alpha^\times M = MK$, and thus

$$M^*QMK = M^*(Q\alpha^\times)M$$

is selfadjoint too. It follows that $M^*QMK = K^*M^*Q^*M = -K^*M^*QM$. Thus M^*QM is a solution of the following Lyapunov equation:

$$(M^*QM)K + K^*(M^*QM) = 0. \tag{2.3.12}$$

Since the eigenvalues of K are in Π^-, equation (2.3.12) has only the trivial solution (cf., Theorem I.5.5 in Gohberg-Goldberg-Kaashoek [1990]), and thus

$$0 = M^*QM = \begin{pmatrix}-X^* & I\end{pmatrix}\begin{pmatrix}0 & I\\ -I & 0\end{pmatrix}\begin{pmatrix}-X\\ I\end{pmatrix} = X - X^*.$$

Hence $X^* = X$.

Part (d). In this part we prove that X in (2.3.9) is positive semi-definite. For this purpose we first compute for $W(\lambda) = G^\#(\lambda)JG(\lambda)$ a right canonical factorization with respect to $i\mathbb{R}$. From (2.3.8) and (2.3.9) it follows that the projection Π of $\mathcal{Y}$ along $\operatorname{Im} P$ onto $\operatorname{Ker} P^\times$ is given by

$$\Pi = \begin{pmatrix}0 & -X\\ 0 & I\end{pmatrix}. \tag{2.3.13}$$

Now put

$$W_+(\lambda) = I + \delta^{-1}\gamma\Pi(\lambda-\alpha)^{-1}\beta, \quad W_-(\lambda) = \delta + \gamma(\lambda-\alpha)^{-1}(I-\Pi)\beta.$$

Then, according to Theorem 1.3.1, we have $W = W_-W_+$ and this factorization is a right canonical factorization with respect to $i\mathbb{R}$. Since Π is given by (2.3.12), we can use the definitions of α, β, γ, and δ to show that

$$W_+(\lambda) = I + (D^*JD)^{-1}(D^*JC + B^*X)(\lambda - A)^{-1}B. \tag{2.3.14}$$

Now, let $G = \Theta R$ be a J-inner-outer factorization of G. Without loss of generality we may assume that $\Theta(\infty) = I$, and hence $R(\infty) = D$. Since

$$W(\lambda) = G^\#(\lambda)JG(\lambda) = R^\#(\lambda)JR(\lambda),$$

we also know that $R(\lambda)$ is a plus factor in a right canonical factorization of W too. Thus R is equal to W_+ up to multiplication on the left by a constant non-singular matrix. By comparing the values at infinity, we see that $R(\lambda) = DW_+(\lambda)$. But then we can use (2.3.14) to show that

$$R(\lambda) = D + (C + JD^{-*}B^*X)(\lambda - A)^{-1}B. \tag{2.3.15}$$

Next, we compute Θ from $\Theta = GR^{-1}$. Formula (2.3.15) and (2.3.11) yield

$$R(\lambda)^{-1} = D^{-1} - D^{-1}(C + JD^{-*}B^*X)(\lambda - K)^{-1}BD^{-1}. \tag{2.3.16}$$

Hence

$$\begin{aligned}\Theta(\lambda) &= (D + C(\lambda-A)^{-1}B)R(\lambda)^{-1}\\ &= I - (C + JD^{-*}B^*X)(\lambda-K)^{-1}BD^{-1} + C(\lambda-A)^{-1}BD^{-1} -\\ &\qquad -C(\lambda-A)^{-1}\{BD^{-1}C + BD^{-1}JD^{-*}B^*X\}(\lambda-K)^{-1}BD^{-1}.\end{aligned}$$

According to (2.3.11)

$$BD^{-1}C + BD^{-1}JD^{-*}B^*X = (\lambda - K) - (\lambda - A), \tag{2.3.17}$$

and thus the J-inner factor Θ is given by

$$\Theta(\lambda) = I - JD^{-*}B^*X(\lambda - K)^{-1}BD^{-1}. \tag{2.3.18}$$

To see that $X \geq 0$, we apply Corollary 2.2.3 to Θ. Since X satisfies (2.3.3), we have

$$XK + K^*X = -(JD^{-*}B^*X)^*J(JD^{-*}B^*X). \tag{2.3.19}$$

Moreover, $XBD^{-1} = -(-JD^{-*}B^*X)^*J$. Thus, by Corollary 2.2.3, the fact that Θ is J-inner yields $X \geq 0$.

Notice that formula (2.3.15) for R is equal to (2.3.4), and formula (2.3.18) for Θ is equal to (2.3.5). Thus, assuming that G admits a J-inner-outer factorization, we have shown that equation (2.3.3) has a solution X satisfying the conditions (i) and (ii). Moreover, we have shown that a J-inner-outer factorization $G = \Theta R$ is obtained by taking R as in (2.3.4) and Θ as in (2.3.5).

Part (e). This part concerns the sufficiency of the conditions. We assume that equation (2.3.3) has a solution X satisfying conditions (i) and (ii), and we prove that G admits a J-inner-outer factorization. To do this we repeat some of the previous arguments in the reverse direction.

Let R and Θ be as in (2.3.4) and (2.3.5). From (ii) it follows that (2.3.17) is valid. Hence, by a straightforward computation, using (2.3.17), one sees that $G = \Theta R$. Obviously, $R \in \mathcal{R}at(H_\infty)$, because the eigenvalues of A are in Π^-. From (2.3.17) it also follows that $R(\cdot)^{-1}$ is given by (2.3.16). According to condition (ii) the eigenvalues of K are in Π^-, and formula (2.3.16) shows that $R(\cdot)^{-1} \in \mathcal{R}at(H_\infty)$. Thus R is outer. Next, we prove that Θ is J-inner. Since X is a solution of (2.3.3), the definition of K in (ii) yields the identity (2.3.19). Also, $XBD^{-1} = -(-JD^{-*}B^*X)^*J$. But then we can apply Corollary 2.3.3 to show that Θ is J-inner, because of condition (i). □

2.4 Notes

Theorem 2.1.1 is taken from Alpay-Gohberg [1982], see also Section 6.1 in Ball-Gohberg-Rodman [1990]. An earlier version of Theorem 2.1.1 for unitary rational matrix functions may be found in Glover [1984]. Theorem 2.2.1, which is a further refinement of Theorem 2.1.1, appears in disguised form in Ball-Gohberg-Rodman [1990] (cf., Theorem 6.2.2 and Corollary 13.1.5 in the latter book). The J-inner-outer factorization theorem in Section 2.3 is the "stable" version of the more general J-inner-outer factorization theorem in Ball-Kaashoek [to appear] (see also Sections IV.2 and IV.3 in Kaashoek-Ran [1991]); the result is also closely related to the J-spectral factorization theorem in Section 3 of Green-Glover-Limebeer-Doyle [1990]. We shall return to J-inner rational matrix functions in the fourth chapter. See also the supplementary notes in Chapter 1 of Dym [1989] for more information about the literature on J-contractive matrix functions.

CHAPTER 3

Analysis of zeros

In this chapter we analyze the zeros of a regular rational matrix function, and the corresponding eigenvectors and generalized eigenvectors. In the first section canonical systems of null functions are used to organize the zero data. In the second section canonical right and left null pairs are introduced, and their connection with Laurent principal parts is clarified. The third section concerns global left and right null pairs. The last section specifies the results for the analytic case.

3.1 Eigenvectors and canonical systems of null functions

Throughout this chapter $W = W(\lambda)$ is an $m \times m$ rational matrix function which is assumed to be regular (i.e., $\det W(\lambda)$ does not vanish identically). Fix a point μ in the complex plane $\mathbb{C}$, and consider the Laurent expansion of W at μ:

$$W(\lambda) = \sum_{j=-q}^{\infty} (\lambda - \mu)^j A_j. \tag{3.1.1}$$

We call μ a (*right*) *zero* of W if there exist vectors $x_0, \ldots, x_q$ in $\mathbb{C}^m$, $x_0 \neq 0$, such that

$$\begin{aligned} &A_{-q}x_0 = 0, \\ &A_{-q}x_1 + A_{-q+1}x_0 = 0, \\ &\quad\vdots \\ &A_{-q}x_q + \cdots + A_0 x_0 = 0. \end{aligned} \tag{3.1.2}$$

In this case x_0 is called a *right eigenvector* of W at μ. If W is analytic at μ, i.e., if $q = 0$, then the $q+1$ identities in (3.1.2) reduce to the more familiar formula $W(\mu)x_0 = 0$. In the definition of a zero we put the word "right" between parentheses; we shall see later (at the end of this section) that this prefix is superfluous.

Let $x_0, \ldots, x_q$ be vectors in $\mathbb{C}^m$, and construct the following vector polynomial:

$$\varphi(\lambda) = x_0 + (\lambda - \mu)x_1 + \cdots + (\lambda - \mu)^q x_q. \tag{3.1.3}$$

The condition $x_0 \neq 0$ is equivalent to $\varphi(\mu) \neq 0$. Furthermore, (3.1.2) is equivalent to the requirement that $W\varphi = W(\cdot)\varphi(\cdot)$ is analytic at μ and has the value zero at μ. It follows that the definitions of a zero and an eigenvector do not depend on the particular choice of the number q in (3.1.1).

The remarks made in the previous paragraph lead to the following definition. A $\mathbb{C}^m$-valued function φ is called a *right null function* of W at μ if φ is analytic in a neighborhood of μ, the vector $\varphi(\mu)$ is nonzero, the function $W\varphi$ is analytic at μ and $W\varphi$ is equal to zero at μ. Notice that in order for a right null function φ of W at μ to exist it is necessary and sufficient that μ is a (right) zero of W, and in this case $\varphi(\mu)$ is a right eigenvector of W at μ.

As an example, let us consider the 2×2 matrix function

$$V(\lambda) = \begin{pmatrix} 1 & \frac{c}{\lambda-\mu} \\ 0 & 1 \end{pmatrix} = \frac{1}{\lambda - \mu} \begin{pmatrix} 0 & c \\ 0 & 0 \end{pmatrix} + \begin{pmatrix} 1 & 0 \\ 0 & 1 \end{pmatrix}. \tag{3.1.4}$$

We take $c \neq 0$, and hence both $V(\lambda)$ and $V(\lambda)^{-1}$ have a pole at μ. For this example, the equations in (3.1.2) reduce to

$$\begin{pmatrix} 0 & c \\ 0 & 0 \end{pmatrix} \begin{pmatrix} x_{01} \\ x_{02} \end{pmatrix} = \begin{pmatrix} 0 \\ 0 \end{pmatrix},$$
$$\begin{pmatrix} 0 & c \\ 0 & 0 \end{pmatrix} \begin{pmatrix} x_{11} \\ x_{12} \end{pmatrix} + \begin{pmatrix} 1 & 0 \\ 0 & 1 \end{pmatrix} \begin{pmatrix} x_{01} \\ x_{02} \end{pmatrix} = \begin{pmatrix} 0 \\ 0 \end{pmatrix},$$

which yield

$$cx_{02} = 0, \quad cx_{12} + x_{01} = 0, \quad x_{02} = 0.$$

Thus μ is a zero of V, any right eigenvector of V at μ is a nonzero multiple of the vector $\binom{1}{0}$, and the right null functions φ of V at μ are given by

$$\varphi(\lambda) = \begin{pmatrix} a \\ 0 \end{pmatrix} + (\lambda - \mu) \begin{pmatrix} b \\ -a/c \end{pmatrix} + \text{terms of higher order}, \tag{3.1.5}$$

where $a \neq 0$ and b are free to choose. In particular, we see that a single eigenvector may be related to many different null functions.

Let μ be a zero of W. Since there may be many different right null functions of W at μ related to the same right eigenvector, it will be convenient to select a canonical set of such functions. Given a right null function φ of W at μ, we define the *order* of φ at μ to be the multiplicity of μ as a zero of $W\varphi$. Thus the order of φ is κ (> 0) if and only if

$$(W\varphi)(\lambda) = (\lambda - \mu)^{\kappa}\psi_{\kappa} + \text{terms of higher order}, \quad \psi_{\kappa} \neq 0.$$

Since $W(\lambda)^{-1}[W(\lambda)\varphi(\lambda)]$ is analytic at μ and $\varphi(\mu) \neq 0$, it follows that the order of φ does not exceed the order of μ as a pole of $W(\lambda)^{-1}$. Thus from the set of all right null functions of W at μ we may choose a null function φ_1 of maximal order. With φ_1 chosen, we pick from all right null functions φ of W at μ, with $\varphi(\mu) \notin \operatorname{span}\{\varphi_1(\mu)\}$, a null function φ_2 of highest order. We proceed in this way by induction. Assume the null functions $\varphi_1, \ldots, \varphi_{k-1}$ are chosen according ot the above method. We choose the next φ_k to be a null function of maximal order among all right null functions φ of W at μ such that $\varphi(\mu) \notin \operatorname{span}\{\varphi_1(\mu), \ldots, \varphi_{k-1}(\mu)\}$. This process stops at the moment when the vectors $\varphi_1(\mu), \ldots, \varphi_s(\mu)$ span the space consisting of the zero vector and the right eigenvectors of W at μ. Any set of null functions obtained in this manner is called a *canonical system of right null functions* of W at μ. The next lemma shows that the orders $\kappa_1 \geq \kappa_2 \geq \cdots \geq \kappa_s$ of the null functions in such a canonical system are uniquely determined by W; the canonical systems by themselves are not unique.

Lemma 3.1.1 *Let $\psi_1, \ldots, \psi_r$ be a set of right null functions of W at μ with orders $\rho_1 \geq \cdots \geq \rho_r$ such that the vectors $\psi_1(\mu), \ldots, \psi_r(\mu)$ are linearly independent, and let $\varphi_1, \ldots, \varphi_s$ be a canonical system of right null functions of W at μ with orders $\kappa_1 \geq \cdots \geq \kappa_s$. Then $s \geq r$ and $\kappa_i \geq \rho_i$ for $i = 1, \ldots, r$. Moreover, $\psi_1, \ldots, \psi_r$ is a canonical system of right null functions of W at μ if and only if $r = s$ and $\kappa_i = \rho_i$ for $i = 1, \ldots, r$.*

Proof Let $\mathrm{Ker}_{\mathcal{R}}(W;\mu)$ be the space consisting of the zero vector and the right eigenvectors of W at μ. Since $s = \dim\ \mathrm{Ker}_{\mathcal{R}}(W;\mu)$ and the set of eigenvectors $\psi_1(\mu),\dots,\psi_r(\mu)$ is independent in $\mathrm{Ker}_{\mathcal{R}}(W;\mu)$, we have $r \le s$.

From the definition of a canonical system we know that φ_1 is a right null function of maximal order. In particular, $\kappa_1 \ge \rho_1$. Fix $1 < k \le r$, and let us prove $\kappa_k \ge \rho_k$. Since the vectors $\psi_1(\mu),\dots,\psi_k(\mu)$ are linearly independent, there exists $1 \le j \le k$ such that $\psi_j(\mu)$ is not in $\mathrm{span}\{\varphi_1(\mu),\dots,\varphi_{k-1}(\mu)\}$. From the choice of φ_k it follows that $\kappa_k \ge \rho_j$. But $1 \le j \le k$ implies that $\rho_j \ge \rho_k$, and therefore $\kappa_k \ge \rho_k$.

If $\psi_1,\dots,\psi_r$ is a canonical system of right null functions of W at μ, then we may interchange the roles of the sets $\psi_1,\dots,\psi_r$ and $\varphi_1,\dots,\varphi_s$. This yields $s \le r$ and $\kappa_k \le \rho_k$ for $k = 1,\dots,s$, and therefore $r = s$ and $\kappa_i = \rho_i$ for $i = 1,\dots,r$.

Finally, assume that $r = s$ and $\kappa_i = \rho_i$ for $i = 1,\dots,r$. We have to show that $\psi_1,\dots,\psi_s$ is a canonical system. Assume not. Then for some k the function ψ_k is not a right null function of maximal order among all right null functions ψ with $\psi(\mu) \notin \mathrm{span}\{\psi_1(\mu),\dots,\psi_{k-1}(\mu)\}$. So we could choose an k-th right null function with order strictly larger than $\rho_k = \kappa_k$. But then we can construct a canonical system of right null functions with orders $\eta_1 \ge \cdots \ge \eta_r$ such that $\eta_k > \kappa_k$. By the result proved in the second paragraph of the proof this is impossible. □

Let $\varphi_1,\dots,\varphi_s$ be a canonical system of right null functions of W at μ. We shall refer to the corresponding orders $\kappa_1 \ge \cdots \ge \kappa_s$ as the *partial (right) null multiplicities* of W at μ. By the preceding lemma this definition does not depend on the particular choice of the canonical system.

Let us consider a diagonal matrix function $D(\lambda) = \mathrm{diag}\left((\lambda-\mu)^{\rho_j}\right)_{j=1}^{m}$, where

$$\rho_1 \ge \cdots \ge \rho_s > 0 \ge \rho_{s+1} \ge \cdots \ge \rho_m. \tag{3.1.6}$$

Let $\varphi(\lambda) = \mathrm{col}\big(f_j(\lambda)\big)_{j=1}^{m}$ be a right null function of D at μ of order κ, say. Since

$$D(\lambda)\varphi(\lambda) = \begin{pmatrix} (\lambda-\mu)^{\rho_1} f_1(\lambda) \\ \vdots \\ (\lambda-\mu)^{\rho_m} f_m(\lambda) \end{pmatrix}, \tag{3.1.7}$$

we see that $f_{s+1}(\mu),\dots,f_m(\mu)$ must be zero and at least one of the numbers $f_1(\mu),\dots,f_s(\mu)$ is non-zero. Moreover, the order κ of φ satisfies

$$\kappa \le \min\ \{\rho_j \mid 1 \le j \le s,\quad f_j(\mu) \ne 0\}. \tag{3.1.8}$$

Let $e_1,\dots,e_m$ be the standard basis of $\mathbb{C}^m$. From (3.1.8) it follows that the functions

$$\eta_1(\lambda) \equiv e_1,\quad \eta_2(\lambda) \equiv e_2,\dots,\eta_s(\lambda) \equiv e_s \tag{3.1.9}$$

form a canonical system of right null functions of D at μ. In particular, $\rho_1 \ge \cdots \ge \rho_s$ are the partial right null multiplicities of D at μ. To prove the above statements, let $\varphi(\lambda) = \mathrm{col}\big(f_j(\lambda)\big)_{j=1}^{m}$ be a right null function of D at μ, and assume that

$$\varphi(\mu) \notin \mathrm{span}\{\eta_1(\mu),\dots,\eta_{k-1}(\mu)\} = \mathrm{span}\{e_1,\dots,e_{k-1}\}.$$

Then at least one of the numbers $f_k(\mu), \ldots, f_s(\mu)$ is non-zero, and hence, by (3.1.8), the order κ of φ is at most ρ_k. On the other hand the order of η_k is precisely ρ_k. Thus the functions in (3.1.9) form a canonical system.

Theorem 3.1.2 *Let $W(\lambda) = E(\lambda)D(\lambda)F(\lambda)$ be a local Smith-McMillan form of W at μ, that is,*

$$D(\lambda) = \begin{pmatrix} (\lambda-\mu)^{\rho_1} & & \\ & \ddots & \\ & & (\lambda-\mu)^{\rho_m} \end{pmatrix}, \tag{3.1.10}$$

$\rho_1 \geq \cdots \geq \rho_m$ are integers, the matrix functions E and F are analytic in a neighborhood of μ and their values at μ are invertible. Choose s such that $\rho_s > 0 \geq \rho_{s+1}$. Then $\rho_1 \geq \cdots \geq \rho_s$ are the partial right null multiplicities of W at μ and the first s columns of $F(\lambda)^{-1}$ form a canonical system of null functions of W at μ.

Proof Let $e_1, \ldots, e_m$ be the standard basis of $\mathbb{C}^m$, and put $\eta_j(\lambda) \equiv e_j$, $j = 1, \ldots, s$. We already know that the functions $\eta_1, \ldots, \eta_s$ form a canonical system of right null functions of D at μ. Put $\psi_j(\lambda) = F(\lambda)^{-1}\eta_j(\lambda)$, $j = 1, \ldots, s$. Then $\psi_1, \ldots, \psi_s$ are the first s columns of $F(\cdot)^{-1}$. Furthermore,

$$W(\lambda)\psi_j(\lambda) = E(\lambda)D(\lambda)\eta_j(\lambda), \quad j = 1, \ldots, s,$$

implies that $\psi_1, \ldots, \psi_s$ is a set of right null functions of W at μ of orders $\rho_1 \geq \cdots \geq \rho_s$ and the vectors $\psi_1(\mu), \ldots, \psi_s(\mu)$ are linearly independent.

Let $\varphi_1, \ldots, \varphi_t$ be a canonical system of right null functions of W at μ with orders $\kappa_1 \geq \cdots \geq \kappa_t$. By Lemma 3.1.1 the result of the previous paragraph shows that

$$t \geq s, \qquad \kappa_j \geq \rho_j \quad (j = 1, \ldots, s).$$

Note that

$$E(\lambda)^{-1}W(\lambda)\varphi_j(\lambda) = D(\lambda)F(\lambda)\varphi_j(\lambda), \quad j = 1, \ldots, t.$$

Thus $F\varphi_1, \ldots, F\varphi_t$ are right null functions of D at μ of orders $\kappa_1 \geq \cdots \geq \kappa_t$ and the vectors $(F\varphi_1)(\mu), \ldots, (F\varphi_t)(\mu)$ are linearly independent. But then, again using Lemma 3.1.1, we must have

$$s \geq t, \qquad \rho_j \geq \kappa_j \quad (j = 1, \ldots, t).$$

So $t = s$ and $\kappa_j = \rho_j$ for $j = 1, \ldots, s$, and $\psi_1, \ldots, \psi_s$ is a canonical system for W at μ. □

The next theorem shows that all canonical systems may be obtained by the method described in Theorem 3.1.2.

Theorem 3.1.3 *Let $\varphi_1, \ldots, \varphi_s$ be a canonical system of right null functions of W at μ with orders $\kappa_1 \geq \cdots \geq \kappa_s$. Then one may choose $\mathbb{C}^m$-valued functions $\varphi_{s+1}, \ldots, \varphi_m$ and non-positive integers $\kappa_{s+1} \geq \cdots \geq \kappa_m$ such that the matrix functions*

$$\Phi(\lambda) := (\varphi_1(\lambda) \ldots \varphi_m(\lambda)), \tag{3.1.11}$$

$$E(\lambda) := W(\lambda)\Phi(\lambda) \begin{pmatrix} (\lambda-\mu)^{-\kappa_1} & & \\ & \ddots & \\ & & (\lambda-\mu)^{-\kappa_m} \end{pmatrix} \tag{3.1.12}$$

are analytic at μ, *and the matrices* $\Phi(\mu)$ *and* $E(\mu)$ *are non-singular. In particular,*

$$W(\lambda) = E(\lambda) \begin{pmatrix} (\lambda-\mu)^{\kappa_1} & & \\ & \ddots & \\ & & (\lambda-\mu)^{\kappa_m} \end{pmatrix} \Phi(\lambda)^{-1}$$

is a local Smith-McMillan form of W *at* μ.

Proof For the sake of this proof, let us call a $\mathbb{C}^m$-valued function φ *admissible* if φ is analytic at μ and the vector $\varphi(\mu)$ is non-zero. Given an admissible function φ, we define the *pseudo-order* of φ to be the integer $\kappa(\varphi)$ given by

$$\kappa(\varphi) = \max\ \{n \in \mathbb{Z} \mid (\lambda-\mu)^{-n} W\varphi \text{ is analytic at } \mu\}.$$

In other words, $\kappa(\varphi) = \kappa$ whenever

$$(W\varphi)(\lambda) = (\lambda-\mu)^{\kappa}\psi_\kappa + \text{terms of higher order}, \quad \psi_\kappa \neq 0.$$

Notice that $\kappa(\varphi) > 0$ if and only if φ is a right null function of W at μ, and in this case $\kappa(\varphi)$ is precisely the order of φ as a null function.

Let $\varphi_1, \dots, \varphi_s$ be a canonical system of right null functions of W at μ. From the construction of a canonical system (see the paragraph preceeding Lemma 3.1.1) and from the remark made at the end of the previous paragraph we see that a canonical system can be extended in a natural way by adding admissible functions. More precisely, we may choose admissible functions $\varphi_{s+1}, \dots, \varphi_m$ such that for $j = s+1, \dots, m$ the function φ_j is an admissible function of maximal pseudo-order among all admissible functions φ such that

$$\varphi_j(\mu) \notin \operatorname{span}\{\varphi_1(\mu), \dots, \varphi_{j-1}(\mu)\}. \tag{3.1.13}$$

Put $\kappa_j = \kappa(\varphi_j)$ for $j = s+1, \dots, m$. The construction of a canonical system and its continuation described above imply that $\kappa_1 \geq \dots \geq \kappa_m$ and that the matrix function $\Phi(\lambda)$ and $E(\lambda)$ defined by (3.1.11) and (3.1.12), respectively, are analytic at μ. Furthermore, since $\varphi_1(\mu) \neq 0$ and (3.1.13) holds for $j = 2, \dots, m$, we have $\det \Phi(\mu) \neq 0$.

It remains to show that $\det E(\mu) \neq 0$. Assume not. Then we may choose a non-zero vector y so that $E(\mu)y = 0$. Without loss of generality we may assume that the first non-zero coordinate of y is normalized to 1, i.e., the vector y has the form

$$y = (0, \dots, 0, 1, y_{r+1}, \dots, y_m)^T.$$

Consider

$$\widetilde{\varphi}_r(\lambda) = \varphi_r(\lambda) + \sum_{i=r+1}^{m} (\lambda-\mu)^{\kappa_r - \kappa_i} y_i \varphi_i(\lambda).$$

Notice that $\widetilde{\varphi}_r$ is an admissible function and $\widetilde{\varphi}_r(\mu) \notin \operatorname{span}\{\varphi_1(\mu), \dots, \varphi_{r-1}(\mu)\}$. Furthermore,

$$(W\widetilde{\varphi}_r)(\lambda) = (\lambda-\mu)^{\kappa_r}\{c_r(\lambda) + \sum_{i=r+1}^{m} y_i c_i(\lambda)\},$$

where $c_1(\lambda), \dots, c_m(\lambda)$ are the columns of $E(\lambda)$. Since $E(\mu)y = 0$, we have

$$c_r(\mu) + \sum_{i=r+1}^{m} y_i c_i(\mu) = 0,$$

and thus $\kappa(\widetilde{\varphi}_r) \geq \kappa_r + 1$. But the latter contradicts the choice of φ_r. So $E(\mu)$ is nonsingular. □

Notice that, as a by-product, the proof of Theorem 3.1.3 tells us how to construct a local Smith-McMillan form for a regular rational matrix function W. Theorem 3.1.3 also shows that μ is a (right) zero of W if and only if $W(\lambda)^{-1}$ has a pole at μ.

So far we considered right zeros, right eigenvectors, right null functions and partial right null multiplicities. These notions have an obvious left analogue. For example, a row vector function ψ of length m is called a *left null function* of W at μ if ψ is analytic in a neighborhood of μ, the vector $\psi(\mu)$ is non-zero, the function ψW is analytic at μ and ψW is equal to zero at μ. By taking transposed matrices, results for right null functions may be transformed into statements about left null functions. In this way one may derive the analogues of Theorems 3.1.2 and 3.1.3 for left null functions. From Theorem 3.1.3 and its analogue for left null functions it follows that μ is a right zero of W if and only if μ is a left zero of W, and in this case the partial right null multiplicities coincide with the partial left null multiplicities. For this reason we shall simply speak about zeros and partial null multiplicities, and omit the prefix "right" or "left".

3.2 Canonical right and left null pairs at a point

Let W be a regular $m \times m$ rational matrix function. In this section we show how one may use canonical systems of right and left null functions of W at μ to construct a mimimal realization of the Laurent principal part of $W(\lambda)^{-1}$ at μ. For this purpose we need the notions of canonical right and left null pairs.

Let φ be a right null function of W at μ of order κ, and consider its Taylor expansion at μ:

$$\varphi(\lambda) = x_0 + (\lambda - \mu)x_1 + (\lambda - \mu)^2 x_2 + \cdots .$$

With φ, its order κ, and the point μ we associate two matrices, namely

$$\begin{aligned} X_{\varphi,\kappa,\mu} &= \begin{pmatrix} x_0 & x_1 & \dots & x_{\kappa-1} \end{pmatrix}, \\ J_{\kappa,\mu} &= \begin{pmatrix} \mu & 1 & & & \\ & \mu & 1 & & \\ & & \ddots & \ddots & \\ & & & \mu & 1 \\ & & & & \mu \end{pmatrix}. \end{aligned}$$

The matrix $X_{\varphi,\kappa,\mu}$ has size $m \times \kappa$ and its j-th column is precisely the j-th coefficient in the Taylor expansion of φ at μ; the matrix $J_{\kappa,\mu}$ is a single Jordan block of size $\kappa \times \kappa$ and with μ as eigenvalue. The sequence $x_0, x_1, \dots, x_{\kappa-1}$ is called a *right Jordan chain* of W at μ. Its first element x_0 is a right eigenvector; the other vectors $x_1, \dots, x_{\kappa-1}$ are referred to as *generalized right eigenvectors*.

The fact that the eigenvector x_0 is nonzero implies that the pair $(X_{\varphi,\kappa,\mu}, J_{\kappa,\mu})$ is a null-kernel pair, i.e.,

$$\bigcap_{\nu=0}^{\kappa-1} \operatorname{Ker} X_{\varphi,\kappa,\mu} J_{\kappa,\mu}^{\nu} = \{0\}. \tag{3.2.1}$$

To see this, note that $J_{\kappa,\mu} = \mu I + J_{\kappa,0}$, where $J_{\kappa,0}$ is the $\kappa \times \kappa$ Jordan block with 0 on the main diagonal. It follows that the space in the left hand side of (3.2.1) does not change if $J_{\kappa,\mu}$ is replaced by $J_{\kappa,0}$. But

$$\begin{pmatrix} X_{\varphi,\kappa,\mu} \\ X_{\varphi,\kappa,\mu}J_{\kappa,0} \\ \vdots \\ X_{\varphi,\kappa,\mu}J_{\kappa,0}^{\kappa-1} \end{pmatrix} = \begin{pmatrix} x_0 & x_1 & \dots & x_{\kappa-1} \\ & x_0 & \dots & x_{\kappa-2} \\ & & \ddots & \vdots \\ & & & x_0 \end{pmatrix}, \tag{3.2.2}$$

where the blank spots denote zero entries. Thus the right hand side of (3.2.2) is a block upper triangular matrix with the main diagonal entries equal to a non-zero column vector. It follows that the matrix in the left hand side of (3.2.2) has a trivial kernel, and therefore (3.2.1) holds.

Now, let $\varphi_1, \dots, \varphi_s$ be a canonical system of right null functions of W at μ, and let $\kappa_1 \geq \dots \geq \kappa_s$ be the corresponding orders. For each φ_j and κ_j we build the pair $(X_{\varphi_j,\kappa_j,\mu}, J_{\kappa_j,\mu})$, and next we build from these pairs a new pair of matrices (X, J) by setting

$$X = \begin{pmatrix} X_{\varphi_1,\kappa_1,\mu} & \dots & X_{\varphi_s,\kappa_s,\mu} \end{pmatrix}, \tag{3.2.3}$$

$$J = \begin{pmatrix} J_{\kappa_1,\mu} & & \\ & \ddots & \\ & & J_{\kappa_s,\mu} \end{pmatrix}. \tag{3.2.4}$$

Thus X is an $m \times N$ matrix, where $N = \kappa_1 + \dots + \kappa_s$ and J is an $N \times N$ Jordan matrix, consisting of s Jordan blocks, with μ as the only eigenvalue. Using the linear independence of the vectors $\varphi_1(\mu), \dots, \varphi_s(\mu)$ and the analogue of the identity (3.2.2) for the pair (X, J), one proves that (X, J) is a null-kernel pair. We call the pair (X, J) the *canonical right null pair* of W at μ associated with the canonical system $\varphi_1, \dots, \varphi_s$.

Notice that the Jordan matrix J in a canonical right null pair is uniquely determined by the partial null multiplicities of W at μ, and hence the matrix J does not depend on the particular choice of the canonical system we started with. However, different canonical systems may lead to different matrices X.

Canonical left null pairs are defined analogously. Since we want to keep the Jordan matrices upper triangular, we cannot just take the transpose. Some additional notation is needed. Let

$$\psi(\lambda) = y_0 + (\lambda - \lambda_0)y_1 + (\lambda - \lambda_0)^2 y_2 + \cdots$$

be a left null function of W at μ of order κ. With ψ, κ, and the point μ we associate the $\kappa \times m$ matrix

$$Y_{\psi,\kappa,\mu} = \begin{pmatrix} y_{\kappa-1} \\ \vdots \\ y_1 \\ y_0 \end{pmatrix}, \tag{3.2.5}$$

and the single Jordan block $J_{\kappa,\mu}$. Since $y_0 \neq 0$, we have

$$\operatorname{Im} \begin{pmatrix} Y_{\psi,\kappa,\mu} & Y_{\psi,\kappa,\mu}J_{\kappa,\mu} & \dots & Y_{\psi,\kappa,\mu}J_{\kappa,\mu}^{\kappa-1} \end{pmatrix} = \mathbb{C}^\kappa, \tag{3.2.6}$$

and hence the pair $(Y_{\psi,\kappa,\mu}, J_{\kappa,\mu})$ is a full-range pair. The identity (3.2.6) is proved by using the same type of argument as was used in the proof of (3.2.1). The sequence of row vectors $y_0, y_1, \dots, y_{\kappa-1}$ in (3.2.5) is called a *left Jordan chain* of W at μ. Its first element is a left eigenvector; the other elements $y_1, \dots, y_{\kappa-1}$ are referred to as *generalized left eigenvectors.*

Next, let $\psi_1, \dots, \psi_s$ be a canonical system of left null functions of W at μ of orders $\kappa_1 \geq \dots \geq \kappa_s$. Put

$$Y = \begin{pmatrix} Y_{\psi_1,\kappa_1,\mu} \\ \vdots \\ Y_{\psi_s,\kappa_s,\mu} \end{pmatrix} \tag{3.2.7}$$

and let J be as in (3.2.4). By using the linear independence of the vectors $\psi_1(\mu), \dots, \psi_s(\mu)$ one shows that the pair (J, Y) is a full-range pair. We call the pair (J, Y) the *canonical left null pair* of W at μ associated with the canonical system $\psi_1, \dots, \psi_s$.

Theorem 3.2.1 *Consider a local Smith-McMillan form of W at μ:*

$$W(\lambda) = E(\lambda) \begin{pmatrix} (\lambda-\mu)^{\kappa_1} & & \\ & \ddots & \\ & & (\lambda-\mu)^{\kappa_m} \end{pmatrix} F(\lambda), \tag{3.2.8}$$

where

$$\kappa_1 \geq \dots \geq \kappa_s > 0 \geq \kappa_{s+1} \geq \dots \geq \kappa_m.$$

Let $\varphi_1, \dots, \varphi_s$ be the first s columns of $F(\cdot)^{-1}$, and let $\psi_1, \dots, \psi_s$ be the first s rows of $E(\cdot)^{-1}$. Then $\varphi_1, \dots, \varphi_s$ is a canonical system of right null functions of W at μ with orders $\kappa_1 \geq \dots \geq \kappa_s$, and $\psi, \dots, \psi_s$ is a canonical system of left null functions of W at μ with order $\kappa_1 \geq \dots \geq \kappa_s$. Furthermore, if (X, J) and (J, Y) are the corresponding canonical right and left null pairs, then $X(\lambda - J)^{-1}Y$ is a minimal realization of the Laurent principal part of $W(\cdot)^{-1}$ at μ.

Proof In view of Theorem 3.1.2 and its analogue for left null pairs, it is sufficient to prove the statement about the Laurent principal part of $W(\cdot)^{-1}$. Put

$$R_i(\lambda) = \varphi_i(\lambda)(\lambda-\mu)^{-\kappa_i}\psi_i(\lambda), \quad i = 1, \dots, s.$$

Formula (3.2.8) implies that the Laurent principal part of $W(\lambda)^{-1}$ is equal to the sum of the Laurent principal parts of $R_i(\lambda)$ for $i = 1, \dots, s$. A straightforward computation shows that the Laurent principal part of $R_i(\lambda)$ at μ is equal to

$$X_{\varphi_i,\kappa_i,\mu}(\lambda - J_{\kappa_i,\mu})^{-1} Y_{\psi_i,\kappa_i,\mu},$$

and thus the Laurent principal part of $W(\lambda)^{-1}$ at μ is equal to

$$\sum_{i=1}^{s} X_{\varphi_i,\kappa_i,\mu}(\lambda - J_{\kappa_i,\mu})^{-1} Y_{\psi_i,\kappa_i,\mu}.$$

Using (3.2.3), (3.2.4), and (3.2.7) the latter expression can be written in a more concise form as $X(\lambda - J)^{-1}Y$. Since the pair (X, J) is a null-kernel pair and the pair (J, Y) is a full-range pair, we know from Section 1.1 that the realization $X(\lambda - J)^{-1}Y$ is minimal. □

We conclude this section with a warning (cf., Example 1.1 in Lerer-Rodman [1994]). Not all information about null functions is retained by a null pair. For example, consider the following 2×2 rational matrix functions:

$$V(\lambda) = \begin{pmatrix} 1 & \frac{1}{\lambda-\mu} \\ 0 & 1 \end{pmatrix}, \qquad \widetilde{V}(\lambda) = \begin{pmatrix} 1 & \frac{2}{\lambda-\mu} \\ 0 & 1 \end{pmatrix}.$$

As we have seen in the previous section,

$$\varphi(\lambda) = \begin{pmatrix} a \\ 0 \end{pmatrix} + (\lambda - \mu) \begin{pmatrix} b \\ -a \end{pmatrix} + \cdots, \quad a \neq 0,$$

is the general form of a right null function of V at μ. Similarly, all right null functions of $\widetilde{V}$ at μ are given by

$$\widetilde{\varphi}(\lambda) = \begin{pmatrix} a \\ 0 \end{pmatrix} + (\lambda - \mu) \begin{pmatrix} b \\ -\frac{1}{2}a \end{pmatrix} + \cdots, \quad a \neq 0,$$

Thus V and $\widetilde{V}$ have no common right null function at μ. On the other hand, the pair $\left(\left(\begin{smallmatrix} 1 \\ 0 \end{smallmatrix}\right), \mu \right)$ is a canonical right null pair for both V and $\widetilde{V}$ at μ.

The phenomenon described in the previous paragraph is typical for the case when the rational matrix function W has a zero and a pole at μ. It does not occur if W is analytic at μ (see Theorem 3.4.1 below).

3.3 Right and left null pairs on a set

In this section we consider canonical null pairs for several zeros at the same time, and we introduce arbitrary null pairs (not necessarily canonical ones). For the latter it will be convenient to work in a coordinate free setting.

Fix a non-empty subset Ω of $\mathbb{C}$, and let $\mu_1, \ldots, \mu_N$ be the different zeros of $W(\lambda)$ in Ω. For each k choose a canonical right null pair (X_k, J_k) of W at μ_k. Next, consider the matrices

$$X = \begin{pmatrix} X_1 & X_2 & \ldots & X_N \end{pmatrix}, \tag{3.3.1}$$

$$J = \begin{pmatrix} J_1 & & & \\ & J_2 & & \\ & & \ddots & \\ & & & J_N \end{pmatrix}. \tag{3.3.2}$$

The matrix J is a Jordan matrix with eigenvalues $\mu_1, \ldots, \mu_N$. Any pair of matrices (X, J) arising in this way is called a *canonical right null pair* of W on Ω.

Now, let $A : \mathcal{X} \to \mathcal{X}$ and $C : \mathcal{X} \to \mathbb{C}^m$ be linear operators, with $\mathcal{X}$ being a finite dimensional inner product space. The pair (C, A) is called a *right null pair* of W on Ω if there exists a canonical right null pair (X, J) of W on Ω and an invertible linear operator $S : \mathcal{X} \to \mathbb{C}^M$, where M is the order of the Jordan matrix J, such that

$$C = XS, \qquad A = S^{-1}JS. \tag{3.3.3}$$

We shall refer to $\mathcal{X}$ as the *main space* of the pair (C, A).

A left null pair of W on Ω is defined in an analogous way. First, one chooses for each zero μ_k a canonical left null pair (J_k, Y_k) of W at μ_k, and next one builds the matrices

$$J = \begin{pmatrix} J_1 & & & \\ & J_2 & & \\ & & \ddots & \\ & & & J_N \end{pmatrix}, \qquad Y = \begin{pmatrix} Y_1 \\ Y_2 \\ \vdots \\ Y_N \end{pmatrix}. \tag{3.3.4}$$

Any pair of matrices (J, Y) arising in this way is called a *canonical left null pair* of W on Ω. A pair (A, B) of linear operators, $A : \mathcal{X} \to \mathcal{X}$ and $B : \mathbb{C}^m \to \mathcal{X}$, with $\mathcal{X}$ a finite dimensional inner product space, is said to be a *left null pair* of W on Ω if for some invertible linear operator S the pair (SAS^{-1}, SB) is a canonical left null pair of W on Ω. Here, as well as in (3.3.3), we identify matrices in a standard way with linear operators acting between complex Euclidean spaces. The space $\mathcal{X}$, appearing above, is called the *main space* of the pair (A, B).

Notice that the eigenvalues of the operator A in a right null pair (C, A) of W on Ω are precisely the zeros $\mu_1, \ldots, \mu_N$ of W in Ω. In particular, all eigenvalues of A belong to Ω. These statements also hold true for the operator A in a left null pair (A, B) of W on Ω.

Theorem 3.3.1 *Consider a pair of linear operators $A : \mathcal{X} \to \mathcal{X}$ and $C : \mathcal{X} \to \mathbb{C}^m$, where $\mathcal{X}$ is a finite dimensional inner product space. The pair (C, A) is a right null pair of W on Ω if and only if the eigenvalues of A belong to Ω and there exists a linear operator $B : \mathbb{C}^m \to \mathcal{X}$ such that $C(\lambda - A)^{-1}B$ is a minimal realization and*

$$W(\lambda)^{-1} - C(\lambda - A)^{-1}B \tag{3.3.5}$$

extends to an analytic function on Ω. In this case, (A, B) is a left null pair of W on Ω.

Proof Let $\mu_1, \ldots, \mu_N$ be the different zeros of W in Ω. For each μ_k let (X_k, J_k) be a canonical right null pair of W at μ_k. According to Theorem 3.1.3 each canonical system of right null functions of W at μ_k arises from a local Smith-McMillan form in the way explained in Theorem 3.1.2. But then Theorem 3.2.1 tells us that there exists a matrix Y_k so that (J_k, Y_k) is a canonical left null pair of W at μ_k and $X_k(\lambda - J_k)^{-1}Y_k$ is a minimal realization of the Laurent principal part of $W(\lambda)^{-1}$ at μ. Build X, J and Y as in (3.3.1), (3.3.2) and (3.3.4). Then

$$X(\lambda - J)^{-1}Y = \sum_{k=1}^{N} X_k(\lambda - J)^{-1}Y_k,$$

and hence the function

$$W(\lambda)^{-1} - X(\lambda - J)^{-1}Y \tag{3.3.6}$$

is analytic on Ω.

Next, let us prove that the realization is minimal. Recall that μ_k is the only eigenvalue of J_k. Thus the spectra of $J_1, \ldots, J_N$ are mutually disjoint. Since $(X_1, J_1), \ldots, (X_N, J_N)$ are null-kernel pairs, it follows that (X, J) is a null-kernel pair too. Indeed, put

$$\mathcal{L} = \bigcap_{\nu=0}^{M-1} \operatorname{Ker} XJ^{\nu},$$

where M is the order of J, and assume that $\mathcal{L}$ contains non-zero vectors. Since $J\mathcal{L} \subset \mathcal{L}$, the operator J has an eigenvector x in $\mathcal{L}$. Let μ be the corresponding eigenvalue. Clearly, $\mu = \mu_k$ for some k. Partition x as a block column vector according to the partitioning of J in (3.2.2). Thus

$$x = \begin{pmatrix} x_1 \\ x_2 \\ \vdots \\ x_N \end{pmatrix}, \qquad Jx = \begin{pmatrix} J_1 x_1 \\ J_2 x_2 \\ \vdots \\ J_N x_N \end{pmatrix}.$$

Notice that $J_i - \mu_k$ is invertible for $i \neq k$. So $(J - \mu_k)x = 0$ implies that $x_i = 0$ for $i \neq k$. But then $XJ^{\nu}x = 0$ yields $X_k J_k^{\nu} x_k = 0$, and hence

$$0 \neq x_k \in \bigcap_{\nu=0}^{M-1} \operatorname{Ker} X_k J_k^{\nu},$$

which contradicts the fact that (X_k, J_k) is a null-kernel pair. We conclude that (X, J) is a null-kernel pair. In an analogous way one shows that (J, Y) is a full-range pair. Thus the realization $X(\lambda - J)^{-1}Y$ is minimal.

Now, assume that (C, A) is a right null pair of W on Ω. Then the eigenvalues of A are in Ω, and there exists a canonical right null pair (X, J) of W on Ω and an invertible operator S such that (3.3.3) holds. Without loss of generality we may assume that (X, J) is the pair appearing in the first paragraph of the proof. Choose Y also as in the first paragraph, and put $B = S^{-1}Y$. Then

$$C(\lambda - A)^{-1}B = X(\lambda - J)^{-1}Y,$$

and we see from (3.3.6) that the function (3.3.5) is analytic on Ω. Since A and J act on spaces of the same dimension and the realization $X(\lambda - J)^{-1}Y$ is minimal, we conclude that the realization $C(\lambda - A)^{-1}B$ is minimal too.

To prove the converse implication, assume that the eigenvalues of A are in Ω, that $C(\lambda - A)^{-1}B$ is a minimal realization, and that the function (3.3.5) is analytic on Ω. Choose X, J and Y as in the first paragraph of the proof. Then

$$C(\lambda - A)^{-1}B - X(\lambda - J)^{-1}Y = Q(\lambda) - R(\lambda), \tag{3.3.7}$$

where $R(\lambda)$ is the function defined by (3.3.5) and $Q(\lambda)$ is the function defined by (3.3.6). Notice that the right hand side of (3.3.7) is analytic on Ω. Since the eigenvalues of A and J are in Ω, the left hand side of (3.3.7) is analytic outside Ω and vanishes at infinity. Thus, by Liouville's theorem,

$$C(\lambda - A)^{-1}B = X(\lambda - J)^{-1}Y.$$

It follows that $C(\lambda - A)^{-1}B$ and $X(\lambda - J)^{-1}Y$ are minimal realizations of the same rational matrix function, and hence, by the state space similarity theorem (see Section 1.1), there exists an invertible linear operator S such that

$$S^{-1}AS = J, \qquad S^{-1}B = Y, \qquad CS = X.$$

Therefore, (C, A) is a right null pair and (A, B) is a left null pair of W on Ω. $\square$

From Theorem 3.3.1 it follows that right (left) null pairs are unique up to similarity. More precisely, the following corollary holds.

Corollary 3.3.2 *Let (C, A) be a right null pair of W on Ω. Then $(\widetilde{C}, \widetilde{A})$ is another right null pair of W on Ω if and only if there exists a (unique) invertible linear operator S such that $\widetilde{C} = CS$ and $\widetilde{A} = S^{-1}AS$.*

Proof If $\widetilde{C} = CS$ and $\widetilde{A} = S^{-1}AS$, then we see directly from the definition in the third paragraph of this section that $(\widetilde{C}, \widetilde{A})$ is a right null pair of W on Ω.

To prove the converse implication, assume that (C, A) and $(\widetilde{C}, \widetilde{A})$ are both right null pairs of W on Ω. By Theorem 3.3.1 we may choose linear operators B and $\widetilde{B}$ such that $C(\lambda - A)^{-1}B$ and $\widetilde{C}(\lambda - \widetilde{A})^{-1}\widetilde{B}$ are minimal realizations, and the functions

$$W(\lambda)^{-1} - C(\lambda - A)^{-1}B, \qquad W(\lambda)^{-1} - \widetilde{C}(\lambda - \widetilde{A})^{-1}B \tag{3.3.8}$$

are both analytic on Ω. Since the eigenvalues of A and $\widetilde{A}$ are in Ω, the difference of the two functions in (3.3.8) is analytic on $\mathbb{C}$ and vanishes at infinity. Hence, by Liouville's theorem,

$$C(\lambda - A)^{-1}B = \widetilde{C}(\lambda - \widetilde{A})^{-1}\widetilde{B},$$

and we can use the state space similarity theorem to show that there exists an invertible linear operator S such that $\widetilde{C} = CS$, $\widetilde{A} = S^{-1}AS$, and $\widetilde{B} = S^{-1}B$. In particular, the pairs (C, A) and $(\widetilde{C}, \widetilde{A})$ are related as desired.

It remains to prove the uniqueness of the similarity transformation S. Assume that T is another invertible linear operator such that $\widetilde{C} = CT$ and $\widetilde{A} = T^{-1}AT$. Then $CA^{\nu}(T - S) = 0$ for each $\nu \geq 0$. Since (C, A) is a null-kernel pair, we conclude that $T = S$. □

Theorem 3.3.1 and Corollary 3.3.2 have analogous versions for left null pairs in which the roles of right and left are interchanged.

In what follows we shall often use the characterization of a null pair in Theorem 3.3.1 as a definition. In order to work with right and left null pairs of W at μ we do not have to remember their construction via the canonical systems of null functions; we may start directly from minimal realizations of the sum of the Laurent principal parts of $W(\lambda)^{-1}$ in Ω. However, one important fact about null pairs, which one could also derive from Theorem 3.3.1, follows immediately from the constructions made in the first two sections of this chapter, namely if $\widetilde{W}(\lambda) = W(\lambda)F(\lambda)$, where F is biregular on Ω (i.e., F is analytic on Ω and $\det F(\lambda)$ does not vanish in Ω), then $\widetilde{W}$ and W have the same left null pairs on Ω. Indeed, multiplying W on the right by a rational matrix function F, which is biregular on Ω, does not change the left null functions of W at each point of Ω. Hence it does not change the canonical left null pairs of W at $\mu_1, \ldots, \mu_N$, and therefore it does not change the left null pairs of W on Ω.

3.4 The analytic case

In this section we consider regular rational matrix functions W that are analytic on Ω, where Ω is a non-empty subset of $\mathbb{C}$. Thus W may have zeros in Ω but no poles. In this case all the information about the singularities of W in Ω is contained in a null pair. The next theorem is the main result of this section.

Theorem 3.4.1 *Let W and $\widetilde{W}$ be regular $m \times m$ rational matrix functions which are analytic on Ω. Then $\widetilde{W}(\lambda) = W(\lambda)F(\lambda)$, with F biregular on Ω, if and only if W and $\widetilde{W}$ have a common left null pair on Ω.*

For the proof of Theorem 3.4.1 we need the following lemma.

Lemma 3.4.2 *Let W be a regular $m \times m$ rational matrix function which is analytic on Ω. If (A, B) is a left null pair of W on Ω, then $(\lambda - A)^{-1}BW(\lambda)$ is analytic on Ω.*

Proof By Theorem 3.3.1 there exists a linear operator C such that $C(\lambda - A)^{-1}B$ is a minimal realization and the function $W(\lambda)^{-1} - C(\lambda - A)^{-1}B$ is analytic on Ω. In particular, (C, A) is a null-kernel pair. Since W is analytic on Ω, we see that the product

$$\{W(\lambda)^{-1} - C(\lambda - A)^{-1}B\}W(\lambda)$$

is also analytic on Ω, and therefore the same holds true for the function $C(\lambda - A)^{-1}BW(\lambda)$. Analyticity on Ω does not change by multiplying by λ. Thus $\lambda C(\lambda - A)^{-1}BW(\lambda)$ is analytic on Ω. Notice that

$$\begin{aligned} \lambda C(\lambda - A)^{-1}BW(\lambda) &= C(\lambda - A + A)(\lambda - A)^{-1}BW(\lambda) \\ &= CBW(\lambda) + CA(\lambda - A)^{-1}BW(\lambda), \end{aligned}$$

and the term $CBW(\lambda)$ is analytic on Ω. It follows that $CA(\lambda - A)^{-1}BW(\lambda)$ is analytic on Ω. Again multiply by λ, and repeat the above argument. We conclude that $CA^2(\lambda - A)^{-1}BW(\lambda)$ is analytic on Ω. Proceeding in this way, we see that the function

$$\lambda \longmapsto \begin{pmatrix} C \\ CA \\ \vdots \\ CA^{n-1} \end{pmatrix} (\lambda - A)^{-1}BW(\lambda) \tag{3.4.1}$$

is analytic on Ω. Here n is the dimension of the main space of the pair (A, B). Since (C, A) is a null-kernel pair, the operator matrix defined by the column in (3.4.1) has a left inverse. By multiplying (3.4.1) from the left with this left inverse, we obtain that $(\lambda - A)^{-1}BW(\lambda)$ is analytic on Ω. □

Proof of Theorem 3.4.1 The necessity of having a common left null pair of Ω follows from the remark made at the end of the previous section. To prove the sufficiency, let us assume that the pair (A, B) is a left null pair of both W and $\widetilde{W}$ on Ω. Put $F(\lambda) = W(\lambda)^{-1}\widetilde{W}(\lambda)$. We have to show that F is analytic on Ω and $\det F(\mu) \neq 0$ for each $\mu \in \Omega$. Since W and $\widetilde{W}$ are regular, the same holds for F. By Theorem 3.3.1 we may choose linear operators C and $\widetilde{C}$ such that

$$\begin{aligned} R(\lambda) &:= W(\lambda)^{-1} - C(\lambda - A)^{-1}B, \\ \widetilde{R}(\lambda) &:= \widetilde{W}(\lambda)^{-1} - \widetilde{C}(\lambda - A)^{-1}B \end{aligned}$$

are analytic on Ω. By Lemma 3.4.2, also $(\lambda - A)^{-1}BW(\lambda)$ and $(\lambda - A)^{-1}B\widetilde{W}(\lambda)$ are both analytic on Ω. Notice that

$$\begin{aligned} F(\lambda) &= R(\lambda)\widetilde{W}(\lambda) + C\{(\lambda - A)^{-1}B\widetilde{W}(\lambda)\}, \\ F(\lambda)^{-1} &= \widetilde{R}(\lambda)W(\lambda) + \widetilde{C}\{(\lambda - A)^{-1}BW(\lambda)\}. \end{aligned}$$

It follows that both $F(\lambda)$ and $F(\lambda)^{-1}$ are analytic on Ω, and hence F is biregular. $\square$

Let W be a regular $m \times m$ rational matrix function which is analytic on Ω, and let (A, B) be a left null pair of W on Ω. From the constructions made in the first two sections of this chapter and Lemma 3.4.2 it follows that the following three conditions are satisfied:

(i) the eigenvalues of A are in Ω and the dimension of the space on which A acts is equal to the sum of the orders of the zeros of $\det W(\lambda)$ in Ω;

(ii) (A, B) is a full-range pair;

(iii) $(\lambda - A)^{-1}BW(\lambda)$ is analytic on Ω.

It turns out that the converse statement is also true, that is, if (A, B) is a pair of linear operators satisfying (i), (ii) and (iii), then (A, B) is a left null pair of W on Ω (see, e.g., Gohberg-Kaashoek-Van Schagen [1993]).

3.5 Notes

Instead of null function the term root function is often used. The notion of a null-function has its origin in the theory of chains of eigenvectors and generalized eigenvectors. The latter has a long history, in algebra as well as in analysis. One of the earliest systematic uses of these chains for operator functions, in the context of infinite dimensional spaces, appears in the work of Keldysh [1951], [1971] on completeness problems (see also Gohberg-Krein [1970], Section 5.9, and Markus [1988]). Null functions were first introduced by S.G. Kreĭn and V.P. Trofimov in Krein-Trofimov [1969] and V.I. Matsaev and Yu.A. Palant (see Palant [1970]). The idea to relate such functions to pairs of matrices appears for the first time in the theory of matrix polynomials (see Lancaster [1977a], [1977b], and Gohberg-Lancaster-Rodman [1978a], [1987b], [1982]). For analytic matrix functions the theory of null functions has been developed systematically in Gohberg-Rodman [1981]. Theorem 3.2.1 has its roots in the paper Gohberg-Sigal [1971]. The present chapter follows closely the set up of Gohberg-Kaashoek-Van Schagen [1993], which concerns also the case of analytic matrix functions and which is extended here to the rational matrix case.

CHAPTER 4

Inverse problems involving null pairs

In this chapter we treat three inverse problems involving null pairs. The first, considered in the first section, is a Lagrange type interpolation problem, and asks for the construction of a regular rational matrix function W in $\mathcal{R}at(H_\infty)$ with a prescribed null pair on the open right half plane. The second section deals with the same problem but now under the additional condition that the interpolant W is required to be J-inner. The last section is devoted to the Nevanlinna-Pick interpolation problem for rational matrix functions in $\mathcal{R}at(H_\infty)$.

4.1 An inverse problem in $\mathcal{R}at(H_\infty)$

Consider a full-range pair (A, B), where $A : \mathcal{X} \to \mathcal{X}$ and $B : \mathbb{C}^m \to \mathcal{X}$, and assume that A has all its eigenvalues in the open right half plane Π^+. In this section we consider the problem of constructing a regular rational matrix function $W \in \mathcal{R}at(H_\infty)$ such that (A, B) is a left null pair of W on Π^+.

To illustrate this problem with an example, let us take

$$A = \begin{pmatrix} s_1 & & & \\ & s_2 & & \\ & & \ddots & \\ & & & s_N \end{pmatrix}, \qquad B = \begin{pmatrix} u_1 \\ u_2 \\ \vdots \\ u_N \end{pmatrix}. \tag{4.1.1}$$

Thus A is a diagonal matrix with scalar diagonal entries, and the entries $u_1, \ldots, u_N$ of B belong to $\mathbb{C}^{1\times m}$. Notice that

$$\begin{aligned} \begin{pmatrix} B & AB & \ldots & A^{N-1}B \end{pmatrix} &= \begin{pmatrix} u_1 & & \\ & \ddots & \\ & & u_N \end{pmatrix} \cdot \\ &\quad \cdot \begin{pmatrix} I_m & s_1 I_m & \ldots & s_1^{N-1} I_m \\ \vdots & \vdots & & \vdots \\ I_m & s_N I_m & \ldots & s_N^{N-1} I_m \end{pmatrix}. \end{aligned} \tag{4.1.2}$$

The second factor in the right hand side of (4.1.2) is the Kronecker product of a Vandermonde matrix with the $m \times m$ identity matrix, and hence the pair (A, B) in (4.1.1) is full-range if and only if all $s_1, \ldots, s_N$ are distinct and the vectors $u_1, \ldots, u_N$ are non-zero. Furthermore, the eigenvalues of A are in Π^+ if and only if $s_1, \ldots, s_N$ are in Π^+. Assume now that the entries $u_1, \ldots, u_N$ are non-zero and that the points $s_1, \ldots, s_N$ are distinct and in Π^+. Then the pair (A, B) in (4.1.1) is a left null pair of $W \in \mathcal{R}at(H_\infty)$ if and only if the following two conditions are satisfied

(a) $u_i W(s_i) = 0, \qquad i = 1, \ldots, N;$

(b) $s_1, \ldots, s_N$ are the only zeros of $\det W(\lambda)$ in Π^+ and the multiplicities of $s_1, \ldots, s_N$ as zeros of $\det W(\lambda)$ are all equal to one.

Thus in this case our inverse problem is a homogeneous left tangential Lagrange type interpolation problem.

It turns out that the problem stated in the first paragraph of this section is always solvable. To state its solution we need a result from system theory. Since the pair (A, B) is full-range, we can find (see, e.g., Section 5.7 in Wonham [1979]) a basis in the main space $\mathcal{X}$ and a basis in $\mathbb{C}^m$ such that $\mathrm{mat}(A)$ and $\mathrm{mat}(B)$, the matrices of A and B with respect to these bases, have the following form:

$$
\mathrm{mat}(A) = \begin{pmatrix}
0 & 1 & & & & & & & & & & & & & \\
 & & \ddots & & & & & & & & & & & & \\
 & & & 1 & & & & & & & & & & & \\
* & * & \ldots & * & * & * & \ldots & * & \ldots & \ldots & * & * & \ldots & * \\
 & & & & 0 & 1 & & & & & & & & \\
 & & & & & & \ddots & & & & & & & \\
 & & & & & & & 1 & & & & & & \\
* & * & \ldots & * & * & * & \ldots & * & \ldots & \ldots & * & * & \ldots & * \\
 & & & & & & & & \ddots & & & & & \\
 & & & & & & & & & \ddots & & & & \\
 & & & & & & & & & & 0 & 1 & & \\
 & & & & & & & & & & & & \ddots & \\
 & & & & & & & & & & & & & 1 \\
* & * & \ldots & * & * & * & \ldots & * & \ldots & \ldots & * & * & \ldots & *
\end{pmatrix},
$$

$$
\mathrm{mat}(B) = \begin{pmatrix}
0 & 0 & \ldots & \ldots & 0 & 0 & \ldots & 0 \\
\vdots & \vdots & & & \vdots & \vdots & & \vdots \\
0 & 0 & & & 0 & 0 & & 0 \\
1 & 0 & & & 0 & 0 & & 0 \\
0 & 0 & & & 0 & 0 & & 0 \\
\vdots & \vdots & & & \vdots & \vdots & & \vdots \\
0 & 0 & & & 0 & 0 & & 0 \\
0 & 1 & & & 0 & 0 & & 0 \\
\vdots & \vdots & & & \vdots & \vdots & & \vdots \\
\vdots & \vdots & & & \vdots & \vdots & & \vdots \\
0 & 0 & & & 0 & 0 & & 0 \\
\vdots & \vdots & & & \vdots & \vdots & & \vdots \\
0 & 0 & & & 0 & 0 & & 0 \\
0 & 0 & & & 1 & 0 & \ldots & 0
\end{pmatrix}.
$$

In $\mathrm{mat}(A)$ the blank spots in the row without $*$'s denote zero entries. The entry 1 in the first column of $\mathrm{mat}(B)$ has the same row index as the first row in $\mathrm{mat}(A)$ with $*$'s, the entry 1 in the second column of $\mathrm{mat}(B)$ has the same row index as the second row with $*$'s in $\mathrm{mat}(A)$, etc. Now, let F be an arbitrary linear operator from $\mathcal{X}$ into $\mathbb{C}^m$. Then mat$(A - BF)$, the matrix of $A - BF$ with respect to the above bases, has the same form as mat(A). In fact, mat(A) and mat$(A-BF)$ differ only in the entries indicated by $*$'s. More precisely, with an appropriate choice of F we can arrange the $*$'s in mat$(A - BF)$ to be what ever we like. In particular, we can choose F so that mat$(A - BF)$ is a direct sum of companion matrices

$$\begin{pmatrix} 0 & 1 & & \\ & & \ddots & \\ & & & 1 \\ * & * & \dots & * \end{pmatrix}$$

of which the entries in the last row may be choosen freely. Hence, the n eigenvalues of $A - BF$, where n is dim $\mathcal{X}$, are free to choose if we allow F to vary. The latter result is in system theory known as the *pole assignment theorem* (see Section 2.2 in Wonham [1979]). We conclude that we always may choose F so that the eigenvalues of $A - BF$ are in the open left half plane Π^-.

Theorem 4.1.1 *Let (A, B) be a full-range pair, and assume that the eigenvalues of A are in Π^+. Choose F so that the eigenvalues of $A - BF$ are in Π^-, and set*

$$G(\lambda) = I - F[\lambda - (A - BF)]^{-1}B. \tag{4.1.3}$$

Then G is a regular rational matrix function in $\mathcal{R}at(H_\infty)$, and (A, B) is a left null pair of G on Π^+. Moreover, all regular rational matrix functions in $\mathcal{R}at(H_\infty)$ with the property that (A, B) is a left null pair of W on Π^+ are given by $W = GR$, where R is outer.

Proof Let G be as in (4.1.3). Obviously, G is regular, and $G \in \mathcal{R}at(H_\infty)$. According to formula (1.3.3) we have

$$G(\lambda)^{-1} = I + F(\lambda - A)^{-1}B. \tag{4.1.4}$$

Put $\alpha = A$, $\beta = B$, and $\gamma = F$. Then $[\alpha, \beta, \gamma, I]$ is a realization of $G(\cdot)^{-1}$. Notice that $\alpha^\times := \alpha - \beta\gamma = A - BF$. Since α has its eigenvalues in Π^+ and $\alpha^\times$ in Π^-, the operators α and $\alpha^\times$ have no common eigenvalues, and we can use the arguments of the last but one paragraph of the proof of Corollary 2.2.3 to show that the realization $[\alpha, \beta, \gamma, I]$ is minimal. In other words, the right hand side of (4.1.4) is a minimal realization. Recall that A has its eigenvalues in Π^+. So we may apply Theorem 3.3.1 to show that (A, B) is a left null pair of G on Π^+. Next, by Theorem 3.4.1, any regular rational matrix function W in $\mathcal{R}at(H_\infty)$ for which (A, B) is a left null pair on Π^+ is of the form $W = GR$, with R outer. □

4.2 J-inner matrix functions in $\mathcal{R}at(H_\infty)$ with prescribed null pairs

Let (A, B) be a full-range pair, and assume that A has its eigenvalues in Π^+. In this section we deal with the problem of finding a J-inner matrix function Θ in $\mathcal{R}at(H_\infty)$ such that (A, B) is a left null pair of Θ on Π^+.

Assume that the above problem has a solution Θ. Let G be defined by (4.1.3), where F is chosen in such a way that the eigenvalues of $A - BF$ are in Π^-. Then, by Theorem 4.1.1, the pair (A, B) is a left null pair of G on Π^+, and we can apply Theorem 3.4.1 to show that $G = \Theta R$, with R outer. In other words, the function G admits a J-inner-outer factorization. Conversely, if G admits a J-inner-outer factorization $G = \Theta R$, say, then, using Theorem 3.4.1 again, the J-inner factor Θ has (A, B) as a left null pair on Π^+. We conclude that the problem stated in the first paragraph of this section has a solution if and only if the matrix function G in (4.1.3) admits a J-inner-outer factorization. We can apply Theorem 2.3.1 to find the necessary and sufficient conditions in order that G admits such a factorization. Also, Theorem 2.3.1 provides a method to construct the factors. In this way we come to the following theorem.

Theorem 4.2.1 *Let (A, B) be a full-range pair, and assume that the eigenvalues of A are in Π^+. Then there exists $\Theta \in \mathcal{R}at(H_\infty)$ such that Θ is J-inner and (A, B) is a left null pair of Θ on Π^+ if and only if the (unique) solution Y of the Lyapunov equation*

$$AY + YA^* = BJB^* \tag{4.2.1}$$

is positive definite. In this case any such Θ is given by

$$\Theta(\lambda) = D - JB^*(\lambda + A^*)^{-1}Y^{-1}BD, \tag{4.2.2}$$

where Y is given by (4.2.1) *and D is an arbitrary J-unitary matrix.*

Proof Define G by (4.1.3), where F is chosen such that the eigenvalues of $A - BF$ are in Π^-.

Assume $\Theta \in \mathcal{R}at(H_\infty)$ is J-inner and has (A, B) as a left null pair on Π^+. From the remarks made in the paragraph preceding the present theorem we know that G admits a J-inner-outer factorization. So we can apply Theorem 2.3.1 to show that the algebraic Riccati equation

$$XA + A^*X - XBJB^*X = 0 \tag{4.2.3}$$

has a solution X satisfying

(i) X is positive semi-definite,

(ii) $K := A - BJB^*X$ has its eigenvalues in Π^-.

From (4.2.3) we see that $\operatorname{Ker} X$ is invariant under A. Condition (ii) implies that A and K coincide on $\operatorname{Ker} X$. But A has its eigenvalues in Π^+, and K in Π^-. So A and K have no common eigenvalues. Thus $\operatorname{Ker} X = \{0\}$, and hence X is invertible. Put $Y = X^{-1}$. Then we see from (4.2.3) that Y satisfies (4.2.1). Since X is positive semi-definite and invertible, we have $X > 0$, and hence the same is true for Y. We conclude that Y is a positive definite solution of (4.2.1). The solution is unique, because A and $-A^*$ have no common eigenvalues.

To prove the converse implication, assume that Y is a positive definite solution of (4.2.1). Put $X = Y^{-1}$. Then X satisfies (4.2.3), and condition (i) in the previous paragraph is fulfilled. Notice that

$$K := A - BJB^*X = X^{-1}(XA - XBJB^*X) = X^{-1}(-A^*)X,$$

and hence K has its eigenvalues in Π^-. So, condition (ii) in the previous paragraph is also fulfilled. Hence, by Theorem 2.3.1, the matrix function G defined in the first paragraph of the proof admits a J-inner-outer factorization. Moreover, a J-inner-outer factor Θ of G is obtained (use (2.3.5)) by taking

$$\Theta(\lambda) = I - JB^*X(\lambda - K)^{-1}B. \tag{4.2.4}$$

The remarks in the paragraph preceding the present theorem imply that (A, B) is a left null pair of Θ on Π^+. Since $K = X^{-1}(-A^*)X$ and $Y = X^{-1}$, the function Θ in (4.2.4) is also given by (4.2.2) with $D = I$.

In a J-inner-outer factorization of G the J-inner factor is unique up to multiplication on the right by a constant J-unitary matrix. We conclude that the general solution of our inverse problem is given by (4.2.2), with D an arbitrary J-unitary matrix. □

In Section 6.2 we shall consider a generalization of Theorem 4.2, and there we shall use a different method (not based on factorization) to obtain a J-inner rational matrix function, not necessarily in $\mathcal{R}at(H_\infty)$, with prescibed zeros and prescribed poles in the open right half plane.

4.3 Left tangential Nevanlinna-Pick interpolation

In this section we solve a left tangential Nevanlinna-Pick interpolation problem, by using the method of the book Ball-Gohberg-Rodman [1990], which reduces the problem to an inverse problem of the type considered in the previous section.

The interpolation problem we shall deal with is the following. Given n distinct points $s_1, \ldots, s_n$ in Π^+, non-zero row vectors $x_1, \ldots, x_n$ in $\mathbb{C}^{1\times m}$, and arbitrary row vectors $y_1, \ldots, y_n$ in $\mathbb{C}^{1\times r}$, find (all) $m \times r$ matrix functions F such that

(TNP1) $F \in \mathcal{R}at(H_\infty)$,
(TNP2) $\sup\{\|F(\lambda)\| \mid \mathrm{Re}\ \lambda \geq 0\} < 1$,
(TNP3) $x_i F(s_i) = y_i, \qquad i = 1, \ldots, n.$

Condition (TNP1) means that F is required to be an H_∞-matrix function, and condition (TNP2) requires the corresponding H_∞-norm to be strictly less than one. The problem of finding an F satisfying (TNP1) and (TNP3) is a Lagrange type interpolation problem which always has a solution. Condition (TNP2) turns the problem into a Nevanlinna-Pick type problem, which may not be solvable.

In order for the problem (TNP1)–(TNP3) to be solvable it is necessary that the following matrix, which is built from the given data, is positive definite:

$$\Lambda = \left[\frac{x_i x_j^* - y_i y_j^*}{s_i + \overline{s_j}}\right]_{i,j=1}^n. \tag{4.3.1}$$

One refers to Λ as the *Pick matrix* associated with the problem (TNP1)–(TNP3). To prove this necessity statement, let H_2^m and H_2^r be the Hilbert spaces of all functions with values in $\mathbb{C}^m$ and $\mathbb{C}^r$, respectively, analytic on Π^+, with entries in the Hardy space H_2 of the right half plane (see Hoffman [1962], Chapter 8), and consider the operators

$$R : H_2^m \to \mathbb{C}^n, \qquad Rf = \sum_{i=1}^n \begin{pmatrix} x_1 f(s_1) \\ \vdots \\ x_n f(s_n) \end{pmatrix};$$

$$\widetilde{R} : H_2^r \to \mathbb{C}^n, \qquad \widetilde{R}f = \sum_{i=1}^n \begin{pmatrix} y_1 f(s_1) \\ \vdots \\ y_n f(s_n) \end{pmatrix}.$$

Assume the problem (TNP1)–(TNP3) has a solution F, say, and let $T : H_2^m \to H_2^r$ be the operator of multiplication by F, that is, $(Tf)(\lambda) = F(\lambda)f(\lambda)$ for $\lambda \in \Pi^+$. By condition (TNP1) the operator T is well-defined and bounded. Condition (TNP2) is equivalent to the requirement that $\|T\| < 1$, and condition (TNP3) yields $RT = \widetilde{R}$. It follows that

$$\widetilde{R}\widetilde{R}^* = RTT^*R^* \leq \|T\|^2 RR^*,$$

hence $RR^* - \widetilde{R}\widetilde{R}^* \geq (1 - \|T\|^2)RR^*$, and thus the operator $RR^* - \widetilde{R}\widetilde{R}^*$ is a positive definite operator on $\mathbb{C}^n$. Next, one computes that

$$R^* : \mathbb{C}^n \to H_2^m, \qquad \left(R^* \begin{pmatrix} z_1 \\ \vdots \\ z_n \end{pmatrix}\right)(\lambda) = \sum_{j=1}^n -\frac{z_j}{\lambda + \overline{s_j}} x_j^*,$$

$$\widetilde{R}^* : \mathbb{C}^n \to H_2^r, \qquad \left(\widetilde{R}^* \begin{pmatrix} z_1 \\ \vdots \\ z_n \end{pmatrix}\right)(\lambda) = \sum_{j=1}^n -\frac{z_j}{\lambda + \overline{s_j}} y_j^*.$$

We conclude that with respect to the standard basis of $\mathbb{C}^n$ the matrix of the operator $RR^* - \widetilde{R}\widetilde{R}^*$ is precisely the matrix Λ in (4.3.1). Hence, in order for the problem (TNP1)–(TNP3) to be solvable it is necessary that $\Lambda > 0$.

Now, let us see how the results of the previous section may be used to solve the (TNP1)–(TNP3) problem. The analysis starts with some heuristic arguments. From earlier experience—for instance, from the classical Nevanlinna [1929] paper or from the more recent papers Adamjan-Arov-Krein [1968], [1971] and Kovalishina-Potapov [1974], [1981] —one expects the solution (assuming it exists) to be given by a linear fractional map

$$F(\lambda) = \big(\Theta_{11}(\lambda)G(\lambda) + \Theta_{12}(\lambda)\big)\big(\Theta_{21}(\lambda)G(\lambda) + \Theta_{22}(\lambda)\big)^{-1}, \tag{4.3.2}$$

where the free parameter G is an arbitrary $m \times r$ matrix function in $\mathcal{R}at(H_\infty)$ such that

$$\sup\{\|G(\lambda)\| \mid \mathrm{Re}\ \lambda \geq 0\} < 1 \tag{4.3.3}$$

and

$$\Theta(\lambda) = \begin{pmatrix} \Theta_{11}(\lambda) & \Theta_{12}(\lambda) \\ \Theta_{21}(\lambda) & \Theta_{22}(\lambda) \end{pmatrix} \in \mathcal{R}at\big(H_\infty^{(m+r)\times(m+r)}\big). \tag{4.3.4}$$

Notice that (4.3.2) implies that

$$\begin{pmatrix} F(\lambda) \\ I \end{pmatrix} = \Theta(\lambda) \begin{pmatrix} G(\lambda) \\ I \end{pmatrix} \big(\Theta_{21}(\lambda)G(\lambda) + \Theta_{22}(\lambda)\big)^{-1}. \tag{4.3.5}$$

Since we want (4.3.3) to yield condition (TNP2), it is natural to assume that the coefficient matrix Θ is J-unitary with respect to the imaginary axis, with the signature matrix J being given by

$$J = \begin{pmatrix} I_m & 0 \\ 0 & -I_r \end{pmatrix}. \tag{4.3.6}$$

By taking $G(\lambda) \equiv 0$ in (4.3.2), one sees that $\Theta_{12}(\cdot)\Theta_{22}(\cdot)^{-1}$ has to satisfy (TNP1), and hence $\Theta_{12}(\cdot)\Theta_{22}(\cdot)^{-1}$ must have all its poles in Π^-. To achieve the latter one requires the coefficient matrix Θ to be J-inner, where J is as in (4.3.6). Finally, since

$$\begin{aligned} x_iF(s_i) - y_i &= \begin{pmatrix} x_i & -y_i \end{pmatrix} \begin{pmatrix} F(s_i) \\ I \end{pmatrix} \\ &= \begin{pmatrix} x_i & -y_i \end{pmatrix} \Theta(s_i) \begin{pmatrix} G(s_i) \\ I \end{pmatrix} \left(\Theta_{21}(s_i)G(s_i) + \Theta_{22}(s_i)\right)^{-1}, \end{aligned}$$

condition (TNP1) will be fulfilled if Θ satisfies the homogeneous interpolation condition

$$\begin{pmatrix} x_i & -y_i \end{pmatrix} \Theta(s_i) = 0 \qquad i = 1, \dots, n. \tag{4.3.7}$$

In other words, (TNP1) holds if for each i the coefficient matrix Θ has a zero at s_i and $\begin{pmatrix} x_i & -y_i \end{pmatrix}$ is a corresponding left eigenvector.

The above suggests the following strategy for solving the (TNP1)–(TNP3) problem:

1. construct Θ as in (4.3.4) such that Θ is J-inner, with J given by (4.3.6), and Θ is a solution of the homogeneous interpolation problem (4.3.7);
2. build F by using the linear fractional representation (4.3.2), where $G \in \mathcal{R}at(H_\infty^{m\times r})$ and satisfies (4.3.3);
3. check that F is a solution of the problem (TNP1)–(TNP3);
4. show that all solutions of the problem (TNP1)–(TNP3) are obtained in this way.

The following theorem shows that this strategy works.

Theorem 4.3.1 *Assume Θ in* (4.3.4) *has the following properties:*
(a) *Θ is J-inner, with J given by* (4.3.6),
(b) *Θ satisfies* (4.3.7).
Then for any $G \in \mathcal{R}at(H_\infty^{m\times r})$ such that (4.3.3) *holds, the matrix $F(\lambda)$ in* (4.3.2) *is well-defined for each $\lambda \in \Pi^+$, and the corresponding matrix function F satisfies* (TNP1)–(TNP3). *Furthermore, if*
(c) *$s_1, \dots, s_n$ are the only zeros of $\det \Theta(\lambda)$ in Π^+, and the multiplicities of $s_1, \dots, s_n$ as zeros of $\det \Theta(\lambda)$ are equal to one,*
then any solution F of (TNP1)–(TNP3) *is given by* (4.3.2) *for some matrix function $G \in \mathcal{R}at(H_\infty^{m\times r})$ satisfying* (4.3.3).

We shall return to Theorem 4.3.1 at the end of this section. First, let us show how Theorem 4.3.1 may be used to solve the (TNP1)–(TNP3) problem. Introduce the matrices

$$A = \begin{pmatrix} s_1 & & \\ & \ddots & \\ & & s_n \end{pmatrix}, \qquad B = \begin{pmatrix} x_1 & -y_1 \\ \vdots & \vdots \\ x_n & -y_n \end{pmatrix}. \tag{4.3.8}$$

Here the blank spots denote zero entries, and thus A is a diagonal matrix. According to our hypotheses, A has its eigenvalues in Π^+. Since the points $s_1, \dots, s_n$ are distinct and the vectors $x_1, \dots, x_n$ are non-zero, we know from the remarks made in the second paragraph of Section 4.1 that (A, B) is a full-range pair. By the final statement in the same second paragraph of Section 4.1, conditions (b) and

(c) in Theorem 3.4.1 are equivalent to the requirement that (A, B) is a left null pair of Θ on Π^+. But then we can apply Theorem 4.2.1 to show that there exists $\Theta \in \mathcal{R}at(H_\infty^{(m+r)\times(m+r)})$ satisfying conditions (a), (b), and (c) in Theorem 4.3.1 if and only if the Lyapunov equation

$$AY + YA^* = B\begin{pmatrix} I_m & 0 \\ 0 & -I_r \end{pmatrix} B^* \tag{4.3.9}$$

has a positive definite solution Y, and in this case one such Θ is given by

$$\Theta(\lambda) = I - \begin{pmatrix} I_m & 0 \\ 0 & -I_r \end{pmatrix} B^*(\lambda + A^*)^{-1} Y^{-1} B. \tag{4.3.10}$$

Let y_{ij} be the (i, j)-th entry of the $n \times n$ matrix Y. From the special form of A and B in (4.3.8) we see that equation (4.3.9) reduces to

$$s_i y_{ij} + y_{ij}\overline{s_j} = x_i x_j^* - y_i y_j^*, \qquad i, j = 1, \ldots, n.$$

Hence Y is precisely the matrix Λ given by (4.3.1). Thus, if Λ in (4.3.1) is positive definite, then $Y = \Lambda$ satisfies (4.3.9), and hence Θ in (4.3.10) is a matrix function in $\mathcal{R}at(H_\infty)$ which satisfies conditions (a), (b) and (c) in Theorem 4.3.1. We have proved the following theorem.

Theorem 4.3.2 *The left tangential Nevanlinna-Pick interpolation problem* (TNP1)–(TNP3) *has a solution if and only if the Pick matrix* Λ *associated with the problem is positive definite. In this case all solutions* F *are given by the linear fractional map*

$$F(\lambda) = (\Theta_{11}(\lambda)G(\lambda) + \Theta_{12}(\lambda))\left(\Theta_{21}(\lambda)G(\lambda) + \Theta_{22}(\lambda)\right)^{-1},$$

where the free parameter G *is an arbitrary* $m \times r$ *matrix function in* $\mathcal{R}at(H_\infty)$ *satisfying* $\sup\{\|G(\lambda)\| \mid \mathrm{Re}\ \lambda \geq 0\} < 1$, *and*

$$\begin{pmatrix} \Theta_{11}(\lambda) & \Theta_{12}(\lambda) \\ \Theta_{21}(\lambda) & \Theta_{22}(\lambda) \end{pmatrix} = \begin{pmatrix} I_m & 0 \\ 0 & I_r \end{pmatrix} - \begin{pmatrix} I_m & 0 \\ 0 & -I_r \end{pmatrix} B^*(\lambda + A^*)^{-1}\Lambda^{-1}B,$$

where A *and* B *are the matrices defined by* (4.3.8).

We conclude this section with the proof of Theorem 4.3.1.

Proof of Theorem 4.3.1 We split the proof into five parts. Throughout the proof Θ is as in (4.3.4), and we assume that Θ satisfies conditions (a) and (b).

Part (a). Let $G \in \mathcal{R}at(H_\infty^{m\times r})$ satisfy the norm constraint (4.3.3), and let F be defined by (4.3.2). In this part we show that F is well-defined and a solution of the problem (TNP1)–(TNP3).

Fix $z \in \overline{\Pi^+}$, the closure of the open right half plane in $\mathbb{C} \cup \{\infty\}$. Since $\Theta(z)$ is J-contractive, where J is given by (4.3.6), we can apply Lemma 2.2.2 to show that $\Theta_{22}(z)$ is non-singular and $\|\Theta_{22}(z)^{-1}\Theta_{21}(z)\| < 1$. It follows that

$$\Theta_{22}(z)^{-1}\Theta_{21}(z)G(z) + I$$

is invertible, and hence the same is true for $\Theta_{21}(z)G(z) + \Theta_{22}(z)$. Since z is an arbitrary point in $\overline{\Pi^+}$, we conclude that $\left(\Theta_{21}(\cdot)G(\cdot) + \Theta_{22}(\cdot)\right)^{-1} \in \mathcal{R}at(H_\infty)$, and thus $F(\cdot)$ in (4.3.2) is a well-defined $m \times r$ matrix function in $\mathcal{R}at(H_\infty)$.

Next, take $z \in i\mathbb{R}$. Since $\Theta(z)$ is a J-unitary matrix, where J is given by (4.3.6), we can use (4.3.5) to show that

$$\begin{aligned} F(z)^*F(z) - I &= \begin{pmatrix} F(z)^* & I \end{pmatrix} \begin{pmatrix} I_m & 0 \\ 0 & -I_r \end{pmatrix} \begin{pmatrix} F(z) \\ I \end{pmatrix} \\ &= X(z)^* \begin{pmatrix} G(z)^* & I \end{pmatrix} \Theta(z)^* \begin{pmatrix} I_m & 0 \\ 0 & -I_r \end{pmatrix} \Theta(z) \begin{pmatrix} G(z) \\ I \end{pmatrix} X(z) \\ &= X(z)^* \begin{pmatrix} G(z)^* & I \end{pmatrix} \begin{pmatrix} I_m & 0 \\ 0 & -I_r \end{pmatrix} \begin{pmatrix} G(z) \\ I \end{pmatrix} X(z) \\ &= X(z)^*(G(z)^*G(z) - I)X(z), \end{aligned}$$

where

$$X(\lambda) = \big(\Theta_{21}(\lambda)G(\lambda) + \Theta_{22}(\lambda)\big)^{-1}.$$

Now, use the fact that G satisfies (4.3.3). It follows that F satisfies condition (TNP2).

To prove that condition (TNP3) is also fulfilled, one uses (4.3.5) and the fact that Θ satisfies the homogeneous interpolation conditions (4.3.7).

The remaining four parts of the proof concern the second part of Theorem 4.3.1. Therefore, in the sequel we assume that Θ also satisfies condition (c) in Theorem 4.3.1, and that F is a solution of the interpolation problem (TNP1)–(TNP3). The aim is to show that F may be written in the form (4.3.2). Define rational matrix functions G_1 and G_2 of sizes $m \times r$ and $r \times r$, respectively, by setting

$$\begin{pmatrix} G_1(\lambda) \\ G_2(\lambda) \end{pmatrix} = \Theta(\lambda)^{-1} \begin{pmatrix} F(\lambda) \\ I \end{pmatrix}. \tag{4.3.11}$$

Part (b). In this part we show that G_1 and G_2 are in $\mathcal{R}at(H_\infty)$. Consider the matrices A and B in (4.3.8). We know that conditions (b) and (c) in Theorem 4.3.1 are equivalent to the requirement that (A, B) is a left null pair of Θ on Π^+. So, there exists a $(m + r) \times n$ matrix C such that $\Theta(\lambda)^{-1} - C(\lambda - A)^{-1}B$ is in $\mathcal{R}at(H_\infty)$. From the special form of A and B in (4.3.8) we deduce that

$$(\lambda - A)^{-1}B \begin{pmatrix} F(\lambda) \\ I \end{pmatrix} = \begin{pmatrix} (\lambda - s_1)^{-1}(x_1F(\lambda) - y_1) \\ \vdots \\ (\lambda - s_n)^{-1}(x_nF(\lambda) - y_n) \end{pmatrix}, \tag{4.3.12}$$

and hence the left hand side of (4.3.12) is also in $\mathcal{R}at(H_\infty)$. It follows that

$$\Theta(\lambda)^{-1} \begin{pmatrix} F(\lambda) \\ I \end{pmatrix} = \{\Theta(\lambda)^{-1} - C(\lambda - A)^{-1}B\} \begin{pmatrix} F(\lambda) \\ I \end{pmatrix} + K(\lambda), \tag{4.3.13}$$

for some K in $\mathcal{R}at(H_\infty)$. The first term in the right hand side of (4.3.13) is the product of two matrix functions in $\mathcal{R}at(H_\infty)$, and hence belongs to $\mathcal{R}at(H_\infty)$. So G_1 and G_2 in (4.3.11) are in $\mathcal{R}at(H_\infty)$.

Part (c). In this part we show that

$$\|G_1(z)G_2(z)^{-1}\| < 1, \quad z \in i\mathbb{R} \cup \{\infty\}. \tag{4.3.14}$$

Fix $z \in i\mathbb{R} \cup \{\infty\}$. Since $\Theta(z)$ is J-unitary, where J is given by (4.3.6), we have

$$\begin{aligned}
0 > F(z)^*F(z) - I &= \begin{pmatrix} F(z)^* & I \end{pmatrix} \begin{pmatrix} I_m & 0 \\ 0 & -I_r \end{pmatrix} \begin{pmatrix} F(z) \\ I \end{pmatrix} \\
&= \begin{pmatrix} G_1(z)^* & G_2(z)^* \end{pmatrix} \Theta(z)^* \begin{pmatrix} I_m & 0 \\ 0 & -I_r \end{pmatrix} \Theta(z) \begin{pmatrix} G_1(z) \\ G_2(z) \end{pmatrix} \\
&= \begin{pmatrix} G_1(z)^* & G_2(z)^* \end{pmatrix} \begin{pmatrix} I_m & 0 \\ 0 & -I_r \end{pmatrix} \begin{pmatrix} G_1(z) \\ G_2(z) \end{pmatrix} \\
&= G_1(z)^*G_1(z) - G_2(z)^*G_2(z).
\end{aligned}$$

If $G_2(z)x = 0$ and $x \neq 0$, then, by the above calculation,

$$0 > x^*G_1(z)^*G_1(z)x = \|G_1(z)x\|^2 \geq 0,$$

which is impossible. Thus $\det G_2(z) \neq 0$. But then (use again the above calculation)

$$I - G_2(z)^{-*}G_1(z)^*G_1(z)G_2(z)^{-1} > 0,$$

which proves (4.3.14).

Part (d). In this part we show that $\det G_2(\lambda) \neq 0$ for $\lambda \in \Pi^+$. To do this we use a winding number argument. Recall (see Part (a)) that $\|\Theta_{22}(z)^{-1}\Theta_{21}(z)\| < 1$ for $z \in i\mathbb{R} \cup \{\infty\}$. It follows that

$$\det(I + t\Theta_{22}(z)^{-1}\Theta_{21}(z)G_1(z)G_2(z)^{-1}) \neq 0, \quad z \in i\mathbb{R} \cup \{\infty\}, \quad 0 \leq t \leq 1. \tag{4.3.15}$$

Hence, by a continuity argument,

$$\text{wno } \det\left(I + \Theta_{22}(\cdot)^{-1}\Theta_{21}(\cdot)G_1(\cdot)G_2(\cdot)^{-1}\right) = \text{wno } \det I = 0.$$

Here wno f is the winding number relative to the origin of the curve parametrized by $z \mapsto f(z)$, where z runs from $-\infty$ to $+\infty$ over $i\mathbb{R}$. Notice that $\Theta_{21}G_1 + \Theta_{22}G_2 = I$, according to (4.3.11), and thus

$$I = \Theta_{22}(z)\{\Theta_{22}(z)^{-1}\Theta_{21}(z)G_1(z)G_2(z)^{-1} + I\}G_2(z), \qquad z \in i\mathbb{R} \cup \{\infty\}.$$

Now, the winding number of a product is the sum of the winding numbers of the separate factors. Therefore,

$$0 = \text{wno } \det I = \text{wno } \det \Theta_{22} + \text{wno } \det G_2.$$

But wno $\det \Theta_{22} = 0$, because $\det \Theta_{22}(\lambda) \neq 0$ for all $\lambda \in \overline{\Pi^+}$, including $\lambda = \infty$. So wno $\det G_2 = 0$, and hence $\det G_2(\lambda) \neq 0$ for $\lambda \in \Pi^+$.

Part (e). Put $G = G_1G_2^{-1}$. From the results proved in Parts (c) and (d) we know that $G \in \mathcal{R}at(H_\infty)$. According to (4.3.14) we have $\|G(z)\| \leq 1$ for $z \in i\mathbb{R} \cup \{\infty\}$, and hence, by the maximum modulus principle, G satisfies (4.3.3). From (4.3.11) we see that

$$\begin{aligned}
F &= \Theta_{11}G_1 + \Theta_{12}G_2 = (\Theta_{11}G + \Theta_{12})G_2 \\
I &= \Theta_{21}G_1 + \Theta_{22}G_2 = (\Theta_{21}G + \Theta_{22})G_2.
\end{aligned}$$

Thus F has the desired representation (4.3.2). □

4.4 Notes

Tangential Nevanlinna-Pick interpolation problems were introduced by Fedchina [1975]. Problems of this type appear in an natural way in optimal control studies and have led to a new branch of control theory which is nowadays called H_∞-*control*; see Francis [1987] and the references therein. A well-readable and simple (scalar) illustration of the connection between interpolation and control problems may be found in Chapter 9 of Doyle-Francis-Tannenbaum [1992]. There are several different methods to deal with the tangential Nevanlinna-Pick interpolation problem; see the books Ball-Gohberg-Rodman [1990], Dym [1989], Foias-Frazho [1990], Helton [1987], Rosenblum-Rovnyak [1985], and the papers in Kovalishina-Potapov [1988] and Gohberg-Sakhnovich (eds) [1994]. The method used in Section 4.3 is taken from Ball-Gohberg-Rodman [1990]. With some modifications (see also Ball-Gohberg-Rodman [1990]) this method applies equally well to the left tangential Nevanlinna-Pick interpolation problem for rational matrix functions that are analytic on the unit circle. Furthermore these interpolation problems have non-stationary analogues. Indeed, if one replaces the input-output system in (1.1.1) by a time-varying system

$$\Sigma \begin{cases} x'(t) &= A(t)x(t) + B(t)u(t), \quad t \geq 0 \\ y(t) &= C(t)x(t) + D(t)u(t), \\ x(0) &= x_0. \end{cases}$$

and rational matrix functions by input-output maps of time-varying systems of the above type, then the notion of point-evaluation for analytic functions extends in a natural way to a generalized point evaluation for input-output maps, and the latter allows one to introduce time-varying non-stationary analogues of the interpolation problem discussed in the present chapter (see the papers in Gohberg (ed.) [1992] and the references therein). Moreover, it turns out that an appropriate time-variant version of the state space method of Section 4.3 can be used to produce solutions of these non-stationary Nevanlinna-Pick interpolation problems in much the same way as they are constructed in Section 4.3 (see Ball-Gohberg-Kaashoek [1992a,b]).

See Kovalishina-Potapov [1988] and Chapter 1 of Dym [1980] for further information about J-inner matrix functions.

CHAPTER 5

Analysis of zeros and poles

In this chapter the theory of null pairs is developed further. Now, the poles are also taken into account. In the first section pole pairs and null-pole subspaces are introduced. The coupling transformation, which intertwines the zero and pole data, and null-pole triples are defined in the second section. A formula for the coupling transformation and one of its main properties are derived in the third section. In the fourth section null-pole triples are described in terms of realization. In the fifth section an alternative construction of a null-pole triple is given. In the final section null-pole triples at a single point are derived from a local Smith-McMillan form.

5.1 Pole pairs and null-pole subspace

Throughout this chapter W is a regular $m \times m$ rational matrix function, and Ω is a non-empty subset of $\mathbb{C}$.

Let $A : \mathcal{X} \to \mathcal{X}$ and $C : \mathcal{X} \to \mathbb{C}^m$ be linear operators, with $\mathcal{X}$ being a finite dimensional inner product space. The pair (C, A) is said to be a *right pole pair* of W on Ω if (C, A) is a right null pair of $W(\lambda)^{-1}$ on Ω. In other words, by Theorem 3.3.1, the pair (C, A) is a right pole pair of W on Ω if and only if the eigenvalues of A are in Ω and there exists a linear operator $B : \mathbb{C}^m \to \mathcal{X}$ such that $C(\lambda - A)^{-1}B$ is a minimal realization and the function

$$W(\lambda) - C(\lambda - A)^{-1}B \tag{5.1.1}$$

is analytic on Ω. In the sequel, we shall refer to $\mathcal{X}$ as the *main space* of the pair (C, A). Left pole pairs are defined in an analogous way.

By definition, the *null-pole subspace* of W on Ω is the set of all functions $W(\cdot)h(\cdot)$, where h is an arbitrary rational $\mathbb{C}^m$-valued function which has no poles in Ω. This set of functions is a linear space which will be denoted by $\mathcal{M}(W; \Omega)$.

To analyze the null-pole subspace, two operations on rational functions will be useful. Given a rational vector-valued or operator-valued function G, we define $\Xi_\Omega G$ to be the sum of the Laurent principal parts of G at the poles of G in Ω. Notice that $\Xi_\Omega G$ is analytic outside Ω and vanishes at infinity. On the other hand,

$$G(\lambda) - (\Xi_\Omega G)(\lambda)$$

is analytic on Ω. These three properties characterize the function $\Xi_\Omega G$ uniquely, and therefore we call $\Xi_\Omega G$ the *singular part* of G on Ω. The second map is the *residue map* which assigns to G the vector or operator $\operatorname{Res}_\Omega G$ which, by definition, is equal to the sum of the residues of G at the poles of G in Ω. It will be convenient to think about Ξ_Ω and $\operatorname{Res}_\Omega$ as linear operators. Notice that (C, A) is a right pole pair of W on Ω if and only if there exist an operator B such that $C(\lambda - A)^{-1}B$ is a minimal realization of the singular part of W on Ω.

In the sequel, the following lemma will be used a number of times.

Lemma 5.1.1 *Let $A : \mathcal{X} \to \mathcal{X}$ be a linear operator acting on the finite dimensional linear space $\mathcal{X}$, and assume that the eigenvalues of A are in Ω. Let f be a rational $\mathcal{X}$-valued function which is analytic on Ω. Then*

$$\Xi_\Omega\{(\lambda - A)^{-1} f(\lambda)\} = (\lambda - A)^{-1} y, \tag{5.1.2}$$

where

$$y = \text{Res}_\Omega (\lambda - A)^{-1} f(\lambda). \tag{5.1.3}$$

Proof It suffices to prove the lemma for the case when Ω is a singleton, $\Omega = \{\mu\}$, say. In this case, μ is the only eigenvalue of A, and hence $A - \mu$ is nilpotent. It follows that

$$(\lambda - A)^{-1} = \left((\lambda - \mu) - (A - \mu)\right)^{-1} = \sum_{\nu=1}^{n} (\lambda - \mu)^{-\nu} (A - \mu)^{\nu - 1},$$

where $n = \dim \mathcal{X}$. Let $f(\lambda) = f_0 + (\lambda - \mu) f_1 + (\lambda - \mu)^2 f_2 + \cdots$ be the Taylor expansion of f at μ. We see that for $k \geq 1$ the coefficient of $(\lambda - \mu)^{-k}$ in the Laurent principal part of $(\lambda - A)^{-1} f(\lambda)$ at μ is equal to $(A - \mu)^{k-1} y$, where

$$y = \sum_{\nu=0}^{n-1} (A - \mu)^\nu f_\nu.$$

It follows that y is precisely the residue of $(\lambda - A)^{-1} f(\lambda)$ at μ, and the Laurent principal part of $(\lambda - A)^{-1} f(\lambda)$ at μ is equal to $(\lambda - A)^{-1} y$. □

The following proposition is the main result of this section. We use subscripts π to emphasis the fact that we are dealing with pole pairs.

Proposition 5.1.2 *Let (C_π, A_π) be a right pole pair of W on Ω, with main space $\mathcal{X}_\pi$, and let $\mathcal{M}(W; \Omega)$ be the null-pole subspace of W on Ω. Then*

$$\Xi_\Omega \mathcal{M}(W, \Omega) = \{C_\pi (\lambda - A_\pi)^{-1} x \mid x \in \mathcal{X}_\pi\}. \tag{5.1.4}$$

Proof We may choose a linear operator B_π so that $C_\pi(\lambda - A_\pi)^{-1} B_\pi$ is a minimal realization and $W(\lambda) - C_\pi (\lambda - A_\pi)^{-1} B_\pi$ is analytic on Ω.

Let h be a $\mathbb{C}^m$-valued rational function which is analytic on Ω. We have

$$\begin{aligned} (\Xi_\Omega W h)(\lambda) &= \Xi_\Omega W(\lambda) h(\lambda) \\ &= \Xi_\Omega \{W(\lambda) - C_\pi (\lambda - A_\pi)^{-1} B_\pi\} h(\lambda) + \Xi_\Omega C_\pi (\lambda - A_\pi)^{-1} B_\pi h(\lambda). \end{aligned}$$

Since $W(\lambda) - C_\pi (\lambda - A_\pi)^{-1} B_\pi$ and $h(\lambda)$ are both analytic on Ω, the term

$$\Xi_\Omega \{W(\lambda) - C_\pi (\lambda - A_\pi)^{-1} B_\pi\} h(\lambda)$$

is identically zero. Next, we use the fact that the eigenvalues of A_π are in Ω. So we may apply Lemma 5.1.1 to show that

$$\Xi_\Omega C_\pi (\lambda - A_\pi)^{-1} B_\pi h(\lambda) = C_\pi \Xi_\Omega (\lambda - A_\pi)^{-1} B_\pi h(\lambda) = C_\pi (\lambda - A_\pi)^{-1} x,$$

where

$$x = \text{Res}_\Omega (\lambda - A_\pi)^{-1} B_\pi h(\lambda). \tag{5.1.5}$$

It follows that

$$(\Xi_\Omega W h)(\lambda) = C_\pi(\lambda - A_\pi)^{-1}x. \tag{5.1.6}$$

Conversely, take $x \in \mathcal{X}_\pi$. We seek a $\mathbb{C}^m$-valued rational function h, analytic on Ω, so that (5.1.6) holds. Since the realization is minimal, the pair (A_π, B_π) is full-range. Hence we may choose vectors $h_0, \ldots, h_{n_\pi - 1}$ in $\mathbb{C}^m$, where $n_\pi = \dim\ \mathcal{X}_\pi$, such that

$$\begin{pmatrix} B_\pi & A_\pi B_\pi & \ldots & A_\pi^{n_\pi - 1} B_\pi \end{pmatrix} \begin{pmatrix} h_0 \\ h_1 \\ \vdots \\ h_{n_\pi - 1} \end{pmatrix} = x. \tag{5.1.7}$$

Put $h(\lambda) = h_0 + \lambda h_1 + \cdots + \lambda^{n_\pi - 1} h_{n_\pi - 1}$. Then h is a $\mathbb{C}^m$-valued rational function, which is analytic on Ω, and according to the result of the second paragraph of the proof, we have

$$(\Xi_\Omega W h)(\lambda) = C_\pi(\lambda - A_\pi)^{-1}y,$$

where $y = \mathrm{Res}_\Omega(\lambda - A)^{-1}B_\pi h(\lambda)$. It remains to show that $y = x$. To do this, we choose a circle γ, with center at zero and with a sufficiently large radius, so that the eigenvalues of A_π are inside γ. Since h is a polynomial, h is analytic at each point of $\mathbb{C}$, and hence

$$\begin{aligned} y &= \mathrm{Res}_\Omega(\lambda - A_\pi)^{-1}B_\pi h(\lambda) = \frac{1}{2\pi i}\int_\gamma (\lambda - A_\pi)^{-1}B_\pi h(\lambda)\, d\lambda \\ &= \sum_{j=0}^{n_\pi - 1}\{\frac{1}{2\pi i}\int_\gamma \lambda^j(\lambda - A_\pi)^{-1}\, d\lambda\}B_\pi h_j = \sum_{j=0}^{n_\pi - 1} A_\pi^j B_\pi h_j = x, \end{aligned}$$

because of (5.1.7). □

5.2 Coupling transformation and null-pole triples

Let W and $\widetilde{W}$ be regular $m \times m$ rational matrix functions, and let Ω be a non-empty subset of $\mathbb{C}$. Let us consider the problem of finding necessary and sufficient conditions that guarantee that $\widetilde{W}(\lambda) = W(\lambda)F(\lambda)$ with F biregular on Ω. Assume such a matrix function F exists. Then necessarily (see the remark made at the end of Section 3.3) the functions W and $\widetilde{W}$ have a common left null pair on Ω. According to Theorem 3.4.1, the latter condition is also sufficient if W and $\widetilde{W}$ are both analytic on Ω. In the general case, when W and $\widetilde{W}$ may have poles in Ω, the identity $W(\lambda)^{-1} = F(\lambda)\widetilde{W}(\lambda)^{-1}$ shows that W and $\widetilde{W}$ also must have a common right pole pair on Ω. The two conditions together are still not sufficient, that is, it may happen that W and $\widetilde{W}$ have a common left null pair and a common right pole pair, both on Ω, and one cannot write $\widetilde{W}(\lambda) = W(\lambda)F(\lambda)$ with F biregular on Ω.

To illustrate this phenomenon on an example, let us consider again the 2×2 rational matrix functions

$$V(\lambda) = \begin{pmatrix} 1 & \frac{1}{\lambda - \mu} \\ 0 & 1 \end{pmatrix}, \quad \widetilde{V}(\lambda) = \begin{pmatrix} 1 & \frac{2}{\lambda - \mu} \\ 0 & 1 \end{pmatrix}.$$

Take $\Omega = \{\mu\}$. Clearly, it is impossible to factor $\widetilde{V}(\lambda)$ as $\widetilde{V}(\lambda) = V(\lambda)F(\lambda)$ with F biregular at μ. On the other hand, $\big(\mu, (0 \quad 1)\big)$ is a left null pair for both V and $\widetilde{V}$ on $\Omega = \{\mu\}$, and $\big(\left(\begin{smallmatrix}1\\0\end{smallmatrix}\right), \mu\big)$ is a right pole pair for both V and $\widetilde{V}$ on $\Omega = \{\mu\}$.

To resolve the problem some additional information is needed. It turns out that there exists a certain coupling between the zero data and the pole data which has to be employed. To see this, we need the following lemma, which may be viewed as a further refinement of Lemma 3.4.2. We use subscripts ζ to emphasis the fact that we are dealing with zeros and null pairs.

Lemma 5.2.1 *Let (A_ζ, B_ζ) be a left null pair of W on Ω, and let f be a $\mathbb{C}^m$-valued function, which is analytic on Ω. If f belongs to the null-pole subspace $\mathcal{M}(W;\Omega)$, then $(\lambda - A_\zeta)^{-1}B_\zeta f(\lambda)$ is analytic on Ω.*

Proof Choose a matrix C_ζ such that $C_\zeta(\lambda - A_\zeta)^{-1}B_\zeta$ is a minimal realization of the singular part of $W(\lambda)^{-1}$ on Ω. Since $f \in \mathcal{M}(W;\Omega)$, the function $W(\lambda)^{-1}f(\lambda)$ is analytic on Ω. Also, $W(\lambda)^{-1} - C_\zeta(\lambda - A_\zeta)^{-1}B_\zeta$ and $f(\lambda)$ are analytic on Ω, and hence the same holds true for the product of these two functions. It follows that $C_\zeta(\lambda - A_\zeta)^{-1}B_\zeta f(\lambda)$ is analytic on Ω. By using the same type of reasoning as in the proof of Lemma 3.4.2, we conclude that the function

$$\lambda \longmapsto \begin{pmatrix} C_\zeta \\ C_\zeta A_\zeta \\ \vdots \\ C_\zeta A_\zeta^{n_\zeta - 1} \end{pmatrix} (\lambda - A_\zeta)^{-1}B_\zeta f(\lambda)$$

is analytic on Ω. Here n_ζ is the dimension of the main space of the pair (A_ζ, B_ζ). Since (C_ζ, A_ζ) is a null-kernel pair, we obtain (as in the proof of Lemma 3.4.2) that $(\lambda - A_\zeta)^{-1}B_\zeta f(\lambda)$ is analytic on Ω. □

Now, let (C_π, A_π) be a right pole pair of W on Ω, and let (A_ζ, B_ζ) be a left null pair of W on Ω. Let $\mathcal{X}_\pi$ and $\mathcal{X}_\zeta$ be the corresponding main spaces. Take $x \in \mathcal{X}_\pi$. According to Proposition 5.1.2 there exists a $\mathbb{C}^m$-valued rational function h, analytic on Ω, such that

$$\big(\Xi W h\big)(\lambda) = C_\pi(\lambda - A_\pi)^{-1}x.$$

Put $f(\lambda) = W(\lambda)h(\lambda) - C_\pi(\lambda - A_\pi)^{-1}x$. Note that f is a $\mathbb{C}^m$-valued rational function, which is analytic on Ω, and

$$C_\pi(\lambda - A_\pi)^{-1}x + f(\lambda) \in \mathcal{M}(W;\Omega). \tag{5.2.1}$$

Thus, given $x \in \mathcal{X}_\pi$, one can always find a $\mathbb{C}^m$-valued rational function f, analytic on Ω, such that (5.2.1) holds. In general, the function f in (5.2.1) is not uniquely determined by x. However, we shall see that the vector

$$\mathrm{Res}_\Omega(\lambda - A_\zeta)^{-1}B_\zeta f(\lambda) \tag{5.2.2}$$

does not depend on the particular choice of f in (5.2.1). Indeed, suppose that $\widetilde{f}$ is another $\mathbb{C}^m$-valued rational function, analytic on Ω, such that (5.2.1) holds with $\widetilde{f}$ in place of f. Then $\widetilde{f} - f$ is analytic on Ω and $\widetilde{f} - f \in \mathcal{M}(W;\Omega)$. But then we can apply Lemma 5.2.1 to show that

$$\mathrm{Res}_\Omega(\lambda - A_\zeta)^{-1}B_\zeta f(\lambda) = \mathrm{Res}_\Omega(\lambda - A_\zeta)^{-1}B_\zeta \widetilde{f}(\lambda).$$

Hence the vector defined by (5.2.2) is uniquely determined by x, and the map

$$x \longmapsto \mathrm{Res}_\Omega(\lambda - A_\zeta)^{-1}B_\zeta f(\lambda) \tag{5.2.3}$$

is well-defined.

Since the right hand side of (5.2.1) is a linear space, the map $\Gamma : \mathcal{X}_\pi \to \mathcal{X}_\zeta$ defined by (5.2.3) is linear. This map Γ is called the *coupling transformation* corresponding to the pole pair (C_π, A_π), the null pair (A_ζ, B_ζ), and the function W on Ω, and one refers to the triple $\{C_\pi, A_\pi; A_\zeta, B_\zeta; \Gamma\}$ as a *null-pole triple of* W *on* Ω. Notice that Γ is uniquely determined by the pairs (C_π, A_π), (A_ζ, B_ζ), and by the function W. The next theorem settles the question raised in the beginning of this section.

Theorem 5.2.2 *Let W and $\widetilde{W}$ be regular $m \times m$ rational matrix functions. Then $\widetilde{W}(\lambda) = W(\lambda)F(\lambda)$ with F biregular on Ω if and only if W and $\widetilde{W}$ have a common null-pole triple.*

For the proof of Theorem 5.2.2 we shall use the following theorem, which describes the null-pole subspace in terms of a null-pole triple.

Theorem 5.2.3 *Let $\{C_\pi, A_\pi; A_\zeta, B_\zeta; \Gamma\}$ be a null-pole triple of W on Ω. Then the null-pole subspace $\mathcal{M}(W;\Omega)$ consists of all functions*

$$g(\lambda) = C_\pi(\lambda - A_\pi)^{-1}x + f(\lambda), \tag{5.2.4}$$

where x is an arbitrary vector in $\mathcal{X}_\pi$, the main space of the pair (C_π, A_π), and f is any $\mathbb{C}^m$-valued rational function, analytic on Ω, satisfying

$$\mathrm{Res}_\Omega(\lambda - A_\zeta)^{-1}B_\zeta f(\lambda) = \Gamma x \tag{5.2.5}$$

Proof Let $\widetilde{g} \in \mathcal{M}(W;\Omega)$. According to Proposition 5.1.2 we may choose $x \in \mathcal{X}_\pi$ such that $(\Xi_\Omega \widetilde{g})(\lambda) = C_\pi(\lambda - A_\pi)^{-1}x$. Let $\widetilde{f}$ be the $\mathbb{C}^m$-valued rational function defined by

$$\widetilde{f}(\lambda) = \widetilde{g}(\lambda) - C_\pi(\lambda - A_\pi)^{-1}x.$$

Then $\widetilde{f}$ is analytic on Ω, and the identity (5.2.4) holds with f and g replaced by $\widetilde{f}$ and $\widetilde{g}$, respectively. Furthermore, from the definition of the coupling transformation it follows that (5.2.5) is satisfied with $\widetilde{f}$ in place of f. Thus $\widetilde{g}$ has the desired representation.

Conversely, let g be as in (5.2.4), where x is an arbitrary vector in $\mathcal{X}_\pi$, and f is a $\mathbb{C}^m$-valued rational function, analytic on Ω, such that (5.2.5) is satisfied. We have to prove that $g \in \mathcal{M}(W;\Omega)$. According to Proposition 5.1.2 there exists $\widetilde{g} \in \mathcal{M}(W;\Omega)$ such that $(\Xi_\Omega \widetilde{g})(\lambda) = C_\pi(\lambda - A_\pi)^{-1}x$. Define $\widetilde{f}$ as in the first paragraph of the proof. It suffices to show that $f - \widetilde{f} \in \mathcal{M}(W;\Omega)$. Since (5.2.5) is fulfilled for both f and $\widetilde{f}$, we have

$$\mathrm{Res}_\Omega(\lambda - A_\zeta)^{-1}B_\zeta\big(f(\lambda) - \widetilde{f}(\lambda)\big) = 0. \tag{5.2.6}$$

Put $h(\lambda) = W(\lambda)^{-1}\{f(\lambda) - \widetilde{f}(\lambda)\}$. Note that h is a $\mathbb{C}^m$-valued rational function. We shall prove that h is analytic on Ω. To do this, we choose a linear operator C_ζ such that $C_\zeta(\lambda - A_\zeta)^{-1}B_\zeta$ is a minimal realization and $W(\lambda)^{-1} - C_\zeta(\lambda - A_\zeta)^{-1}B_\zeta$ is analytic on Ω. Recall that f and $\widetilde{f}$ are also analytic on Ω. It follows that

$$\{W(\lambda)^{-1} - C_\zeta(\lambda - A_\zeta)^{-1}B_\zeta\}\{f(\lambda) - \widetilde{f}(\lambda)\}$$

is analytic on Ω as well. Furthermore, since the eigenvalues of A_ζ are in Ω, we can apply Lemma 5.1.1 to show that

$$\Xi_\Omega(\lambda - A_\zeta)^{-1}B_\zeta\{f(\lambda) - \widetilde{f}(\lambda)\} = (\lambda - A_\zeta)^{-1}y,$$

where y is the vector given by the left hand side of (5.2.6). Thus $y = 0$, and we see that

$$C_\zeta(\lambda - A_\zeta)^{-1}B_\zeta\{f(\lambda) - \widetilde{f}(\lambda)\}$$

is also analytic on Ω. It follows that $W(\lambda)^{-1}\{f(\lambda) - \widetilde{f}(\lambda)\}$ is analytic on Ω. □

Proof of Theorem 5.2.2 Assume W and $\widetilde{W}$ have a common null-pole triple on Ω. Then we see from Theorem 5.2.3 that $\mathcal{M}(W;\Omega) = \mathcal{M}(\widetilde{W};\Omega)$. The latter property implies that $\widetilde{W}(\lambda) = W(\lambda)F(\lambda)$ with F biregular on Ω. Indeed, put $F(\lambda) = W(\lambda)^{-1}\widetilde{W}(\lambda)$, and let y be an arbitrary vector in $\mathbb{C}^m$. Then $\widetilde{g}(\lambda) := \widetilde{W}(\lambda)y$ is in $\mathcal{M}(\widetilde{W};\Omega) = \mathcal{M}(W;\Omega)$, and hence there exists a $\mathbb{C}^m$-valued rational function h, analytic on Ω, such that $\widetilde{g}(\lambda) = W(\lambda)h(\lambda)$. But then

$$F(\lambda)y = W(\lambda)^{-1}\widetilde{W}(\lambda)y = W(\lambda)^{-1}\widetilde{g}(\lambda) = h(\lambda),$$

and we see that F is analytic on Ω. Since $F(\lambda)^{-1} = \widetilde{W}(\lambda)W(\lambda)^{-1}$, we can apply the same reasoning, with W and $\widetilde{W}$ interchanged, to show that $F(\lambda)^{-1}$ is analytic on Ω. Thus F has the desired properties.

Conversely, assume that $\widetilde{W}(\lambda) = W(\lambda)F(\lambda)$, where F is biregular on Ω. Then $\widetilde{W}$ and W have a common left null pair on Ω, (A_ζ, B_ζ) say, and a common right pole pair on Ω, (C_π, A_π) say. Furthermore, W and $\widetilde{W}$ have the same null-pole subspace on Ω. But, if the pairs (C_π, A_π) and (A_ζ, B_ζ) are fixed, then the corresponding coupling transformation is uniquely determined by the null-pole subspace of W on Ω. Thus W and $\widetilde{W}$ have a common null-pole triple on Ω. □

Let us compute a null-pole triple on $\Omega = \{\mu\}$ for the 2×2 rational matrix function

$$V(\lambda) = \begin{pmatrix} 1 & \frac{c}{\lambda-\mu} \\ 0 & 1 \end{pmatrix}, \tag{5.2.7}$$

where $c \neq 0$. We already know that we may take

$$(A_\zeta, B_\zeta) = \big(\mu, \begin{pmatrix} 0 & 1 \end{pmatrix}\big), \qquad (C_\pi, A_\pi) = \Big(\begin{pmatrix} 1 \\ 0 \end{pmatrix}, \mu\Big),$$

which do not depend on the choice of c. Notice that $\mathcal{X}_\pi = \mathcal{X}_\zeta = \mathbb{C}$, and hence the coupling transformation reduces to a complex number. Take $x \in \mathbb{C}$. In this case

$$C_\pi(\lambda - A_\pi)^{-1}x = \frac{1}{\lambda - \mu}\begin{pmatrix} 1 \\ 0 \end{pmatrix} x,$$

and hence

$$C_\pi(\lambda - A_\pi)^{-1}x + \begin{pmatrix} 0 \\ 1/c \end{pmatrix} = V(\lambda)\begin{pmatrix} 0 \\ 1/c \end{pmatrix} \in \mathcal{M}(V;\{\mu\}).$$

We conclude that

$$\Gamma x = \operatorname{Res}_{\{\mu\}}(\lambda - A_\zeta)^{-1}B_\zeta\begin{pmatrix} 0 \\ 1/c \end{pmatrix} x = \operatorname{Res}_{\{\mu\}}\frac{1}{\lambda - \mu}\begin{pmatrix} 0 & 1 \end{pmatrix}\begin{pmatrix} 0 \\ 1/c \end{pmatrix} x = \frac{1}{c}x.$$

Thus $\{\binom{1}{0}, \mu; \mu, (0 \quad 1); \frac{1}{c}\}$ is a null-pole triple of V on $\Omega = \{\mu\}$.

A null-pole triple of W on Ω is essentially unique. The freedom one has in the choice of such a triple is described by the next proposition.

Proposition 5.2.4 *Let $\{C_\pi, A_\pi; A_\zeta, B_\zeta; \Gamma\}$ be a null-pole triple of W on Ω, and let $\mathcal{X}_\pi$ and $\mathcal{X}_\zeta$ be the corresponding main spaces. Let $E : \mathcal{X}_\pi \to \widetilde{\mathcal{X}}_\pi$ and $F : \mathcal{X}_\zeta \to \widetilde{\mathcal{X}}_\zeta$ be invertible linear operators. Then*

$$\{C_\pi E^{-1}, EA_\pi E^{-1}; FA_\zeta F^{-1}, FB_\zeta; F\Gamma E^{-1}\} \tag{5.2.8}$$

is again a null-pole triple of W on Ω, and all null-pole triples of W on Ω are obtained in this way.

Proof Let $\{\widetilde{C}_\pi, \widetilde{A}_\pi; \widetilde{A}_\zeta, \widetilde{B}_\zeta; \widetilde{\Gamma}\}$ be another null-pole triple of W on Ω. Write $\widetilde{\mathcal{X}}_\pi$ and $\widetilde{\mathcal{X}}_\zeta$ for the corresponding main spaces. By Corollary 3.3.2 and its analogue for left null pairs, there exist invertible linear operators $E : \mathcal{X}_\pi \to \widetilde{\mathcal{X}}_\pi$ and $F : \mathcal{X}_\zeta \to \widetilde{\mathcal{X}}_\zeta$ such that

$$\widetilde{C}_\pi = C_\pi E^{-1}, \quad \widetilde{A}_\pi = EA_\pi E^{-1}, \quad \widetilde{B}_\zeta = FB_\zeta, \quad \widetilde{A}_\zeta = FA_\zeta F^{-1}.$$

Now, let x be an arbitrary element of $\mathcal{X}_\pi$. We may choose a $\mathbb{C}^m$-valued rational function f, analytic on Ω, such that (5.2.1) holds. It follows that

$$\widetilde{C}_\pi(\lambda - \widetilde{A}_\pi)^{-1}Ex + f(\lambda) \in \mathcal{M}(W; \Omega),$$

and hence, using the definition of the coupling transformation, we see that

$$\widetilde{\Gamma}Ex = \mathrm{Res}_\Omega(\lambda - \widetilde{A}_\zeta)\widetilde{B}_\zeta f(\lambda) = F\{\mathrm{Res}_\Omega(\lambda - A_\zeta)^{-1}B_\zeta f(\lambda)\} = F\Gamma x.$$

So $\widetilde{\Gamma} = F\Gamma E^{-1}$, and we conclude that the triple $\{\widetilde{C}_\pi, \widetilde{A}_\pi; \widetilde{A}_\zeta, \widetilde{B}_\zeta; \widetilde{\Gamma}\}$ is of the form (5.2.8). The fact that all null-pole triples of W on Ω are of the form (5.2.8) is proved in a similar way. □

5.3 More about the coupling transformation

In this section we derive a formula for the coupling transformation in a null-pole triple, and we discuss one of its main properties.

Theorem 5.3.1 *Let $\{C_\pi, A_\pi; A_\zeta, B_\zeta; \Gamma\}$ be a null-pole triple of W on Ω. Put $n_\pi = \dim \mathcal{X}_\pi$, where $\mathcal{X}_\pi$ is the main space of the pair (C_π, A_π). We may choose a linear operator B_π such that $C_\pi(\lambda - A_\pi)^{-1}B_\pi$ is a minimal realization of the singular part of W on Ω, and we may choose linear operators $\Lambda_0, \dots, \Lambda_{n_\pi - 1}$ from $\mathcal{X}_\pi$ into $\mathbb{C}^m$ such that*

$$\sum_{j=0}^{n_\pi - 1} A_\pi^j B_\pi \Lambda_j = I_{\mathcal{X}_\pi}. \tag{5.3.1}$$

Put $P(\lambda) = \Lambda_0 + \lambda\Lambda_1 + \cdots + \lambda^{n_\pi - 1}\Lambda_{n_\pi - 1}$. Then the coupling transformation Γ is given by

$$\Gamma = \mathrm{Res}_\Omega(\lambda - A_\zeta)^{-1}B_\zeta W(\lambda)P(\lambda). \tag{5.3.2}$$

5.4 Null-pole triples from minimal realizations

Throughout this section W is a *bicausal*rational $m \times m$ matrix function, that is, W is proper (analytic at infinity) and $W(\infty)$ is non-singular. We shall construct a null-pole triple for W on Ω starting from a minimal realization of W.

Theorem 5.4.1 *Let $W(\lambda) = D + C(\lambda - A)^{-1}B$ be a minimal realization, and assume D is non-singular. Write $A^\times = A - BD^{-1}C$, and let Q and $Q^\times$ be the Riesz projections of A and $A^\times$, respectively, corresponding to the eigenvalues in Ω. Consider the following partitionings:*

$$\begin{aligned}
A &= \begin{pmatrix} A_0 & 0 \\ 0 & A_1 \end{pmatrix} : \operatorname{Im} Q \dot{+} \operatorname{Ker} Q \to \operatorname{Im} Q \dot{+} \operatorname{Ker} Q, \\
B &= \begin{pmatrix} B_0 \\ B_1 \end{pmatrix} : \mathbb{C}^m \to \operatorname{Im} Q \dot{+} \operatorname{Ker} Q, \\
C &= \begin{pmatrix} C_0 & C_1 \end{pmatrix} : \operatorname{Im} Q \dot{+} \operatorname{Ker} Q \to \mathbb{C}^m, \\
A^\times &= \begin{pmatrix} A_0^\times & 0 \\ 0 & A_1^\times \end{pmatrix} : \operatorname{Im} Q^\times \dot{+} \operatorname{Ker} Q^\times \to \operatorname{Im} Q^\times \dot{+} \operatorname{Ker} Q^\times, \\
BD^{-1} &= \begin{pmatrix} B_0^\times \\ B_1^\times \end{pmatrix} : \mathbb{C}^m \to \operatorname{Im} Q^\times \dot{+} \operatorname{Ker} Q^\times, \\
-D^{-1}C &= \begin{pmatrix} C_0^\times & C_1^\times \end{pmatrix} : \operatorname{Im} Q^\times \dot{+} \operatorname{Ker} Q^\times \to \mathbb{C}^m,
\end{aligned}$$

and set

$$\Gamma_0 = Q^\times | \operatorname{Im} Q : \operatorname{Im} Q \to \operatorname{Im} Q^\times. \tag{5.4.1}$$

Then the triple $\{C_0, A_0; A_0^\times, B_0^\times; \Gamma_0\}$ is a null-pole triple of W on Ω.

Proof It is straightforward to check that $C_0(\lambda - A_0)^{-1}B_0$ and $C_0^\times(\lambda - A_0^\times)B_0^\times$ are minimal realizations of the singular parts of W and $W(\cdot)^{-1}$ on Ω. It follows that the pair (C_0, A_0) is a right pole pair of W on Ω, and the pair $(A_0^\times, B_0^\times)$ is a left null pair of W on Ω. It remains to prove that the corresponding coupling transformation Λ is equal to the operator Γ_0 defined by (5.4.1). To do this we apply Theorem 5.3.1.

According to Theorem 5.3.1, the coupling transformation Λ is given by

$$\Lambda = \operatorname{Res}_\Omega(\lambda - A_0^\times)^{-1}B_0^\times W(\lambda)P(\lambda), \tag{5.4.2}$$

where $P(\lambda)$ is an operator polynomial, $P(\lambda) = P_0 + \lambda P_1 + \ldots \lambda^{n-1}P_{n-1}$, whose coefficients satisfy the identity

$$\sum_{j=0}^{n-1} A_0^j B_0 P_j = I_{\mathcal{X}_0}. \tag{5.4.3}$$

Here $\mathcal{X}_0 = \operatorname{Im} Q$ and $n = \dim\ \mathcal{X}_0$. First, observe that

$$(\lambda - A_0^\times)^{-1}B_0^\times = Q^\times(\lambda - A^\times)^{-1}BD^{-1}.$$

Next, notice that

$$\begin{aligned}
(\lambda - A^\times)^{-1}BD^{-1}W(\lambda) &= (\lambda - A^\times)^{-1}B + (\lambda - A^\times)^{-1}BD^{-1}C(\lambda - A)^{-1}B \\
&= (\lambda - A^\times)^{-1}B + (\lambda - A^\times)^{-1}\{(\lambda - A^\times) - (\lambda - A)\}(\lambda - A)^{-1}B \\
&= (\lambda - A)^{-1}B.
\end{aligned}$$

Since the singular part of $(\lambda - A)^{-1}B$ on Ω is the function $Q(\lambda - A)^{-1}B = (\lambda - A_0)^{-1}B_0$, it follows that Λ in (5.4.2) satisfies

$$\begin{aligned}\Lambda &= \operatorname{Res}_\Omega Q^\times(\lambda - A^\times)^{-1}BD^{-1}W(\lambda)P(\lambda) = \operatorname{Res}_\Omega Q^\times(\lambda - A)^{-1}BP(\lambda)\\ &= \operatorname{Res}_\Omega Q^\times Q(\lambda - A)^{-1}BP(\lambda) = \Gamma_0\{\operatorname{Res}_\Omega(\lambda - A_0)^{-1}B_0P(\lambda)\}.\end{aligned}$$

From the remark made at the end of Section 5.3 (see formula (5.3.7)) we know that $\operatorname{Res}_\Omega(\lambda - A_0)^{-1}B_0P(\lambda) = I_{\mathcal{X}_0}$, and hence $\Lambda = \Gamma_0$. □

Let W be bicausal rational $m \times m$ matrix function, and put $D = W(\infty)$. Let (C_π, A_π) be a right pole pair of W on Ω, and let (A_ζ, B_ζ) be a left null pair of W on Ω. Theorem 5.4.1 provides an alternative way to construct the coupling transformation corresponding (C_π, A_π), (A_ζ, B_ζ) and W, namely as follows. Write

$$W(\lambda) = D + W_1(\lambda) + W_2(\lambda), \quad W(\lambda)^{-1} = D^{-1} + W_1^\times(\lambda) + W_2^\times(\lambda),$$

where W_1 and $W_1^\times$ have their poles in Ω, and the poles of W_2 and $W_2^\times$ are in $\mathbb{C}\backslash\Omega$. We may choose operators B_π and C_ζ such that $C_\pi(\lambda - A_\pi)^{-1}B_\pi$ is a minimal realization of W_1 and $C_\zeta(\lambda - A_\zeta)^{-1}B_\zeta$ is a minimal realization of $W_1^\times$. Next, choose minimal realizations $\widehat{C}_\pi(\lambda - \widehat{A}_\pi)^{-1}\widehat{B}_\pi$ and $\widehat{C}_\zeta(\lambda - \widehat{A}_\zeta)^{-1}\widehat{B}_\zeta$ of W_2 and $W_2^\times$, respectively. Then

$$W(\lambda) = D + \begin{pmatrix} C_\pi & \widehat{C}_\pi \end{pmatrix} \left(\lambda - \begin{pmatrix} A_\pi & 0 \\ 0 & \widehat{A}_\pi \end{pmatrix}\right)^{-1} \begin{pmatrix} B_\pi \\ \widehat{B}_\pi \end{pmatrix}, \tag{5.4.4}$$

$$W(\lambda)^{-1} = D^{-1} + \begin{pmatrix} C_\zeta & \widehat{C}_\zeta \end{pmatrix} \left(\lambda - \begin{pmatrix} A_\zeta & 0 \\ 0 & \widehat{A}_\zeta \end{pmatrix}\right)^{-1} \begin{pmatrix} B_\zeta \\ \widehat{B}_\zeta \end{pmatrix}. \tag{5.4.5}$$

Since D is non-singular, it follows from (5.4.4) that $W(\lambda)^{-1}$ is also given by

$$W(\lambda)^{-1} = D^{-1} - C^\times(\lambda - A^\times)^{-1}B^\times, \tag{5.4.6}$$

where

$$\begin{aligned} A^\times &= \begin{pmatrix} A_\pi & 0 \\ 0 & \widehat{A}_\pi \end{pmatrix} - \begin{pmatrix} B_\pi \\ \widehat{B}_\pi \end{pmatrix} D^{-1} \begin{pmatrix} C_\pi & \widehat{C}_\pi \end{pmatrix}, \\ B^\times &= \begin{pmatrix} B_\pi \\ \widehat{B}_\pi \end{pmatrix} D^{-1}, \qquad C^\times = D^{-1}\begin{pmatrix} C_\pi & \widehat{C}_\pi \end{pmatrix}. \end{aligned}$$

The realizations (5.4.4)–(5.4.6) are all minimal. In particular, (5.4.5) and (5.4.6) are minimal realizations of the same rational matrix function. But then we can apply the state space similarity theorem (Theorem 1.1.3) to show that there exists an invertible linear operator T such that

$$TA^\times T^{-1} = \begin{pmatrix} A_\zeta & 0 \\ 0 & \widehat{A}_\zeta \end{pmatrix}, \quad TB^\times = \begin{pmatrix} B_\zeta \\ \widehat{B}_\zeta \end{pmatrix}, \quad -C^\times T^{-1} = \begin{pmatrix} C_\zeta & \widehat{C}_\zeta \end{pmatrix}. \tag{5.4.7}$$

We may partition T as follows

$$T = \begin{pmatrix} T_{11} & T_{12} \\ T_{21} & T_{22} \end{pmatrix} : \mathcal{X}_\pi \dot{+} \widehat{\mathcal{X}}_\pi \to \mathcal{X}_\zeta \dot{+} \widehat{\mathcal{X}}_\zeta. \tag{5.4.8}$$

Here $\mathcal{X}_\pi$, $\widehat{\mathcal{X}}_\pi$, $\mathcal{X}_\zeta$, $\widehat{\mathcal{X}}_\zeta$ are the respective spaces on which the operators A_π, $\widehat{A}_\pi$, A_ζ, $\widehat{A}_\zeta$ act.

Corollary 5.4.2 *The operator T_{11} in (5.4.8) is the coupling transformation corresponding to the pairs (C_π, A_π), (A_ζ, B_ζ) and W on Ω.*

Proof We apply Theorem 5.4.1. The various operators introduced in Theorem 5.4.1 provide minimal realizations for the matrix functions W_1, $W_1^\times$, W_2 and $W_2^\times$. It follows that there exist invertible operators E_0, E_1, F_0, F_1 such that the following identities hold:

$$\begin{array}{lll} A_\pi = E_0 A_0 E_0^{-1}, & B_\pi = E_0 B_0, & C_\pi = C_0 E_0^{-1}, \\ \widehat{A}_\pi = E_1 A_1 E_1^{-1}, & \widehat{B}_\pi = E_1 B_1, & \widehat{C}_\pi = C_1 E_1^{-1}, \\ A_\zeta = F_0 A_0^\times F_0^{-1}, & B_\zeta = F_0 B_0^\times, & C_\zeta = C_0^\times F_0^{-1}, \\ \widehat{A}_\zeta = F_1 A_1^\times F_1^{-1}, & \widehat{B}_\zeta = F_1 B_1^\times, & \widehat{C}_\zeta = C_1^\times F_1^{-1}. \end{array}$$

Put

$$\begin{aligned} E &= \begin{pmatrix} E_0 & 0 \\ 0 & E_1 \end{pmatrix} : \operatorname{Im} Q \dot{+} \operatorname{Ker} Q \to \mathcal{X}_\pi \dot{+} \widehat{\mathcal{X}}_\pi, \\ F &= \begin{pmatrix} F_0 & 0 \\ 0 & F_1 \end{pmatrix} : \operatorname{Im} Q^\times \dot{+} \operatorname{Ker} Q^\times \to \mathcal{X}_\zeta \dot{+} \widehat{\mathcal{X}}_\zeta. \end{aligned}$$

Let T be as in (5.4.8). By the uniqueness statement in the state space similarity theorem, the operator $F^{-1}TE$ is an identity operator. It follows that $T_{11} = F_0 \Gamma_0 E_0^{-1}$. But then we can apply Proposition 5.2.4 and Theorem 5.4.1 to show that T_{11} is the desired coupling transformation. □

From (5.4.7) it readily follows that the coupling transformation T_{11} in Corollary 5.4.2 satisfies the Sylvester equation

$$T_{11} A_\pi - A_\zeta T_{11} = B_\zeta C_\pi. \tag{5.4.9}$$

Indeed, in (5.4.7) bring T^{-1} to the right hand side, and consider the $(1,1)$-entries. In this way one sees that

$$T_{11}(A_\pi - B_\pi D^{-1} C_\pi) = A_\zeta T_{11}, \qquad T_{11} B_\pi D^{-1} = B_\zeta,$$

which implies (5.4.9). [1]

5.5 Another construction of a null-pole triple

Let $W(\lambda)$ be a regular $m \times m$ rational matrix function, and let Ω be a non-empty subset of $\mathbb{C}$. In this section we present an alternative way to construct a null-pole triple of W on Ω.

First we fix some notation. By $\mathcal{R}$ we denote the linear space of all $\mathbb{C}^m$-valued rational functions. Notice that the map Ξ_Ω acts as a linear projection on $\mathcal{R}$, and hence we may decompose $\mathcal{R}$ as

$$\mathcal{R} = \mathcal{H}(\Omega) \dot{+} \mathcal{P}(\Omega), \tag{5.5.1}$$

where

$$\mathcal{H}(\Omega) = \{ f \in \mathcal{R} \mid f \text{ analytic on } \Omega \} = \operatorname{Ker} \Xi_\Omega, \quad \mathcal{P}(\Omega) = \{ \Xi_\Omega f \mid f \in \mathcal{R} \} = \operatorname{Im} \Xi_\Omega.$$

[1]The next two sections are not necessary for understanding the results of Chapter 6.

By M we denote the operator of multiplication by W acting on $\mathcal{R}$. Since W is regular, M is invertible. We partition M and M^{-1} as 2×2 operator matrices relative to the direct sum decomposition (5.5.1), as follows

$$M = \begin{pmatrix} T & K \\ H & U \end{pmatrix} : \mathcal{H}(\Omega)\dot{+}\mathcal{P}(\Omega) \to \mathcal{H}(\Omega)\dot{+}\mathcal{P}(\Omega), \tag{5.5.2}$$

$$M^{-1} = \begin{pmatrix} T^{\times} & K^{\times} \\ H^{\times} & U^{\times} \end{pmatrix} : \mathcal{H}(\Omega)\dot{+}\mathcal{P}(\Omega) \to \mathcal{H}(\Omega)\dot{+}\mathcal{P}(\Omega). \tag{5.5.3}$$

Next, let $V : \mathcal{R} \to \mathcal{R}$ denote the operator of multiplication by λ (where λ is the independent variable). Obviously, V leaves $\mathcal{H}(\Omega)$ invariant, and hence relative to (5.5.1) the operator V partitions as:

$$V = \begin{pmatrix} S & S_0 \\ 0 & S_* \end{pmatrix} : \mathcal{H}(\Omega)\dot{+}\mathcal{P}(\Omega) \to \mathcal{H}(\Omega)\dot{+}\mathcal{P}(\Omega). \tag{5.5.4}$$

In particular, $S_* : \mathcal{P}(\Omega) \to \mathcal{P}(\Omega)$ is defined by

$$S_* = \Xi_\Omega V|\mathcal{P}(\Omega).$$

The fact that $\lambda W(\lambda) = W(\lambda)\lambda$ implies that V commutes with M and M^{-1}, which among others yields:

$$HS = S_*H, \qquad H^{\times}S = S_*H^{\times}. \tag{5.5.5}$$

Finally, $\tau : \mathbb{C}^m \to \mathcal{H}(\Omega)$ will be the map which assigns to the vector $y \in \mathbb{C}^m$ the function identically equal to y, that is, $(\tau y)(\lambda) = y$.

To construct a null-pole triple of W on Ω we need the spaces

$$\mathcal{X}_\pi = \operatorname{Im} H, \qquad \mathcal{X}_\zeta = \operatorname{Im} H^{\times}. \tag{5.5.6}$$

The two identities in (5.5.5) show that $\mathcal{X}_\pi$ and $\mathcal{X}_\zeta$ are invariant under the operator S_*. The next theorem is the main result of this section.

Theorem 5.5.1 *Let $\mathcal{X}_\pi$ and $\mathcal{X}_\zeta$ be as in* (5.5.6), *and consider the following operators:*

$$\begin{aligned} A_\pi &= S_*|\mathcal{X}_\pi : \mathcal{X}_\pi \to \mathcal{X}_\pi, & C_\pi &= \operatorname{Res}_\Omega|\mathcal{X}_\pi : \mathcal{X}_\pi \to \mathbb{C}^m, \\ A_\zeta &= S_*|\mathcal{X}_\zeta : \mathcal{X}_\zeta \to \mathcal{X}_\zeta, & B_\zeta &= H^{\times}\tau : \mathbb{C}^m \to \mathcal{X}_\zeta, \\ \Gamma &= -U^{\times}|\mathcal{X}_\pi : \mathcal{X}_\pi \to \mathcal{X}_\zeta. \end{aligned}$$

Then the triple $\{C_\pi, A_\pi; A_\zeta, B_\zeta; \Gamma\}$ is a null-pole triple of W on Ω.

For the proof of Theorem 5.5.1 we need to make some preparations. By interchanging the rows in (5.5.2) and the columns in (5.5.3) we see that

$$\begin{pmatrix} H & U \\ T & K \end{pmatrix}^{-1} = \begin{pmatrix} K^{\times} & T^{\times} \\ U^{\times} & H^{\times} \end{pmatrix}. \tag{5.5.7}$$

It follows that the operators H and $H^{\times}$ are matricially coupled (see Bart-Gohberg-Kaashoek [1984] or Section III.4 in Gohberg-Goldberg-Kaashoek [1990], where this notion is defined for bounded operators). The fact that H and $H^{\times}$ are related as in

(5.5.7) implies (cf., Gohberg-Goldberg-Kaashoek [1990], Corollary III.4.3) among other things that

$$\begin{aligned} \operatorname{Ker} H &= T^\times \operatorname{Ker} H^\times, \qquad \operatorname{Ker} H^\times = T \operatorname{Ker} H, &(5.5.8)\\ \operatorname{Im} H &= \{f \in \mathcal{P}(\Omega) \mid U^\times f \in \operatorname{Im} H^\times\}, &(5.5.9)\\ \operatorname{Im} H^\times &= \{f \in \mathcal{P}(\Omega) \mid U f \in \operatorname{Im} H\}. &(5.5.10)\end{aligned}$$

From (5.5.9) we see that $U^\times(\operatorname{Im} H) \subset \operatorname{Im} H^\times$, and thus the map Γ in Theorem 5.5.1 maps $\mathcal{X}_\pi$ into $\mathcal{X}_\zeta$ as desired.

Lemma 5.5.2 *The null-pole subspace $\mathcal{M}(W;\Omega)$ consists of all functions $g = f+x$, where x is an arbitrary element of $\mathcal{X}_\pi$ and f is any element of $\mathcal{H}(\Omega)$ satisfying*

$$H^\times f + U^\times x = 0. \tag{5.5.11}$$

Proof Notice that $\mathcal{M}(W;\Omega) = M\mathcal{H}(\Omega)$. Take $g \in \mathcal{M}(W;\Omega)$. Thus $g \in M\mathcal{H}(\Omega)$, and hence $g = Th+Hh$ for some $h \in \mathcal{H}(\Omega)$. It follows that $x := Hh \in \mathcal{X}_\pi$, and $g = f+x$ with $f = Th \in \mathcal{H}(\Omega)$. Since $M^{-1}M = I$, we have $H^\times T + U^\times H = 0$, and thus

$$H^\times f + U^\times x = (H^\times T + U^\times H)h = 0,$$

which proves (5.5.11).

Conversely, let x be an arbitrary element of $\mathcal{X}_\pi$, and let f be any element of $\mathcal{H}(\Omega)$ satisfying (5.5.11). We have to show that $g = f + x \in \mathcal{M}(W,\Omega)$. From the first identity in (5.5.6) we see that $x = H\widetilde{h}$ for some $\widetilde{h} \in \mathcal{H}(\Omega)$. But then we know from the first paragraph of this proof that $H^\times T\widetilde{h} + U^\times x = 0$. The latter identity together with (5.5.11) yields $H^\times(T\widetilde{h} - f) = 0$. Now use the second identity in (5.5.8). We see that $T\widetilde{h} - f = Tk$ for some $k \in \operatorname{Ker} H \subset \mathcal{H}(\Omega)$. So, $f = T(\widetilde{h} - k)$ and $x = H\widetilde{h} = H(\widetilde{h} - k)$, where $\widetilde{h} - k \in \mathcal{H}(\Omega)$. Thus

$$g = f + x = \begin{pmatrix} T \\ H \end{pmatrix} (\widetilde{h} - k) = M(\widetilde{h} - k) \in \mathcal{M}(W,\Omega),$$

which proves the lemma. □

Lemma 5.5.3 *Let $C(\lambda - A)^{-1}B$ be an arbitrary realization of the singular part of W on Ω, and let $\mathcal{X}$ be the corresponding state space. Consider the operators*

$$\begin{aligned} &\Phi : \mathcal{H}(\Omega) \to \mathcal{X}, \qquad \Phi h = \operatorname{Res}_\Omega (\lambda - A)^{-1} B h(\lambda); &(5.5.12)\\ &\Psi : \mathcal{X} \to \mathcal{P}(\Omega), \qquad (\Psi x)(\lambda) = C(\lambda - A)^{-1} x. &(5.5.13)\end{aligned}$$

Then $\Psi\Phi = H$, where H is as in (5.5.2).

Proof Fix $h \in \mathcal{H}(\Omega)$ and a complex number μ not in Ω. From (5.5.12) and the resolvent equation (see Gohberg-Goldberg-Kaashoek [1990], Section I.2) we see that

$$(\Psi\Phi h)(\mu) = \operatorname{Res}_\Omega C(\mu - A)^{-1}(\lambda - A)^{-1} B h(\lambda) = (A) - (B), \tag{5.5.14}$$

where

$$(A) = \operatorname{Res}_\Omega \frac{1}{\lambda - \mu} C(\mu - A)^{-1} B h(\lambda), \quad (B) = \operatorname{Res}_\Omega \frac{1}{\lambda - \mu} C(\lambda - A)^{-1} B h(\lambda).$$

Since $\mu \notin \Omega$ and h is analytic on Ω, we have $(A) = 0$. To analyze the term (B) notice that the poles of the rational functions

$$W(\lambda) - C(\lambda - A)^{-1}B, \qquad W(\lambda)h(\lambda) - (\Xi_\Omega Wh)(\lambda)$$

are outside Ω. Also, μ is outside Ω. It follows that

$$(B) = \mathrm{Res}_\Omega \frac{1}{\lambda - \mu} W(\lambda)h(\lambda) = \mathrm{Res}_\Omega \frac{1}{\lambda - \mu}(\Xi_\Omega Wh)(\lambda).$$

Finally, note that $(\Xi_\Omega Wh)$ is analytic outside Ω and $(\Xi_\Omega Wh)$ vanishes at infinity. This implies that

$$\mathrm{Res}_\Omega \frac{1}{\lambda - \mu}(\Xi_\Omega Wh)(\lambda) = -(\Xi_\Omega Wh)(\mu) = -(Hh)(\mu).$$

Therefore, $(B) = -(Hh)(\mu)$, which proves the lemma. □

Notice that the identity $H = \Psi\Phi$ in Lemma 5.5.3 implies that $\dim \mathcal{X} \geq \operatorname{rank} H$. It follows that H is a finite rank operator. Furthermore, we see that the state space dimension of an arbitrary realization of the singular part of W on Ω is at least rank H. By applying Lemma 5.5.3 to $W(\cdot)^{-1}$, we see that the analogous statements for $H^\times$ also hold. In particular, the spaces $\mathcal{X}_\pi$ and $\mathcal{X}_\zeta$ in (5.5.6) are finite dimensional.

Proof of Theorem 5.5.1 We split the proof into three parts.

Part (a). In this part we show that the pair (C_π, A_π) is a right pole pair of W on Ω. Define

$$B_\pi : \mathbb{C}^m \to \mathcal{X}_\pi, \qquad B_\pi = H\tau. \tag{5.5.15}$$

It suffices to show that $C_\pi(\lambda - A_\pi)^{-1}B_\pi$ is a minimal realization of $\Xi_\Omega W$. First, we prove that

$$C_\pi(\mu - A_\pi)^{-1}x = x(\mu), \qquad x \in \mathcal{X}_\pi. \tag{5.5.16}$$

To understand the right hand side of (5.5.16), recall that $\mathcal{X}_\pi = \operatorname{Im} H$, and hence each element of $\mathcal{X}_\pi$ is a $\mathbb{C}^m$-valued rational function which has its poles in Ω. Take $x \in \mathcal{X}_\pi$. Let μ be a complex number outside Ω; thus μ is not an eigenvalue of A_π. From the definitions of A_π and C_π in Theorem 5.5.1 we see that

$$C_\pi(\mu - A_\pi)^{-1}x = \mathrm{Res}_\Omega \left(\begin{pmatrix} 0 & I \end{pmatrix} (\mu - V)^{-1} \begin{pmatrix} 0 \\ x \end{pmatrix} \right) = \mathrm{Res}_\Omega \frac{1}{\mu - \lambda} x(\lambda).$$

Since $x \in \mathcal{X}_\pi \subset \mathcal{P}(\Omega)$, the function x has all its poles in Ω and vanishes at infinity. Also, recall that μ is outside Ω. Hence the usual residue calculus yields

$$\mathrm{Res}_\Omega \frac{1}{\mu - \lambda} x(\lambda) = x(\mu),$$

which proves (5.5.16).

From (5.5.16) and the definition of B_π in (5.5.15) it is clear that

$$C_\pi(\lambda - A_\pi)^{-1}B_\pi u = (H\tau u)(\lambda) = \Xi_\Omega W(\lambda)u.$$

Thus $C_\pi(\lambda - A_\pi)^{-1}B_\pi$ is a realization of the singular part of W on Ω. Since $\dim \mathcal{X}_\pi = \operatorname{rank} H$, we know from the remarks made after the proof of Lemma 5.5.3 that $C_\pi(\lambda - A_\pi)^{-1}B_\pi$ is a minimal realization.

Part (b). In this part we show that (A_ζ, B_ζ) is a left null pair of W on Ω. Define

$$C_\zeta : \mathcal{X}_\zeta \to \mathbb{C}^m, \qquad C_\zeta = \operatorname{Res}_\Omega |\mathcal{X}_\zeta. \tag{5.5.17}$$

By applying the result of Part (a) to $W(\cdot)^{-1}$ in place of W, we see that $C_\zeta(\lambda - A_\zeta)^{-1}B_\zeta$ is a minimal realization of the singular part of $W(\cdot)^{-1}$ on Ω. Thus (A_ζ, B_ζ) has the desired property.

Part (c). To complete the proof we have to compute the coupling transformation associated with the pole pair (C_π, A_π), the null pair (A_ζ, B_ζ) and W. Define

$$\begin{array}{ll} \Phi_\zeta : \mathcal{H}(\Omega) \to \mathcal{X}_\zeta, & \Phi_\zeta h = \operatorname{Res}_\Omega(\lambda - A_\zeta)^{-1}B_\zeta h(\lambda), \\ \Psi_\zeta : \mathcal{X}_\zeta \to \mathcal{P}(\Omega), & (\Psi_\zeta x)(\lambda) = C_\zeta(\lambda - A_\zeta)^{-1}x, \end{array}$$

where C_ζ is defined by (5.5.17). Since $C_\zeta(\lambda - A_\zeta)^{-1}B_\zeta$ is a realization of the singular part of $W(\cdot)^{-1}$ on Ω, we can apply Lemma 5.5.3 to show that $\Psi_\zeta\Phi_\zeta = H^\times$. On the other hand, as in Part (a), one shows that Ψ_ζ is the identity map on $\mathcal{X}_\zeta$. We conclude that $\Phi_\zeta = H^\times$.

Now, take $x \in \mathcal{X}_\pi$. According to Lemma 5.5.2 there exists $f \in \mathcal{H}(\Omega)$ such that $x + f$ belongs to the null-pole subspace $\mathcal{M}(W; \Omega)$. Furthermore, x and f satisfy the identity (5.5.11). Using (5.5.16) we see that

$$C_\pi(\lambda - A_\pi)^{-1}x + f(\lambda) \in \mathcal{M}(W; \Omega),$$

and thus, by definition, the action of the coupling transformation Γ on x is given by

$$\Gamma x = \operatorname{Res}_\Omega(\lambda - A_\zeta)^{-1}B_\zeta f(\lambda) = \Phi_\zeta f = H^\times f.$$

But then, since x and f satisfy (5.5.11), we see that $\Gamma x = -U^\times x$, as desired. □

In conclusion, let us show that the Sylvester equation (5.3.6) appears in a natural way for the null-pole triple defined in Theorem 5.5.1. Indeed, since M in (5.5.2) commutes with the operator V in (5.5.4), we have

$$S_* U^\times - U^\times S_* = H^\times S_0.$$

For $x \in \mathcal{X}_\pi$, we have $S_0 x = \tau C_\pi x$. Indeed, by (5.5.16),

$$\begin{aligned} (Vx)(\lambda) &= \lambda C_\pi(\lambda - A_\pi)^{-1}x = C_\pi(\lambda - A_\pi + A_\pi)(\lambda - A_\pi)^{-1}x \\ &= C_\pi x + C_\pi(\lambda - A_\pi)^{-1}A_\pi x, \end{aligned}$$

and therefore $(S_0 x)(\lambda) = C_\pi x$. Thus, for $x \in \mathcal{X}_\pi$, we have

$$\begin{aligned} \Gamma A_\pi x - A_\zeta \Gamma x &= -U^\times S_* x + S_* U^\times x = (S_* U^\times - U^\times S_*)x \\ &= H^\times S_0 x = H^\times \tau C_\pi x = B_\zeta C_\pi x, \end{aligned}$$

which yields the identity (5.3.6).

5.6 Null-pole triples and the Smith-McMillan form

In this section Ω consists of one point μ, say. In this case we speak about null-pole triples at μ (rather than on $\{\mu\}$). Our aim is to derive such a triple from a Smith-McMillan form of W at μ. We begin with a general result which yields another construction of the coupling transformation.

Theorem 5.6.1 *Let (C_π, A_π) be a right pole pair of W at μ, and let (A_ζ, B_ζ) be a left null pair of W at μ. Write $\mathcal{X}_\pi$ and $\mathcal{X}_\zeta$ for the corresponding main spaces. The associate coupling transformation Γ may be constructed as follows. Choose a positive integer r such that $\widetilde{W}(\lambda) = (\lambda-\mu)^r W(\lambda)$ is analytic at μ, and let $(\widetilde{A}_\zeta, \widetilde{B}_\zeta)$ be a left null pair of $\widetilde{W}$ at μ. Put $\mathcal{L} = \operatorname{Im}(\widetilde{A}_\zeta - \mu)^r$, and set*

$$A_0 := \widetilde{A}_\zeta|\mathcal{L} : \mathcal{L} \to \mathcal{L}, \qquad B_0 := (\widetilde{A}_\zeta - \mu)^r \widetilde{B}_\zeta : \mathbb{C}^m \to \mathcal{L}.$$

Then there exists a unique invertible linear operator $F : \mathcal{X}_\zeta \to \mathcal{L}$ such that

$$A_\zeta = F^{-1}A_0F, \qquad B_\zeta = F^{-1}B_0, \tag{5.6.1}$$

and the coupling transformation Γ is given by $\Gamma = F^{-1}\Gamma_0$, where

$$\Gamma_0 x = -\sum_{\substack{i\geq 0, j\geq 0\\ i+j=r-1}} (\widetilde{A}_\zeta - \mu)^i \widetilde{B}_\zeta C_\pi (A_\pi - \mu)^j x \in \mathcal{L} \tag{5.6.2}$$

for each $x \in \mathcal{X}_\pi$.

Proof We split the proof into two parts.

Part (a). In this part we show that the pair (A_0, B_0) is a left null pair of W at μ. Choose a linear operator $\widetilde{C}_\zeta$ such that $\widetilde{C}_\zeta(\lambda-\widetilde{A}_\zeta)^{-1}\widetilde{B}_\zeta$ is a minimal realization of the Laurent principal part of $\widetilde{W}(\lambda)^{-1}$ at μ, and let $C_0 : \mathcal{L} \to \mathbb{C}^m$ be the restriction of $\widetilde{C}_\zeta$ to $\mathcal{L}$. It suffices to show that $C_0(\lambda - A_0)^{-1}B_0$ is a minimal realization of the Laurent principal part of $W(\lambda)^{-1}$ at μ.

Notice that

$$(\lambda-\mu)^r\widetilde{C}_\zeta(\lambda-\widetilde{A}_\zeta)^{-1}\widetilde{B}_\zeta = \widetilde{C}_\zeta(\lambda-\widetilde{A}_\zeta)^{-1}(\widetilde{A}_\zeta-\mu)^r\widetilde{B}_\zeta + H(\lambda),$$

where $H(\lambda)$ is analytic at μ. It follows that

$$\begin{aligned} W(\lambda)^{-1} - C_0(\lambda-A_0)^{-1}B_0 &= (\lambda-\mu)^r\widetilde{W}(\lambda)^{-1} - \widetilde{C}_\zeta(\lambda-\widetilde{A}_\zeta)^{-1}(\widetilde{A}_\zeta-\mu)^r\widetilde{B}_\zeta \\ &= (\lambda-\mu)^r\{\widetilde{W}(\lambda)^{-1} - \widetilde{C}_\zeta(\lambda-\widetilde{A}_\zeta)^{-1}\widetilde{B}_\zeta\} + H(\lambda), \end{aligned}$$

and hence $W(\lambda)^{-1} - C_0(\lambda-A_0)^{-1}B_0$ is analytic at μ. Since $\widetilde{A}_\zeta$ has μ as its only eigenvalue, the same holds for A_0, and we may conclude that $C_0(\lambda-A_0)^{-1}B_0$ is a realization of the Laurent principal part of $W(\lambda)^{-1}$ at μ.

Next, we show that $C_0(\lambda-A_0)^{-1}B_0$ is a minimal realization. Notice that

$$C_0A_0^j = \widetilde{C}_\zeta\widetilde{A}_\zeta^j|\mathcal{L}, \qquad j \geq 0. \tag{5.6.3}$$

The pair $(\widetilde{C}_\zeta, \widetilde{A}_\zeta)$ is a null-kernel pair. Thus (5.6.3) implies that the same is true for (C_0, A_0). Let $\widetilde{\mathcal{X}}_\zeta$ be the main space of the pair $(\widetilde{A}_\zeta, \widetilde{B}_\zeta)$, and put $n = \dim \widetilde{\mathcal{X}}_\zeta$.

Since $(\widetilde{A}_\varsigma, \widetilde{B}_\varsigma)$ is a full-range pair, we have

$$\begin{aligned}\operatorname{Im}\left(B_0 \quad A_0B_0 \quad \dots \quad A_0^{n-1}B_0\right) &= (\widetilde{A}-\mu)^r\{\operatorname{Im}\left(\widetilde{B}_\varsigma \quad \widetilde{A}_\varsigma\widetilde{B}_\varsigma \quad \dots \quad \widetilde{A}_\varsigma^{n-1}\widetilde{B}_\varsigma\right)\}\\ &= (\widetilde{A}-\mu)^r\widetilde{\mathcal{X}}_\varsigma = \mathcal{L},\end{aligned}$$

and hence the pair (A_0, B_0) is a full-range pair too. Thus $C_0(\lambda - A_0)^{-1}B_0$ is a minimal realization.

We know now that (A_0, B_0) is a left null pair of $W(\lambda)^{-1}$ at μ. But then we can apply the analogue of Corollary 3.3.2 for left null pairs to show that there exists a unique invertible linear operator F such that (5.6.1) holds.

Part (b). In this part we prove the formula for the coupling transformation Γ. Take $x \in \mathcal{X}_\pi$, and choose a $\mathbb{C}^m$-valued function f, analytic at μ, such that

$$g(\lambda) := C_\pi(\lambda - A_\pi)^{-1}x + f(\lambda) \in \mathcal{M}(W;\{\mu\}).$$

In other words, $g(\lambda) = W(\lambda)h(\lambda)$ for some $\mathbb{C}^m$-valued function h which is analytic at μ. It follows that $(\lambda-\mu)^r g(\lambda) = \widetilde{W}(\lambda)h(\lambda)$, and hence we can apply Lemma 3.4.2 to show that $(\lambda-\mu)^r(\lambda-\widetilde{A}_\varsigma)^{-1}\widetilde{B}_\varsigma g(\lambda)$ is analytic at μ. Thus

$$\operatorname{Res}_{\{\mu\}}(\lambda-\mu)^r(\lambda-\widetilde{A}_\varsigma)^{-1}\widetilde{B}_\varsigma\{C_\pi(\lambda-A_\pi)^{-1}x + f(\lambda)\} = 0.$$

Since f is analytic at μ, we have

$$\begin{aligned}\operatorname{Res}_{\{\mu\}}(\lambda-\mu)^r(\lambda-\widetilde{A}_\varsigma)^{-1}\widetilde{B}_\varsigma f(\lambda) &= \operatorname{Res}_{\{\mu\}}(\lambda-\widetilde{A}_\varsigma)^{-1}(\widetilde{A}_\varsigma-\mu)^r\widetilde{B}_\varsigma f(\lambda)\\ &= \operatorname{Res}_{\{\mu\}}(\lambda-A_0)^{-1}B_0 f(\lambda)\\ &= F\{\operatorname{Res}_{\{\mu\}}(\lambda-A_\varsigma)^{-1}B_\varsigma f(\lambda)\}\\ &= F\Gamma x.\end{aligned}$$

Here we used (5.6.1), and the definition of the coupling transformation.

Next, notice that

$$\operatorname{Res}_{\{\mu\}}(\lambda-\mu)^r(\lambda-\widetilde{A}_\varsigma)^{-1}\widetilde{B}_\varsigma C_\pi(\lambda-A_\pi)^{-1}x = \sum_{\substack{i\geq 0, j\geq 0\\ i+j=r-1}} (\widetilde{A}_\varsigma-\mu)^i\widetilde{B}_\varsigma C_\pi(A_\pi-\mu)^j x.$$

We conclude that

$$F\Gamma x = -\sum_{\substack{i\geq 0, j\geq 0\\ i+j=r-1}} (\widetilde{A}_\varsigma-\mu)^i\widetilde{B}_\varsigma C_\pi(A_\pi-\mu)^j x. \tag{5.6.4}$$

In particular, the right hand side of (5.6.4) belongs to the space $\mathcal{L}$. □

Now, consider a local Smith-McMillan form of W at μ:

$$W(\lambda) = E(\lambda)\begin{pmatrix}(\lambda-\mu)^{\kappa_1} & & \\ & \ddots & \\ & & (\lambda-\mu)^{\kappa_m}\end{pmatrix}F(\lambda). \tag{5.6.5}$$

Here the matrix functions $E(\lambda)$ and $F(\lambda)$ are analytic at μ, and the matrices $E(\mu)$ and $F(\mu)$ are non-singular. Assume that the indices in (5.6.5) satisfy the following inequalities:

$$\kappa_1 \geq \cdots \geq \kappa_s > 0 = \kappa_{s+1} = \cdots = \kappa_{m-t} = 0 > \kappa_{m-t+1} \geq \cdots \geq \kappa_m = -r. \tag{5.6.6}$$

In what follows $\psi_1(\lambda), \ldots, \psi_m(\lambda)$ will be the rows of $E(\lambda)^{-1}$ and $\varphi_m(\lambda), \ldots, \varphi_1(\lambda)$ its columns (ordered in the reverse direction), i.e.,

$$E(\lambda)^{-1} = \begin{pmatrix} \psi_1(\lambda) \\ \vdots \\ \psi_m(\lambda) \end{pmatrix}, \qquad E(\lambda)^{-1} = \begin{pmatrix} \varphi_m(\lambda) & \ldots & \varphi_1(\lambda) \end{pmatrix}. \tag{5.6.7}$$

From Theorem 3.2.1 we know that $\varphi_1, \ldots, \varphi_s$ is a canonical system of left null functions of W at μ, with orders $\kappa_1 \geq \cdots \geq \kappa_s$. Let (J_ζ, Y_ζ) be the corresponding canonical left null pair of W at μ. Since

$$W(\lambda)^{-1} = F(\lambda)^{-1} \begin{pmatrix} (\lambda-\mu)^{-\kappa_1} & & \\ & \ddots & \\ & & (\lambda-\mu)^{-\kappa_m} \end{pmatrix} E(\lambda)^{-1}$$

is a Smith-McMillan form of $W(\lambda)^{-1}$ at μ, we can also apply Theorem 3.2.1 to show that $\varphi_1, \ldots, \varphi_t$ is a canonical system of right null functions of $W(\cdot)^{-1}$ at μ, with orders $-\kappa_m \geq \cdots \geq -\kappa_{m-t+1}$. Let (X_π, J_π) be the corresponding canonical right null pair of $W(\cdot)^{-1}$ at μ. Then (X_π, J_π) is a right pole pair of $W(\cdot)$ at μ.

Next, we construct the coupling transformation Γ corresponding to the pole pair (X_π, J_π), the null pair (J_ζ, Y_ζ), and W at μ. For this purpose, put $\widetilde{W}(\lambda) = (\lambda-\mu)^r W(\lambda)$, where $r = -\kappa_m$, and set

$$\widetilde{s} = \max\ \{j \mid \kappa_j + r > 0\}.$$

Notice that

$$\widetilde{W}(\lambda) = E(\lambda) \begin{pmatrix} (\lambda-\mu)^{\kappa_1+r} & & \\ & \ddots & \\ & & (\lambda-\mu)^{\kappa_m+r} \end{pmatrix} F(\lambda)$$

is a Smith-McMillan form of $\widetilde{W}$ at μ. Hence, by Theorem 3.2.1, the functions $\psi_1, \ldots, \psi_{\widetilde{s}}$ form a canonical system of left null functions of $\widetilde{W}$ at μ, with orders $\kappa_1 + r \geq \cdots \geq \kappa_{\widetilde{s}} + r$. Let $(\widetilde{J}_\zeta, \widetilde{Y}_\zeta)$ be the corresponding canonical left null pair of $\widetilde{W}$ at μ.

We are now in the position to apply Theorem 5.6.1. Notice that

$$\mathbb{C}^{\kappa_1+r} \oplus \cdots \oplus \mathbb{C}^{\kappa_{\widetilde{s}}+r}$$

is the main space of the pair $(\widetilde{J}_\zeta, \widetilde{Y}_\zeta)$, and $\mathbb{C}^{\kappa_1} \oplus \cdots \oplus \mathbb{C}^{\kappa_s}$ is the corresponding space for the pair (J_ζ, Y_ζ). Put $\mathcal{L} = \operatorname{Im}(\widetilde{J}_\zeta - \mu)^r$, and set

$$A_0 = \widetilde{J}_\zeta | \mathcal{L} : \mathcal{L} \to \mathcal{L}, \qquad B_0 = (\widetilde{J}_\zeta - \mu)^r \widetilde{Y}_\zeta : \mathbb{C}^m \to \mathcal{L}.$$

By Theorem 5.6.1 there exists a unique invertible linear operator T such that

$$J_\zeta = T^{-1} A_0 T, \qquad Y_\zeta = T^{-1} B_0. \tag{5.6.8}$$

To find this invertible map T, define

$$P : \mathbb{C}^{\kappa_1+r} \oplus \cdots \oplus \mathbb{C}^{\kappa_{\widetilde{s}}+r} \to \mathbb{C}^{\kappa_1} \oplus \cdots \oplus \mathbb{C}^{\kappa_s}$$

to be the operator which assigns to $\widetilde{x}_1 + \cdots + \widetilde{x}_{\widetilde{s}}$ the vector $x_1 + \cdots + x_s$, where for $1 \le j \le s$ the vector x_j is the vector in $\mathbb{C}^{\kappa_j}$ consisting of the first κ_j entries of the vector $\widetilde{x}_j \in \mathbb{C}^{\kappa_j+r}$. From Section 3.2 we know that

$$\widetilde{J}_\zeta - \mu = \begin{pmatrix} J_{\kappa_1+r,0} & & \\ & \ddots & \\ & & J_{\kappa_{\widetilde{s}}+r,0} \end{pmatrix},$$

where for each $1 \le j \le \widetilde{s}$ the matrix $J_{\kappa_j+r,0}$ is a single Jordan block of order $\kappa_j + r$ with zeros on the main diagonal. Hence P maps $\mathcal{L} = \operatorname{Im}(\widetilde{J}_\zeta - \mu)^r$ in a one-one way onto $\mathbb{C}^{\kappa_1} \oplus \cdots \oplus \mathbb{C}^{\kappa_s}$, and it follows that the operator T in (5.6.8) is precisely the map $(P|\mathcal{L})^{-1}$. But then we can use the second part of Theorem 5.6.1 to show that the coupling transformation Γ corresponding to (X_π, J_π), (J_ζ, Y_ζ) and W at μ is given by

$$\Gamma = -P\Big(\sum_{\substack{i\ge 0, j\ge 0\\ i+j=r-1}} (\widetilde{J}_\zeta - \mu)^i \widetilde{Y}_\zeta X_\pi (J_\pi - \mu)^j \Big). \tag{5.6.9}$$

Formula (5.6.9) can be made more explicit by using the Taylor expansions at μ of the functions ψ_i and φ_j for $i = 1, \ldots, s$ and $j = 1, \ldots, t$, respectively. Put

$$\begin{aligned} \psi_i(\lambda) &= \sum_{\nu=0}^{\kappa_i+r-1} (\lambda - \mu)^\nu y_{i\nu} + \text{terms of higher order},\\ \varphi_j(\lambda) &= \sum_{\nu=0}^{\rho_j-1} (\lambda - \mu)^\nu x_{j\nu} + \text{terms of higher order}. \end{aligned}$$

Here $\rho_j = -\kappa_{m-j+1} \le r$ for $j = 1, \ldots, t$. With these Taylor coefficients we build the following matrices

$$R_i = \begin{pmatrix} y_{i,\kappa_i+r-1} & y_{i,\kappa_i+r-2} & \cdots & y_{i,\kappa_i} \\ \vdots & \vdots & & \vdots \\ y_{i,r} & y_{i,r-1} & \cdots & y_{i,1} \end{pmatrix} : \mathbb{C}^{mr} \to \mathbb{C}^{\kappa_i}, \quad i = 1, \ldots, s, \tag{5.6.10}$$

$$Q_j = \begin{pmatrix} 0 & \cdots & \cdots & 0 \\ \vdots & & & \vdots \\ 0 & & & 0 \\ 0 & & & x_{j,0} \\ \vdots & & & \vdots \\ 0 & x_{j0} & \cdots & x_{j,\rho_j-2} \\ x_{j0} & x_{j1} & \cdots & x_{j,\rho_j-1} \end{pmatrix} : \mathbb{C}^{\rho_j} \to \mathbb{C}^{mr}, \quad j = 1, \ldots, t, \tag{5.6.11}$$

The final result is the following theorem.

Theorem 5.6.2 *Consider the Smith-McMillan form of W at μ in* (5.6.5), *and assume that the inequalities* (5.6.6) *are satisfied. Partition $E(\lambda)^{-1}$ as in* (5.6.7). *Let (X_π, J_π) be the right pole pair of W at μ determined by the canonical system $\varphi_1, \ldots, \varphi_t$, and let (J_ζ, Y_ζ) be the left null pair determined by the canonical system $\psi_1, \ldots, \psi_s$. Then the coupling transformation Γ corresponding to the pole pair (X_π, J_π), the null pair (J_ζ, X_ζ), and W at μ is given by*

$$\Gamma = -\begin{pmatrix} R_1 \\ \vdots \\ R_s \end{pmatrix} \begin{pmatrix} Q_1 & \cdots & Q_t \end{pmatrix}, \tag{5.6.12}$$

where R_i and Q_j are given by (5.6.10) *and* (5.6.11), *respectively.*

Proof Rewrite (5.6.9) in the form $\Gamma = -PRQ$, where

$$\begin{aligned} R &= \begin{pmatrix} \widetilde{Y}_\zeta & (\widetilde{J}_\zeta - \mu)\widetilde{Y}_\zeta & \cdots & (\widetilde{J}_\zeta - \mu)^{r-1}\widetilde{Y}_\zeta \end{pmatrix}, \\ Q &= \begin{pmatrix} X_\pi (J_\pi - \mu)^{r-1} \\ \vdots \\ X_\pi (J_\pi - \mu) \\ X_\pi \end{pmatrix}. \end{aligned}$$

Next, use the special form of the pairs $(\widetilde{J}_\zeta, \widetilde{Y}_\zeta)$ and (X_π, J_π) to derive (5.6.12) by a direct computation. □

5.7 Notes

The coupling transformation, which intertwines the null and pole data, appears for the first time explicitly in the work of Ball-Ran [1987a,b]; see also Ball-Cohen-Ran [1988]. In Ball-Ran [1987a] the theory is developed in terms of Fourier coefficients with all null-pole triples over the disc or the exterior of the disc, the paper Ball-Ran [1987b] works in terms of local Laurent coefficients and chains, and in Ball-Cohen-Ran [1988] null and pole pairs of general sets are considered and the analysis is in terms of general contour integrals. Theorem 5.2.2 is taken from Ball-Gohberg-Rodman [1987] (see also Theorem 5.1 in Gohberg-Kaashoek [1987]). Theorem 5.3.1 seems to be new. Theorem 5.4.1 is Lemma 5.2 in Gohberg-Kaashoek [1987]. The construction of the null-pole triple in Section 5.5 is inspired by the restricted shift realization in system theory; the null-pole triple constructed in Theorem 5.5.1 is up to a similarity equal to the null-pole triple appearing in Theorem 14.1.2 of Ball-Gohberg-Rodman [1990]. Formulas (5.6.10), (5.6.11), and Theorem 5.6.2 are taken from Section 12.5 in Ball-Gohberg-Rodman [1990]; the proofs, which are based here on Theorem 5.6.1, seem to be new. Following Ball-Ran [1987b], where augmented chains are introduced, the pair $(\widetilde{A}_\zeta, \widetilde{B}_\zeta)$ in Theorem 5.6.1 could be called an *augmented left null pair* of W at μ. Theorem 5.6.1 gives then a formula for the coupling transformation in terms of a pole pair and an augmented null pair, just as Theorem 12.5.2 in Ball-Gohberg-Rodman [1990] gives the coupling transformation in terms of pole chains and augmented null chains. An extension to non-regular (possibly non-square) rational matrix functions of the theory developed in the present section may be found in Ball-Gohberg-Rakowski [1994] and in the earlier papers Ball-Rakowski [1990], [1991].

CHAPTER 6

Inverse problems involving null-pole triples

In this chapter we treat three inverse problems involving null-pole triples. The problems are of the same type as those considered in Chapter 4, but here we do not restrict the attention to matrix functions in $\mathcal{R}at(H_\infty)$; the interpolants may have zeros as well as poles in Π^+. In the first section we seek regular rational matrix functions that have a prescribed null-pole triple on a given set. The second section deals with existence and construction of a J-inner rational matrix function with a prescribed null-pole triple on Π^+. The last section concerns the bitangential Nudelman interpolation problem for rational matrix functions.

6.1 Rational matrix functions with prescribed null-pole triples

Let W be a regular rational $m \times m$ matrix function, and let

$$\tau = \{C_\pi, A_\pi; A_\zeta, B_\zeta; \Gamma\} \tag{6.1.1}$$

be a null-pole triple of W on Ω. The latter implies that

$$\begin{array}{ccc} A_\pi : \mathcal{X}_\pi \to \mathcal{X}_\pi, & C_\pi : \mathcal{X}_\pi \to \mathbb{C}^m, & A_\zeta : \mathcal{X}_\zeta \to \mathcal{X}_\zeta, \\ B_\zeta : \mathbb{C}^m \to \mathcal{X}_\zeta, & \Gamma : \mathcal{X}_\pi \to \mathcal{X}_\zeta & \end{array} \tag{6.1.2}$$

are linear operators acting between finite dimensional inner product spaces, satisfying the following properties:

1. (C_π, A_π) is a null-kernel pair and the eigenvalues of A_π are in Ω;
2. (A_ζ, B_ζ) is a full-range pair and the eigenvalues of A_ζ are in Ω;
3. $\Gamma A_\pi - A_\zeta \Gamma = B_\zeta C_\pi$.

In the sequel we call τ in (6.1.1) an *Ω-admissible Sylvester data set* if conditions (1), (2), (3) above are fulfilled. The prefix Ω will be omitted whenever it is not directly relevant. As before we refer to Γ as the *coupling transformation* of the data set τ. The word "Sylvester" refers to the fact that Γ satisfies the Sylvester equation in (3). In what follows we keep the integer m in (6.1.2) fixed.

In this section we consider the following inverse problem. Given an Ω-admissible Sylvester data set τ, find a regular rational matrix function W such that τ is a null-pole triple of W on Ω. We shall see that this problem is always solvable, and we shall construct a solution. The case when the coupling transformation is invertible is of special interest.

Recall that a rational matrix function W is called bicausal if W is proper and $W(\infty)$ is invertible.

Theorem 6.1.1 *Let $\tau = \{C_\pi, A_\pi; A_\zeta, B_\zeta; \Gamma\}$ be an admissible Sylvester data set. Then there exists a bicausal rational matrix function W such that τ is a null-pole triple of W on $\mathbb{C}$ if and only if Γ is invertible. In this case all such W are* *given by*

$$W(\lambda) = D + C_\pi(\lambda - A_\pi)^{-1}\Gamma^{-1}B_\zeta D, \tag{6.1.3}$$

where D is an arbitrary invertible matrix of order m. Moreover the inverse of W in (6.1.3) *is given by*

$$W(\lambda)^{-1} = D^{-1} - D^{-1}C_\pi\Gamma^{-1}(\lambda - A_\zeta)^{-1}B_\zeta. \tag{6.1.4}$$

Proof Assume τ is a null-pole triple of a bicausal rational matrix function W on $\mathbb{C}$. We have to prove that Γ is invertible. To do this we construct another null-pole triple of W on $\mathbb{C}$ by applying the construction of Theorem 5.4.1. Choose a minimal realization $D + C(\lambda - A)^{-1}B$ of W. Since $\Omega = \mathbb{C}$, in this case the Riesz projections Q and $Q^\times$ appearing in Theorem 5.4.1 are equal to the identity operator on the state space $\mathcal{X}$ of the realization. But then we can apply Proposition 5.2.4 to show that $\Gamma = F^{-1}I_\mathcal{X}E$ for some invertible linear operators E and F. In particular, Γ is invertible.

To prove the converse, assume Γ is invertible. Define W by (6.1.3), where D is some invertible $m \times m$ matrix. We claim that the realization $[A_\pi, \Gamma^{-1}B_\zeta D, C_\pi, D]$ is minimal. Indeed, by definition the pair (C_π, A_π) is a null-kernel pair. Furthermore, the Sylvester equation for Γ yields

$$A_\pi = \Gamma^{-1}A_\zeta\Gamma + \Gamma^{-1}B_\zeta C_\pi. \tag{6.1.5}$$

It follows that $(A_\pi, \Gamma^{-1}B_\zeta D) = (\Gamma^{-1}A_\zeta\Gamma + \Gamma^{-1}B_\zeta C_\pi, \Gamma^{-1}B_\zeta D)$, and hence the pair $(A_\pi, \Gamma^{-1}B_\zeta D)$ is similar to the pair $(A_\zeta + B_\zeta C_\pi\Gamma^{-1}, B_\zeta D)$. Since (A_ζ, B_ζ) is a full-range pair, the same holds true for the pair $(A_\zeta + B_\zeta C_\pi\Gamma^{-1}, B_\zeta D)$, and hence by the above similarity the pair $(A_\pi, \Gamma^{-1}B_\zeta D)$ is also full-range. Thus the realization (6.1.3) is minimal.

From (6.1.5) we conclude that

$$\begin{aligned} W(\lambda)^{-1} &= D^{-1} - D^{-1}C_\pi(\lambda - A_\pi + \Gamma^{-1}B_\zeta C_\pi)^{-1}\Gamma^{-1}B_\zeta \\ &= D^{-1} - D^{-1}C_\pi(\lambda - \Gamma^{-1}A_\zeta\Gamma)^{-1}\Gamma^{-1}B_\zeta \\ &= D^{-1} - D^{-1}C_\pi\Gamma^{-1}(\lambda - A_\zeta)^{-1}B_\zeta, \end{aligned}$$

which proves (6.1.4). The result of the previous paragraph and the preceding calculation show that (6.1.4) is a minimal realization too.

Since (6.1.3) and (6.1.4) are both minimal realizations, we can apply Theorem 3.3.1 to show that (A_ζ, B_ζ) is a left null pair of W on $\mathbb{C}$ and (C_π, A_π) is a right null pair of $W(\cdot)^{-1}$ on $\mathbb{C}$. By definition the latter implies that (C_π, A_π) is a right pole pair of W on $\mathbb{C}$. A straightforward application of Corollary 5.4.2, using (6.1.5), yields that Γ is the coupling transformation corresponding to the pairs (C_π, A_π), (A_ζ, B_ζ), and W on $\mathbb{C}$.

Finally, let $\widetilde{W}$ be another bicausal rational matrix function which has τ as its null pole triple on $\mathbb{C}$. Then, by Theorem 5.2.2, we have $\widetilde{W} = WR$, where R is biregular on $\mathbb{C}$. Both W and $\widetilde{W}$ are biregular at infinity. Hence R is also biregular at infinity. But then, by Liouville's theorem, R must be an invertible constant matrix. Thus $\widetilde{W}$ is of the form (6.1.3) with D replaced by some invertible constant matrix. $\square$

Corollary 6.1.2 *Assume that W is biregular on the imaginary axis and at infinity, and let $\tau = \{C_\pi, A_\pi; A_\zeta, B_\zeta; \Gamma\}$ be a null-pole triple of W on Π^+. Then W admits a right canonical factorization with respect to the imaginary axis if and only if the coupling transformation is invertible. In this case, such a canonical factorization $W = W_- W_+$ may be obtained by taking*

$$W_-(\lambda) = I + C_\pi(\lambda - A_\pi)^{-1}\Gamma^{-1}B_\zeta, \tag{6.1.6}$$

$$W_+(\lambda) = \{I - C_\pi\Gamma^{-1}(\lambda - A_\zeta)^{-1}B_\zeta\}W(\lambda). \tag{6.1.7}$$

Proof Assume Γ is invertible. Define W_- by (6.1.6). According to the last part of Theorem 6.1.1, the inverse $W_-(\lambda)^{-1}$ is given by the first factor in the right hand side of (6.1.7). Since the eigenvalues of A_π and A_ζ are in Π^+, both $W_-(\cdot)$ and $W_-(\cdot)^{-1}$ are analytic on $\overline{\Pi^-}$ and at infinity. In other words, W_- has no poles on Re $\lambda \leq 0$ and at infinity, and $\det W_-(\lambda) \neq 0$ on Re $\lambda \leq 0$ and at infinity. Furthermore, we see from the first part of Theorem 6.1.1 that W and W_- have the same null-pole triple on Π^+. So, by Theorem 5.2.2, we have $W = W_-R$, with R biregular on Π^+. On the other hand, W and W_- are both biregular on $i\mathbb{R} \cup \{\infty\}$, and thus the same is true for R. Put $W_+ = R$. Then $W = W_-W_+$ is the desired canonical factorization.

To prove the converse, assume that $W = W_-W_+$ is a right canonical factorization of W with respect to the imaginary axis. Theorem 5.2.2 shows that τ is a null-pole triple of W_- on Π^+. Since W_- has no singularities (poles or zeros) on $\overline{\Pi^-}$ and at infinity, we conclude that actually τ is a null-pole triple of W_- on $\mathbb{C}$. Thus, by Theorem 6.1.1, its coupling transformation Γ is invertible. □

To solve the inverse problem stated in the second paragraph of this section we have to consider the case when the coupling transformation is singular. To treat this case we use the notion of a complement of a Sylvester data set.

Let $\tau = \{C_\pi, A_\pi; A_\zeta, B_\zeta; \Gamma\}$ be an Ω-admissible Sylvester data set, and let $\tau_0 = \{C_\pi^{(0)}, A_\pi^{(0)}; A_\zeta^{(0)}, B_\zeta^{(0)}; \Gamma_0\}$ be an Ω_0-admissible Sylvester data set. We say that τ_0 is a *complement* of τ if $\Omega \cap \Omega_0 = \emptyset$ and

$$\det\begin{pmatrix} \Gamma & \Gamma_{10} \\ \Gamma_{01} & \Gamma_0 \end{pmatrix} \neq 0,$$

where Γ_{10} and Γ_{01} are the unique solutions of

$$\Gamma_{10}A_\pi^{(0)} - A_\zeta\Gamma_{10} = B_\zeta C_\pi^{(0)}, \qquad \Gamma_{01}A_\pi - A_\zeta^{(0)}\Gamma_{01} = B_\zeta^{(0)}C_\pi. \tag{6.1.8}$$

Since the sets Ω and Ω_0 are required to be disjoint, $A_\pi^{(0)}$ and A_ζ have no common eigenvalues and the same is true for A_π and $A_\zeta^{(0)}$. Thus, indeed, the two equations in (6.1.8) are uniquely solvable (cf., Theorem I.4.1 in Gohberg-Goldberg-Kaashoek [1990]).

Theorem 6.1.3 *Let $\tau = \{C_\pi, A_\pi; A_\zeta, B_\zeta; \Gamma\}$ be an Ω-admissible Sylvester data set. Then there exists a bicausal matrix function W such that τ is a null-pole triple of W on Ω if and only if τ has a complement $\tau_0 = \{C_\pi^{(0)}, A_\pi^{(0)}; A_\zeta^{(0)}, B_\zeta^{(0)}; \Gamma_0\}$, say. In this case,*

$$W(\lambda) = I + \begin{pmatrix} C_\pi & C_\pi^{(0)} \end{pmatrix}\begin{pmatrix} \lambda - A_\pi & 0 \\ 0 & \lambda - A_\pi^{(0)} \end{pmatrix}^{-1}\begin{pmatrix} \Gamma & \Gamma_{10} \\ \Gamma_{01} & \Gamma_0 \end{pmatrix}^{-1}\begin{pmatrix} B_\zeta \\ B_\zeta^{(0)} \end{pmatrix} \tag{6.1.9}$$

is a bicausal rational matrix function which has τ as its null-pole triple on Ω. Moreover,

$$W(\lambda)^{-1} = I - \begin{pmatrix} C_\pi & C_\pi^{(0)} \end{pmatrix} \begin{pmatrix} \Gamma & \Gamma_{10} \\ \Gamma_{01} & \Gamma_0 \end{pmatrix}^{-1} \cdot \cdot \begin{pmatrix} \lambda - A_\zeta & 0 \\ 0 & \lambda - A_\zeta^{(0)} \end{pmatrix}^{-1} \begin{pmatrix} B_\zeta \\ B_\zeta^{(0)} \end{pmatrix}. \tag{6.1.10}$$

Here Γ_{10} and Γ_{01} are the unique solutions of the two equations in (6.1.8).

Proof Assume τ is a null-pole triple of W on Ω, where W is a bicausal rational matrix function. We use the constructions made in the paragraph after the proof of Theorem 5.4.1. Thus we may assume that $W(\lambda)$ and $W(\lambda)^{-1}$ are given by (5.4.4) and (5.4.5). Recall that the realizations in (5.4.4) and (5.4.5) are minimal. Let T be the unique invertible linear operator determined by (5.4.7), and partition T as in (5.4.8). From Corollary 5.4.2 we know that $\Gamma = T_{11}$. By the minimality of the realizations in (5.4.4) and (5.4.5), the pair $(\widehat{C}_\pi, \widehat{A}_\pi)$ appearing in (5.4.4) is a null-kernel pair and the pair $(\widehat{A}_\zeta, \widehat{B}_\zeta)$ in (5.4.5) is a full-range pair. The identities in (5.4.7) show that $T_{22}\widehat{A}_\pi - \widehat{A}_\zeta T_{22} = \widehat{B}_\zeta \widehat{C}_\pi$ and hence

$$\widehat{\tau} = \{\widehat{C}_\pi, \widehat{A}_\pi; \widehat{A}_\zeta, \widehat{B}_\zeta; T_{22}\} \tag{6.1.11}$$

is an admissible Sylvester data set. The eigenvalues of $\widehat{A}_\pi$ and $\widehat{A}_\zeta$ are in $\mathbb{C}\backslash\Omega$, and thus $\widehat{\tau}$ is $\widehat{\Omega}$-admissible, where $\widehat{\Omega}$ is disjoint from Ω. The identities in (5.4.7) also yield

$$T_{12}\widehat{A}_\pi - A_\zeta T_{12} = B_\zeta \widehat{C}_\pi, \qquad T_{21}A_\pi - \widehat{A}_\zeta T_{21} = \widehat{B}_\zeta C_\pi.$$

From these identities and the fact that T in (5.4.8) is invertible, we conclude that $\widehat{\tau}$ in (6.1.11) is a complement of τ.

To prove the converse, let $\tau_0 = \{C_\pi^{(0)}, A_\pi^{(0)}; A_\zeta^{(0)}, B_\zeta^{(0)}; \Gamma_0\}$ be an Ω_0-admissible Sylvester data set, and assume that τ_0 is a complement of τ. Consider the pairs

$$\left\{ \begin{pmatrix} C_\pi & C_\pi^{(0)} \end{pmatrix}, \begin{pmatrix} A_\pi & 0 \\ 0 & A_\pi^{(0)} \end{pmatrix} \right\} \left\{ \begin{pmatrix} A_\zeta & 0 \\ 0 & A_\zeta^{(0)} \end{pmatrix}^{-1}, \begin{pmatrix} B_\zeta \\ B_\zeta^{(0)} \end{pmatrix} \right\}. \tag{6.1.12}$$

Since the sets Ω and Ω_0 are disjoint it is straightforward to check that the first pair in (6.1.12) is a null-kernel pair and the second is a full-range pair. Recall that Γ and Γ_0 satisfy the Sylvester equations corresponding to τ and τ_0, respectively. These facts, together with the identities (6.1.8) imply that

$$\begin{pmatrix} \Gamma & \Gamma_{10} \\ \Gamma_{01} & \Gamma_0 \end{pmatrix} \begin{pmatrix} A_\pi & 0 \\ 0 & A_\pi^{(0)} \end{pmatrix} - \begin{pmatrix} A_\zeta & 0 \\ 0 & A_\zeta^{(0)} \end{pmatrix} \begin{pmatrix} \Gamma & \Gamma_{10} \\ \Gamma_{01} & \Gamma_0 \end{pmatrix} = \begin{pmatrix} B_\zeta \\ B_\zeta^{(0)} \end{pmatrix} \begin{pmatrix} C_\pi & C_\pi^{(0)} \end{pmatrix},$$

and hence

$$\tau' = \left\{ \begin{pmatrix} C_\pi & C_\pi^{(0)} \end{pmatrix}, \begin{pmatrix} A_\pi & 0 \\ 0 & A_\pi^{(0)} \end{pmatrix}; \begin{pmatrix} A_\zeta & 0 \\ 0 & A_\zeta^{(0)} \end{pmatrix}, \begin{pmatrix} B_\zeta \\ B_\zeta^{(0)} \end{pmatrix}; \begin{pmatrix} \Gamma & \Gamma_{10} \\ \Gamma_{01} & \Gamma_0 \end{pmatrix} \right\}$$

is an admissible Sylvester data set (which is called the *direct sum* of τ and τ_0). The coupling transformation of τ' is invertible. So we can apply Theorem 6.1.1 to τ'.

Let W be defined by (6.1.9). The second part of Theorem 6.1.1 shows that $W(\cdot)^{-1}$ is given by (6.1.10). From the proof of Theorem 6.1.1 we know that

$$\left[\begin{pmatrix} A_\pi & 0 \\ 0 & A_\pi^{(0)} \end{pmatrix}, \begin{pmatrix} \Gamma & \Gamma_{10} \\ \Gamma_{01} & \Gamma_0 \end{pmatrix}^{-1} \begin{pmatrix} B_\zeta \\ B_\zeta^{(0)} \end{pmatrix}, \begin{pmatrix} C_\pi & C_\pi^{(0)} \end{pmatrix}, I\right]$$

is a minimal realization of W, and

$$\left[\begin{pmatrix} A_\zeta & 0 \\ 0 & A_\zeta^{(0)} \end{pmatrix}, \begin{pmatrix} B_\zeta \\ B_\zeta^{(0)} \end{pmatrix}, \begin{pmatrix} C_\pi & C_\pi^{(0)} \end{pmatrix} \begin{pmatrix} \Gamma & \Gamma_{10} \\ \Gamma_{01} & \Gamma_0 \end{pmatrix}^{-1}, I\right]$$

is a minimal realization of $W(\cdot)^{-1}$. Put

$$A^\times = \begin{pmatrix} A_\pi & 0 \\ 0 & A_\pi^{(0)} \end{pmatrix} - \begin{pmatrix} \Gamma & \Gamma_{10} \\ \Gamma_{01} & \Gamma_0 \end{pmatrix}^{-1} \begin{pmatrix} B_\zeta \\ B_\zeta^{(0)} \end{pmatrix} \begin{pmatrix} C_\pi & C_\pi^{(0)} \end{pmatrix},$$

and set

$$T = \begin{pmatrix} \Gamma & \Gamma_{10} \\ \Gamma_{01} & \Gamma_0 \end{pmatrix}.$$

From the Sylvester equation corresponding to τ' we derive that

$$TA^\times T^{-1} = \begin{pmatrix} A_\zeta & 0 \\ 0 & A_\zeta^{(0)} \end{pmatrix}.$$

Furthermore,

$$T \begin{pmatrix} \Gamma & \Gamma_{10} \\ \Gamma_{01} & \Gamma_0 \end{pmatrix}^{-1} \begin{pmatrix} B_\zeta \\ B_\zeta^{(0)} \end{pmatrix} = \begin{pmatrix} B_\zeta \\ B_\zeta^{(0)} \end{pmatrix},$$

$$\begin{pmatrix} C_\pi & C_\pi^{(0)} \end{pmatrix} T^{-1} = \begin{pmatrix} C_\pi & C_\pi^{(0)} \end{pmatrix} \begin{pmatrix} \Gamma & \Gamma_{10} \\ \Gamma_{01} & \Gamma_0 \end{pmatrix}^{-1}.$$

But then we see from Corollary 5.4.2 that τ is a null-pole triple of W on Ω. □

According to Theorem 6.1.3, to solve the inverse problem stated in the second paragraph of this section, we have to construct complements of admissible Sylvester data sets. Here below we shall show that this can always be done.

Let $\tau = \{C_\pi, A_\pi; A_\zeta, B_\zeta; \Gamma\}$ be an Ω-admissible Sylvester data set. We assume that the corresponding spaces are as in (6.1.2). To construct a complement of τ we need some additional notation. Let M be a complement of $\operatorname{Ker}\Gamma$ in $\mathcal{X}_\pi$, and let K be a complement of $\operatorname{Im}\Gamma$ in $\mathcal{X}_\zeta$. Thus

$$\mathcal{X}_\pi = \operatorname{Ker}\Gamma \dot{+} M, \qquad \mathcal{X}_\zeta = \operatorname{Im}\Gamma \dot{+} K. \tag{6.1.13}$$

We write ρ_π for the projection of $\mathcal{X}_\pi$ onto $\operatorname{Ker}\Gamma$ along M, and ρ_ζ for the projection of $\mathcal{X}_\zeta$ onto K along $\operatorname{Im}\Gamma$. The canonical embedding of $\operatorname{Ker}\Gamma$ into $\mathcal{X}_\pi$ will be denoted by η_π, and we let η_ζ be the embedding of K into $\mathcal{X}_\zeta$.

We claim that $(C_\pi\eta_\pi, \rho_\pi A_\pi \eta_\pi)$ is a null-kernel pair and the pair $(\rho_\zeta A_\zeta \eta_\zeta, \rho_\zeta B_\zeta)$ is a full-range pair. To prove this we use the Sylvester equation. Assume

$$\bigcap_{n=0}^{\infty} \operatorname{Ker} C_\pi \eta_\pi (\rho_\pi A_\pi \eta_\pi)^n \tag{6.1.14}$$

is non-trivial. Notice that the space (6.1.14) is invariant under $\rho_\pi A_\pi \eta_\pi$. Hence there exists $x \neq 0$ in $\operatorname{Ker}\Gamma$ such that $C_\pi x = 0$ and $\rho_\pi A_\pi x = \lambda x$ for some λ in $\mathbb{C}$. But then we can use the Sylvester equation to show that

$$\Gamma A_\pi x = \Gamma A_\pi x - A_\zeta \Gamma x = B_\zeta C_\pi x = 0,$$

and thus $\rho_\pi A_\pi x = A_\pi x$. It follows that $C_\pi x = 0$ and $A_\pi x = \lambda x$, and therefore $x \neq 0$ is in $\operatorname{Ker} C_\pi A_\pi^n$ for each n, which contradicts the properties of (C_π, A_π). Hence $(C_\pi \eta_\pi, \rho_\pi A_\pi \eta_\pi)$ is a null-kernel pair. In a similar way one shows that $(\rho_\zeta A_\zeta \eta_\zeta, \rho_\zeta B_\zeta)$ is a full-range pair.

Fix a set Ω_0 in $\mathbb{C}$ which is disjoint with Ω. The result of the previous paragraph allows us to choose (cf., the third paragraph of Section 4.1) linear operators

$$G : \mathbb{C}^m \to \operatorname{Ker}\Gamma, \qquad F : K \to \mathbb{C}^m \tag{6.1.15}$$

in such a way that the operators

$$S = \rho_\pi A_\pi \eta_\pi - G C_\pi \eta_\pi : \operatorname{Ker}\Gamma \to \operatorname{Ker}\Gamma, \quad T = \rho_\zeta A_\zeta \eta_\zeta - \rho_\zeta B_\zeta F : K \to K \tag{6.1.16}$$

have their eigenvalues in Ω_0. Observe that $\rho_\pi A_\pi - G C_\pi - S \rho_\pi$ is zero on $\operatorname{Ker}\Gamma$ and $A_\zeta \eta_\zeta - B_\zeta \Gamma - \eta_\zeta T$ maps K into $\operatorname{Im}\Gamma$. So we may choose linear operators

$$A_{21} : \mathcal{X}_\zeta \to \operatorname{Ker}\Gamma, \qquad A_{12} : K \to \mathcal{X}_\pi \tag{6.1.17}$$

such that the following identities hold:

$$\rho_\pi A_\pi - G C_\pi - S \rho_\pi = A_{21}\Gamma, \qquad A_\zeta \eta_\zeta - B_\zeta F - \eta_\zeta T = \Gamma A_{12}. \tag{6.1.18}$$

Notice that (6.1.18) fixes only $A_{21}(I - \rho_\zeta)$ and $(I - \rho_\pi)A_{12}$, and hence without loss of generality we may assume that

$$A_{21}\eta_\zeta - \rho_\pi A_{12} = GF. \tag{6.1.19}$$

Furthermore, A_π and T have no common eigenvalues, because the sets Ω and Ω_0 are disjoint. For the same reason A_ζ and S have no common eigenvalues. It follows that the Lyapunov equations

$$A_\pi X - XT = A_{12}, \qquad Y A_\zeta - SY = A_{21} \tag{6.1.20}$$

are uniquely solvable (cf., Theorem I.4.1 in Gohberg-Goldberg-Kaashoek [1990]. We are now ready to state the final result of this section.

Theorem 6.1.4 *Let $\tau = \{C_\pi, A_\pi; A_\zeta, B_\zeta; \Gamma\}$ be an Ω-admissible Sylvester data set, and let $\Omega_0 \subset \mathbb{C}$ be a non-empty set disjoint with Ω. Choose operators G and F as in* (6.1.15) *such that the operators S and T defined by* (6.1.16) *have their eigenvalues in Ω_0. Next, choose operators A_{21} and A_{12} as in* (6.1.17) *such that the identities* (6.1.18) *and* (6.1.19) *are satisfied, and let $X : K \to X_\pi$ and $Y : \mathcal{X}_\zeta \to \operatorname{Ker}\Gamma$ be the unique solutions of the Lyapunov equations* (6.1.20). *Then*

$$\tau_0 = \{-C_\pi X - F, T; S, -YB_\zeta + G, Y\Gamma X - Y\eta_\zeta - \rho_\pi X\} \tag{6.1.21}$$

is a complement of the Sylvester data set τ.

Proof First we show that $(-C_\pi X - F, T)$ is a null-kernel pair. Assume not. Then

$$\bigcap_{n=0}^{\infty} \operatorname{Ker}(C_\pi X + F)T^n \tag{6.1.22}$$

contains non-zero vectors. Since the space (6.1.22) is invariant under T, there exists $0 \neq x \in K$ such that $(C_\pi X + F)x = 0$ and $Tx = \lambda x$ for some $\lambda \in \mathbb{C}$. The second identity in (6.1.18) and the Sylvester equation for τ yield

$$\begin{aligned}(A_\zeta - \lambda)x &= B_\zeta F x + \Gamma A_{12} x = -B_\zeta C_\pi X x + \Gamma A_{12} x \\ &= -\Gamma A_\pi X x + A_\zeta \Gamma X x + \Gamma A_{12} x.\end{aligned}$$

From the first identity in (6.1.20) we obtain that

$$\Gamma A_{12} x = \Gamma A_\pi X x - \Gamma X T x = \Gamma A_\pi X x - \lambda \Gamma X x.$$

Thus $(A_\zeta - \lambda)(x - \Gamma X x) = 0$. Now, recall that the eigenvalues of T are in Ω_0, and those of A_ζ in Ω. But Ω and Ω_0 are disjoint. It follows that $x = \Gamma X x \in \operatorname{Im}\Gamma$. Thus, $x \in K \cap \operatorname{Im}\Gamma$, and hence $x = 0$, which is impossible. We conclude that $(-C_\pi X - F, T)$ is a null-kernel pair.

In a similar way, using duality, one shows that $(S, -YB_\zeta + G)$ is a full-range pair.

Next, we check that $\Gamma_0 = Y\Gamma X - Y\eta_\zeta - \rho_\pi X$ satisfies the Sylvester equation

$$\Gamma_0 T - S\Gamma_0 = (-YB_\zeta + G)(-C_\pi X - F). \tag{6.1.23}$$

Using the Sylvester equation for τ and the identities (6.1.18)–(6.1.20) we may rewrite the right hand side of (6.1.23) as follows:

$$\begin{aligned}(-YB_\zeta + G)(-C_\pi X - F) &= YB_\zeta C_\pi X - GC_\pi X + YB_\zeta F - GF = \\ &= Y(\Gamma A_\pi - A_\zeta \Gamma)X - (\rho_\pi A_\pi X - S\rho_\pi X - A_{21}\Gamma X) + \\ &\quad + (YA_\zeta \eta_\zeta - Y\eta_\zeta T - Y\Gamma A_{12}) - GF \\ &= Y\Gamma(A_\pi X - A_{12}) - (YA_\zeta - A_{21})\Gamma X - \\ &\quad - \rho_\pi(A_\pi X - A_{12}) + (YA_\zeta - A_{21})\eta_\zeta + S\rho_\pi X - Y\eta_\zeta T \\ &= Y\Gamma X T - SY\Gamma X - \rho_\pi X T + SY\eta_\zeta + S\rho_\pi X - Y\eta_\zeta T.\end{aligned}$$

From the definition of Γ_0 we see that the last expression in the above calculation is precisely $\Gamma_0 T - S\Gamma_0$, and hence (6.1.23) is proved.

We have now shown that τ_0 is an Ω_0-admissible Sylvester data set. It remains to prove that τ_0 is a complement of τ. We already know that $\Omega_0 \cap \Omega = \emptyset$. Let Γ_{10} and Γ_{01} be the unique solutions of the equations

$$\Gamma_{10} T - A_\zeta \Gamma_{10} = -B_\zeta(C_\pi X + F), \quad \Gamma_{01} A_\pi - S\Gamma_{01} = (-YB_\zeta + G)C_\pi. \tag{6.1.24}$$

We have to prove that the operator

$$\widetilde{\Gamma} = \begin{pmatrix} \Gamma & \Gamma_{10} \\ \Gamma_{01} & Y\Gamma X - Y\eta_\zeta - \rho_\pi X \end{pmatrix} \tag{6.1.25}$$

is invertible. To do this we first show that the unique solution of the equations (6.1.24) are given by

$$\Gamma_{10} = -\Gamma X + \eta_\zeta, \qquad \Gamma_{01} = -Y\Gamma + \rho_\pi. \tag{6.1.26}$$

This is done by direct checking. Indeed,

$$\begin{aligned}(-\Gamma X + \eta_\zeta)T \quad & -A_\zeta(-\Gamma X + \eta_\zeta) = -\Gamma XT + \eta_\zeta T + A_\zeta \Gamma X - A_\zeta \eta_\zeta = \\ & = \Gamma(A_{12} - A_\pi X) + \eta_\zeta T + (\Gamma A_\pi - B_\zeta C_\pi)X - A_\zeta \eta_\zeta \\ & = \Gamma A_{12} + \eta_\zeta T - A_\zeta \eta_\zeta - B_\zeta C_\pi X \\ & = -B_\zeta(C_\pi X + F).\end{aligned}$$

The second identity in (6.1.26) is proved in a similar way.

By using (6.1.26) in (6.1.25) we see that the operator $\widetilde{\Gamma}$ in (6.1.25) may be rewritten as

$$\widetilde{\Gamma} = \begin{pmatrix} I & 0 \\ -Y & I \end{pmatrix} \begin{pmatrix} \Gamma & \eta_\zeta \\ \rho_\pi & 0 \end{pmatrix} \begin{pmatrix} I & -X \\ 0 & I \end{pmatrix}. \tag{6.1.27}$$

The middle term in the right hand side of (6.1.27) is invertible. In fact,

$$\begin{pmatrix} \Gamma & \eta_\zeta \\ \rho_\pi & 0 \end{pmatrix}^{-1} = \begin{pmatrix} \Gamma^+ & \eta_\pi \\ \rho_\zeta & 0 \end{pmatrix},$$

where Γ^+ is the generalized inverse of Γ defined by $\Gamma^+\Gamma = I - \eta_\pi\rho_\pi$ and $\Gamma\Gamma^+ = I - \eta_\zeta\rho_\zeta$. Thus $\widetilde{\Gamma}$ is invertible, and τ_0 in (6.1.21) is a complement of τ. □

The complement τ_0 in Theorem 6.1.4 has the additional property that

$$\operatorname{rank}\begin{pmatrix} \Gamma & \Gamma_{10} \\ \Gamma_{01} & \Gamma_0 \end{pmatrix}$$

is as small as possible among all complements of τ. For this reason τ_0 is called a *minimal complement*. These minimal complements are of special interest in factorization problems (see Gohberg-Kaashoek-Ran [1988], Section 4).

By combining Theorems 6.1.3 and 6.1.4 we see that the inverse problem stated in the second paragraph of this section is always solvable, and both theorems together provide a way to construct a solution.

6.2 *J*-inner matrix functions with prescribed null-pole triples

Let $\tau = \{C_\pi, A_\pi; A_\zeta, B_\zeta; \Gamma\}$ be an Π^+-admissible Sylvester data set. In this section we consider the problem to find a J-inner rational matrix function Θ which has τ as its null-pole triple on Π^+.

Let us first consider the direct problem. Let Θ be a J-unitary rational matrix function, and assume that Θ is proper and has no pole on the imaginary axis (the latter two conditions are automatically fulfilled if Θ is J-inner). We shall compute a null-pole triple of Θ on Π^+ by using the method of Section 5.4 (in combination with Theorem 2.2.1).

Let $\Theta(\lambda) = D + C(\lambda - A)^{-1}B$ be a minimal realization with state space $\mathcal{X}$. Since Θ has no poles on the imaginary axis, we know from Corollary 1.1.4 that the space $\mathcal{X}$ decomposes into a direct sum of two A-invariant subspaces $\mathcal{X}_1$ and $\mathcal{X}_2$, say, such that the eigenvalues of $A|\mathcal{X}_1$ are in Π^- and those of $A|\mathcal{X}_2$ in Π^+. Without loss of generality we may assume (change the inner product in $\mathcal{X}$ if necessary) that the spaces $\mathcal{X}_1$ and $\mathcal{X}_2$ are perpendicular. Thus $\mathcal{X} = \mathcal{X}_1 \oplus \mathcal{X}_2$. Put $A_1 = A|\mathcal{X}_1$ and $A_2 = A|\mathcal{X}_2$, and partition B and C according to the direct sum decomposition of $\mathcal{X}$ as follows:

$$B = \begin{pmatrix} B_1 \\ B_2 \end{pmatrix} : \mathbb{C}^m \to \mathcal{X}_1 \oplus \mathcal{X}_2, \qquad C = \begin{pmatrix} C_1 & C_2 \end{pmatrix} : \mathcal{X}_1 \oplus \mathcal{X}_2 \to \mathbb{C}^m.$$

Then

$$\Theta = D + \begin{pmatrix} C_1 & C_2 \end{pmatrix} \begin{pmatrix} \lambda - A_1 & 0 \\ 0 & \lambda - A_2 \end{pmatrix}^{-1} \begin{pmatrix} B_1 \\ B_2 \end{pmatrix}. \tag{6.2.1}$$

Since Θ is J-unitary, we have $\Theta(\lambda)^{-1} = J\Theta(-\overline{\lambda})^* J$, and hence

$$\Theta(\lambda)^{-1} = JD^*J + \begin{pmatrix} JB_1^* & JB_2^* \end{pmatrix} \begin{pmatrix} \lambda + A_1^* & 0 \\ 0 & \lambda + A_2^* \end{pmatrix}^{-1} \begin{pmatrix} -C_1^* J \\ -C_2^* J \end{pmatrix}. \tag{6.2.2}$$

Notice that the eigenvalues of $-A_1^*$ are in Π^+ and those of $-A_2^*$ are in Π^-. The realizations (6.2.1) and (6.2.2) are both minimal, and hence (C_2, A_2) is a right pole pair of Θ on Π^+ and $(-A_1^*, -C_1^* J)$ is a left null pair of Θ on Π^+. To find the corresponding coupling transformation we apply Theorem 2.1.1.

According to Theorem 2.1.1 there exists a (unique) invertible linear operator H on $\mathcal{X}_1 \oplus \mathcal{X}_2$ such that $H = H^*$ and

$$H \begin{pmatrix} A_1 & 0 \\ 0 & A_2 \end{pmatrix} + \begin{pmatrix} A_1^* & 0 \\ 0 & A_2^* \end{pmatrix} H = - \begin{pmatrix} C_1^* \\ C_2^* \end{pmatrix} J \begin{pmatrix} C_1 & C_2 \end{pmatrix}, \tag{6.2.3a}$$

$$H \begin{pmatrix} B_1 \\ B_2 \end{pmatrix} = - \begin{pmatrix} C_1^* \\ C_2^* \end{pmatrix} JD. \tag{6.2.3b}$$

Partition H according to the decomposition $\mathcal{X}_1 \oplus \mathcal{X}_2$ as follows:

$$H = \begin{pmatrix} H_{11} & H_{12} \\ H_{21} & H_{22} \end{pmatrix} : \mathcal{X}_1 \oplus \mathcal{X}_2 \to \mathcal{X}_1 \oplus \mathcal{X}_2. \tag{6.2.4}$$

Since H is selfadjoint, we have $H_{12}^* = H_{21}$. We claim that H_{12} is the desired coupling transformation. To see this, put

$$T = \begin{pmatrix} H_{21} & H_{22} \\ H_{11} & H_{12} \end{pmatrix} = \begin{pmatrix} 0 & I \\ I & 0 \end{pmatrix} H.$$

Then

$$T \left\{ \begin{pmatrix} A_1 & 0 \\ 0 & A_2 \end{pmatrix} - \begin{pmatrix} B_1 \\ B_2 \end{pmatrix} D^{-1} \begin{pmatrix} C_1 & C_2 \end{pmatrix} \right\} T^{-1} = \begin{pmatrix} -A_2^* & 0 \\ 0 & -A_1^* \end{pmatrix},$$

$$T \begin{pmatrix} B_1 \\ B_2 \end{pmatrix} D^{-1} = - \begin{pmatrix} C_2^* \\ C_1^* \end{pmatrix} J, \quad -D^{-1} \begin{pmatrix} C_1 & C_2 \end{pmatrix} T^{-1} = \begin{pmatrix} JB_2^* & JB_1^* \end{pmatrix}.$$

Hence we may apply Corollary 5.4.2 to conclude that H_{12} is the coupling transformation corresponding to the pairs (C_2, A_2), $(-A_1^*, -C_1^* J)$ and Θ on Π^+. So we have shown that

$$\{C_2, A_2; -A_1^*, -C_1^* J; H_{12}\}. \tag{6.2.5}$$

is a null-pole triple of Θ on Π^+.

One can reconstruct the selfadjoint operator H in (6.2.3) from the data appearing in the null-pole triple (6.2.5). To see this we partition H as in (6.2.4). The operator H_{12} is the coupling transformation in (6.2.5) and $H_{21} = H_{12}^*$. Furthermore, H_{11} and H_{22} satisfy

$$H_{11}A_1 + A_1^*H_{11} = -C_1^*JC_1, \qquad H_{22}A_2 + A_2^*H_{22} = -C_2^*JC_2. \tag{6.2.6}$$

The first equation in (6.2.6) follows by comparing the $(1,1)$-block entries in the first identity in (6.2.3); in a similar way the $(2,2)$-block entries in the first identity in (6.2.3) yield the second equation in (6.2.6). Recall that A_1 has the eigenvalues in Π^-, and hence A_1 and $-A_1^*$ have no common eigenvalue. It follows that the first equation in (6.2.6) is uniquely solvable. Thus H_{11} is uniquely determined by the null pair in (6.2.5). Similarly, H_{22} is uniquely determined by the pole pair in (6.2.5). In fact,

$$H_{11} = \int_0^\infty e^{tA_1^*}C_1^*JC_1e^{tA_1}\,dt, \qquad H_{22} = -\int_0^\infty e^{-tA_2^*}C_2^*JC_2e^{-tA_2}\,dt. \tag{6.2.7}$$

Recall that (by Theorem 2.2.2) the operator H in (6.2.3) is positive definite if Θ is J-inner. The above analysis leads to the following theorem.

Theorem 6.2.1 *Let $\tau = \{C_\pi, A_\pi; A_\zeta, B_\zeta; \Gamma\}$ be a Π^+-admissible Sylvester data set with corresponding main spaces $\mathcal{X}_\pi$ and $\mathcal{X}_\zeta$. Put*

$$S_1 = \int_0^\infty e^{-tA_\zeta}B_\zeta JB_\zeta^*e^{-tA_\zeta^*}\,dt, \qquad S_2 = -\int_0^\infty e^{-tA_\pi^*}C_\pi^*JC_\pi e^{-tA_\pi}\,dt, \tag{6.2.8}$$

and consider

$$\Lambda = \begin{pmatrix} S_1 & \Gamma \\ \Gamma^* & S_2 \end{pmatrix} : \mathcal{X}_\zeta \oplus \mathcal{X}_\pi \to \mathcal{X}_\zeta \oplus \mathcal{X}_\pi. \tag{6.2.9}$$

Then there exists a J-inner rational matrix function Θ which has τ as its null-pole triple on Π^+ if and only if the operator Λ is positive definite. In this case any such Θ is given

$$\Theta(\lambda) = D + \begin{pmatrix} -JB_\zeta^* & C_\pi \end{pmatrix} \begin{pmatrix} \lambda + A_\zeta^* & 0 \\ 0 & \lambda - A_\pi \end{pmatrix}^{-1} \Lambda^{-1} \begin{pmatrix} B_\zeta \\ -C_\pi^*J \end{pmatrix} D, \tag{6.2.10}$$

where D is an arbitrary J-unitary matrix.

Proof Let Θ be J-inner with respect to the imaginary axis, and assume that τ is a null-pole triple of Θ in Π^+. We may assume that Θ is given by (6.2.1). Thus, by the analysis preceding the present theorem, the triple (6.2.5) is also a null-pole triple of Θ on Π^+. But then we can apply Proposition 5.2.4 to show that there exist invertible linear operators $E : \mathcal{X}_\pi \to \mathcal{X}_2$ and $F : \mathcal{X}_\zeta \to \mathcal{X}_1$ such that

$$\begin{gathered} C_\pi E^{-1} = C_2, \qquad EA_\pi E^{-1} = A_2, \qquad FA_\zeta F^{-1} = -A_1^*, \\ FB_\zeta = -C_1^*J, \qquad F\Gamma E^{-1} = H_{12}. \end{gathered} \tag{6.2.11}$$

It follows that $FS_1F^* = H_{11}$ and $E^{-*}S_2E^{-1} = H_{22}$, where H_{11} and H_{22} are given in (6.2.7). This yields

$$\begin{pmatrix} F & 0 \\ 0 & E^{-*} \end{pmatrix} \Lambda \begin{pmatrix} F^* & 0 \\ 0 & E^{-1} \end{pmatrix} = H, \tag{6.2.12}$$

where H is the operator appearing in (6.2.3). Since Θ is J-inner we know (from Theorem 2.2.2) that H is positive definite. By (6.2.12), this implies that Λ is positive definite.

To prove the converse implication, assume that Λ is positive definite. Put

$$\begin{gathered} C_\pi^{(0)} = -JB_\zeta^*, \qquad A_\pi^{(0)} = -A_\zeta^*, \qquad A_\zeta^{(0)} = -A_\pi^*, \\ B_\zeta^{(0)} = -C_\pi^* J, \qquad \Gamma_0 = \Gamma^*. \end{gathered} \tag{6.2.13}$$

Obviously, $(C_\pi^{(0)}, A_\pi^{(0)})$ is a null-kernel pair, and $(A_\zeta^{(0)}, B_\zeta^{(0)})$ is a full-range pair. Furthermore,

$$\Gamma_0 A_\pi^{(0)} - A_\zeta^{(0)}\Gamma_0 = -\Gamma^* A_\zeta^* + A_\pi^*\Gamma^* = (\Gamma A_\pi - A_\zeta \Gamma)^* = C_\pi^* B_\zeta^* = B_\zeta^{(0)} C_\pi^{(0)}.$$

Since the eigenvalues of $A_\pi^{(0)}$ and $A_\zeta^{(0)}$ are in Π^-, we conclude that the set of operators $\tau_0 = \{C_\pi^{(0)}, A_\pi^{(0)}; A_\zeta^{(0)}, B_\zeta^{(0)}; \Gamma_0\}$ is a Π^--admissible Sylvester data set.

We claim that τ_0 is a complement of τ. To prove this we have to find the (unique) solutions of the equations

$$\Gamma_{10}A_\pi^{(0)} - A_\zeta \Gamma_{10} = B_\zeta C_\pi^{(0)}, \qquad \Gamma_{01}A_\pi - A_\zeta^{(0)}\Gamma_{01} = B_\zeta^{(0)} C_\pi,$$

and check that the operator

$$\widetilde{\Gamma} = \begin{pmatrix} \Gamma & \Gamma_{10} \\ \Gamma_{01} & \Gamma_0 \end{pmatrix} \tag{6.2.14}$$

is invertible. From the definition of S_1 and S_2 in (6.2.8) it follows that

$$A_\zeta S_1 + S_1 A_\zeta^* = B_\zeta J B_\zeta^*, \qquad A_\pi^* S_2 + S_2 A_\pi^* = -C_\pi^* J C_\pi.$$

Using (6.2.13), we see that $\Gamma_{10} = S_1$ and $\Gamma_{01} = S_2$. Thus

$$\widetilde{\Gamma} = \begin{pmatrix} \Gamma & S_1 \\ S_2 & \Gamma^* \end{pmatrix} = \Lambda \begin{pmatrix} 0 & I \\ I & 0 \end{pmatrix}, \tag{6.2.15}$$

and hence $\widetilde{\Gamma}$ is invertible.

Since τ_0 is a complement of τ we may use Theorem 6.1.3, formula (6.2.15) and the identities in (6.2.13) to show that

$$W(\lambda) = I + \begin{pmatrix} -JB_\zeta^* & C_\pi \end{pmatrix} \begin{pmatrix} \lambda + A_\zeta^* & 0 \\ 0 & \lambda - A_\pi \end{pmatrix}^{-1} \Lambda^{-1} \begin{pmatrix} B_\zeta \\ -C_\pi^* J \end{pmatrix} D \tag{6.2.16}$$

is a regular rational matrix function which has τ as its null-pole triple on Π^+. From the proof of Theorem 6.1.3 we know that the realization (6.2.16) is minimal. Thus we can apply Theorem 2.2.2 to show that W is J-inner with respect to the imaginary axis.

If D is J-unitary, then $\Theta(\lambda) = W(\lambda)D$ is J-inner and has τ as its null-pole triple on Π^+. It remains to prove that all solutions Θ are of this form. To do this, assume Θ is J-inner and has τ as its null-pole triple on Π^+. Then, by Theorem

5.2.2, we have $\Theta = WR$, where R is outer. Thus $R = W(\cdot)^{-1}\Theta$ is J-inner and has no poles nor zeros in Π^+. But then we can repeat the analysis preceding the present theorem (for R in place of Θ) to show that $R(\lambda)$ is a constant J-unitary matrix, and hence Θ is of the desired form. $\square$

6.3 Bitangential Nudelman interpolation

In this section we show how one may treat the bitangential Nudelman interpolation problem for rational matrix functions by reducing it to an inverse problem of the type considered in the previous section.

For the bitangential Nudelman interpolation problem the given data form a collection of seven linear operators

$$\omega = (\widetilde{C}, C, A; Z, B, \widetilde{B}; \Gamma), \tag{6.3.1}$$

where

$$\begin{array}{lll} A : \mathcal{X} \to \mathcal{X}, & C : \mathcal{X} \to \mathbb{C}^r, & \widetilde{C} : \mathcal{X} \to \mathbb{C}^m, \\ Z : \mathcal{Y} \to \mathcal{Y}, & B : \mathbb{C}^m \to \mathcal{Y}, & \widetilde{B} : \mathbb{C}^r \to \mathcal{Y}, \\ \Gamma : \mathcal{X} \to \mathcal{Y}. & & \end{array}$$

Here $\mathcal{X}$ and $\mathcal{Y}$ are finite dimensional inner product spaces. The operators A and Z are assumed to have their eigenvalues in the open right half plane Π^+. Furthermore, (C, A) is required to be a null-kernel pair, and (Z, B) a full-range pair.

Given these data the corresponding interpolation problem is to find a rational $m \times r$ matrix function F such that

(BTN1) $F \in \mathcal{R}at(H_\infty)$,
(BTN2) $\sup\{\|F(\lambda)\| \mid \mathrm{Re}\ \lambda \geq 0\} < 1$,
(BTN3) $\mathrm{Res}_{\Pi^+}(\lambda - Z)^{-1}BF(\lambda) = \widetilde{B}$,
(BTN4) $\mathrm{Res}_{\Pi^+}\, F(\lambda)C(\lambda - A)^{-1} = \widetilde{C}$,
(BTN5) $\mathrm{Res}_{\Pi^+}(\lambda - Z)^{-1}BF(\lambda)C(\lambda - A)^{-1} = \Gamma$.

If $F \in \mathcal{R}at(H_\infty)$ satisfies the interpolation conditions (BTN3)–(BTN5), then necessarily

$$Z\Gamma - \Gamma A = \widetilde{B}C - B\widetilde{C}. \tag{6.3.2}$$

Therefore in the sequel we shall assume that the operators in (6.3.1) satisfy the compatability condition (6.3.2), and in this case we shall refer to the septet ω as an *admissible interpolation data set.*

To illustrate the problem, let us make the following choices for the operators in (6.3.1):

$$A = \begin{pmatrix} t_1 & & & \\ & t_2 & & \\ & & \ddots & \\ & & & t_n \end{pmatrix}, \quad C = \begin{pmatrix} u_1 & u_2 & \dots & u_n \end{pmatrix},$$

$$\widetilde{C} = \begin{pmatrix} \widetilde{u}_1 & \widetilde{u}_2 & \dots & \widetilde{u}_n \end{pmatrix}, \quad (6.3.3)$$

$$Z = \begin{pmatrix} s_1 & & & \\ & s_2 & & \\ & & \ddots & \\ & & & s_k \end{pmatrix}, \quad B = \begin{pmatrix} x_1 \\ x_2 \\ \vdots \\ x_k \end{pmatrix}, \quad \widetilde{B} = \begin{pmatrix} \widetilde{x}_1 \\ \widetilde{x}_2 \\ \vdots \\ \widetilde{x}_k \end{pmatrix}, \quad (6.3.4)$$

$$\Gamma = (\gamma_{ij})_{i=1,j=1}^{k\ \ n}. \quad (6.3.5)$$

Here A and Z are diagonal matrices. The diagonal entries $t_1, \dots, t_n$ and $s_1, \dots, s_k$ are points in Π^+ such that $t_i \neq t_j$ and $s_i \neq s_j$ whenever $i \neq j$. The entries of C and $\widetilde{C}$ are column vectors from $\mathbb{C}^{r\times 1}$ and $\mathbb{C}^{m\times 1}$, respectively, those of B and $\widetilde{B}$ are row vectors from $\mathbb{C}^{1\times m}$ and $\mathbb{C}^{1\times r}$, respectively, and the entries of Γ are complex numbers. Furthermore, the entries of C and B are all assumed to be nonzero vectors. It follows that A and Z have the eigenvalues in Π^+, the pair (C, A) is a null-kernel pair, and (Z, B) is a full-range pair. For this case equation (6.3.2) is satisfied if and only if

$$\gamma_{ij} = \frac{\widetilde{x}_i u_j - x_i \widetilde{u}_j}{s_i - t_j} \quad \text{if} \quad s_i \neq t_j, \qquad \widetilde{x}_i u_j - x_i \widetilde{u}_j = 0 \quad \text{if} \quad s_i = t_j. \quad (6.3.6)$$

For the operators in (6.3.3)–(6.3.5) the interpolation conditions in the bitangential Nudelman interpolation problem take the following form:

(BTN3)' $x_i F(s_i) = \widetilde{x}_i, \qquad i = 1, \dots, k,$
(BTN4)' $F(t_j) u_j = \widetilde{u}_j, \qquad j = 1, \dots, n,$
(BTN5)' $x_i F'(s_i) u_j = \gamma_{ij}$, whenever $s_i = t_j$.

Notice that for (BTN3)' and (BTN4)' to admit a solution the data have to be compatible, which in this case means that the second condition in (6.3.6) has to be satisfied. Condition (BTN5)' involves only those γ_{ij} corresponding to $s_i = t_j$; the other γ_{ij} are uniquely determined from the data in (BTN3)' and (BTN4)' by the first part of (6.3.6).

Conversely, the interpolation condition (BTN3)'–(BTN5)' may be rewritten in the form (BTN3)–(BTN5) by taking $A, C, \widetilde{C}, Z, B, \widetilde{B}$ as in (6.3.3) and (6.3.4), and by defining Γ in (6.3.5) by using the given γ_{ij} in (BTN5)' and by choosing the others according to the formula in the first part of (6.3.6). Then (6.3.2) is satisfied if and only if the second part of (6.3.6) holds.

The following theorem shows that the bitangential Nudelman interpolation problem (BTN1)–(BTN5) reduces to an inverse problem of the type considered in the previous section.

Theorem 6.3.1 *Let* $\omega = (\widetilde{C}, C, A; Z, B, \widetilde{B}; \Gamma)$ *be an admissible interpolation data set. Then*

$$\tau = \left\{ \begin{pmatrix} \widetilde{C} \\ C \end{pmatrix}, A; Z, \begin{pmatrix} B & -\widetilde{B} \end{pmatrix}; \Gamma \right\}$$

is a Π^+*-admissible Sylvester data set, and the interpolation problem* (BTN1) – (BTN5) *has a solution if and only if there exists a* J*-inner rational* $(m+r)\times(m+r)$ *matrix function* Θ *which has* τ *as its null-pole triple. Here*

$$J = \begin{pmatrix} I_m & 0 \\ 0 & -I_r \end{pmatrix}. \tag{6.3.7}$$

Furthermore, if Θ *is such a* J*-inner rational matrix function, then all solutions* F *of the interpolation problem* (BTN1)–(BTN5) *are given by*

$$F(\lambda) = \big(\Theta_{11}(\lambda)G(\lambda) + \Theta_{12}(\lambda)\big)\big(\Theta_{21}(\lambda)G(\lambda) + \Theta_{22}(\lambda)\big)^{-1}, \tag{6.3.8}$$

where the free parameter G *is an arbitrary* $m \times r$ *matrix function in* $\mathcal{R}at(H_\infty)$ *satisfying* $\sup\{\|G(\lambda)\| \mid Re\ \lambda \geq 0\} < 1$, *and*

$$\begin{pmatrix} \Theta_{11}(\lambda) & \Theta_{12}(\lambda) \\ \Theta_{21}(\lambda) & \Theta_{22}(\lambda) \end{pmatrix}$$

is the block partitioning of $\Theta(\lambda)$ *according to the partitioning of* J *in* (6.3.7).

Theorem 6.3.1 is proved in a way similar to the proof of Theorem 4.3.1. For the details we refer to the book Ball-Gohberg-Rodman [1990], Section 18.6. A straightforward application of Theorems 6.2.1 and 6.3.1 yield the following solution of the problem (BTN1)–(BTN5).

Theorem 6.3.2 *Let* $\omega = (\widetilde{C}, C, A; Z, B, \widetilde{B}; \Gamma)$ *be an admissible interpolation data set. Put*

$$S_1 = \int_0^\infty e^{-tZ}(BB^* - \widetilde{B}\widetilde{B}^*)e^{-tZ^*}\,dt, \qquad S_2 = \int_0^\infty e^{-tA^*}(C^*C - \widetilde{C}^*\widetilde{C})e^{-tA}\,dt,$$

and set

$$\Lambda = \begin{pmatrix} S_1 & \Gamma \\ \Gamma^* & S_2 \end{pmatrix}. \tag{6.3.9}$$

Then the bitangential Nudelman interpolation problem (BTN1)–(BTN5) *is solvable if and only if* Λ *is positive definite. Furthermore, in this case all solutions* F *are given by*

$$F(\lambda) = \big(\Theta_{11}(\lambda)G(\lambda) + \Theta_{12}(\lambda)\big)\big(\Theta_{21}(\lambda)G(\lambda) + \Theta_{22}(\lambda)\big)^{-1}, \tag{6.3.10}$$

where

$$\begin{pmatrix} \Theta_{11}(\lambda) & \Theta_{12}(\lambda) \\ \Theta_{21}(\lambda) & \Theta_{22}(\lambda) \end{pmatrix} = \begin{pmatrix} I_m & 0 \\ 0 & I_r \end{pmatrix} + \begin{pmatrix} -B^* & \widetilde{C} \\ -\widetilde{B}^* & C \end{pmatrix} \begin{pmatrix} \lambda + Z^* & 0 \\ 0 & \lambda - A \end{pmatrix}^{-1} \cdot \Lambda^{-1} \begin{pmatrix} B & -\widetilde{B} \\ -\widetilde{C}^* & C^* \end{pmatrix}, \tag{6.3.11}$$

and the free parameter G *in* (6.3.10) *is an arbitrary* $m \times r$ *matrix function in* $\mathcal{R}at(H_\infty)$ *satisfying the norm constraint* $\sup\{\|G(\lambda)\| \mid Re\ \lambda \geq 0\} < 1$.

The operators S_1 and S_2 in the previous theorem may also be defined as the unique solutions of the Lyapunov equations

$$ZS_1 + S_1Z^* = BB^* - \widetilde{B}\widetilde{B}^*, \qquad A^*S_2 + S_2A = C^*C - \widetilde{C}^*\widetilde{C}.$$

It follows that for the operators $A, C, \widetilde{C}, Z, B, \widetilde{B}$ in (6.3.3) and (6.3.4) both S_1 and S_2 are Pick matrices, namely

$$S_1 = \left(\frac{x_i x_j^* - \widetilde{x}_i \widetilde{x}_j^*}{s_i + \overline{s_j}}\right)_{i,j=1}^{k}, \qquad S_2 = \left(\frac{u_i^* u_j - \widetilde{u}_i^* \widetilde{u}_j}{\overline{t_i} + t_j}\right)_{i,j=1}^{n}.$$

Hence, in this case the $(m+r) \times (m+r)$ matrix Λ is explicitly given, and to see whether or not the corresponding interpolation problem is solvable one has to check only that the eigenvalues of Λ are positive.

Let us illustrate Theorem 6.3.2 on an example. Consider the problem to find a rational 2×2 matrix function $F \in \mathcal{R}at(H_\infty)$ satisfying the norm constraint

$$\sup\{\|F(\lambda)\| \mid \operatorname{Re}\, \lambda \geq 0\} < 1 \tag{6.3.12}$$

and such that the following interpolation conditions are fulfilled:

$$\begin{pmatrix} 1 & 1 \end{pmatrix} F(\frac{1}{2}) = \begin{pmatrix} a & b \end{pmatrix}, \quad F(\frac{1}{2}) \begin{pmatrix} 1 \\ 0 \end{pmatrix} = \begin{pmatrix} 0 \\ a \end{pmatrix}, \quad \begin{pmatrix} 1 & 1 \end{pmatrix} F'(\frac{1}{2}) \begin{pmatrix} 1 \\ 0 \end{pmatrix} = c. \tag{6.3.13}$$

Here a, b, and c are assumed to be real numbers. To solve this bitangential interpolation problem we apply Theorem 6.3.2 with

$$\begin{aligned} A &= (\tfrac{1}{2}), & B &= \begin{pmatrix} 1 & 1 \end{pmatrix}, & \widetilde{B} &= \begin{pmatrix} a & b \end{pmatrix}, \\ Z &= (\tfrac{1}{2}), & C &= \begin{pmatrix} 1 & 0 \end{pmatrix}, & \widetilde{C} &= \begin{pmatrix} 0 \\ a \end{pmatrix}, \\ \Gamma &= (c). \end{aligned}$$

In this case the matrix Λ in (6.3.9) is given by

$$\Lambda = \begin{pmatrix} 2 - a^2 - b^2 & c \\ c & 1 - a^2 \end{pmatrix},$$

and hence our problem is solvable if and only if

$$|a| < 1, \qquad a^2 + b^2 + c^2(1 - a^2)^{-1} < 2. \tag{6.3.14}$$

Assume the conditions (6.3.14) are fulfilled, then Theorem 6.3.2 tells us (take $G(\lambda) \equiv 0$ in (6.3.10)) that a particular solution of our problem is obtained by taking $F(\lambda) = \Theta_{12}(\lambda)\Theta_{22}(\lambda)^{-1}$, where

$$\Theta_{12}(\lambda) = \begin{pmatrix} -1 & 0 \\ -1 & a \end{pmatrix} \begin{pmatrix} \lambda + \frac{1}{2} & 0 \\ 0 & \lambda - \frac{1}{2} \end{pmatrix}^{-1} \begin{pmatrix} 2 - a^2 - b^2 & c \\ c & 1 - a^2 \end{pmatrix}^{-1} \begin{pmatrix} -a & -b \\ 1 & 0 \end{pmatrix},$$

$$\begin{aligned} \Theta_{22}(\lambda) &= \begin{pmatrix} 1 & 0 \\ 0 & 1 \end{pmatrix} + \begin{pmatrix} -a & 1 \\ -b & 0 \end{pmatrix} \begin{pmatrix} \lambda + \frac{1}{2} & 0 \\ 0 & \lambda - \frac{1}{2} \end{pmatrix}^{-1} \cdot \\ &\quad \cdot \begin{pmatrix} 2 - a^2 - b^2 & c \\ c & 1 - a^2 \end{pmatrix}^{-1} \begin{pmatrix} -a & -b \\ 1 & 0 \end{pmatrix}. \end{aligned}$$

Taking $a = b = \frac{1}{2}$ one sees that our norm constraint bitangential interpolation problem has a solution if and only if $|c| < \frac{3}{4}\sqrt{2}$, and one computes (e.g., by using the computer program Maple V) that in this case a particular solution is given by

$$F(\lambda) = \frac{1}{d(\lambda)} \begin{pmatrix} \left(\frac{3}{2} + 4c\right)\lambda - 2c - \frac{3}{4} & \frac{3}{2}\lambda + \frac{5}{4} \\ \left(\frac{9}{2} + 5c\right)\lambda - \frac{3}{2}c + \frac{5}{4} & \left(\frac{3}{2} + c\right)\lambda + \frac{1}{2}c + \frac{3}{4} \end{pmatrix},$$

where

$$d(\lambda) = (\frac{9}{2} - 4c^2)\lambda + (\frac{15}{2} + 4c)\lambda + c^2 + \frac{17}{8}.$$

For $a = 0$, $b = 1$, $c = 0$ a direct computation yields that

$$F(\lambda) = \frac{1}{2\lambda + 3}\begin{pmatrix} 0 & 2 \\ 0 & 2 \end{pmatrix}$$

satisfies the interpolation conditions

$$\begin{pmatrix} 1 & 1 \end{pmatrix} F(\frac{1}{2}) = \begin{pmatrix} 0 & 1 \end{pmatrix}, \quad F(\frac{1}{2})\begin{pmatrix} 1 \\ 0 \end{pmatrix} = \begin{pmatrix} 0 \\ 0 \end{pmatrix}, \quad \begin{pmatrix} 1 & 1 \end{pmatrix} F'(\frac{1}{2})\begin{pmatrix} 0 \\ 1 \end{pmatrix} = 0,$$

and meets the norm constraint (6.3.12).

6.4 Notes

The first solution of the inverse problem in Section 6.1 appears in Ball-Ran [1987b]. Theorem 6.1.1 is a preliminary version which appears in Gohberg-Kaashoek -Lerer-Rodman [1984]. Corollary 6.1.2, and Theorems 6.1.3 and 6.1.4 are taken from Gohberg-Kaashoek-Ran [1988]. If one allows for a bit more freedom in the choice of the operators A_{21}, A_{12} in (6.1.19) and in the choice of the coupling transformation in (6.1.21), then Theorem 6.1.4 yields modulo similarity transformations all minimal complements (see Gohberg-Kaashoek-Ran [1988] for further details). Minimal complements have also been constructed using different types of realizations (see Gohberg-Kaashoek-Ran [1990] and Section II.3 in Groenewald [1993]). Extensions of the results of Section 6.1 to non-regular rational (possibly non-square) matrix functions may be found in Ball-Gohberg-Rakowski [1994] and in the earlier papers Ball-Rakowski [1990], [1991]. Theorem 6.2.1 appears implicitly in Ball-Gohberg-Rodman [1990]. The sketch of the solution of the bitangential Nudelman interpolation in Section 6.3 follows Ball-Gohberg-Rodman [1990]. Bitangential interpolation problems of the type considered in Section 6.3 but with the norm constraint (BTN2) replaced by other types of constraints are also of interest and may be treated by the state space method too. See, e.g., Antoulas-Ball-Kang (Kim)-Willems [1990], where the norm constraint is replaced by a minimal degree condition, and Ball-Kim [1994], which concerns bitangential interpolation in the class of symmetric rational matrix functions. For recursive methods and algorithms to solve rational matrix function interpolation problems via a state space approach we refer to Gohberg-Olshevsky [1994] and Sayed-Kailath-Lev-Ari-Constantinescu [1994]. The non-stationary version of the bitangential Nudelman interpolation problem of Section 6.3 is solved in Ball-Gohberg-Kaashoek [1995], [1993]; see also Ball-Gohberg-Kaashoek [1994] for the solution of the non-stationary continuous time analogue.

In the present text the state space theory has been illustrated mainly on factorization and interpolation problems. Some of the other applications, related to Wiener-Hopf and Toeplitz operators, have already been listed in the notes at the end of the first chapter. In conclusion we mention linear transport theory (see Kaper-Lekkerkerker-Hejtmanek [1982], Greenberg-van der Mee-Protopopescu [1987], Gohberg-Kaashoek [1991a]), and scattering theory (see Alpay-Gohberg [1993], [1994]) as two other areas where the state space method has been used successfully.

Bibliography

Adamjan, V.M., Arov, D.Z. and Kreĭn, M.G. [1968] *Infinite Hankel and generalized Carathéodory-Fejèr and I. Schur problems* Funktsional. Anal. i Prilozhen **2**,1–17. (Russian).

Adamjan, V.M., Arov, D.Z. and Kreĭn, M.G. [1971] *Infinite block Hankel matrices and their connections with the interpolation problem* Akad. Nauk Armenia SSR Isv. Mat. **6**; English transl. in: Amer. Math. Soc. Transl. Ser. 2 **111** (1978), Providence, RI .

Alpay, D. and Gohberg, I. [1982] *Unitary rational matrix functions* in: Topics in interpolation theory of matrix-valued functions (I. Gohberg, ed.), OT 33, Birkhäuser Verlag, Basel(1988), 175–222.

Alpay, D. and Gohberg, I. [1993] *Inverse spectral problem for differential operators with rational scattering matrix functions* J. Diff. Eqs, to appear.

Alpay, D. and Gohberg, I. [1994] *Inverse spectral problems for difference operators with rational scattering matrix function* Integral Equations and Operator Theory **20**, 125–170.

Antoulas, A.C., Ball, J.A., Kang (Kim), J. and Willems, J.C. [1990] *On the solution of the minimal interpolation problem* Lin. Alg. Appl. **137/138**, 511–573.

Ball, J.A. and Cohen, N. [1991] *De Branges-Rovnyak operator models and systems theory: a survey* in: Topics in Matrix and Operator Theory (H. Bart, I. Gohberg and M.A. Kaashoek, eds.), OT 50, Birkhäuser Verlag, Basel, pp. 93–136.

Ball, J.A., Cohen, N. and Ran, A.C.M. [1988] *Inverse spectral problems for regular improper rational matrix functions*, in: Topics in interpolation theory of rational matrix-valued functions (I.Gohberg, ed.), OT 33, Birkhäuser Verlag, Basel, pp. 123–173.

Ball, J.A., Gohberg, I. and Kaashoek, M.A. [1992a] *Nevanlinna-Pick interpolation for time-varying input-output maps: The discrete case*, in: Time-variant systems and interpolation (I. Gohberg, ed.), OT 56, Birkhäuser Verlag, Basel, pp. 1–51.

Ball, J.A., Gohberg, I. and Kaashoek, M.A. [1992b] *Nevanlinna-Pick interpolation for time-varying input-output maps: The continuous time case*, in: Time-variant systems and interpolation (I. Gohberg, ed.), OT 56, Birkhäuser Verlag, Basel, pp. 52–89.

Ball, J.A., Gohberg, I. and Kaashoek, M.A. [1993] *Bitangential interpolation for input-output operators of time-varying systems: The discrete time case*, in: New aspects in interpolation and completion theories (I. Gohberg, ed.), OT 64, Birkhäuser Verlag, pp. 33–72.

Ball, J.A., Gohberg, I. and Kaashoek, M.A. [1994] *Bitangential interpolation for input-output maps of time-varying systems: the continuous time case* Integral Equations and Operator Theory **20**, 1–43.

Ball, J.A., Gohberg, I. and Kaashoek, M.A. [1995] *Two-sided Nudelman interpolation for input-output operators of discrete time-varying systems* Integral Equations and Operator Theory, **21**, 174–211.

Ball, J.A., Gohberg, I. and Rakowski, M. [1994] *Reconstruction of a rational nonsquare matrix function from local data* Integral Equations and Operator Theory **20**, 249–305.

Ball, J.A., Gohberg, I. and Rodman, L. [1987] *Minimal factorization of meromorphic matrix functions in terms of local data* Integral Equations and Operator Theory **10**, 437–465.

Ball, J.A., Gohberg, I. and Rodman, L. [1990] *Interpolation of Rational Matrix Functions*, Birkhäuser Verlag, Basel.

Ball, J.A., Gohberg, I. and Rodman, L. [1993] *Two-sided tangential interpolation of real rational matrix functions*, in: New aspects in interpolation and completion theories, (I. Gohberg, ed.), OT 64, Birkhäuser Verlag, Basel, pp. 73–102.

Ball, J.A. and Kaashoek, M.A. [to appear] *Indefinite inner-outer factorization of rational matrix functions*, to appear.

Ball, J.A., Kaashoek, M.A., Groenewald, G. and Kim, J. [1994] *Common reduced rational matrix functions with given null-pole data in the complex plane* Linear Alg. Appl. **203/204**, 67–110.

Ball, J.A. and Kim, J. [1994] *Bitangential interpolation problems for symmetric rational matrix functions*, to appear.

Ball, J.A. and Rakowski, M. [1990] *Zero-pole structure of nonregular rational matrix functions*, in: Extension and interpolation of linear operators and matrix functions, OT 47, Birkhäuser Verlag, Basel, pp. 137–193.

Ball, J.A. and Rakowski, M. [1991] *Null-pole subspace of rectangular rational matrix functions* Lin. Alg. Appl. **159**, 81–120.

Ball, J.A. and Ran, A.C.M. [1987a] *Global inverse spectral problems for rational matrix functions* Lin. Alg. Appl. **86**, 237–382.

Ball, J.A. and Ran, A.C.M. [1987b] *Local inverse spectral problems for rational matrix functions* Integral Equations and Operator Theory **10**, 349–415.

Bart, H., Gohberg, I. and Kaashoek, M.A. [1979] *Minimal Factorization of Matrix and Operator Functions*, Birkhäuser Verlag, Basel.

Bart, H., Gohberg, I. and Kaashoek, M.A. [1982a] *Wiener-Hopf integral equations, Toeplitz matrices and linear systems*, in: Toeplitz Centennial (I. Gohberg, ed.), OT 4, Birkhäuser Verlag, Basel, pp. 85–135.

Bart, H., Gohberg, I. and Kaashoek, M.A. [1982b] *Convolution equations and linear systems* Integral Equations and Operator Theory **5**, 283–340.

Bart, H., Gohberg, I. and Kaashoek, M.A. [1984] *The coupling method for solving integral equations*, in: Topics in Operator Theory and Networks, (H. Dym and I. Gohberg, eds.), OT 12, Birkhäuser Verlag, Basel, pp. 39–73.

Bart, H., Gohberg, I. and Kaashoek, M.A. [1986a] *Explicit Wiener-Hopf factorization and realization*, in: Constructive methods of Wiener-Hopf factorization, (I. Gohberg and M.A. Kaashoek, eds.), Birkhäuser Verlag, Basel, pp. 235–316.

Bart, H., Gohberg, I. and Kaashoek, M.A. [1986b] *Invariants for Wiener-Hopf equivalence of analytic operator functions*, Constructive methods of Wiener-Hopf factorization, (I. Gohberg and M.A. Kaashoek, eds.), OT 21, Birkhäuser Verlag, Basel, pp. 317–355.

Bart, H., Gohberg, I. and Kaashoek, M.A. [1986c] *Wiener-Hopf factorization, inverse Fourier transforms and exponentially dichotomous operators* Functional Analysis **68**, 1–42.

Ben-Artzi, A., Gohberg, I. and Kaashoek, M.A. [1993] *Time-varying generalizations of invertibility and Fredholm theorems for Toeplitz operators* Indag. Math. **4**, 385–405.

Ben-Artzi, A., Gohberg, I. and Kaashoek, M.A. [1994] *Exponentially dominated infinite block matrices of finite Kronecker rank* Integral Equations and Operator Theory **18**, 30–77.

Böttcher, A. and Silbermann, B. [1991] *Analysis of Toeplitz Operators*, Springer-Verlag, Berlin.

Callier, F.M. and Desoer, C.A. [1991] *Linear System Theory*, Springer-Verlag, Berlin.

Doyle, J.C., Francis, B.A. and Tannenbaum, A.R. [1992] *Feedback Control Theory*, Macmillan, New York.

Dym, H. [1989] *J contractive Matrix Functions, Reproducing Kernel Hilbert Spaces and Interpolation*, CBMS Regional Conference Series 71, Amer. Math. Soc., Providence, RI.

Fedchina, I.P. [1975] *Nevanlinna-Pick problem with multiple points* Doklady Akad. Nauk Arm. SSR **61**, 214–218. (Russian).

Foias, C. and Frazo, A.E. [1990] *The Commutant Lifting Approach to Interpolation Problems*, OT 44, Birkhäuser Verlag, Basel.

Francis, B.A. [1987] *A Course in H_∞ Control Theory*, Springer-Verlag.

Glover, K. [1984] *All optimal Hankel-norm approximations of linear multivariable systems and the L^∞-erro bounds* Int.J. Control **39**, 1115–1193.

Gohberg, I. (ed.) [1992] *Time-variant Systems and Interpolation*, OT 56, Birkhäuser Verlag, Basel.

Gohberg, I. and Fel'dman, I.A. [1974] *Convolution Equations and Projection Methods for their Solution*, Transl. Math. Monographs 41, Amer. Math. Soc., Providence, R.I.

Gohberg, I., Goldberg, S. and Kaashoek, M.A. [1990] *Classes of Linear Operators* I, Birkhäuser Verlag, Basel.

Gohberg, I., Goldberg, S. and Kaashoek, M.A. [1993] *Classes of Linear Operators* II, Birkhäuser Verlag, Basel.

Gohberg, I. and Kaashoek, M.A. [1987] *An inverse spectral problem for rational matrix functions and minimal divisibility* Integral Equations and Operator Theory **10**, 437-465.

Gohberg, I. and Kaashoek, M.A. [1988a] *Regular rational matrix functions with prescribed pole and zero structure*, in: Topics in interpolation theory of rational matrix functions (I. Gohberg, ed.), OT 33, Birkhäuser Verlag, Basel, pp. 109-122.

Gohberg, I. and Kaashoek, M.A. [1988b] *Block Toeplitz operators with rational symbols*, in: Contributions to operator theory and its applications (I. Gohberg, J.W. Helton and L. Rodman, eds.), OT 35, Birkhäuser Verlag, Basel, pp. 385–440.

Gohberg, I. and Kaashoek, M.A. [1991a] *The Wiener-Hopf method for the transport equation: a finite dimensional version*, in: Modern mathematical methods in transport theory (W. Greenberg and J. Polewczak, eds.), OT 51, Birkhäuser Verlag, Basel, pp. 20–33.

Gohberg, I. and Kaashoek, M.A. [1991b] *The state space method for solving singular integral equations*, in: Mathematical System Theory. The influence of Kalman (A.C. Antoulas, ed.), Springer Verlag, Basel, pp. 509–523.

Gohberg, I., Kaashoek, M.A., Lerer, L. and Rodman, L. [1984] *Minimal divisors of rational matrix functions with prescribed zero and pole structure*, in: Topics in Operator Theory, Systems and Networks, The Rehovot Workshop, (H. Dym and I. Gohberg, eds.), OT 12, Birkhäuser Verlag, Basel, pp. 241–275.

Gohberg, I., Kaashoek, M.A. and Ran, A.C.M. [1988] *Interpolation problems for rational matrix functions with incomplete data and Wiener-Hopf factorization* in: Topics in interpolation theory of rational matrix functions (I. Gohberg, ed.), OT 33, Birkhäuser Verlag, Basel, pp. 73-108.

Gohberg, I., Kaashoek, M.A. and Ran, A.C.M. [1990] *Regular rational matrix functions with prescribed null and pole data except at infinity* Lin.Alg.Appl **137/138**, 387–412.

Gohberg, I., Kaashoek, M.A. and van Schagen, F. [1993] *On the local theory of regular analytic matrix functions* Lin.Alg.Appl **182**, 9-25.

Gohberg, I.C. and Kreĭn, M.G. [1960] *Systems of integral equations on a half line with kernels depending on a difference of arguments* Amer. Math. Soc. Transl. **14**, 217–187.

Gohberg, I.C. and Kreĭn, M.G. [1970] *Introduction to the Theory of Linear Nonselfadjoint Operators in Hilbert Space*, Transl. Math. Monographs 24, Amer. Math. Soc., Providence, RI.

Gohberg, I. and Krupnik, N. [1992a] *One-dimensional Linear Singular Integral Operators I, Introduction*, Birkhäuser Verlag, Basel.

Gohberg, I. and Krupnik, N. [1992b] *One-dimensional Linear Singular Integral Operators II, General Theory and Applications*, Birkhäuser Verlag, Basel.

Gohberg, I., Lancaster, P. and Rodman, L. [1978a] *Spectral analysis of matrix polynomials. I. Canonical forms and divisors* Lin. Alg. Appl. **20**, 1–44.

Gohberg, I., Lancaster, P. and Rodman, L. [1978b] *Spectral analysis of matrix polynomials. II. The resolvent form and spectral divisors* Lin. Alg. Appl. **21**, 65–88.

Gohberg, I., Lancaster, P. and Rodman, L. [1982] *Matrix Polynomials*, Academic Press, New York.

Gohberg, I., Lancaster, P. and Rodman, L. [1986] *Invariant Subspaces of Matrices with Applications*, Wiley, New York.

Gohberg, I. and Olshevsky, V. [1994] *Fast state space algorithms for matrix Nehari and Nehari-Takagi interpolation problems* Integral Equations and Operator Theory **20**, 44–83.

Gohberg, I. and Rodman, L. [1981] *Analytic matrix functions with prescribed local data* J. d'Analyse Math. **40**, 90–128.

Gohberg, I. and Sakhnovich, L.A. (eds) [1994] *Matrix and Operator Valued Functions*, The Vladimir Petrovich Potapov Memorial Volume, OT 72, Birkhäuser Verlag, Basel.

Gohberg, I.C. and Sigal, E.I. [1971] *On operator generalizations of the logarithmic residue theorem and the theorem of Rouché* Math. USSR Sbornik **13**, 603–625.

Green, M. [1987] *Factorization Techniques in Multivariable Phase Problems*, Ph. D. thesis, Australian National University, Department of Systems Engineering, Canberra, Australia.

Green, M., Glover, K., Limebeer, D. and Doyle, J. [1990] *A J-spectral factorization approach to $\mathcal{H}_\infty$ control* Siam J. Control Opt. **28**, 1350–1371.

Greenberg, W., van der Mee, C. and Protopopescu, V. [1987] *Boundary Value Problems in Abstract Kinetic Theory*, OT 23, Birkhäuser Verlag, Basel.

Groenewald, G.J. [1993] *Wiener-Hopf Factorization of Rational Matrix Functions in Terms of Realizations: An Alternative Version*, Ph. D. thesis, Vrije Universiteit, Amsterdam.

Helton, J.W. [1987] *Operator Theory, Analytic Functions, Matrices and Electrical Engineering*, CBMS Regional Conference Series 68, Amer. Math. Soc., Providence, RI.

Hoffman, K. [1962] *Banach Spaces of Analytic Functions*, Prentice-Hall, Inc., Englewood Cliffs, N.Y.

Kaashoek, M.A. [1984] *Minimal factorization, linear systems and integral operators*, in: Operators and Function Theory (S.C. Power, ed.), D. Reidel Publishing Company, Dordrecht, pp. 41–68.

Kaashoek, M.A. and Ran, A.C.M. [1991] *State Space Methods for H_∞-control Problems*, notes of a course for the Dutch Graduate Network on Systems and Control, Spring Semester 1991.

Kailath, T. [1980] *Linear Systems*, Prentice Hall, Englewood Cliffs, N.J.

Kalman, R.E. [1960] *Contributions to the theory of optimal control* Bol. Soc. Matem. Mexicana **5**, 102–119.

Kalman, R.E. [1961] *On the general theory of control systems*, in: Automatic and remote control, Butterworth, London, pp. 481–492.

Kalman, R.E., Falb, P.L. and Arbib, M.A. [1969] *Topics in Mathematical System Theory*, McGraw-Hill, New York.

Kaper, H.G., Lekkerkerker, C.G. and Hejtmanek, J. [1982] *Spectral Methods in Linear Transport Theory*, OT 5, Birkhäuser Verlag, Basel.

Keldysh, M.V. [1951] *On eigenvalues and eigenfunctions of some classes of nonselfadjoint equations* DAN SSSR **77**, 11–14. (Russian).

Keldysh, M.V. [1971] *On the completeness of eigenfunctions of some classes of nonselfadjoint linear operators* Russian Math. Surveys **24**, 15–44.

Kovalishina, I.V. and Potapov, V.P. [1974] *An indefinite metric in the Nevanlinna-Pick problem* Transl. Akad. Nauk. Armyn. SSR Dokl. **59**, 17–22; also in: Amer. Math. Soc. Transl. Series 2 **138** (1988), 15–19.

Kovalishina, I.V. and Potapov, V.P. [1981] *The radii of a Weyl disk in the matrix Nevanlinna-Pick problem*, Theory of operators in function spaces and its applications (V.A. Marchenko, ed.), Naukova Dumka, Kiev, pp. 25–49; in: Amer. Math. Soc. Transl. Series 2 **138** (1988), 37–54.

Kovalishina, I.V. and Potapov, V.P. [1988] *Seven Papers Translated from the Russian*, Amer. Math. Soc. Transl. 138, Amer. Math. Soc., Providence, RI.

Kreĭn, S.G. and Trofimov, V.P. [1969] *Holomorphic operator-valued functions of several variables* Functional Analysis Appl. **3**, 330–331.

Kuijper, A.B. [1992] *The State Space Method for Integro-differential Equations of Convolution Type*, Ph. D. thesis, Vrije Universiteit, Amsterdam.

Lancaster, P. [1977a] *A fundamental theorem of lambda-matrices with applications—I. Ordinary differential equations with constant coefficients* Lin. Alg. Appl. **18**, 189–211.

Lancaster, P. [1977b] *A fundamental theorem of lambda-matrices with applications—II. Difference equations with constant coefficients* Lin. Alg. Appl. **18**, 213–222.

Lerer, L. and Rodman, L. [1994] *Common zero structure of rational matrix functions*, preprint.

Markus, A.S. [1988] *Introduction to the Spectral Theory of Polynomial Operator Pencils*, Transl. Math. Monographs 71, Amer. Math. Soc., Providence, RI.

Mikhlin, S.G. and Prössdorf, S. [1986] *Singular Integral Operators*, Akademie-Verlag.

Nevanlinna, R. [1929] *Über beschränkte analytisch Funktionen* Ann. Acad. Sci. Fenn. Ser. A I Math **32**, no. 7.

Palant, Ju. A. [1970] *On a method of testing for the multiple completeness of a system of eigenvectors and associated vectors of a polynomial operator bundle* Vestnik Har'kov Gos. Univ. **34**, 1–13. (Russian).

Roozemond, L. [1986] *Canonical pseudo-spectral factorization and Wiener-Hopf integral equations*, in: Constructive methods of Wiener-Hopf factorization, (I. Gohberg and M.A. Kaashoek, eds.), OT 21, Birkhäuser Verlag, Basel, pp. 127–156.

Roozemond, L. [1987] *Systems of Non-normal and First Kind Wiener-Hopf Equations*, Ph. D. thesis, Vrije Universiteit, Amsterdam.

Rosenblum, M. and Rovnyak, J. [1985] *Hardy Classes and Operator Theory*, Oxford Univ. Press, Oxford.

Sayed, A.H., Kailath, T., Lev-Ari, H. and Constantinescu, T. [1994] *Recursive solutions of rational interpolation problems via fast matrix factorization* Integral Equations and Operator Theory **20**, 84–118.

Sontag, E.D. [1990] *Mathematical Control Theory*, Texts in Applied Mathematics, 6, Springer Verlag, Berlin.

Vreugdenhil, R. [1990] *Spectral Theory of Selfadjoint Wiener-Hopf and Toeplitz Operators with Rational Symbols*, Ph. D. thesis, Vrije Universiteit, Amsterdam.

Vreugdenhil, R. [1991] *Spectral theory of selfadjoint Winer-Hopf operators with rational symbols*, in: Topics in matrix and operator theory (H. Bart, I. Gohberg, and M.A. Kaashoek, eds), OT 50, Birkhäuser Verlag, Basel, pp. 317–370.

Vreugdenhil, R. [1994] *The resolution of the identity for selfadjoint Toeplitz operators with rational matrix symbol* Integral Equations and Operator Theory **20**, 449–490.

Wonham, W.M. [1979] *Linear Multivariable Control: A Geometric Approach*, Springer Verlag, Berlin.

Index

A

adjoint, 149–150
admissible
  function, 267
  interpolation data set, 320
  Sylvester data set, 309
angle operator, 153
angular operator, 78
antispace, 144

B

basis function, 65
Bergman space, 6
bicausal, 297
bicontraction, 150, 160, 162, 165, 186–187
  characterization of, 162, 169
  matrix extensions, 188, 190
biregular
  at a point, 236
  on a set, 236
  on the right half plane, 258
Block Toeplitz matrix, 35
Bognár–Krámli factorization, 150, 152, 158, 159, 207, 223
boundary operator, 102

C

C^*–algebra, 49
Calkin algebra, and Kreĭn spaces, 222
canonical factorization
  right, 243–4
canonical left null pair, 270, 272
  at a point, 270
  on a set, 272
canonical right null pair
  at a point, 269
  on a set, 271
canonical system
  of right null functions, 264
Carathéodory–Fejér problem, 181
Cauchy singular integral, 5, 25
Cayley transform, 100
characteristic function, 185, 196
  of a colligation, 99
  of a symmetric operator, 101
choice sequence, 182
closed graph theorem, 147
closely (inner, outer) connected, 196
coefficient space, 192
colligation, 99, 196
  coisometric, 196–199
  isometric, 196–197, 202
  unitary, 196–197, 203
collocation method, 67
column extensions, 166–167, 174–175, 183
commutant, 209
  of a positive operator, 224–225
compact operator, 17, 221
compact perturbation, see perturbation
complement, 311
complement, closed, 151
complementation, 206
completion of operator matrices, 171–172, 175–176, see also, row, column, matrix, isometric extensions
connecting operator, for a colligation, 196
continuity, on a Kreĭn space, 147
contraction, 150, 160, 162
  adjoint, 168–170
  continuous extensions, 163–165
  mapping of negative subspaces, 161

contraction (Cont.)
matrix extensions, 174
control
H_∞–, 287
controllable, see also closely inner connected
convergence, in a Kreĭn space, 147
coupling transformation
in a null–pole triple, 292
of a data set, 309
critical point, 82
of finite type, 85
regular, 84
singular, 84
critical points, and definitizable operators, 218
Curie point, 62

D
defect number, 88
defect operator, 158–159
defect space, 158, 166–167
definite space, 144, 152
definitizing polynomial, 80, 215
definitizable, 215, 217, 219, 221
dilation, 240
direct sum, 312
double commutant, 209
dual
of a matrix function, 252
dual pair of subspaces, 213

E
eigenvector, 263
right eigenvector, 263
essential range, 9
essential uniqueness, 161, 187, 198, 210
exposed arc, 221
extension theorems, see row, column, matrix, isometric
external coefficient, 238

F
factorization
J–inner–outer, 258
right canonical, 243, 244
feedthrough coefficient, 238
final space, of a partial isometry, 156
finite rank operator, 18
finite section method, 14, 40, 46, 66
Fisher–Hartwig conjecture, 64
Fredholm alternative, 18
Fredholm operator, 26
full range completeness, 113
full–range pair, 241
functional calculus, 217–218
fundamental decomposition, 145, 154
fundamental symmetry, 77, 145

G
Galerkin–Petrov method, 65
generalized left eigenvector, 270
generalized resolvent, 89
generalized right eigenvector, 268
Ginzberg–Potapov transform, 254
Gram matrix, 151
graph representation, 153–155

H
H_∞–control, 287
half range completeness, 113
Hankel matrix, 17
Hankel operator, 17
hermitian analytic kernel, 192
Hilbert matrix, 25
Hilbert space, 144
Hilbert–Schmidt operator, 59
homomorphism, 209
hyperinvariant subspace, 210–212, 221, 223

I
indefinite space, 144, 152
index, 26
index formulas, 159, 160, 166–168, 175–176, 180, 186, 203
indices, of a Kreĭn space, 146
of an operator, 158, 171
initial space, of a partial isometry, 156
inner product space, 144
inner, see J-inner, also closely (inner, outer) connected
input coefficient, 238
input vector, 237
interface condition, 118
invariant subspace, 210
invariant subspace, see also hyperinvariant and maximal dual pair
Ising model, 62

isometric operator, 150, 157, 160
  and bicontractions, 168
  continuous extensions, 163–165
  isometric extensions of, 172, 177, 180
isomorphism, 145, 157, 161
isotropic part, 77

J

J–contractive, 253
J–inner, 253
J–inner–outer factorization, 258
Jordan chain
  left, 270
  right, 268
J–unitary
  matrix, 252
  matrix function, 252
Julia operator, 158–159, 162–163, 165–167, 171, 182, 184, 209
  generalization, 209
  norm estimates, 213

K

kernel, 79, 192–193, see also reproducing kernel
  $\mathbb{C}^n$–kernel, 79
  hermitian, 79, 99
  $\mathfrak{K}$–kernel, 99
  nonnegative, 79
  with negative, positive squares, 79, 99
kernel of a contraction, 161
Kreĭn space, 77, 145
  completion, 79
  continuous operators on, 149
  invariance of index, 146
Kreĭn subspace, 151

L

Laguerre polynomial, 4
Laurent matrix, 5
left null pair, 272
  augmented, 308
linear fractional representation, 186, 189, 205
linear system, see colligation
linearization, 95, 101, 128
local principle, 30
Lomonosov's theorem, 221

M

main operator, 238
main space
  of a pair, 271, 272, 288
  of a realization, 240
mapping, of negative subspaces, 161
matrix
  block Toeplitz, 35
  Hankel, 17
  Hilbert, 25
  Laurent, 5
  paired, 6
  Toeplitz, 4, 8
matrix extensions, 165–167, 171–172, 175–176, 182
matrix representation, 3
maximal dual pair of subspaces, 213–214, 219
maximal positive, negative, etc., 153–154, 162, 164, 169
minimal realization, 240
Minkowski space, 147–148, 151, 152, 163, 172
multiplier problem, 44

N

negative index, 146, 158
negative operator, 150, 158
negative space, 144, 152
negative squares, 192, 203
negative subspace, invariant, 211, 213
neutral operator, 172, 180
neutral space, 144, 152
Nevanlinna function, 82, 91
  generalized, 94
nondegenerate space, 144
nontrivial subspace, see proper subspace
norm, 145
  equivalence, 146
  estimates, 212–213, 221–222
  operator, 149
null function
  left, 268
  right, 263
null–kernel pair, 241
null pair
  canonical left, 270, 272
  canonical right, 269, 271

null pair (Cont.)
  left, 272
  right, 271
null–pole subspace, 288
null–pole triple, 292

O

observable, 241, see also closely outer connected
operator, 79, 146, 149, 150
  compact, 17
  contractive, 79
  definitizable, 80
  finite rank, 18
  Fredholm, 26
  Hankel, 17
  isometric, 79
  nonnegative, 80
  positive definite, 16
  quasi-uniformly positive, 80
  selfadjoint, 16, 79
  simple symmetric, 93
  symmetric, 79
  Toeplitz, 8, 39
  trace class, 59
  unitary, 79
  Wiener–Hopf, 4, 39
order, 264
orthogonal, 144
  companion, 151
  complement, 151
orthogonal companion, 77
outer, 258, see also closely (inner, outer) connected
output coefficient, 238
output vector, 237

P

π_κ–space, 77
paired matrix, 6
partial isometry, 150, 155–156
partial null multiplicities, 265
pencil, operator, 128
perturbation, compact, 221, 223, 224
Phillips' theorem, 154, 213
Pick matrix, 281
piecewise continuous function, 31, 44
piecewise quasicontinuous function, 34
point spectrum, 209, 215
polar factorization, 207
pole, 240
pole assignment theorem, 279
pole pair
  right, 288
polynomial, definitizing, 80, 215
Pontryagin space, 77, 146, 160, 164–165, 176, 192–193
Pontryagin's theorem, 164, 165
positive definite operator, 16
positive index, 146, 158
positive operator, 150, 158, 211–212, 214, 222, 224– 225
positive space, 144, 152
positive subspace, invariant, 211, 213
projection method, 64
projection operator, 150, 152, 156
projection theorem, 151
proper
  rational matrix function, 238
proper subspace, 210
pseudo–order, 267
pseudospectrum, 52

Q

quasicontinuous function, 34
quasisimilar, 223–224

R

realization, 240
  minimal, 240
realization of differential expression
  maximal, 87
  minimal, 88
  selfadjoint, 89
realization problem, 198, 202, 203
Redheffer transform, 254
reducing subspace, 210
regular subspace, see Kreĭn subspace
relation, 199
reproducing kernel, 66, 193, 195, 203
residue map, 288
resolvent, 209
  norm estimates, 221–222
Riesz functional calculus, 209, 217
Riesz idempotents, 215
Riesz index, 215
Riesz representation theorem, 147

right null pair, 271
right pole pair, 288
row extensions, 165–166, 174–175, 182

S

Schatten class, 221
Schur algorithm, 181–182, 187, 189–191
Schur class, 181–182, 187, 189, 204
  generalized, 192, 204–205
Schur function, generalized, 99
Schur parameters, 181, 187, 190
selfadjoint extension, 88
  canonical, 88
  minimal, 88
  with exit space, 88
selfadjoint operator, 16, 150, 207, 213–214, 224–225, see also Bognár–Krámli factorization
  products, 223–224
shift operator, 157
Shmul′yan's theorem, 164–165, 176, 200
signature matrix, 252
signature operator, see fundamental symmetry
similar, 240
singular integral equation, 5
singular part, 288
Smith–McMillan form, 266
spaces $\mathfrak{H}(S)$ and $\mathfrak{D}(S)$, 196, 203
spectral function, 83, 217
spectral measure, 91
spectral theorem, 217
spectrum, 7, 209, 215
stable, 258
stable sequence, 15
state operator, 238
state space method, 244
state space
  of a system, 237
  of a realization, 240
  similarity theorem, 241
state vector, 237
Štraus extension, 93, 101
strictly negative, positive space, 144, 152
strong convergence, 15
strong topology for a Kreĭn space, 147
subspace
  maximal nonnegative etc., 78
  projection complete, 77
  regular, 77
  uniformly positive, 78
symbol, 8, 39, 242
Szegő's limit theorem, 59

T

test function, 65
Titchmarch–Weyl coefficient, 90, 91, 93, 96, 97, 106, 107, 115, 131
Toeplitz matrix, 4, 8, 185–187, 190, 204
Toeplitz operator, 8, 39
topology
  on a Kreĭn space, 147
  on spaces of operators, 149
trace class operator, 59
transfer function, 238, see also characteristic function
two–by–two extensions, 167–168, 174–175

U

uniformly negative, positive, 152, 154
unitary operator, 150, 158–159

V

W

Weyl's alternative, 89
Weyl's theorem, 221
Wiener–Hopf operator, 4, 39
winding number, 9
Wold decomposition, 157

X

Y

Z

zero, 263
  right zero, 263